# Does Your Seismic Data Give You the Edge?

AF737558

The unparalleled imaging qualities of Western's Dual Sensor OBC data have helped identify prospects previously difficult to discern on vintage or even contemporary streamer data in the Gulf of Mexico.

Whether your objective is a farm-in, lease-sale evaluation, or drillsite selection, you need the edge — Western Geophysical's Dual Sensor[SM] ocean-bottom cable (OBC) data, the best-quality seismic data available, anywhere.

Numerous new phases of Western's OBC data are now available in the central Gulf of Mexico shelf area. These new phases are based on the same unique orthogonal acquisition geometry and Dual Sensor processing services that established Western's original Mega Survey™ data as the benchmark in seismic quality.

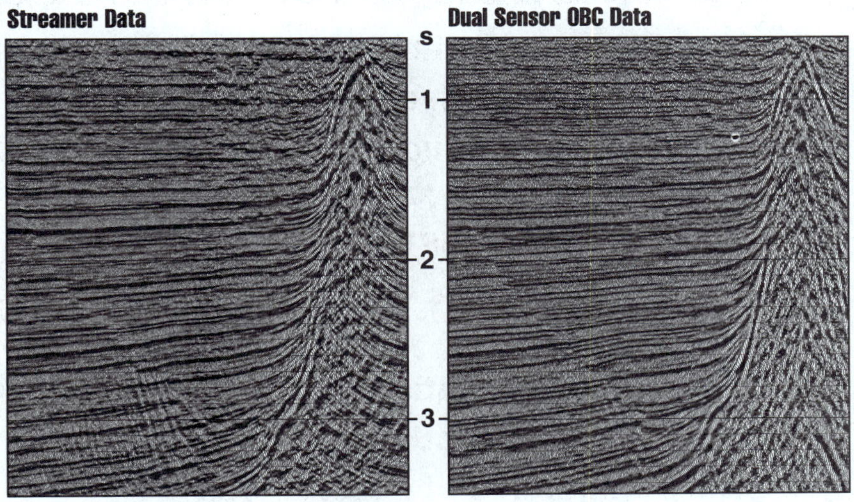


| OBC Parameters | Advantages for Razor-Sharp Imaging |
|---|---|
| 6000-m continuous offset | Expanded bandwidth due to Dual Sensor ghost notch removal and good midpoint distribution |
| 11-sec record length | Improved S/N ratio due to absence of wave action and tow noise |
| 25 m x 25 m natural cell size | Consistent fold, azimuth, and offset distribution eliminates the need for flexbinning |
| 60 fold | Superior statics solution due to surface-consistent geometry, producing inline and crossline coupling |
| Dual Sensor receivers | Minimal impact from surface obstructions |

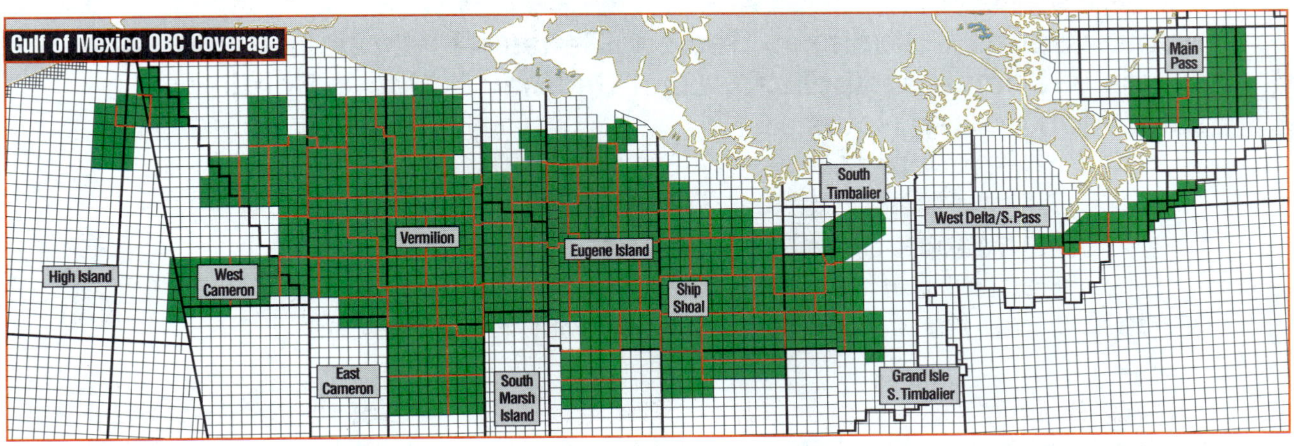


## Value for the Future

### Are you ready for the demands of the future?

**3-D AVO and Prestack Depth Migration** — The consistent fold, azimuth and offset distribution, as well as the surface-consistency of the orthogonal geometry make Western's Dual Sensor OBC data highly suitable for enhanced special processing.

**4-D Ready** — The continuity of coverage due to the absence of impact from obstructions and feathering, and superior source-receiver positioning make Western's OBC data ideal as a baseline for time-lapse reservoir monitoring.

For more information, contact the Data Sales Department today.

| | |
|---|---|
| **Houston** | 713-963-2500 |
| **New Orleans** | 504-523-6781 |
| **Dallas** | 972-490-9832 |
| **E-mail** | data.sales@waii.com |
| **Web Site** | www.waii.com |

WESTERN ATLAS

Western Geophysical

# To the Members

## Gulf of Mexico Theme Issue

Some of the few remaining exploration frontiers for the petroleum industry are the deep-water portions of continental margins. This special issue includes nine papers that describe a part of one deep-water area in the northern Gulf of Mexico. The purpose of these papers is to illustrate, for AAPG membership, the geology of an active deep-water exploration and production setting. The special issue is the first in a series of issues that will contain collected papers on specific themes. This format, an alternative to publishing a book, was conceived by former AAPG Elected Editor Kevin Biddle and is supported by the current AAPG Elected Editor. Let us know your opinion of this volume and any ideas you have for future theme issues.

This series of papers grew out of research from the Gulf of Mexico Industrial Research Consortium at the University of Colorado from June 1992 to May 1996. Initial work on this consortium began in November 1990, when Bob Graebner at Halliburton made the seismic database available to Roy Kligfield and Paul Weimer. During 1991 and early 1992, Kligfield and Weimer developed the infrastructure necessary to conduct the consortium by purchasing the proper software and hardware (viz. CogniSeis, GeoQuest, Landmark, Platte River Associates), writing the research proposal to solicit petroleum company support (25 companies joined), and hiring additional expertise (Mark Rowan, David Knapp).

The research consortium had three original purposes: (1) to study sequence stratigraphy, salt tectonics, and their interaction in the Green Canyon and Ewing Bank area, (2) to develop a close working relationship between the University of Colorado and the petroleum industry, and (3) to generate student research projects, (2 Ph.D. and 8 M.S. degrees were awarded). As the consortium acquired more data sets, biostratigraphic and petroleum systems studies were added to the project, along with two additional research scientists (Tomas Villamil, Claudia Arango).

A follow-up research consortium is being conducted by Mark Rowan, Bruce Trudgill, and Paul Weimer. This new consortium focuses on salt-sediment interaction in different areas of the Gulf of Mexico, petroleum systems, and biostratigraphy.

We began discussions in 1994 with AAPG to find a venue to publish this series of papers. We present these papers in the order that we integrated the different disciplines to analyze the petroleum systems of this area. Although the petroleum companies that explore deep-water margins have more modern databases, these papers represent a rare opportunity for AAPG members to learn about the geology of an active exploration and production area in a deep-water setting.

Neil Hurley
AAPG Elected Editor

Paul Weimer
University of Colorado

The *AAPG Bulletin* (ISSN 0149-1423) is published monthly by the American Association of Petroleum Geologists. Offices are located at 1444 South Boulder Avenue, Tulsa, Oklahoma 74119-3604. Periodicals postage paid at Tulsa, Oklahoma, and at additional mailing offices. Copyright © 1998 by the American Association of Petroleum Geologists. All rights reserved. Phone (918) 584-2555, fax (918) 560-2632, e-mail: bulletin@aapg.org. The *AAPG Bulletin* is also available on CD-ROM through Datapages, Inc., P.O. Box 702708, Tulsa, Oklahoma 74170. Phone (918) 496-7777, fax (918) 496-3756.

**SUBSCRIPTIONS**

Nonmember subscriptions are $140 U.S.; $165 non-U.S. by surface mail; $220 non-U.S. by Airmail. AAPG member subscriptions are included in the dues. Single copies: $8 member, $12 nonmember. *Prices subject to change without notice.* Claims for nonreceipt of any issue, if reported within three months of the date of publication, will be filled gratis. Back issues, if available, may be ordered from the Association. Price list on request. POSTMASTER: send change of address to *AAPG Bulletin,* P.O. Box 979, Tulsa, Oklahoma 74101-0979.

# Applications of 3-D Seismic Data to Exploration and Production

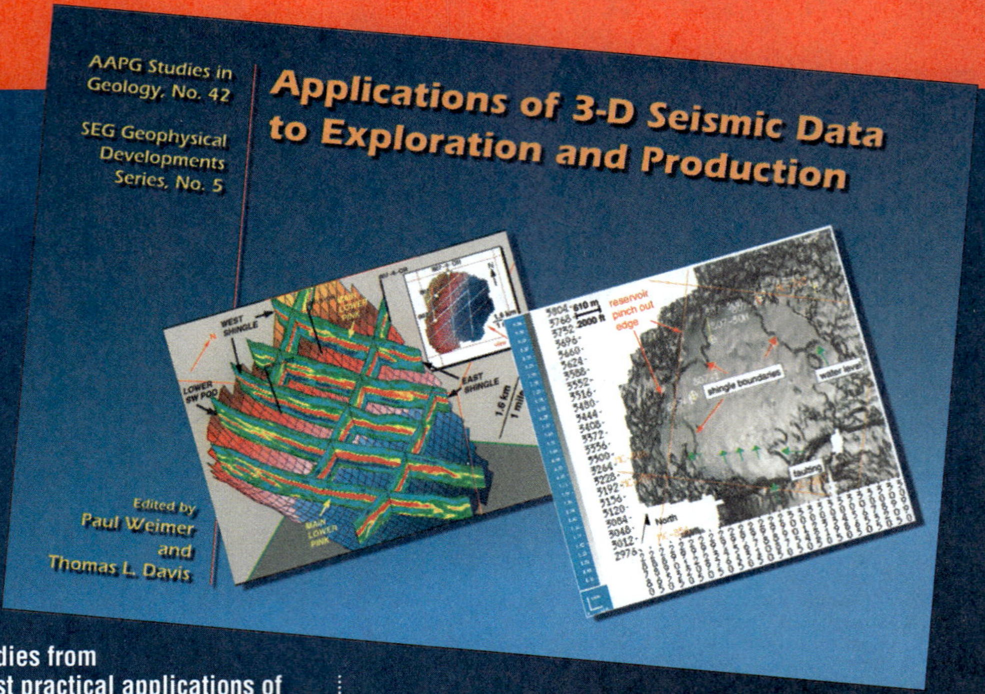


**Edited by P. Weimer and T. L. Davis**

**AAPG Studies in Geology 42.**

**Thirty profusely illustrated case studies from around the world demonstrate the latest practical applications of 3-dimensional seismic data. This book covers fluvial-deltaic, eolian, deep-water clastic, carbonate, and structural reservoirs. Special emphasis is placed on the application of 3-D data to development drilling, reservoir characterization, and reservoir management. This atlas is designed to confirm 3-D seismic interpretation in drilling and production—for despite its great impact on the geosciences community, relatively few published case studies have documented 3-D's importance. This book fills that gap. Extensive use of color illustrations and an 11x17-inch format enhance the presentation. 270 p. 11x17 inches. Wire-o binding. Color throughout. ISBN: 0-89181-050-1**

**Catalog # 534-02**
**AAPG Member price: $94 List price: $127**

"I am enthusiastic about this book! Every geophysical interpreter should learn new applications from it. It's simply packed with important ideas that have been applied profitably."

—R. E. Sheriff
*The Leading Edge*, Nov. 1996

"Hydrocarbon geologists will naturally find [the book] useful, but geoscientists in other areas of application ... and university-based geoscientists should also find it illuminating."

—T. Elliot
*Geological Magazine*, May 1997

## *Order Form*

Please send me _____ copies of **Applications of 3-D Seismic Data to Exploration and Production** $94 each, member price; $127 for nonmembers, plus shipping/handling: (surface) $5.75 in U.S.; $6.75 to Canada/Mexico; $7.25 outside North America. Call for air charges or special handling.

534-02 Member # __|__|__|__|__|__|__|__

Name ______________________________

Address ______________________________

City ______________________________

State ____________ ZIP ____________ Country ____________

Phone ____________ Fax: ____________

E:mail ______________________________

☐ Payment enclosed ☐ VISA ☐ MasterCard ☐ American Express

Account Number ______________________________

Name (please print) ______________________________

**Save time, use our toll-free number to order:**

**1-800-364-AAPG**
**(U.S.A. Only)**

Please have your AAPG member number ready, along with your credit card and catalog order numbers.

AAPG Bookstore, P.O. Box 979, Tulsa, OK 74101-0979 USA
Fax: (918) 560-2652 • Phone: (918) 584-2555
E-mail: bookstore@aapg.org

Outside the U.S., contact our distributors:

**Australia**
Australian Mineral Foundation
Phone: (08) 379-0444

**Canada**
Canadian Society of Petroleum Geologists
Phone: (403) 264-5610

**United Kingdom**
Geological Society
Phone: 01225-445046 • Fax: 01225-442836

# NEW FIELD SEMINAR

## ON LOCATION in the Bahamas....

## Sequence Stratigraphy and Reservoir Distribution in a Modern Carbonate Platform

**Led by: Gregor P. Eberli, G. Michael Grammer, P. Mitch Harris, Donald F. McNeill, and Peter K. Swart**

### June 27- July 1, 1998

Four days of studying and snorkeling on Great Bahama Bank highlight this exciting new field seminar. Participants will have an opportunity to observe and assess the lateral variability in these environments and the potential for creating heterogeneities in the reservoir facies.

Also included is a one-day seismic and core workshop in Miami where participants will study the intimate relationship between sea-level controlled deposition, diagenesis, and petrophysics in carbonates.

This environment provides a unique opportunity to assess the sequence stratigraphic distribution of facies and diagenetic modification in platform carbonate reservoirs. In addition, the facies belts displayed on Great Bahama Bank illustrate the depositional heterogeneities that occur in ancient hydrocarbon reservoirs.

**Cost** for the course is $2200. Included in the cost are field transportation on chartered boat, most meals, and snorkeling equipment.

**Limit:** 25 participants
**Content:** 3.5 CEUs

**To register:**
Contact AAPG Education Department
918-560-2650
E-mail: cammy@aapg.org

# DEDICATION

## Robert J. Graebner

Robert Graebner is a man on a mission. In his quest for better ways of finding and producing petroleum, he has worked tirelessly for the past 50 years to integrate different geologic and geophysical disciplines and reservoir engineering. He has been an innovator in the development and application of three-dimensional (3-D) seismic data, and later in reservoir monitoring. Friendly, outgoing, with a wry chuckle, Bob is truly a pleasure to interact with. His generosity, friendship, and insight are renowned in our industry. He has demonstrated unparalleled leadership and guidance for many people throughout the past five decades, with a pervasive sense of optimism about the future of our industry.

Bob was born in Kansas, but calls his home Colorado Springs, Colorado. After starting his undergraduate education at the University of Colorado, Bob served in the U.S. Army during World War II. After the war, Bob finished his B.S. degree in engineering physics at the University of Colorado in 1948, and joined Geophysical Services, Inc. (GSI). He took a leave of absence and completed his M.S. degree in physics at the University of Colorado in 1954. GSI was merged with Halliburton in 1989. Bob then served in many capacities for Halliburton, culminating as chief geophysicist in 1988–1993. At the end of 1993, Bob "retired."

Bob was especially influential in the development and application of 3-D seismic data, beginning in the 1960s and up to today. He was one of the leaders in the development of the first 3-D seismic survey, acquired in 1966 across the Bell Lake field in Lea County, New Mexico. He has been a major advocate for removing the many barriers between different disciplines in petroleum geology and encouraging a more integrated approach in the application of reservoir sedimentology, reservoir geophysics, petrophysics, and reservoir engineering.

Bob has served the SEG and AAPG in many capacities, including president of SEG during 1986–1987. He was the spring SEG Distinguished Lecturer in 1993, focusing on the future use of reservoir monitoring and four-dimensional seismic data. He received honorary membership from the SEG in 1995, the Mines Medal (distinguished service) from the Colorado School of Mines in 1992, and the Distinguished Alumni award from the University of Colorado in 1997. He served on the AAPG Geophysics and Integration Committee from 1993 to the present.

At a stage in his career when most people are resting on their laurels, Bob continues to work enthusiastically with industry and academic groups. Most importantly, Bob continues to be a leader and visionary in geosciences by encouraging joint industry-academic research. Through GSI and Halliburton, he has supported university research consortia at the University of Colorado, Colorado School of Mines, Columbia University, University of Houston, MIT, Stanford, and the University of Texas. Industry has benefited from these consortia, both from the research results and from the students who are trained with modern skills. He has served on the departmental advisory committees to the University of Colorado, Colorado School of Mines, and Southern Methodist University. He long ago recognized the growing vacuum created by the lack of interested students in petroleum-related geosciences, and the long-term impact this will have on the future health of our industry. He has worked hard to rectify this problem.

The entire success of our applied research program here at the University of Colorado is predicated on the initial support that we received from Bob Graebner. All of the seismic data presented in these papers are the result of Bob Graebner's interest in the proposed research at the University of Colorado, and the data that he made available to develop the industrial research consortium during the early 1990s. He believed in us at a time when we had little success generating support from other parties.

Thus, with great gratitude, pride, and pleasure, we dedicate these papers to Bob Graebner. He has provided an immeasurable service for our research program and for many others. Our industry is better because of the selfless dedication of many individuals such as Bob. For our industry to remain vital well into the twenty-first century will require the actions of more people with Bob's leadership and vision.

**Paul Weimer**
**University of Colorado**

# Evaluating the Petroleum Systems of the Northern Deep Gulf of Mexico Through Integrated Basin Analysis: An Overview[1]

**Paul Weimer,[2] Mark G. Rowan,[2] Barry C. McBride,[3] and Roy Kligfield[4]**

## ABSTRACT

Exploration and development activity has increased significantly during the past 5 years in the northern deep Gulf of Mexico. This activity has been caused by several factors, including significant discoveries in deep water (>1500-ft water depth), outstanding reservoir performance in some of these discoveries, expiration of 10-year leases originally purchased in the mid-1980s, innovative production techniques, and new federal royalty relief. Exploration and production has occurred in three general exploration subprovinces: present shelf, deep water, and the subsalt that extends from shelf into upper slope. Each subprovince consists of slightly different geology and, subsequently, different economic scenarios.

This paper introduces the geologic setting for a portion of the outer shelf and upper to middle slope region in the northern Gulf of Mexico. The following eight papers demonstrate how the petroleum systems of the deep Gulf of Mexico can be analyzed by using an integrated approach. This issue of the *Bulletin* includes papers that describe the petroleum geology of the northern Green Canyon and Ewing Bank region: petroleum fields and discoveries, sequence stratigraphy, biostratigraphy, three-dimensional seismic stratigraphic interpretation, structural geology using restorations, interaction of salt tectonics and sedimentation, and geothermal modeling and path migration prediction.



[1]Manuscript received December 12, 1996; revised manuscript received October 16, 1997; final acceptance November 19, 1997.

[2]Energy and Minerals Applied Research Center, Department of Geological Sciences, University of Colorado, Boulder, Colorado 80309-0399.

[3]Energy and Minerals Applied Research Center, Department of Geological Sciences, University of Colorado, Boulder, Colorado 80309-0399. Present address: HS Resources, 1999 Broadway, Suite 3600, Denver, Colorado 80202.

[4]Energy and Minerals Applied Research Center, Department of Geological Sciences, University of Colorado, Boulder, Colorado 80309-0399. Present address: SignalSoft Corp., 2045 Broadway, Boulder, Colorado 80302.

We thank Kevin Biddle (former AAPG elected editor) and Ken Wolgemuth (AAPG publications manager) for their support and input in reviewing these papers. This series of papers reflects research that was done in the Gulf of Mexico Research Consortium at the University of Colorado during 1992–1996. Two Ph.D. and eight M.S. degrees were awarded for research done related to this project. We thank the following companies and their many representatives for their support and input: Agip, Amoco, Anadarko, BHP, BP Exploration, Burlington Resources, CXY Energy, Chevron USA, CNG, Conoco, Enterprise, Exxon, Marathon, Mobil, Occidental, PanCanadian, Pennzoil, Petrobras, Phillips, Shell, Texaco, Total, Union Pacific, and Unocal. Their support has been crucial for the success of this research. We extend special thanks to Halliburton Geophysical (now Western Geophysical), specifically to Bob Graebner, John Anderson, and Gary White, for the seismic data used in this project. We are indebted to PaleoData Inc. and Art Waterman for use of their biostratigraphic data. Digital logs were kindly provided by Marathon (Tricia Santogrossi). Software donations from GeoQuest, CogniSeis Development (Pat Poe, Robert Ratliff), Landmark Graphics (Thom Fisher), Platte River Associates (Jay and China Leonard), and Zeh Graphics (Jerry Martin) were essential to the success of our study. Special thanks to Ryan Crow, Jennifer Crews, and Peter Varnai for their help in drafting the figures. We thank Kevin Biddle, Bryant Korn, Clint Moore, and Kevin Schofield for their comments on drafts of this paper, although we alone are responsible for the interpretations presented here. Acknowledgment is made to the Donors of the Petroleum Research Fund, administered by the American Chemical Society, for partial support of this work.

## INTRODUCTION

The northern Gulf of Mexico is the most active major exploration province in the United States today, comprising three primary subprovinces that overlap (Figure 1): the deep-water subprovince (>1500 ft present water depths); the subsalt subprovince that extends from shallow water depths (100 ft) to several thousand feet; and the suprasalt subprovince in shallow water depths (0–1500 ft). Each of these areas has slightly different geology, different economics, and different methods of evaluation; consequently, companies of various sizes typically are involved with different areas.

In the northern deep Gulf of Mexico, exploration drilling activity has grown from eight and nine wells in 1991 and 1992, respectively, to more than 60 wells in 1995 and 70 in 1996. The large increase in drilling and leasing activity has been caused by at least seven factors. (1) Several deep-water reservoirs have been brought on line during the past few years. The petroleum industry has been pleasantly surprised by the extremely high

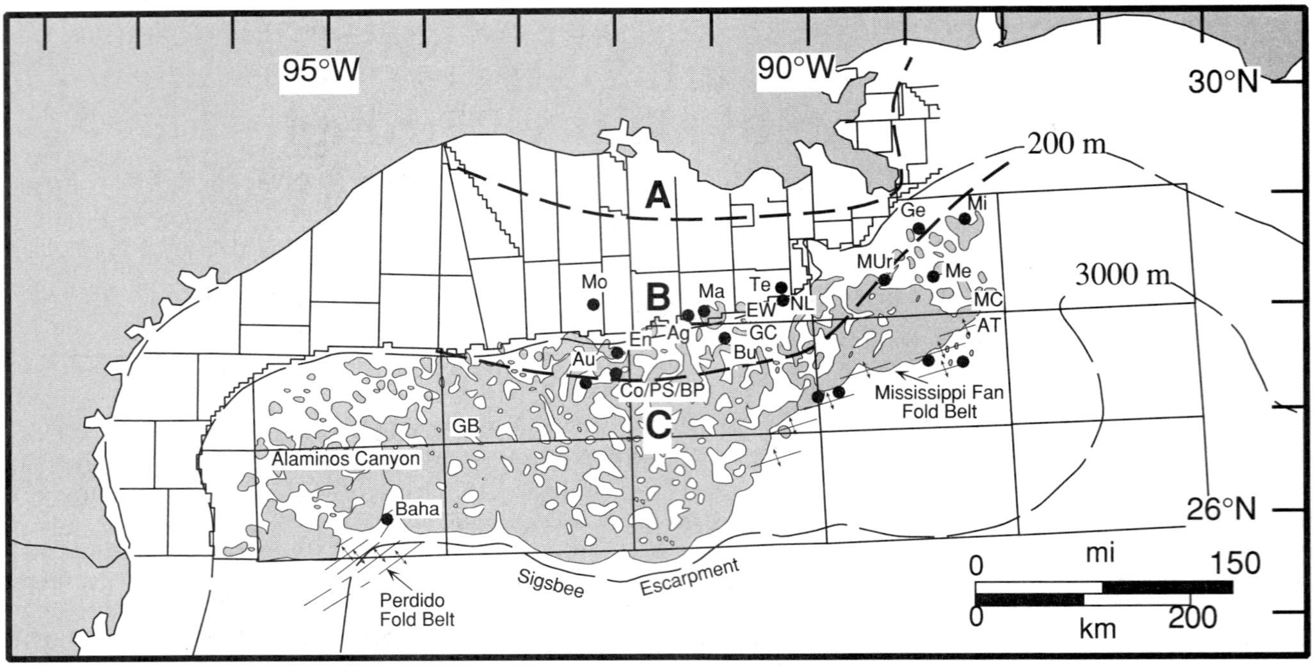


Figure 1—Map showing the northern Gulf of Mexico Basin and the areas of the three main petroleum subprovinces today: shallow water (A), subsalt (B), and deep water (C). Locations of key discoveries are shown: Mi = Mickey (subsalt), Ge = Gemini (subsalt), En = Enchilada (subsalt), Co = Cougar (subsalt), Mo = Monazite (subsalt), Te = Teak (subsalt), Ma = Mahogany (subsalt), NL = North Lobster (subsalt), Co/PS/BP = Conger/PennState/Balpate (subsalt), MUr = Mars-Ursa, Au = Auger, Bu = Bullwinkle, Me = Mensa. Protraction (lease) area abbreviations: MC = Mississippi Canyon, EW = Ewing Bank, GC = Green Canyon, AT = Atwater Valley, GB = Garden Banks. Shaded area shows the distribution of salt in the slope (after Diegel et al., 1995).

Figure 2—Location map of the study area in the northern Gulf of Mexico. Study area includes northern Green Canyon, Ewing Bank, and the southernmost portion of Eugene Island, Ship Shoal, South Timbalier, and Grand Isle protraction areas.

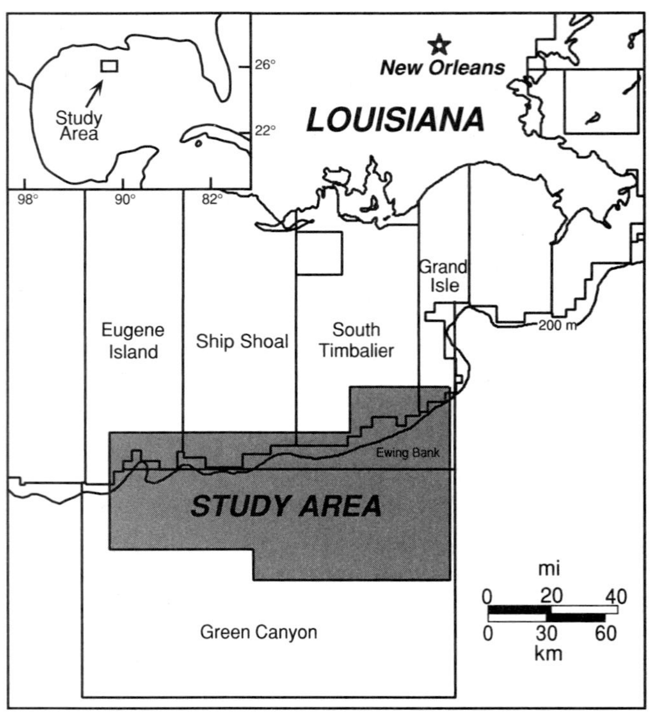


sustainable rates of production from these reservoirs (e.g., Auger, Pfeiffer et al., 1995; Mars, Haines, 1996b; Bullwinkle, Holman and Robertson, 1994) (Figure 1). These rates clearly have demonstrated that deep-water sands are excellent reservoirs capable of sustainable production, thus dramatically increasing the economics for potential deep-water projects. (2) Many 10-year leases were originally purchased during the mid-1980s, and as the lease expiration dates approach, drilling activity has increased to evaluate the prospects. (3) The growing infrastructure for pipelines and production facilities has pushed development projects into increasingly deeper water; therefore, deeper water production is becoming more economic. (4) Innovative subsea production techniques also have made development of these fields more economic. For example, instead of setting traditional production platforms, subsea tiebacks to existing infra-

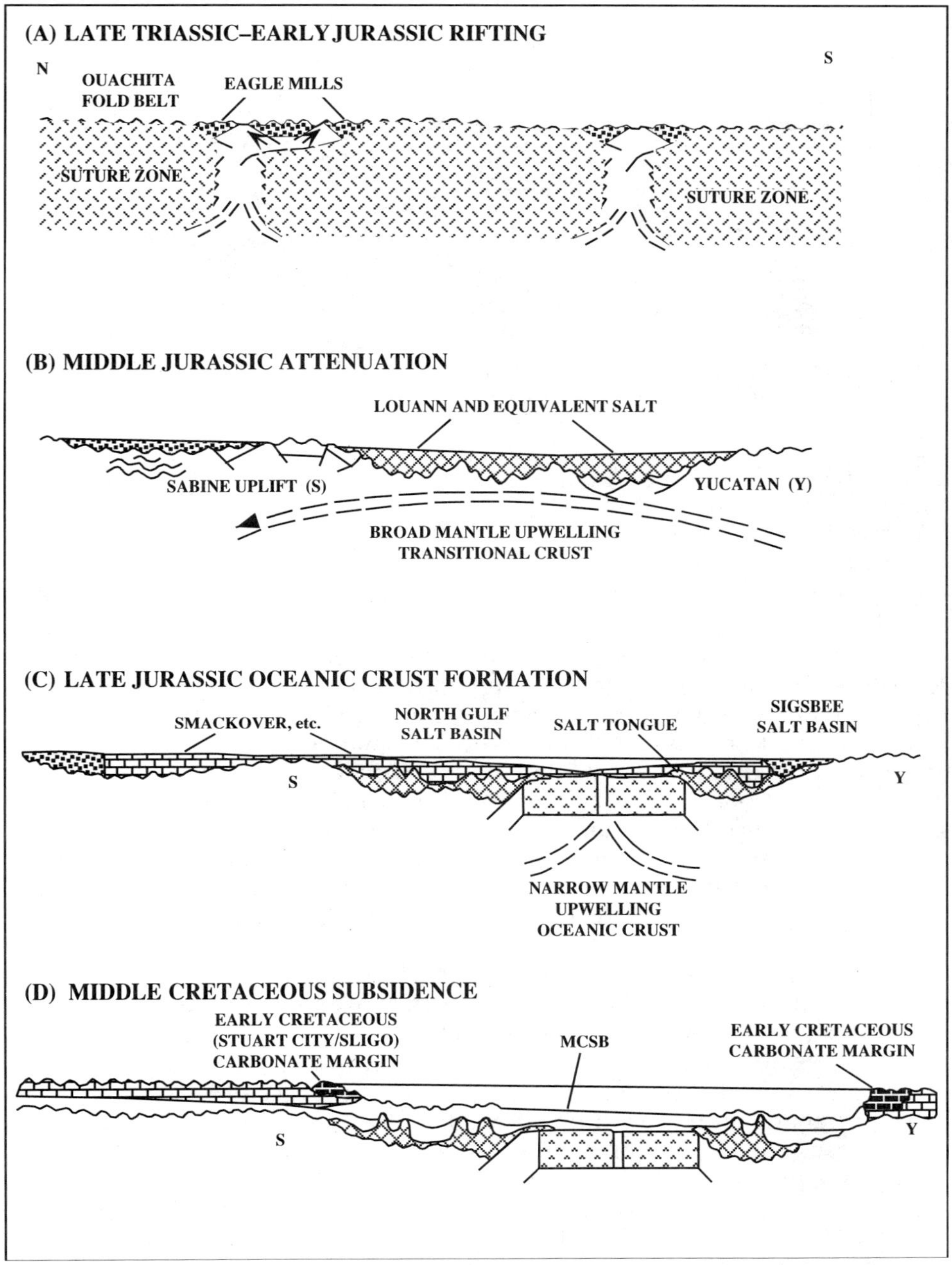


**Figure 3—Schematic cross section illustrating the early evolution of the Gulf of Mexico Basin: (A) Late Triassic, (B) Middle Jurassic, (C) Late Jurassic, and (D) middle Cretaceous sequence boundary. After Buffler (1991).**

structures are being used for the development of the giant Mensa gas field in 5400 ft of water, and for the Troika oil field in 2761 ft of water (Figure 1). Several one- and two-well fields are also being tied back to nearby platforms. (5) A large industry-sponsored engineering consortium called Deep Star (Burton, 1997) was formed to address many engineering and technical problems associated with subsea production techniques. (6) The increased drilling activity has led to the development of new drillships that can drill in deeper water depths. During 1996, the record water depth of 7612 ft for exploration drilling was set in the BAHA prospect in Alaminos Canyon 600 (Trudgill et al., 1995; Brown, 1996). Four additional wells have been drilled in the Atwater Valley protraction area in greater than 6000 ft of water into the Mississippi Fan fold belt (Figure 1), and four wells have been drilled in the Mississippi Canyon protraction area in water depths greater than 6000 ft. The number of drillships in the Gulf of Mexico that can drill in water depths greater than 6000 ft will increase from one to four during the next year. Despite the high operational costs and capital expenditures,

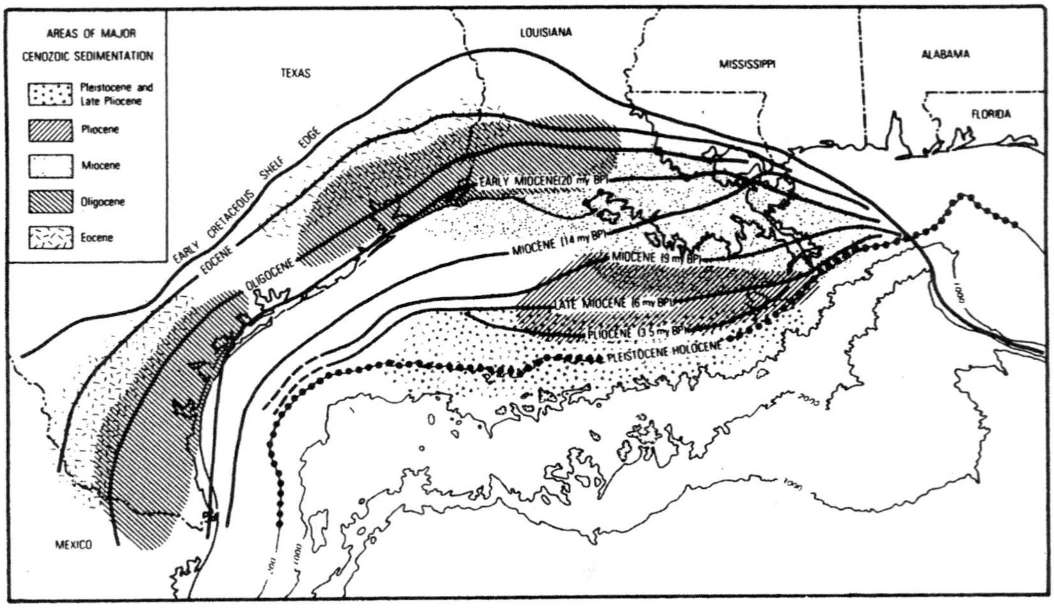


**Figure 4—Map showing the regional distribution of marginal marine depocenters from the early Miocene–Pleistocene in the northern Gulf of Mexico. Modified from Martin (1978).**

both independent and major oil companies hold leases in the deep water; only majors are currently operating in the ultradeep water arena (>6000 ft). (7) Royalty relief passed by U.S. Congress last year has increased the economics of deep-water prospects by deferring royalties on production.

Exploration in the subsalt subprovince has resulted in eight discoveries in the present outer shelf and upper slope area (Mickey and Gemini in Mississippi Canyon; Teak and North Lobster in South Timbalier; Mahogany in Ship Shoal; Monazite in Viosca Knoll; and Cougar, Enchilada, Conger, PennState, and Balpate in Garden Banks) (Figure 1) (Montgomery and Moore, 1997). To date, only the Mahogany discovery is being developed, but the Gemini, Mickey, and Enchilada discoveries have delineation drilling programs underway. Potential reservoirs are primarily in Miocene–Pliocene turbidite systems similar to the deep-water targets. These subsalt discoveries have refocused exploration efforts in shallower water depths in the northern Gulf Basin, where several subsalt prospects have been drilled or will be drilled in the next few years (Haines, 1996a; Montgomery and Moore, 1997; Ratcliff and Weber, 1997). In addition, many subsalt prospects in greater than 3000 ft of water were purchased in the past few OCS lease sales.

The subsalt subprovince has developed because of the evolution of several new concepts or techniques in geosciences. Concepts for salt tectonics have changed considerably during the past decade [see the many references in Jackson et al. (1995) and Travis et al. (1995)]. Seismic acquisition boats now using multiple streamers have allowed for better seismic imaging of the shallow salt bodies. New seismic processing techniques, especially prestack depth migration, have allowed for the clear imaging of the base of salt and better resolution of the structure and stratigraphy of the subsalt region (e.g., Ratcliff and Weber, 1997). The subsalt play in shallower waters has been driven by the smaller major and large independent oil companies, partly because the major oil companies focused their efforts in the deep-water subprovince during the late 1980s and early 1990s.

In the mature suprasalt subprovince of shallower water depths (<300 ft), the smaller major and independent oil companies have rejuvenated new exploration and development projects. Exploration in this subprovince is primarily driven by the economic use of three-dimensional (3-D) seismic data, which allows better exploitation of bypassed pay and identification of potential deeper pays. Companies with producing fields are increasingly having greater success with stepout exploration. With a well-developed production infrastructure, this area has received a dramatic increase in leasing activity during the past three OCS Louisiana lease sales of May 1995, April 1996, and March 1997. Geoscientists with critical skills increasingly are leaving major companies and working for independent companies. Clearly, there is a miniboom in the northern Gulf of Mexico, where the deep-water, subsalt, and shallow-water subprovinces will remain active exploration and production areas for the next several decades.

All of this activity has refocused the petroleum companies' efforts on reevaluating all aspects of the petroleum systems of the northern Gulf of Mexico, which present some of the most complex and challenging areas in the world. The success of

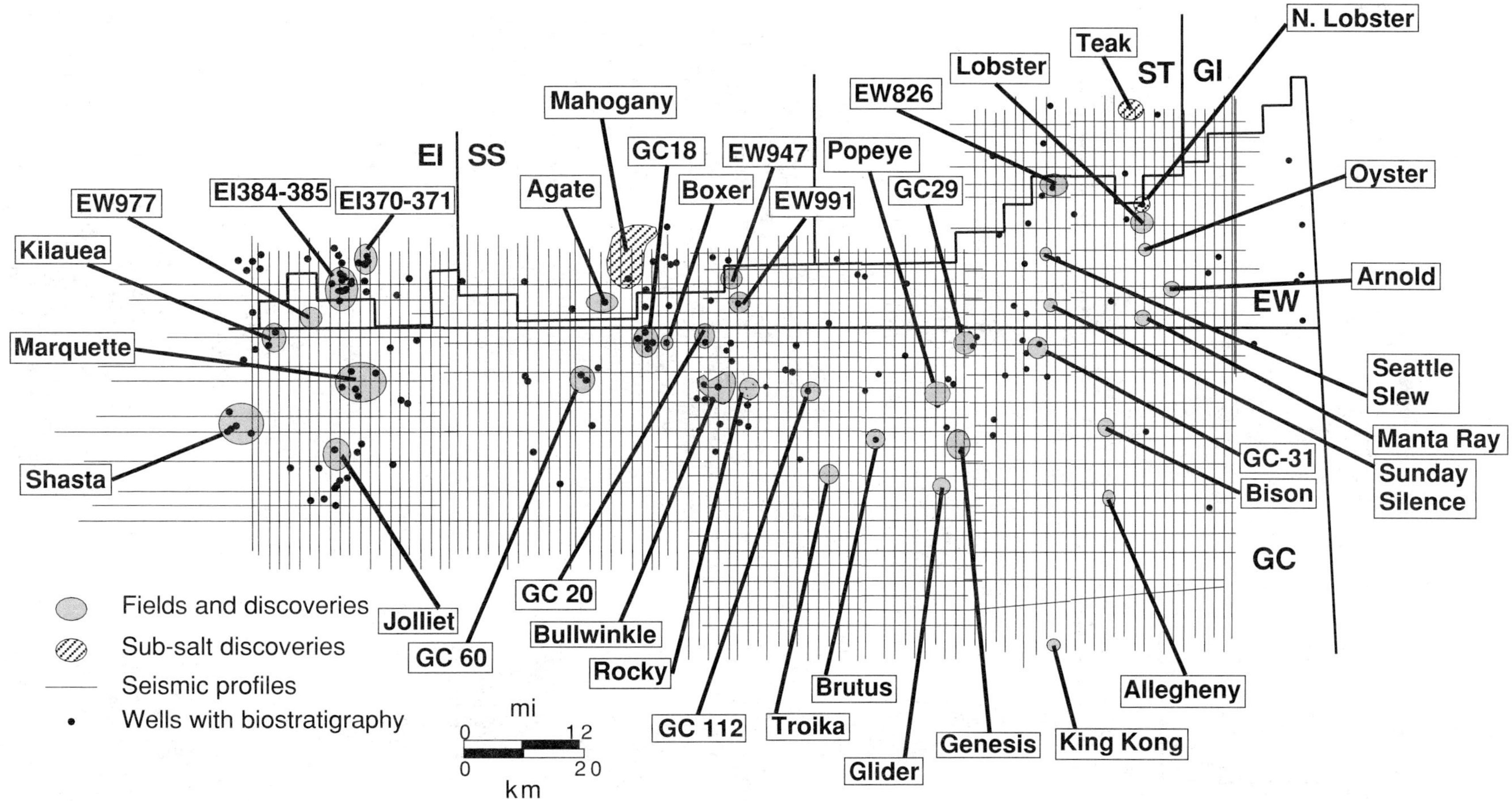


Figure 5—Database used for this study. Dark lines indicate distribution of two-dimensional (2-D) seismic data. Wells used for this study are shown by black dots. Wells with biostratigraphy are shown by different dots. Fields and discoveries are shown.

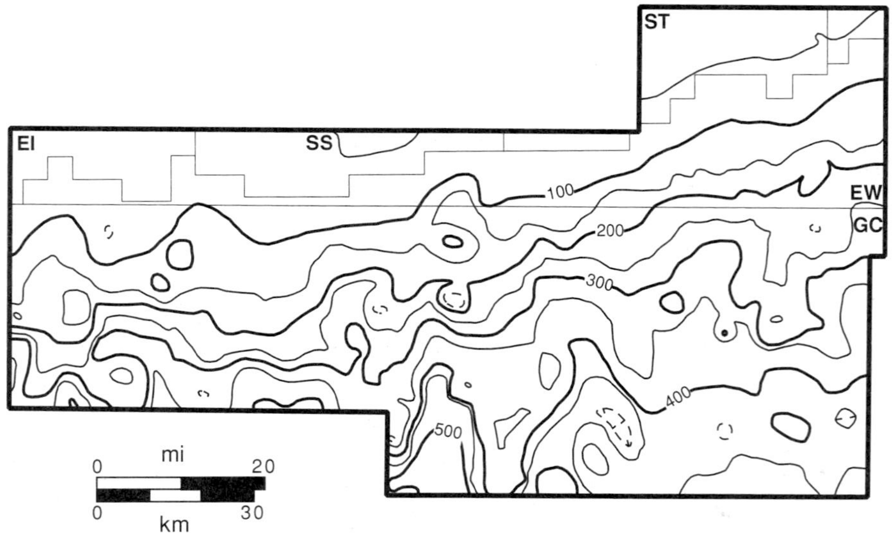


**Figure 6—Map showing bathymetry of the study area. Contour interval is 50 m. See Figure 2 for location of study area.**

the area reflects the confluence of a number of factors. The area consists of complex intraslope basins, where allochthonous salt has formed a series of bathymetric highs, and turbidite systems flow through the adjacent lows, depositing sand. Source rocks are interpreted to be maturing below allochthonous salt, which retards the timing of source rock maturation and controls the flow paths of petroleum. Migration pathways and areas of potential petroleum accumulation can be predicted only after an integrated structural and stratigraphic interpretation, along with detailed geothermal maturation modeling of subsalt source rocks. Successful exploration efforts have developed, especially in the deep-water subprovince, largely due to the persistence of a few companies that developed a good understanding of the regional geologic framework and then developed successful strategies based on their knowledge.

This issue of the *AAPG Bulletin* includes eight papers that address the many components of the petroleum systems for a portion of the northern Gulf of Mexico (Figure 2). This introductory paper briefly reviews the regional geologic setting of the Gulf of Mexico, the local setting of the study area, the database used for the papers, and the significance of each paper in the issue.

## REGIONAL GEOLOGIC SETTING

### Northern Gulf of Mexico

The general geologic evolution of the northern Gulf of Mexico is briefly reviewed here to introduce the province. The Gulf of Mexico Basin is one of the most thoroughly studied sedimentary basins in the world, primarily because it is a major petroleum province. The extensive geological and geophysical databases have allowed for the detailed documentation of the basin's original evolution. The general trends of the Gulf Basin are presented in a series of summaries (e.g., see Salvador, 1987, 1991, and many papers therein; Winker and Buffler, 1988; Buffler, 1991). The early evolution of the Gulf Basin is shown in a schematic cross section in Figure 3. The basin began to form during two rifting episodes: Late Triassic (represented by the Eagle Mills deposits in eastern Texas) and Middle Jurassic (represented by half grabens in the deep Gulf of Mexico) (Figure 3). An extremely important geologic event associated with the second period of rifting was the deposition of thick Callovian (Louann) salt across the basin (Figure 3). This salt, when remobilized later, played a critical role in the maturation, migration, and entrapment of petroleum in the northern Gulf Basin. Oceanic spreading then occurred for about 10 m.y. during the Late Jurassic (Figure 3C).

Early postrift sedimentation (Middle Jurassic to Early Cretaceous) consisted primarily of shallow-marine platform carbonates rimming the northern Gulf basin (underlain by thick continental crust), with periodic influx of siliciclastic sediments (Figure 3C, D). These formed important reservoirs in the onshore basin. In the deep Gulf of Mexico, post-breakup thermal subsidence of highly attenuated continental and oceanic crust caused the region to subside more rapidly than the shelf areas to the north, so that by the time of the middle Cretaceous sequence boundary (MCSB, Figure 3D), the deep Gulf rested in more than 1 km water depth. The deep Gulf stratigraphy for the Middle Jurassic–Lower Cretaceous section is interpreted to consist of deep-water shales and marls and to be the source rocks for the deep-water Gulf of Mexico province.

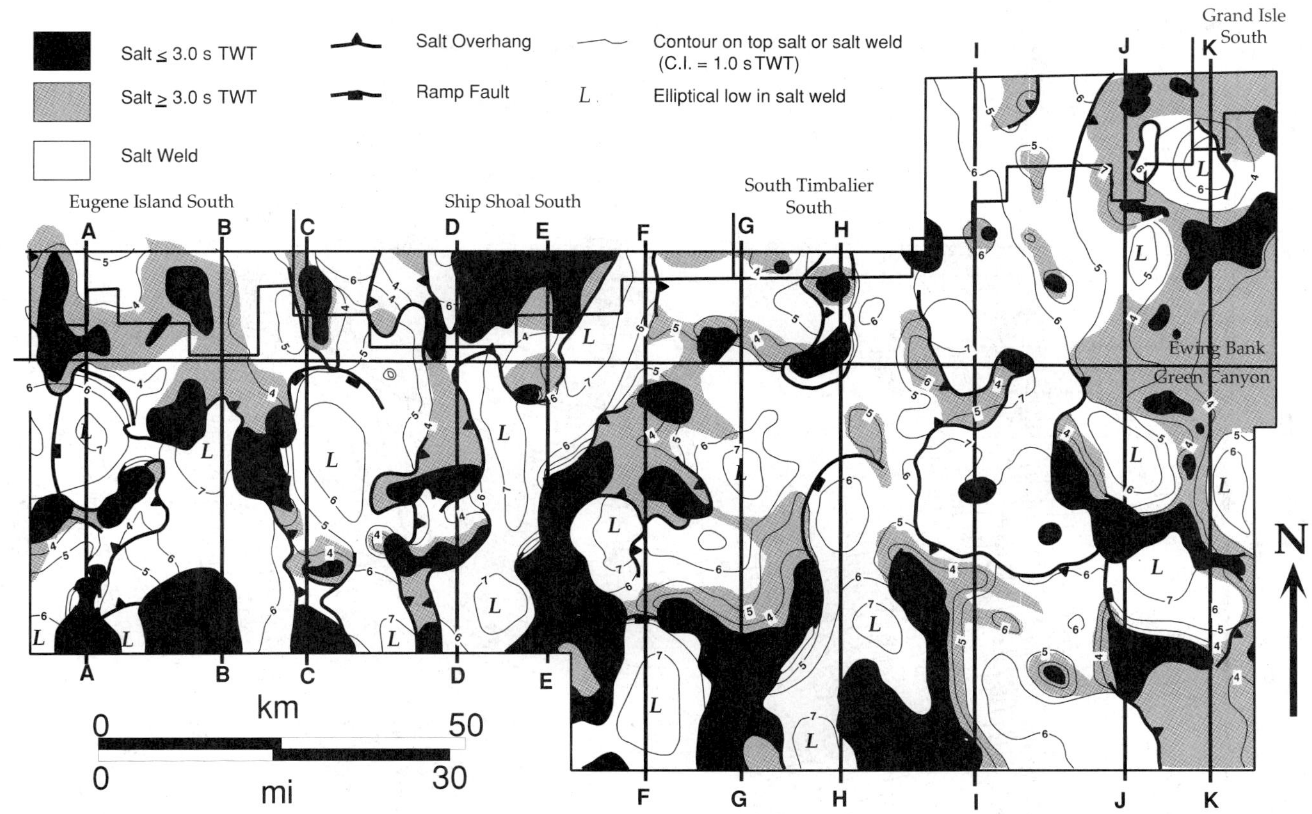


**Figure 7—Structure contour map (in two-way traveltime) of the top salt or equivalent salt weld. The salt systems are characterized by large relief, with different levels of salt either separated by overhangs or connected by ramps in the salt weld. Top salt contours shallower than 3.0 s not shown. After Rowan (1995). Location of Figures 8 (profile F) and 9 (profile I) as shown, along with other profile locations.**

By the early Cenozoic, regionally extensive shallow-water to continental siliciclastic depocenters became established in the northwestern Gulf Basin, reflecting the drainage of sediments from the Laramide orogeny in the United States and Mexico (Martin, 1978). These deposits form important reservoirs in both the onshore and offshore Basin. During the Miocene, these depocenters switched to the east, so that the sediments from the entire central portion of North America were draining into the Louisiana area in the north-central Gulf Basin (Figure 4). The coeval bathyal sediments consist of turbidite systems deposited in a series of minibasins on top of and between allochthonous salt bodies. These turbidite systems are the reservoirs for the northern deep-water Gulf subprovince (see many references in Weimer et al., 1994a). The extensive loading of Cenozoic sediments caused salt to deform from the original autochthonous layer into one or more allochthonous levels. Allochthonous salt geometries vary considerably across the Gulf Basin; these geometries and their evolution are summarized in several papers in Jackson et al. (1995) and Travis et al. (1995).

## Northern Green Canyon–Ewing Bank Study Area

The study area for the papers in this issue is located offshore central Louisiana in the present upper slope and outer shelf, including northern Green Canyon, Ewing Bank, and the southern parts of Grand Isle, South Timbalier, Ship Shoal, and Eugene Island protraction areas (Figure 2). The northern part of the study area is a developing petroleum region; the southern part of the study area is a frontier exploration region. The entire area includes 35 fields or discoveries, primarily in turbidite-related reservoirs (Figure 5). Three subsalt discoveries have been made in the area, with several more prospects being tested within the next few years. Good potential for future exploration exists both in subsalt opportunities and in the deeper water portions of the area.

Water depths within the study area range from 200 ft to the north to 4500 ft to the south (Figure 6). The modern bathymetry is highly irregular, reflecting the complex salt tectonics that have created a series of salt highs and intraslope basins

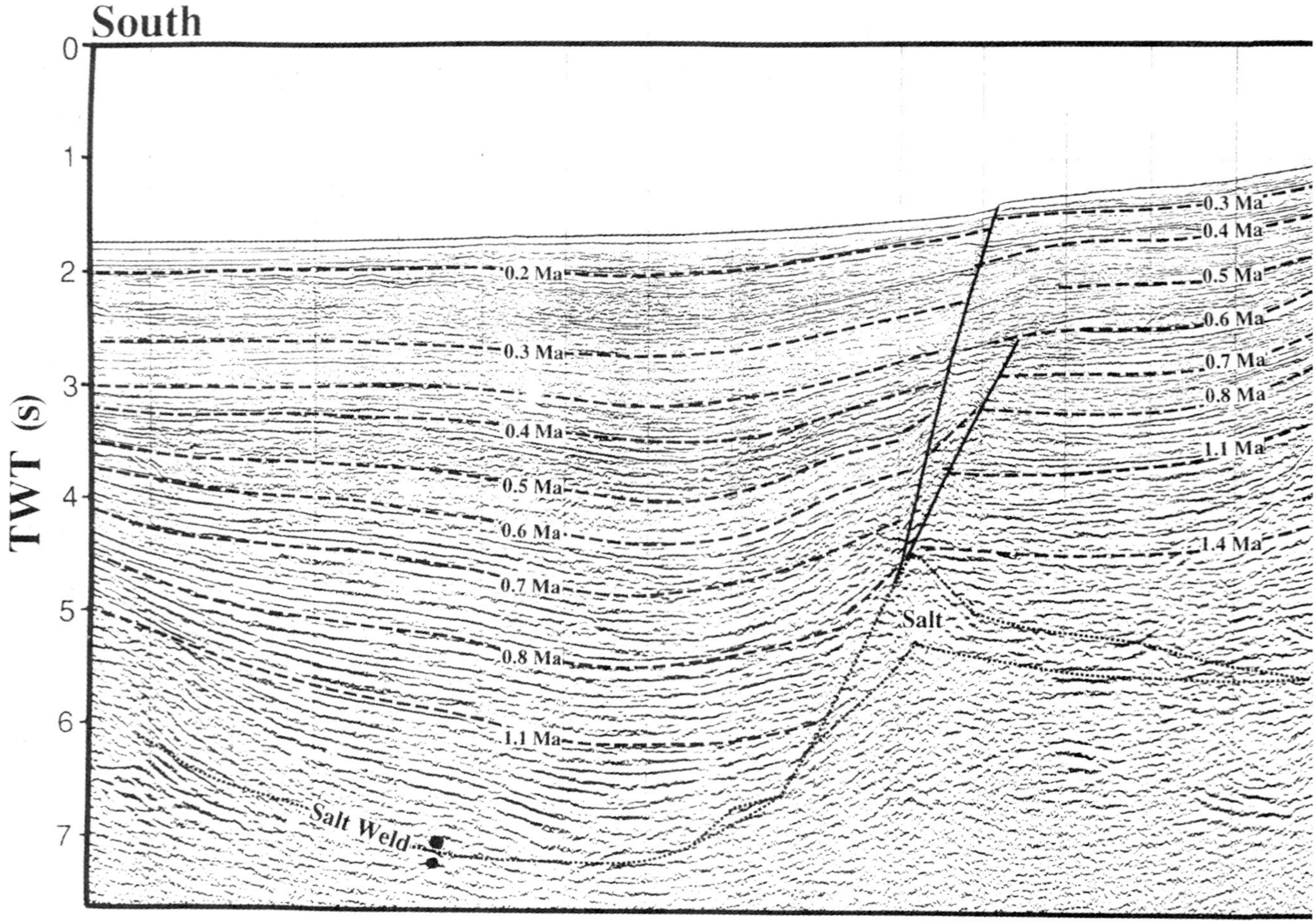


**Figure 8—Regional seismic profile across the western part of the study area illustrating sequences 1.9–0.4 Ma. Profile crosses (at the north) the southern toe of a shallow salt sheet that appears to have experienced some contraction and extension, a thick shallow salt sheet/massif, and counterregional growth sediments forming a deep minibasins (south). See Figure 7 for location of the profile (profile F). For additional regional profiles, see Weimer et al. (1998, this issue) and McBride et al. (1998a, this issue).**

throughout the Pliocene–Pleistocene. The regional setting of the study area is illustrated in Figure 7, which is a structure contour map in seconds two-way traveltime (s TWT) of the top salt or salt weld, and in two regional seismic profiles (Figures 8, 9). Details of the structure of the area are included in Rowan (1995) and McBride et al. (1998a, this issue).

The most prominent feature of the salt system is its large relief, with the tops of the shallowest diapirs at less that 1.0 s TWT and the deepest parts of the salt weld at more than 7.0 s TWT (Figures 7–9). The difference does not represent the thickness of salt in diapirs that are rising from a common deep source level; instead, there is significant topographic relief on both the weld and the base salt (Figures 7–9). The area consists of both salt-stock canopies and subhorizontal salt sheets, both with a variety of diapir styles. Faults include both normal (both basinward dipping and counterregional) and thrust faults.

The Pliocene–Pleistocene sediments consist primarily of bathyal turbidite systems deposited in a series of intraslope minibasins (Figures 7–9). These systems are up to 20,000 ft thick in the deeper basins. Sediments were deposited in several different environments, including channels, lobes, overbanks, slides, and large mud-dominated unchanneled settings (Weimer et al., 1994b, 1998b, this issue). Shifts in the shelf depocenter that fed these turbidite systems greatly affected the distribution of sediments. The best reservoir facies are in sheet and amalgamated sheet sands (lobes). All fields or discoveries consist of combined structural/stratigraphic traps.

## Database

We used an integrated data set from three primary sources (Figure 5). First, we interpreted 6300 mi

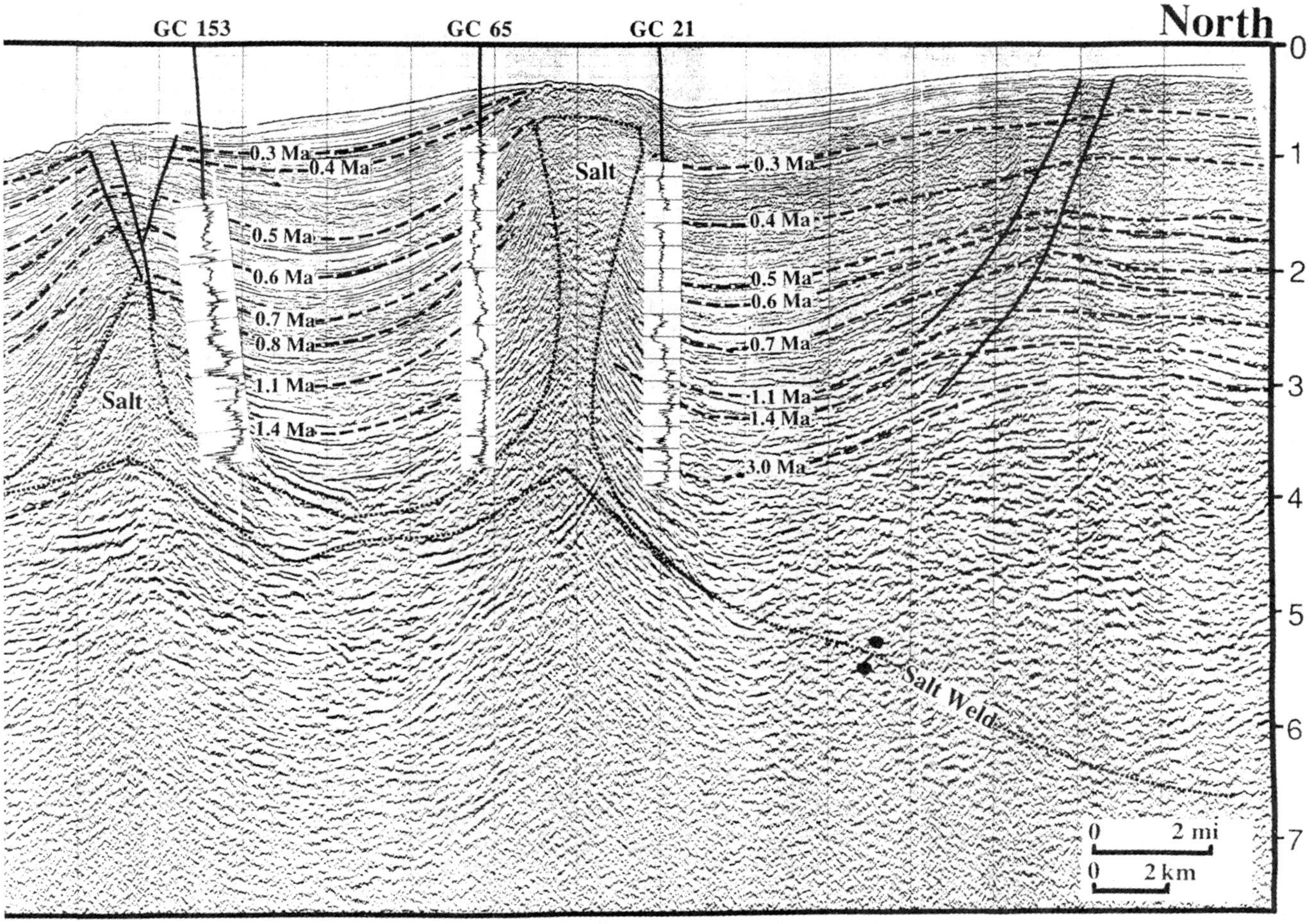


Figure 8—Continued.

of 48- and 96-fold two-dimensional (2-D) seismic data; these consist of 4300 mi dip direction (north-south) and 2000 mi strike direction (east-west). Line spacing is 1 mi in dip direction and 2 to 3 mi in strike direction. These seismic data were collected between 1984 and 1986 as a regional 3-D data set, and were reprocessed as 2-D lines during 1988 and 1989. Several subregional 3-D surveys also were interpreted for details in the stratigraphy and structure.

Second, we used digital logs from 185 of the approximately 440 wells in the area. Synthetic seismograms and time-based logs were generated for all 185 wells.

Third, the biostratigraphic database consists of foraminifera reports from 180 wells, calcareous nannoplankton reports from 83 wells, and high-resolution foraminifera reports from 60 wells. This biostratigraphic data set has been essential to age-dating the sequences for correlating abundance curves to seismic profiles to identify condensed sections, and for establishing the paleoecology of each sequence.

## Objectives of These Papers

The papers in this issue of the *Bulletin* illustrate how the petroleum systems for one part of the northern deep Gulf of Mexico (Figure 2) were interpreted by integrating studies that used modern concepts in sequence stratigraphy, structural geology, source rock maturation modeling, and petroleum migration analysis. Note that none of these disciplines, by themselves, can unravel the highly complex petroleum systems. All of these disciplines must be integrated to help direct exploration. The specific results, in part, are unique to this area because its geologic evolution is slightly different than other areas in the northern Gulf of Mexico; however, these papers do document the systematic approaches and techniques necessary for doing integrated studies.

This study presents eight papers that illustrate our approach to this integrated basin analysis. Weimer et al. (1998a, this issue) illustrate the seismic stratigraphic and wireline-log expression of the producing turbidite systems of 35 discoveries or

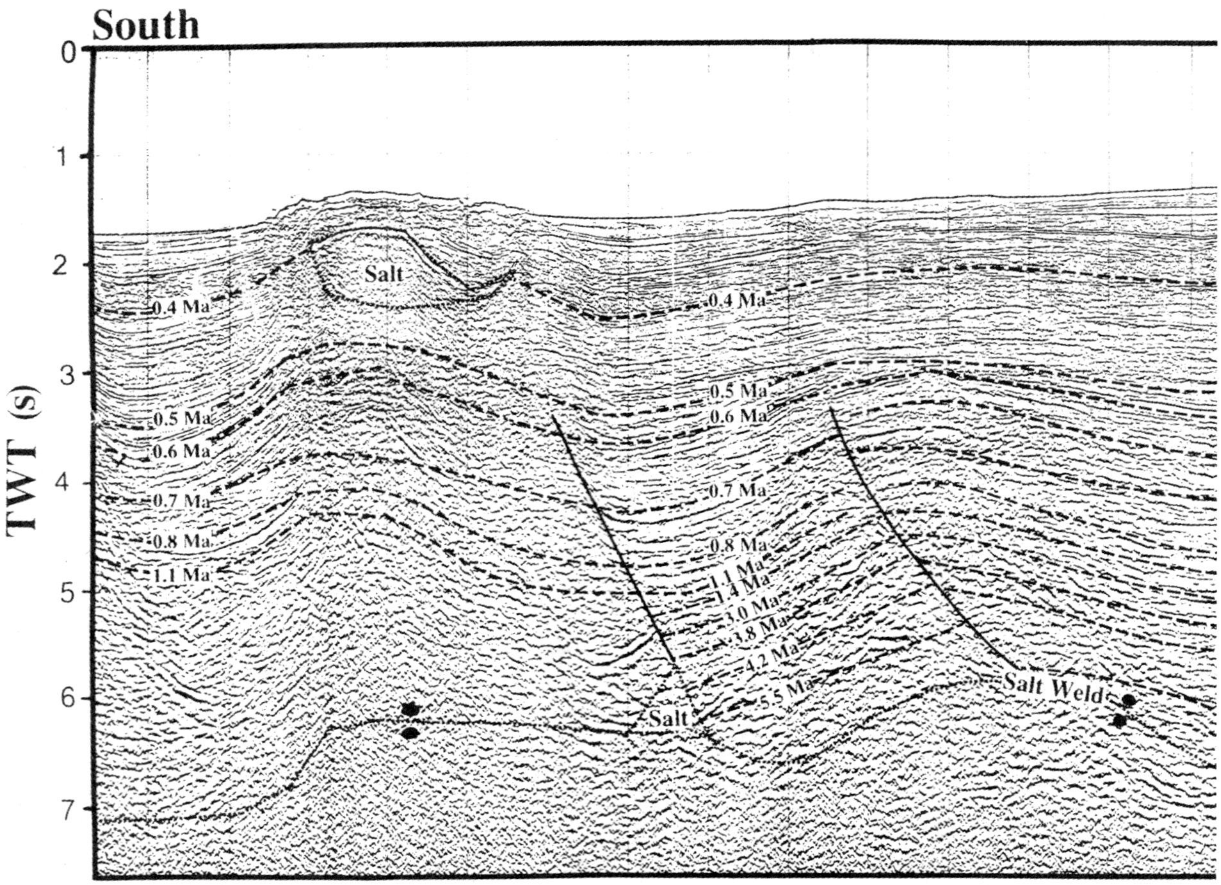


**Figure 9—Regional seismic profile across the central part of the study area illustrating sequences 5.5–0.4 Ma. At the north, the profile crosses diapirs within a shallow salt sheet, which appears to have experienced contraction; minibasins bounded by normal faults; and a shallow, isolated salt sheet (south). A separate weld surface exists for each of the salt systems. The deepest salt weld extends the length of the section. Several salt rollers exist along the deep salt weld/base salt surface and are located under the major normal faults that define the minibasin. See Figure 7 for location of the profile (profile H). For additional regional profiles, see Weimer et al. (1998, this issue) and McBride et al. (1998a, this issue).**

fields, and some of their production statistics. This compilation provides a useful summary of the variety of trapping mechanisms and field information for deep-water reservoirs. Weimer et al. (1998, this issue) systematically describe the sequence stratigraphy of the Pliocene–Pleistocene sediments in the study area. These workers show how temporal and spatial changes in sedimentation occurred in the minibasins, where major sands were deposited, and present a model of how sands are deposited in these sorts of basins. Villamil et al. (1998, this issue) present three biostratigraphic concepts that help in basin analysis: how the accuracy of correlations can be evaluated through graphic correlation, how estimates of the paleobathymetry of the basin can be improved by integrating paleoecology with lithostratigraphy and palinspastic restorations, and how biostratigraphic condensation changes throughout one minibasin in three and four dimensions. Varnai (1998, this issue) interprets the sequence stratigraphy of three minibasins using a subregional 3-D seismic data set, and demonstrates the evolution of turbidite systems throughout a 2.4 m.y. time interval. This 3-D interpretation allows for far greater definition of the turbidite elements than the regional stratigraphic study. McBride et al. (1998a, this issue) use 2-D structural restorations to illustrate how the salt geometry changes through time, how the suprasalt sedimentary basins develop and evolve through time, and how rates of these processes vary. McBride (1998, this issue) shows a structural restoration of a megaregional profile from the northern (updip) edge of the salt basin (onshore Louisiana) to the southern (downdip) edge (Sigsbee Escarpment). The restoration shows some of the ways in which the northern Gulf of

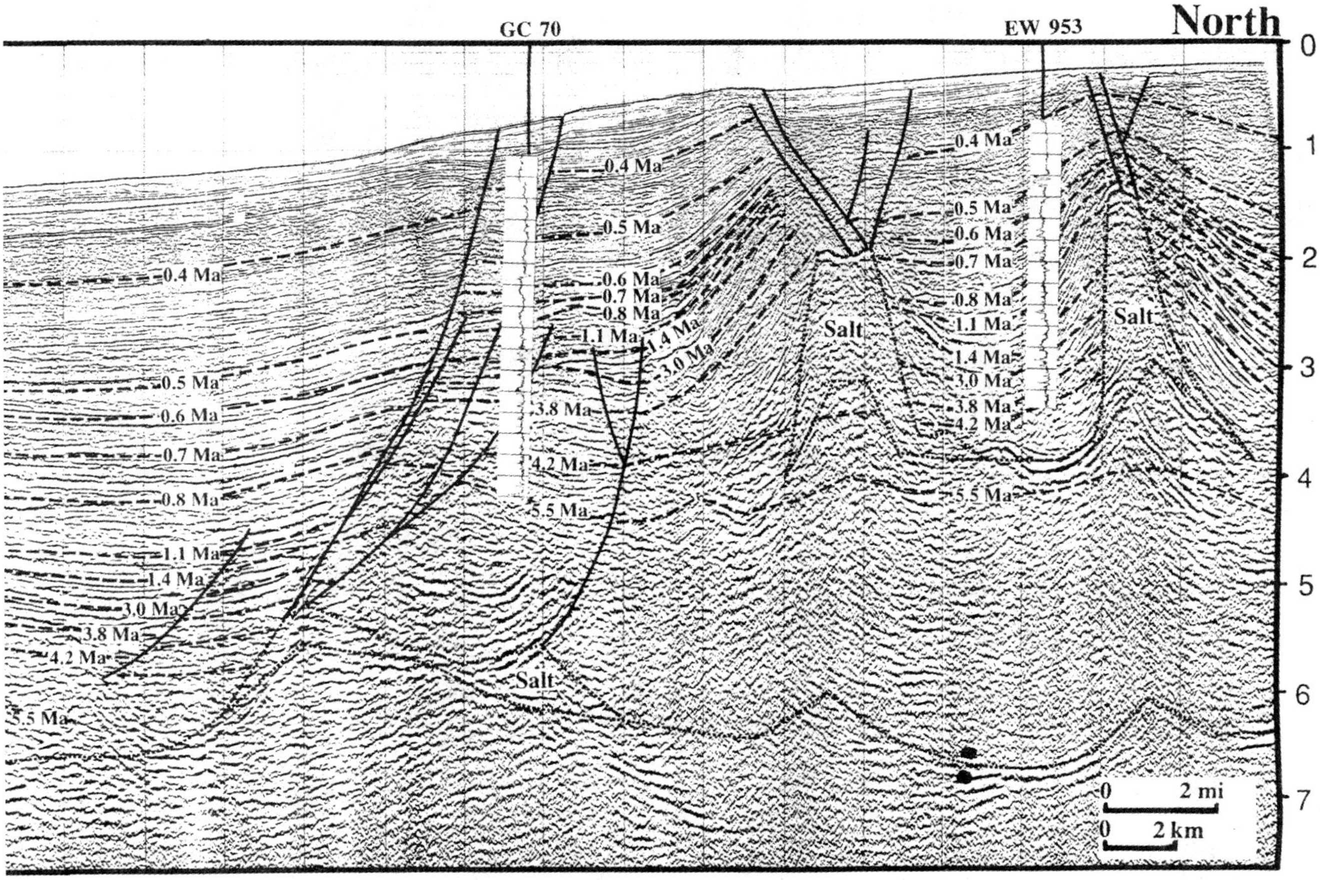


Figure 9—Continued.

Mexico province evolved through time, illustrating the changing position of salt features, the different structural/stratigraphic provinces along this profile, and the various timing of emplacement of salt sheets. By integrating the results of the detailed sequence stratigraphic and structural studies, Rowan and Weimer (1998, this issue) document how the evolving salt impacted the sea-floor paleobathymetry, which in turn influenced the location of sediment pathways and depocenters. The manner in which salt evolved also produced different styles in minibasins, which is one of many factors that affect lithofacies distribution. McBride et al. (1998b, this issue) show that by integrating all of the previous studies with thermal maturation modeling, a comprehensive analysis of the petroleum systems can be made. Special emphasis is placed on the timing of peak generation of source rocks and the determination of which areas receive concentrated petroleum flow through time, which in turn controls the development of major petroleum accumulations. In a related paper, Dixon and Weimer (1998, in press) describe the evolution of the Pleistocene eastern Mississippi Fan in the northwest deep Gulf of Mexico. This fan is coeval to the larger western Mississippi Fan, and reflects deposition in the unconfined deep basin, in contrast to the turbidite systems deposited in the intraslope basins.

These eight papers provide the opportunity to view different aspects of the petroleum systems of one area in the northern deep Gulf of Mexico. The papers can be read separately, but are best read in the order presented here to follow the petroleum systems logic. We hope that these papers serve as a catalyst for further development of exploration concepts in the deep Gulf of Mexico and similar sedimentary basins around the world, especially as deep-water drilling activity increases globally.

## REFERENCES CITED

Brown, D., 1996, Industry watches deep Gulf test: AAPG Explorer, v. 17, no. 6, p. 8, 9, 29.

Buffler, R. T., 1991, Early evolution of the Gulf of Mexico Basin, *in* D. Goldthwaite, ed., An introduction to central Gulf Coast

geology: New Orleans Geological Society, p. 1-15.
Burton, C., 1997, Resources limitations force new operating concepts: World Oil, Deepwater Technology, August, p. 9-14.
Diegel, F. A., D. C. Schuster, J. F. Karlo, R. C. Shoup, and P. R. Tauvers, 1995, Cenozoic structural evolution and tectono-stratigraphic framework of the northern Gulf Coast continental margin, *in* M. P. A. Jackson, D. G. Robert, and S. Snelson, eds., Salt tectonics: a global perspective: AAPG Memoir 65, p. 109-151.
Dixon, B. T., and P. Weimer, in press, Sequence stratigraphy and depositional history of the eastern Mississippi Fan (Pleistocene), northern deep Gulf of Mexico: AAPG Bulletin, v. 82, no. 6.
Haines, L., 1996a, Sub-salt: Oil and Gas Investor, February, v. 16, p. 23-31.
Haines, L., 1996b, Going deep: Oil and Gas Investor, September, v. 16, p. 24-32.
Holman, W. E., and S. S. Robertson, 1994, Field development, depositional model and production performance of the turbidite "J" sands at the prospect Bullwinkle, Green Canyon 65 fields, outer shelf Gulf of Mexico, *in* P. Weimer, A. H. Bouma, and B. F. Perkins, eds., Submarine fans and turbidite systems: GCS-SEPM 15th Annual Research Conference, p. 139-150.
Jackson, M. P. A., D. G. Robert, and S. Snelson, eds., 1995, Salt tectonics: a global perspective: AAPG Memoir 65, 454 p.
Martin, R. G., 1978, Northern and eastern Gulf of Mexico continental margin: stratigraphic and structural framework: AAPG Studies in Geology 7, p. 21-42.
McBride, B. C., 1998, The evolution of allochthonous salt along a megaregional profile across the northern Gulf of Mexico Basin: AAPG Bulletin, v. 82, no. 5B, p. 1037-1054.
McBride, B. C., M. R. Rowan, P. Weimer, 1998a, The evolution of allochthonous salt systems, northern Green Canyon and Ewing Bank (offshore Louisiana), northern Gulf of Mexico: AAPG Bulletin, v. 82, no. 5B, p. 1013-1036.
McBride, B. C., P. Weimer, M. G. Rowan, 1998b, The effect of allochthonous salt on the petroleum systems of northern Green Canyon and Ewing Bank (offshore Louisiana), northern Gulf of Mexico: AAPG Bulletin, v. 82, no. 5B, p. 1083-1112.
Montgomery, S. L., and D. Moore, 1997, Subsalt play, Gulf of Mexico: a review: AAPG Bulletin, v. 81, p. 871-896.
Pfeiffer, D. S., D. T. McGee, and J. C. Crump, 1995, Auger's geology guidepoint to deepwater U.S. Gulf potential: Offshore, May, p. 22-28.
Ratcliff, D., and D. Weber, 1997, Geophysical imaging of subsalt geology: The Leading Edge, v. 16, p. 115-142.
Rowan, M. G., 1995, Structural styles and evolution of allochthonous salt, central Louisiana outer shelf and upper slope, *in* M. P. A. Jackson, D. G. Robert, and S. Snelson, eds., Salt tectonics: a global perspective: AAPG Memoir 65, p. 199-228.
Rowan, M. G., and P. Weimer, 1998, Salt-sediment interaction, northern Green Canyon and Ewing Bank (offshore Louisiana), northern Gulf of Mexico: AAPG Bulletin, v. 82, no. 5B, p. 1055-1082.
Salvador, A., 1987, Late Triassic-Jurassic paleogeography and origin of the Gulf of Mexico: AAPG Bulletin, v. 71, p. 419-451.
Salvador, A., ed., 1991, The Gulf of Mexico Basin: Geological Society of America, Geology of North America, v. J, 568 p.
Travis, C. J., H. Harrison, M. R. Hudec, B. C. Vendeville, F. J. Peel, and B. F. Perkins, eds., 1995, Salt, sediment and hydrocarbons: GCS-SEPM Foundation 16th Annual Research Conference, 308 p.
Trudgill, B., et al., 1995, The geological evolution of the deep-water Perdido foldbelt, Alaminos Canyon, northwestern deep Gulf of Mexico: Transactions of the Gulf Coast Association of Geological Societies, v. 45, p. 573-579.
Varnai, P., Three-dimensional seismic stratigraphic expression of Pliocene-Pleistocene turbidite systems, northern Green Canyon (offshore Louisiana), northern Gulf of Mexico: AAPG Bulletin, v. 82, no. 5B, p. 986-1012.
Villamil, T., C. Arango, P. Weimer, A. Waterman, M. G. Rowan, P. Varnai, A. Pulham, and J. R. Crews, 1998, Biostratigraphic techniques for analyzing benthic biofacies, stratigraphic condensation, and key surface identification, Pliocene and Pleistocene sediments, northern Green Canyon and Ewing Bank (offshore Louisiana), northern Gulf of Mexico: AAPG Bulletin, v. 82, no. 5B, p. 961-985.
Weimer, P., A. H. Bouma, and B. F. Perkins, eds., 1994a, Submarine fans and turbidite systems: GCS-SEPM Foundation 15th Annual Research Conference, 440 p.
Weimer, P., P. Varnai, Z. Acosta, F. M. Budhijanto, R. Martinez, A. Navarro, M. G. Rowan, B. McBride, and T. Villamil, 1994b, Sequence stratigraphy of a Neogene turbidite system, northern Green Canyon and Ewing Bank, northern Gulf of Mexico, *in* P. Weimer, A. H. Bouma, and B. F. Perkins, eds., Submarine fans and turbidite systems: GCS-SEPM Foundation 15th Annual Research Conference, p. 383-396.
Weimer, P., J. Crews, R .S. Crow, and P. Varnai, 1998a, Atlas of petroleum fields and discoveries, northern Green Canyon, Ewing Bank and southern Ship Shoal and South Timbalier areas (offshore Louisiana), northern Gulf of Mexico: AAPG Bulletin, v. 82, no. 5B, p. 878-917.
Weimer, P., et al., 1998b, Sequence stratigraphy of Pliocene and Pleistocene tubidite systems, northern Green Canyon and Ewing Bank (offshore Louisiana), northern Gulf of Mexico: AAPG Bulletin, v. 82, no. 5B, p. 918-960.
Winker, C. D., and R. T. Buffler, 1988, Paleogeographic evolution of early deep-water Gulf of Mexico and margins, Jurassic to middle Cretaceous (Comanchean): AAPG Bulletin, v. 72, p. 318-345.

## ABOUT THE AUTHORS

### Paul Weimer

Paul Weimer is an associate professor and director of the Energy and Minerals Applied Research Center at the University of Colorado at Boulder. His research interests are in sequence stratigraphy and basin analysis, three-dimensional seismic interpretation, and reservoir geology.

### Mark G. Rowan

Mark Rowan is a graduate of Caltech (1976), University of California, Berkeley (1982), and the University of Colorado (1991), where he is currently an assistant research professor in structural geology. He also has worked in the petroleum industry, both as an exploration geologist for Sohio and as a consultant with Geo-Logic Systems and Alastair Beach Associates. His research interests focus on the three-dimensional geometry and evolution of both extensional and contractional structures, with a recent emphasis on salt tectonics.

### Barry C. McBride

Barry McBride is a geologist at HS Resources Inc. He received his Ph.D. in geology at the University of Colorado, his M.S. degree in geology from Western Michigan University, and his B.S. degree in geology from Furman University. From 1988 to 1993 he worked for Mobil Exploration and Producing U.S. Inc. and Mobil Global Interpretation Support as an exploration geologist. His research interests focus on complex structures, petroleum exploration, and petroleum systems.

### Roy Kligfield

Roy Kligfield received his B.A. (1972), M.Phil. (1976), and Ph.D. (1978) degrees from Columbia Universiy. He taught at the University of Colorado from 1981 to 1992. Kligfield's research and teaching career initially focused on the structural geology and tectonics of mountain belts and collisional basins, with fieldwork in the Alps, Himalayas, Apennines, and Basin and Range during the 1970s and 1980s. Beginning about 1980, he helped pioneer the development and applications of computer methods for subsurface structural and stratigraphic interpretation—especially the creation of balanced cross sections and their 3-D visualization. As part of this effort, he was a founding partner of Geo-Logic Systems, and later general manager of the geological division of CogniSeis Development. He is currently vice president of Development at SignalSoft Corp., as well as adjunct associate professor in the Department of Geological Sciences at the University of Colorado.

# Atlas of Petroleum Fields and Discoveries, Northern Green Canyon, Ewing Bank, and Southern Ship Shoal and South Timbalier Areas (Offshore Louisiana), Northern Gulf of Mexico[1]

**Paul Weimer,[2] Jennifer R. Crews,[2] Ryan S. Crow,[2] and Peter Varnai[3]**


## ABSTRACT

Thirty-five fields and discoveries from the northern Green Canyon, Ewing Bank, and southern Ship Shoal and South Timbalier protraction areas (offshore Louisiana) are characterized by a series of seismic profiles, wireline logs, and summaries of their production characteristics. All fields or discoveries consist of combined structural/stratigraphic traps. Reservoirs occur in various elements of turbidite systems—primarily in sheet sands, channel fill, and thin beds. Best production rates occur in sheet sand facies. Development facilities vary from fixed platforms to subsea tiebacks to existing fields.


## INTRODUCTION

Although the northern deep Gulf of Mexico is the major exploration province in the United States today, few examples have been published on the setting of the fields. This paper presents the seismic stratigraphic and wireline-log expression of 35 fields or discoveries in the northern Green Canyon, Ewing Bank, and southern Ship Shoal and South Timbalier protraction areas (Figure 1 for location, Figure 2 for index map, Figures 3–37 for data of fields, Table 1 for Catalog). Tables with the figures summarize the salient characteristics of each field. The purpose of the paper is to provide the AAPG membership with an overview of selected fields from this important province. This paper is designed to complement the AAPG *Treatise of Petroleum Geology* series on different kinds of petroleum fields (Beaumont and Foster, 1990–1992).

All the fields in the study area consist of combined structural and stratigraphic traps. Salt and/or faults play a critical role in the trapping mechanisms. Using the reservoir classification of Chapin et al. (1994), the depositional environments of the producing sands consist of sheet and amalgamated sheet sands, channel-fill sands, or thin sands encased in overbank shales. These latter sands are known as thin-bedded pay (Shew et al., 1994) or low-contrast, low-resistivity pay. An important aspect of these fields is that although a fair amount of sand is encased in shales, as indicated on the well logs, only one sand or series of sand bodies gets charged with petroleum. Clearly, a complex migration history occurs in these fields after the trapped petroleum is expelled from beneath the salt sheets (McBride et al., 1998, this issue). The fields are oil, gas, and combined producers. The reserves of the fields are quite variable, ranging from a few million barrels



[1]Manuscript received December 16, 1996; revised manuscript received January 13, 1998; final acceptance January 14, 1998.

[2]Energy and Minerals Applied Research Center, Department of Geological Sciences, University of Colorado, Boulder, Colorado 80309-0399.

[3]Energy and Minerals Applied Research Center, Department of Geological Sciences, University of Colorado, Boulder, Colorado 80309-0399. Present address: Exxon Exploration Company, P.O. Box 4480, Houston, Texas 77210.

This paper was compiled as part of a laboratory exercise for a graduate course, co-taught by Roger Slatt and Paul Weimer, entitled, "Petroleum Geology of Turbidite Systems." In compiling the information in the paper, we have spoken with many people who kindly supplied us with nonproprietary information about the fields. We are grateful to the following people for their help: Robert Bunch (Mobil), Wayne Camp (Anadarko), John Dombrowski (Texaco), Ben Funderburk (Forest), Peter Gale (British-Borneo), Gary Guthrie (Marathon), William Holman (Newfield), Mike Jobe (Walter Oil), Todd Jones (Newfield), Michael Joseph (Texaco), Alan Kornacki (Shell), David Lawrence (Shell), Mike Mahaffie (Shell), Vinod Mathur (Mobil), William Mills (BP), Clint Moore (Anadarko), John Pike (Tathum), John Rafalowski (Chevron), David Robertson (Forest), Tricia Santogrossi (Vastar), William Schneider (Conoco), Kevin Schofield (Conoco), Roger Shew (Shell), Mike Smith (CNG), Mike Strickler (Texaco), Scott Tinker (Marathon), Brian Wallick (Mobil), David Walz (CNG), Tom Wilson (Shell), and Kris Wojcik (Shell). We thank Mark Rowan for his help with the structural interpretations shown here. We extend special thanks to Halliburton Geophysical (now Western Geophysical), specifically to Bob Graebner, John Anderson, and Gary White, for the seismic data used in this project. We are indebted to PaleoData Inc. and Art Waterman for use of their biostratigraphic data. Digital logs were kindly provided by Marathon (Tricia Santogrossi). We thank Davis Ratcliff and Diamond Geophysical, SEPM, and the Gulf Coast Section SEPM Foundation for their permission to reprint figures shown here. We thank Kevin Biddle (former AAPG elected editor), and AAPG reviewers Brad Prather and Denise Stone for their reviews of an earlier version of this paper, although we alone are responsible for the interpretations presented here. Acknowledgment is made to the donors of the Petroleum Research Fund, administered by the American Chemical Society, for partial support of this work.

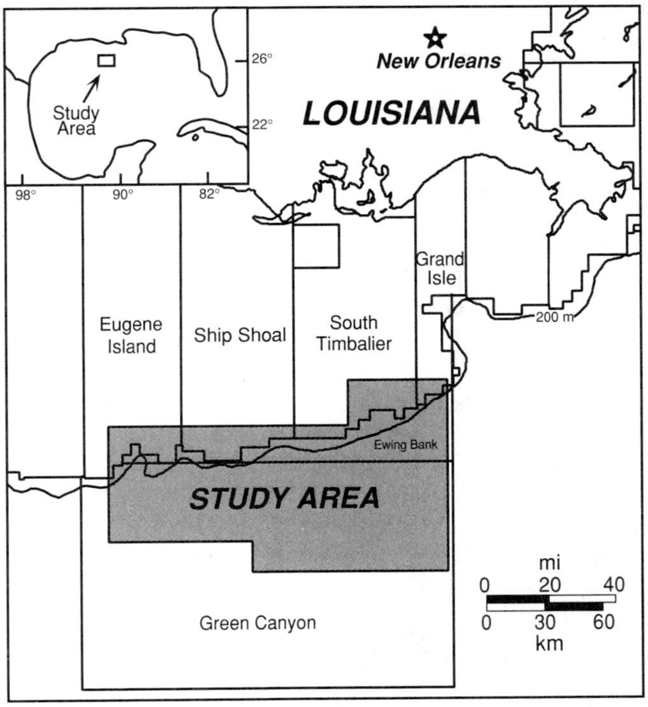


**Figure 1—Location of study area in northern Gulf of Mexico. Area includes northern Green Canyon, Ewing Bank, and southernmost portions of Eugene Island, Ship Shoal, South Timbalier, and southeastern Grand Island.**

**Table 1. Catalog of the 35 Fields Represented in Figures 3–37**

| Figure Number | Field Name | Block Number |
|---|---|---|
| 3 | King Kong | GC 472, 473 |
| 4 | Allegheny | GC 253, 254, 297, 298 |
| 5 | Bison | GC 166 |
| 6 | Green Canyon 31 | GC 31 |
| 7 | Green Canyon 29 | GC 29 |
| 8 | Popeye | GC 72, 73, 116 |
| 9 | Genesis | GC 161, 205 |
| 10 | Glider | GC 248 |
| 11 | Manta Ray | EW 1006 |
| 12 | Arnold | EW 963 |
| 13 | Oyster | EW 917 |
| 14 | Lobster | EW 873 |
| 15 | North Lobster | EW 947, ST 308 |
| 16 | Teak | ST 260 |
| 17 | Ewing Bank 826 | EW 826 |
| 18 | Seattle Slew | EW 914 |
| 19 | Sunday Silence | EW 1003 |
| 20 | Brutus | GC 158 |
| 21 | Troika | GC 200, 201, 244, 245 |
| 22 | Rocky | GC 110 |
| 23 | Bullwinkle | GC 65 |
| 24 | Stellaria | GC 112, 113 |
| 25 | Green Canyon 20 | GC 20 |
| 26 | Boxer | GC 19 |
| 27 | Green Canyon 18 | GC 18, EW 988 |
| 28 | Ewing Bank 947 | EW 947 |
| 29 | Mahogany | SS 349, 359 |
| 30 | Agate | SS 361 |
| 31 | Green Canyon 60 | GC 60 |
| 32 | Marquette | GC 52, 53 |
| 33 | Jolliet | GC 184 |
| 34 | Shasta | GC 136 |
| 35 | Kilauea | GC 6 |
| 36 | Ewing Bank 977 | EW 977 |
| 37 | Ewing Bank 991 | EW 991 |

and/or a few billion cubic feet of gas to greater than 200 million barrels. Production facilities include fixed-leg platforms, tension-leg platforms, and subsea tiebacks to existing platforms. The companies involved include major oil companies and large independents. Clearly, the economics for development in this area have changed considerably during the past decade as a production infrastructure has developed.

A few caveats are necessary regarding the data that are presented here. (1) All seismic profiles shown here are two-dimensional (2-D) seismic data reprocessed in 1989 from 3-D seismic data collected in 1984–1986. Newer seismic data, especially three-dimensional (3-D) data, can image the reservoirs far better than the seismic data shown here. (2) No detailed reservoir maps (e.g., structure contour, sand percentage, amplitude extraction) are shown because we lack detailed 3-D data and details of the development wells; additionally, the 2-D seismic line spacing is too great to map the reservoir in detail. (3) The reservoir characteristics listed in the summary tables are derived from a number of public sources. Much of the detailed field information remains proprietary. (4) Because this is an active exploration area, well logs from eight discoveries have not yet been released. (5) The interpretation of depositional environments is based on an integrated approach. We do not have access to core information so that we can be more precise in our interpretations of subenvironments within the overall turbidite systems. Consequently, we have simplified the depositional interpretations. (6) We have high-resolution biostratigraphic data for some of the producing wells illustrated here; consequently, some of the interpretation of ages and location of sequence boundaries, systems tracts boundaries, and condensed sections are based on regional correlations (Weimer et al., 1998, this issue). (7) We did not include summaries of fields or discoveries that were on the edge of our areas where we did not have enough seismic data to illustrate the fields (e.g., Morpeth, Ewing Bank 921/922/964; Fuji, Green Canyon 506/507).

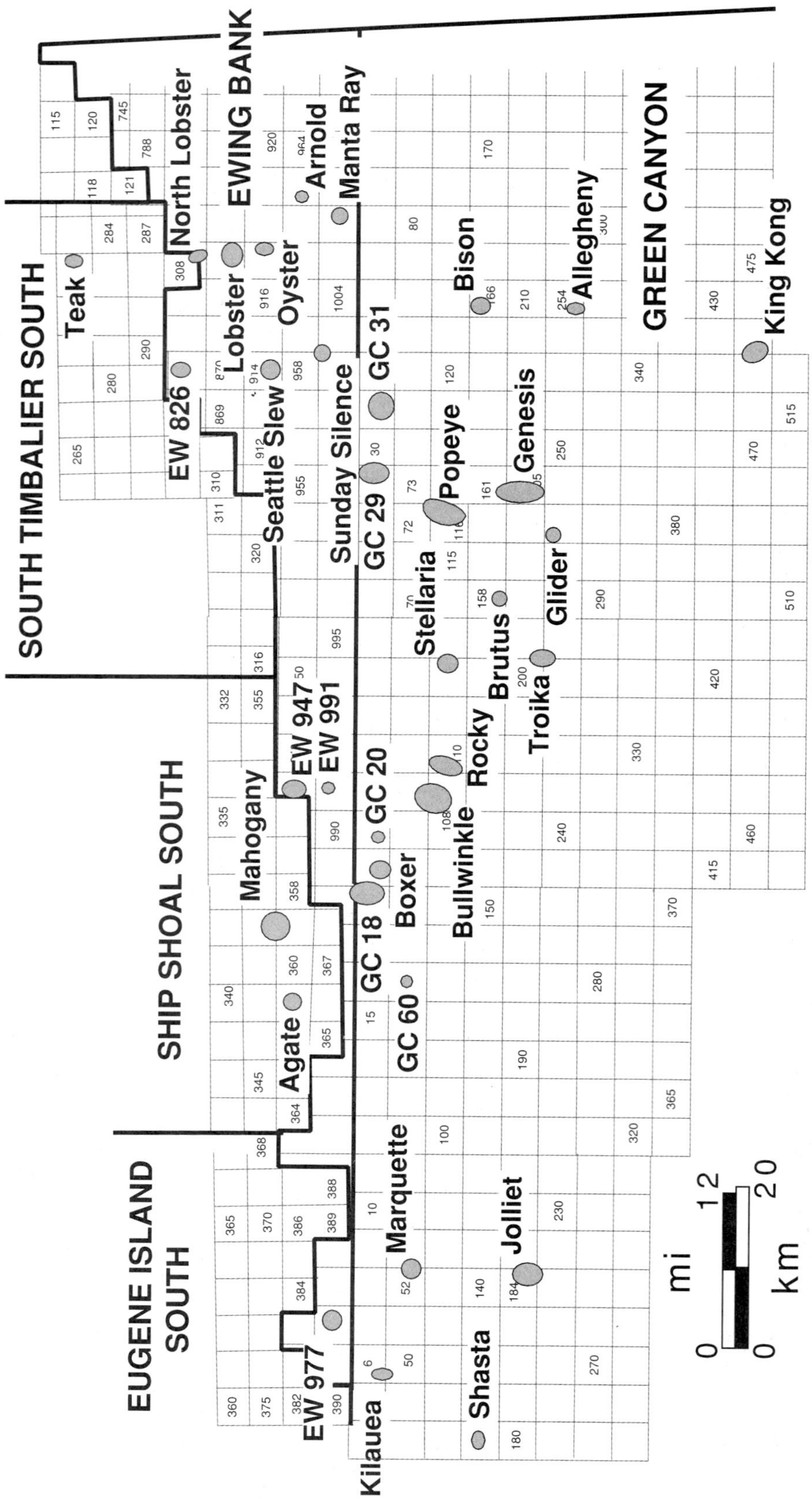


Figure 2—Location map of fields and discoveries in the study area. GC = Green Canyon; EW = Ewing Bank.

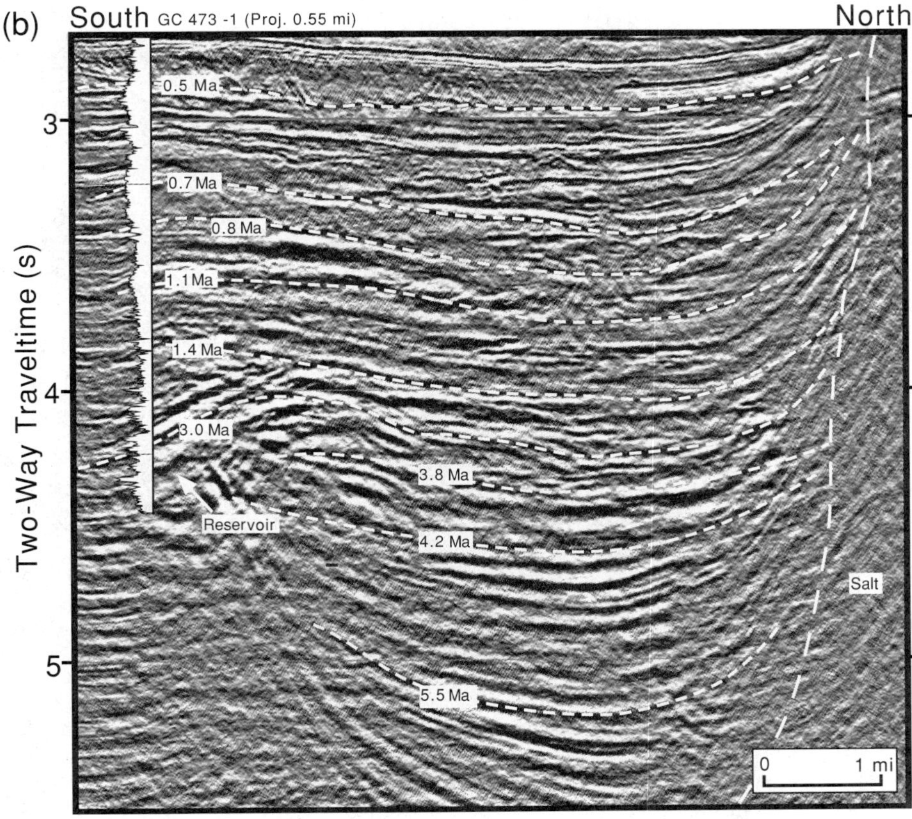


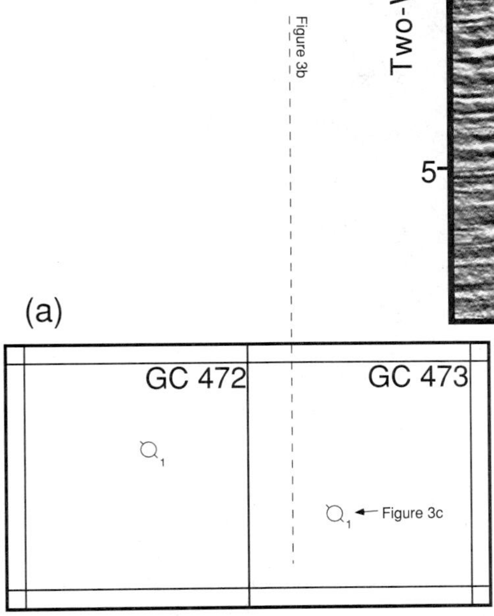


## Green Canyon 472, 473
## King Kong

**Partners:** British Borneo 50%, Shell 50%
**Discovery Date:** 1989
**Type of Production Facility:** Subsea tieback planned
**Water Depth:** 3700 feet
**Number of Wells:** 2
**Age:** early Pliocene
**Status:** Under evaluation

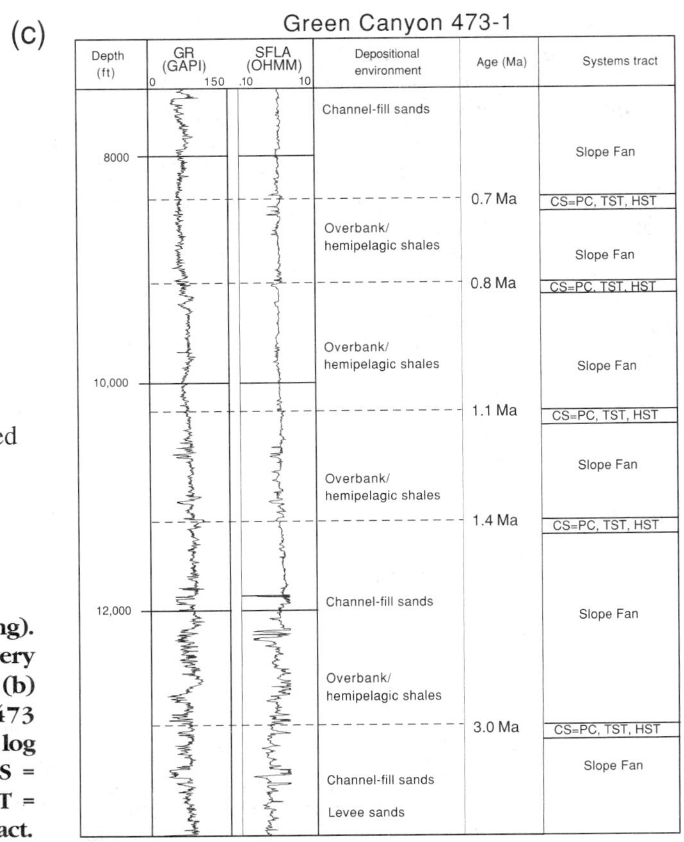


**Figure 3—Green Canyon 472-473 discovery (King Kong). (a) Location map of the Green Canyon 472-473 discovery (King Kong). Also shown are locations for (b) and (c). (b) Seismic profile across the Green Canyon 472-473 discovery (King Kong). (c) Gamma-ray and resistivity log from the GC 473-1 well, King Kong discovery. CS = condensed section, PC = prograding complex, TST = transgressive systems tract, HST = highstand systems tract.**

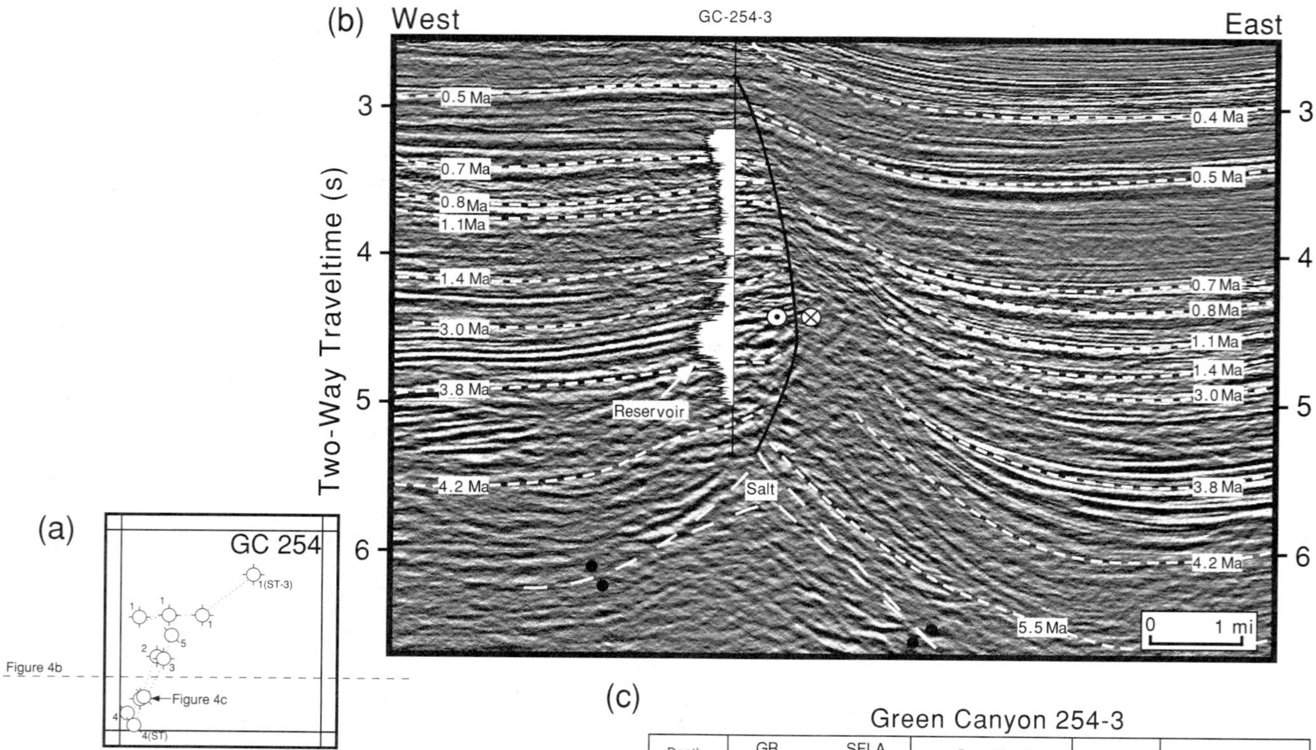


**Green Canyon 253, 254, 297, 298**
**Allegheny**

**Partners:** British Borneo 60%, Reading & Bates 40%
**Discovery Date:** November 1991
**Production anticipated to begin:** third quarter 1999
**Type of production facility:** Sea Star mini-tension leg platform with subsea satellite wells
**Water Depth:** 3115'
**Reserves:** 80 million barrels
**IP:** 25,000 BOPD, 35 MMSCF/day
**Number of reservoirs:** 5
**Depositional environment:** amalgamated sheets sands (Basin-floor fan)
**Number of wells:** 5
**Age:** early Pliocene
**Producing Interval:** 11,500'–15,200'
**Seals upper:** shale
**lateral:** shale
**Size acres:** 500
**API:** 26-44 degrees
**Average GOR:** 1350 SCF/Bbl
**Status:** Appraisal Drilling
**Source of information:** Haines, 1996; R. Bunch, pers. comm., 1996; press release, P. Gale, British Borneo, Inc., pers. comm., 1998

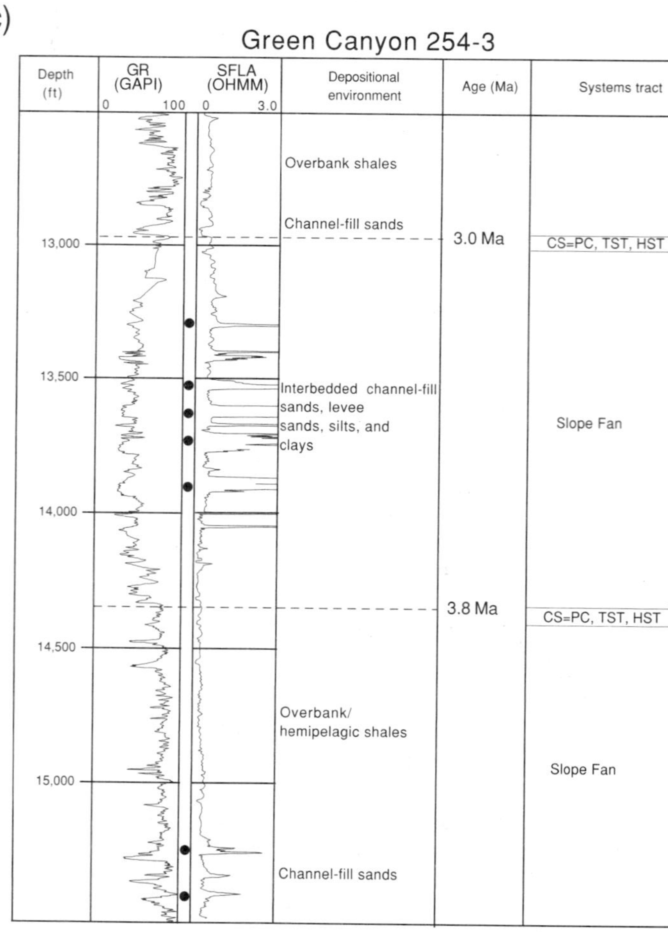


**Figure 4—Green Canyon 254 field (Allegheny). (a) Location map of the Green Canyon 254 field (Allegheny). Also shown are locations for (b) and (c). (b) Seismic profile across the Green Canyon 254 field (Allegheny field). (c) Gamma-ray and resistivity log from the GC 254-3 well, Allegheny field. CS = condensed section, PC =prograding complex, TST = transgressive systems tract, HST = highstand systems tract.**

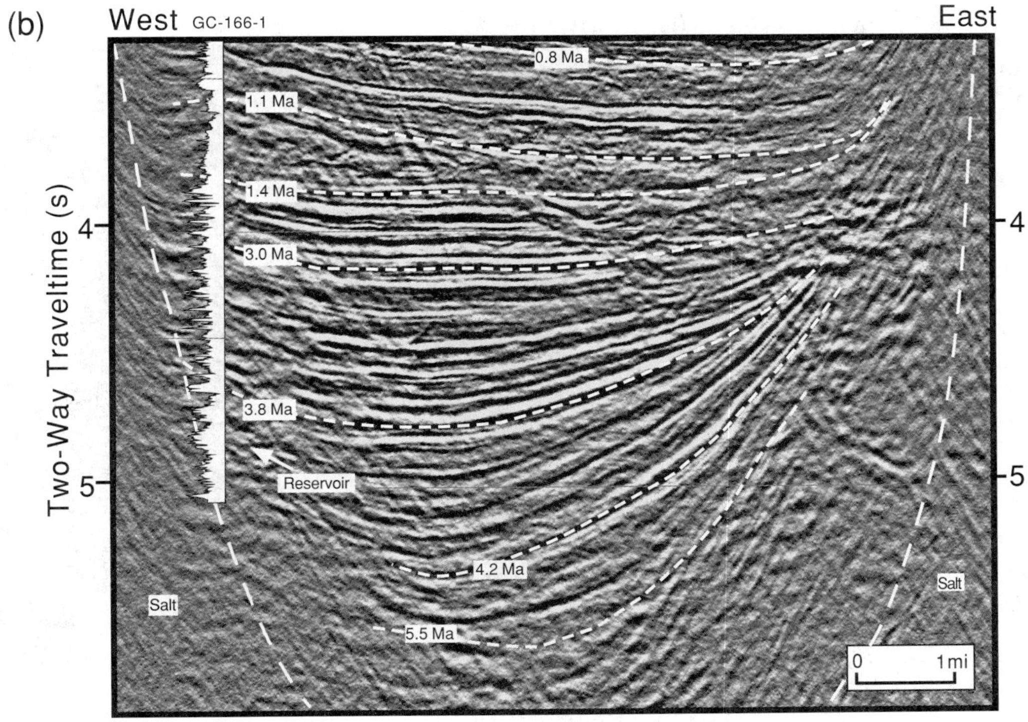


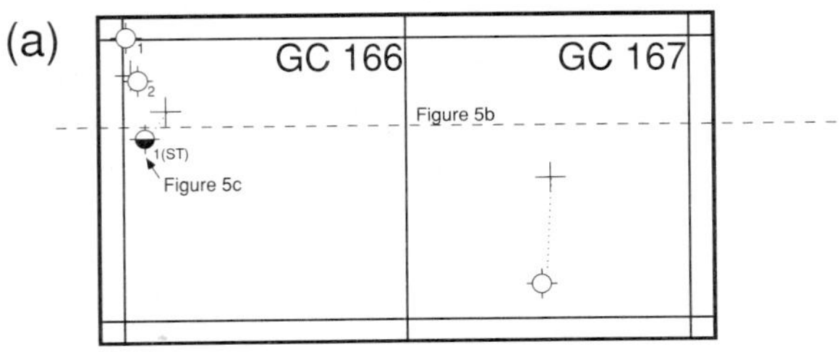


**Green Canyon 166**
**Bison**

**Partners:** Exxon (originally), now BP (100%)
**Discovery Date:** March, 1986
**Water Depth:** 2381 feet
**Number of Wells:** 5
**Age:** lower Pliocene
**Producing Interval:** 16,800; 17,050-17,150 feet
**Status:** Initial discovery was not developed, block relinquished, and leased in 1995 by BP. Currently under evaluation.

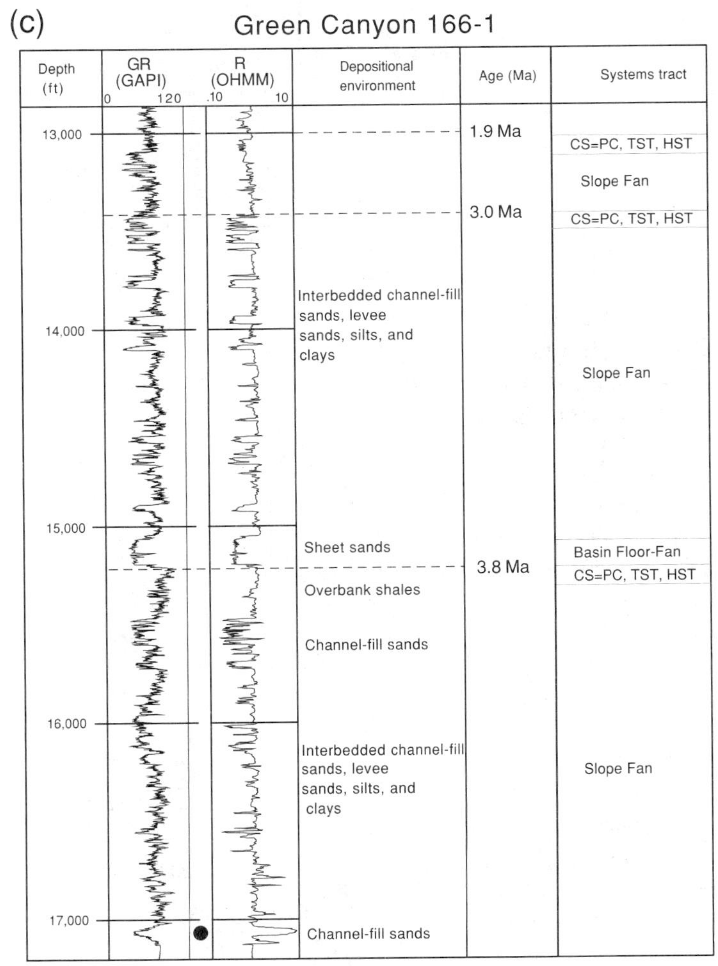


**Figure 5—Green Canyon 166 Discovery (Bison). (a) Location map of the Green Canyon 166 Discovery (Bison). Also shown are locations for (b) and (c). (b) Seismic profile across the Green Canyon 166 Discovery (Bison). (c) Gamma-ray and resistivity log from the GC 166-1 well, Bison discovery. CS = condensed section, PC = prograding complex, TST = transgressive systems tract, HST = highstand systems tract.**

**Figure 6—Green Canyon 31 field. (a) Location map of the Green Canyon 31 field. Also shown are locations for (b) and (c). (b) Seismic profile across the Green Canyon 31 field. (c) Gamma-ray and resistivity log from the GC 31-1 well. CS = condensed section, PC = prograding complex, TST = transgressive systems tract, HST = highstand systems tract.**

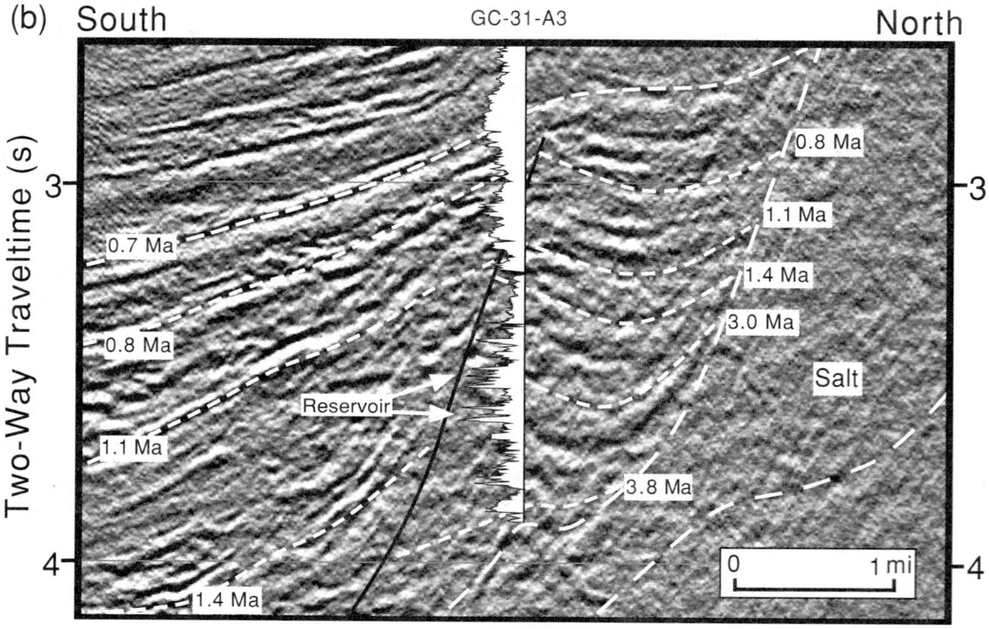


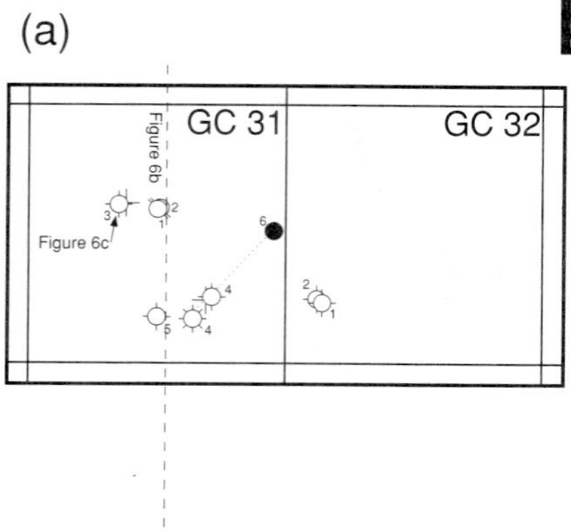


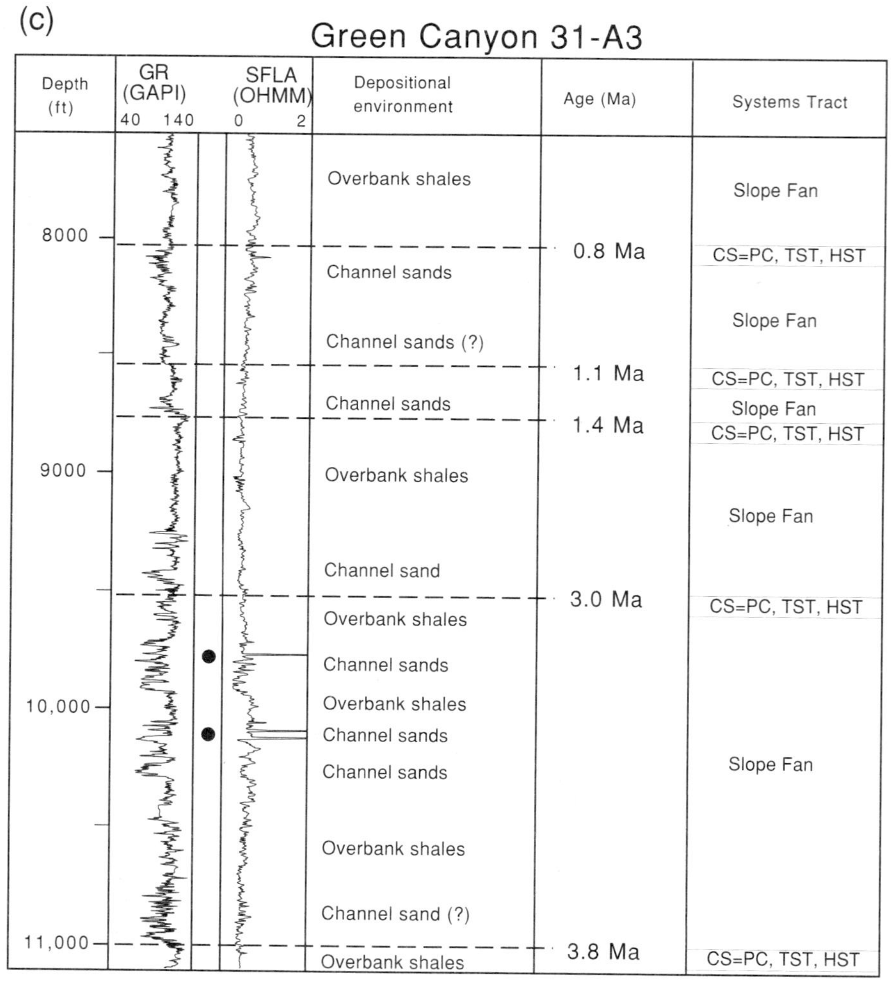


### Green Canyon 31

**Partners:** Placid
**Discovery Date:** February 1985
**Production began:** 1989
**Type of production facility:** subsea tieback to GC 29, abandoned
**Water Depth:** 2050 feet
**Number of reservoirs completed:** 1
**Gross thickness:** 75 feet
**Depositional environment:** channel-fill
**Number of wells:** 6
**Age:** early Pleistocene
**Producing Intervals:** 9700–9900, 10,100–10,150 feet
**Seals upper:** shale
**lateral:** shale
**Status:** produced minimum oil, abandoned
**Source of information:** Mike Smith, pers. comm., 1996

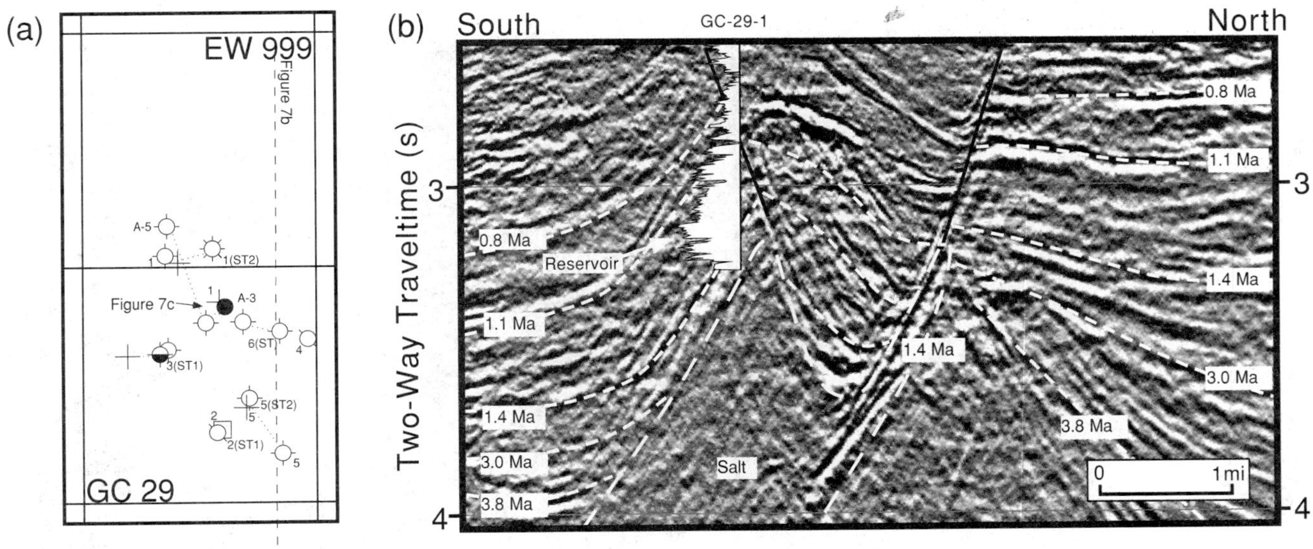


**Green Canyon 29**

**Partners:** Placid
**Discovery Date:** May 1985
**Production began:** 1989
**Type of production facility:** Floating production platform, abandoned
**Water Depth:** 1519 feet
**Reserves:** 0.5 million barrels
**Number of reservoirs completed:** 1
**Reservoir Depth:** 8074
**Gross thickness:** 75 feet
**Gross pay thickness:** 10 feet
**Drive mechanism:** Unknown
**Average Porosity:** 33.0
**Reservoir Temperature:** 166°F
**Reservoir Pressure(BHP):** 3769
**Water Saturation:** 40.0
**API gravity:** 35.0
**Gas Gravity:** 0.60
**Gas-oil ratio:** 8440
**Yield:** 27.4
**Cumulative oil-cond production:** 90
**Cumulative gas production:** 1598
**Oil reserves:** 0
**Oil recovery efficiency:** 0
**Gas recovery efficiency:** 0.10
**Depositional environment:** channel-fill
**Number of wells:** 10
**Age:** early Pleistocene
**Producing Interval:** 9400–9470 feet
**Trap:** Sediments overlying dome
**Seals upper:** shale
  **lateral:** shale
**Status:** produced 0.5 million barrels, abandoned
**Source of information:** Mike Smith, pers. comm., 1996; MMS, 1997

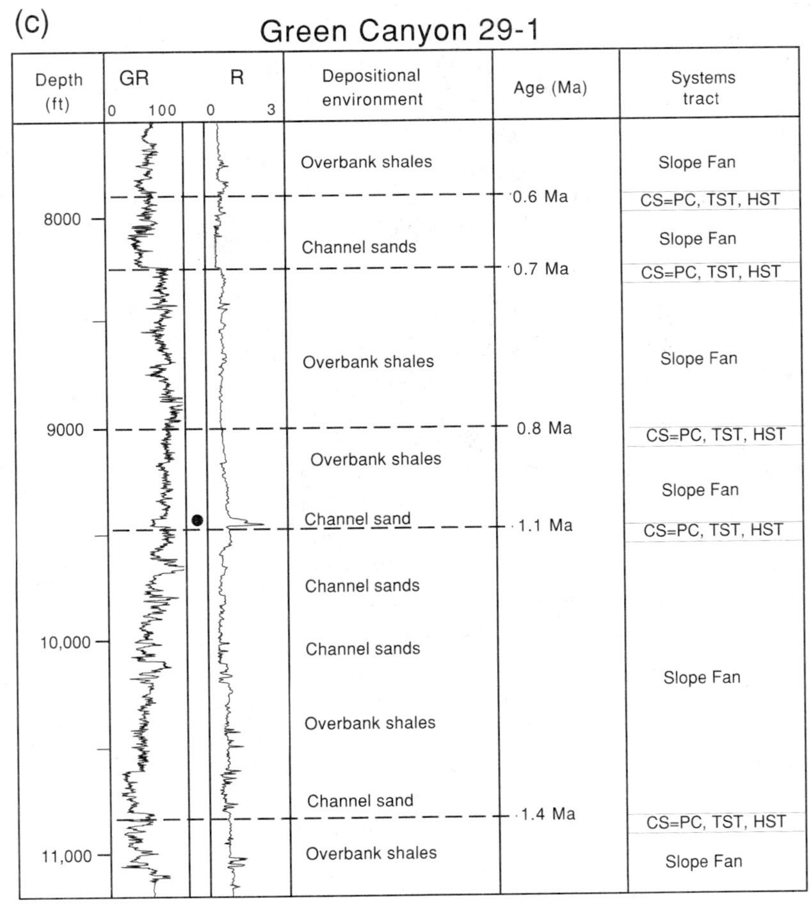


**Figure 7—Green Canyon 29 field. (a) Location map of the Green Canyon 29 field. Also shown are locations for (b) and (c). (b) Seismic profile across the Green Canyon 29 field. (c) Gamma-ray and resistivity log from the GC 29-1 well. CS = condensed section, PC = prograding complex, TST = transgressive systems tract, HST = highstand systems tract.**

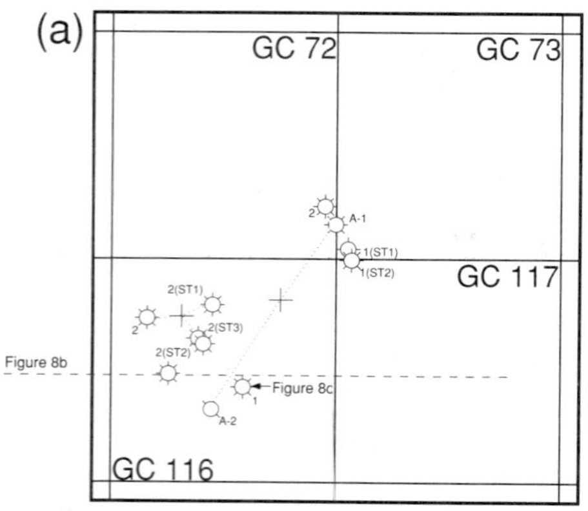


**Green Canyon 72, 73, 116, 117**
**Popeye**

Note: there are two separate fields being produced together. One is in GC 72, 73; the second one is in GC 116, 117.

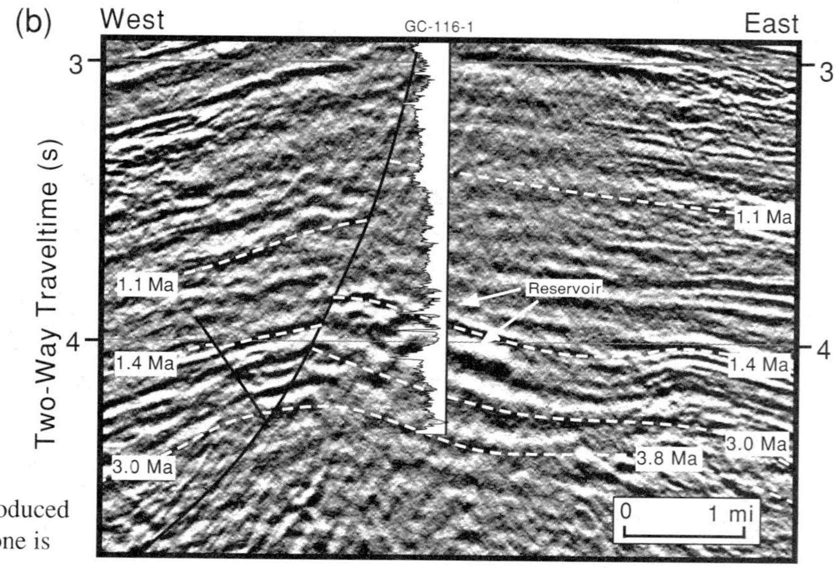


**Partners:** Shell Offshore, Inc. 37.5%, CNG 37.5%, Mobil 12.5%, BP 12.5%
**Discovery Date:** January 1985
**Production began:** January 1996
**Type of production facility:** Subsea tieback to Cougar platform, South Timbalier 300
**Water Depth:** 2100 feet
**Reserves:** 320 Bcfg, 7 MMbc
**Number of reservoirs completed:** 4
**Gross pay thickness:** 21–37 feet
**Average Porosity:** 29%
**Reservoir Temperature:** 194°F
**Reservoir Pressure (BHP):** 8208–8461
**Water Saturation:** 17.0
**Depositional environment:** channel-fill, thin-bedded levee, sheets
**Number of wells:** 9
**Age:** upper Pliocene, lower Pleistocene
**Producing Interval:** 11,250'–11,350', 12,100'–12,400'
**API Gravity:** 30.0
**Gas Gravity:** 0.60
**Gas-oil ratio:** 0-1000
**Yield:** 0.1
**Source of information:** Haines, 1996; Shew, 1997

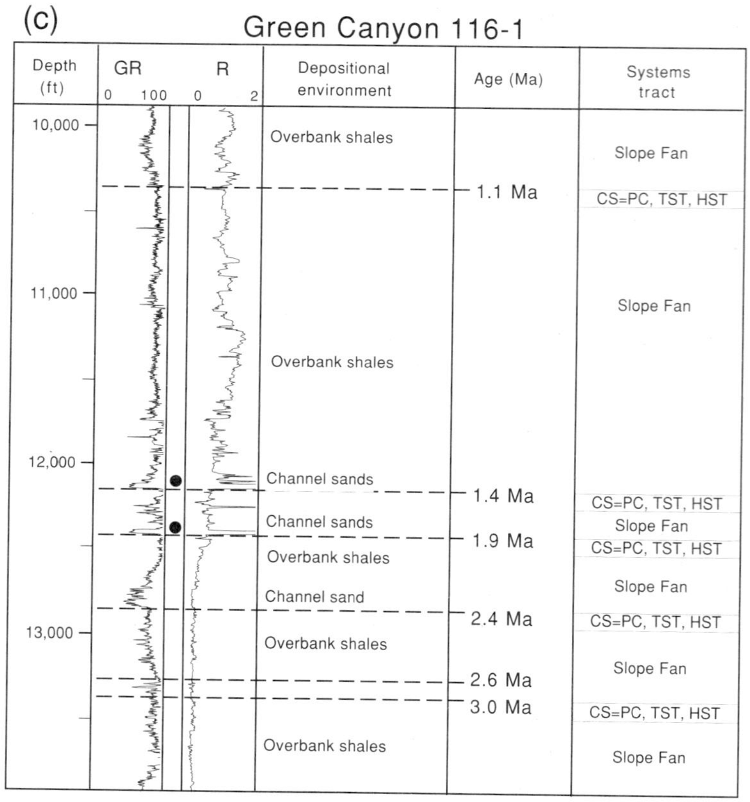


**Figure 8—Green Canyon 72, 73, 116 field (Popeye). Note that there are two separate fields being produced together. The first field is GC 72, 73, and the second field is GC 116, 117. (a) Location map of the Green Canyon 72, 73, 116 field (Popeye). Also shown are locations for (b) and (c). (b) Seismic profile across the Green Canyon 116 field (Popeye). (c) Gamma-ray and resistivity log of the GC-116-1 well from the Popeye field. CS = condensed section, PC = prograding complex, TST = transgressive systems tract, HST = highstand systems tract.**

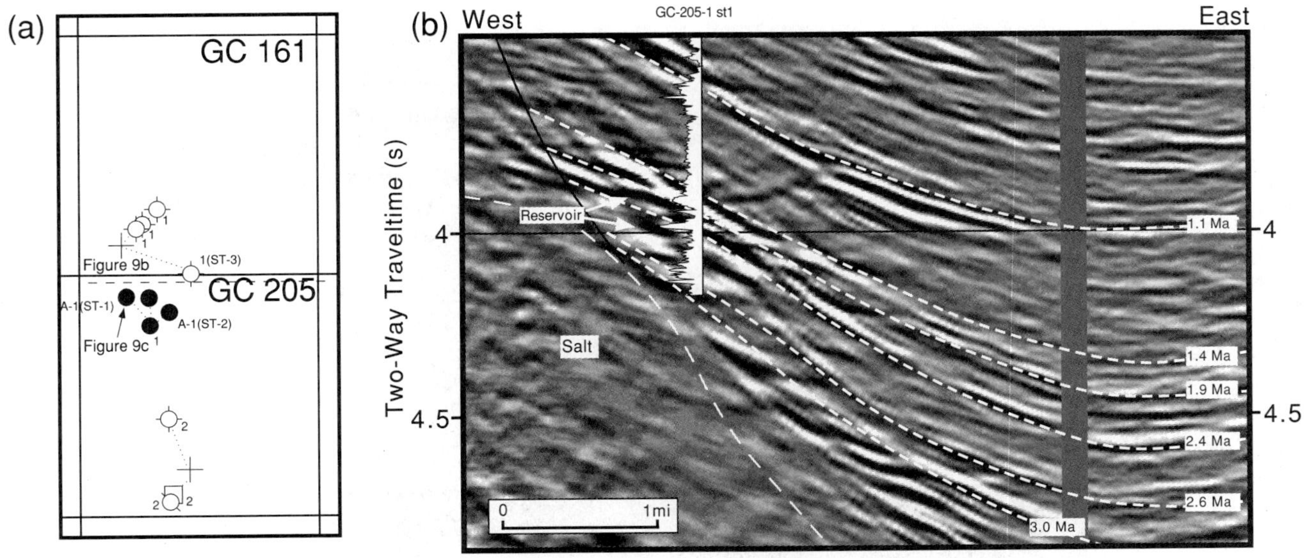


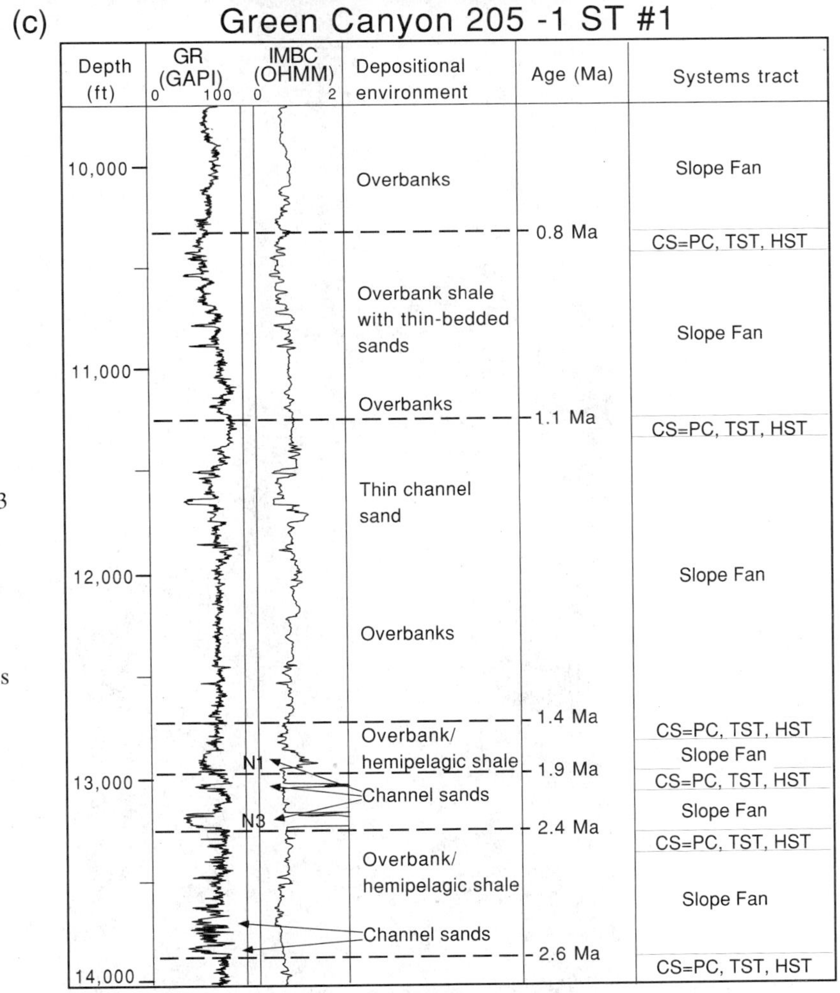


**Figure 9—Green Canyon 161/205 field (Genesis). (a) Location map of the Green Canyon 161/205 field (Genesis). Also shown are locations for (b) and (c). (b) Seismic profile across the Green Canyon 161/205 field (Genesis). (c) Gamma-ray and resistivity log from the GC 205-1 well, Genesis field. CS = condensed section, PC = prograding complex, TST = transgressive systems tract, HST = highstand systems tract.**

### Green Canyon 205, 161
### Genesis

**Partners:** Chevron 56.67%, Exxon 38.38%, Fina 4.95%
**Discovery Date:** June 1988
**Production began:** Anticipated last quarter, 1998
**Type of production facility:** Floating Production Facility, 20 slot
**Water Depth:** 2580 feet
**Water Depth:** 2630 feet
**Reserves:** 160 Million BBOE
**Anticipated production rates:** 55,000 bblspd, 72 million cubic ft/day
**IP:** N-3 8600 barrels/day 24/64" choke, N-1 2700 BBOE
**Number of reservoirs completed:** 6, N-1 and N-3 main reservoirs
**Depths:** 12,900–13,200'
**Gross Thickness:** 400 feet
**Average Porosity:** 27–35%
**Average Permeability:** 0.2–4 darcies
**Depositional environment:** amalgamated channels and sheets
**Number of wells:** 10 to date, from five locations
**Age:** late Pliocene
**Trap configuration:** sand pinchout on flank of salt feature
**Seals: upper:** shale
**lateral:** shale/salt
**Secondary recovery:** N/A
**Oil sulfur** (3% elemental)
**Initial GOR:** 1000–1200
**Viscosity:** 0.6 –1.0 cp@150° F.
**Temperature:** 130–160°F
**Oil Column Thickness:** 1100–1500 feet
**Status:** Development drilling
**Source of information:** Rafalowski et al., 1994; John Rafalowski, pers. comm., 1996

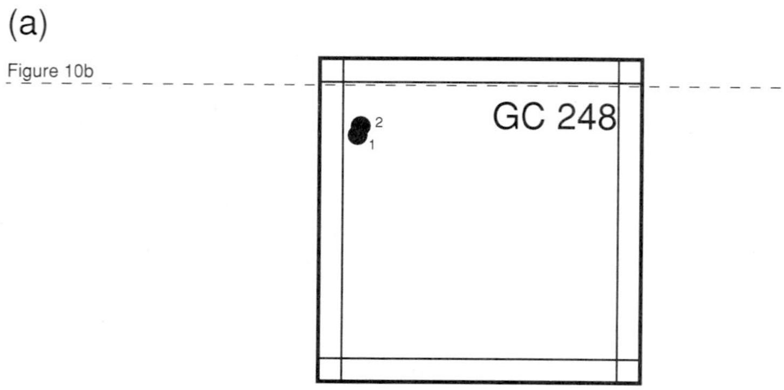


**Glider, GC 248**
**Discovery Date:** January 1997
**Operator:** Shell 100%
**Water Depth:** 3300 feet
**Number of Wells:** 2
**Age:** early Pleistocene (?)
**Status:** development plan underway

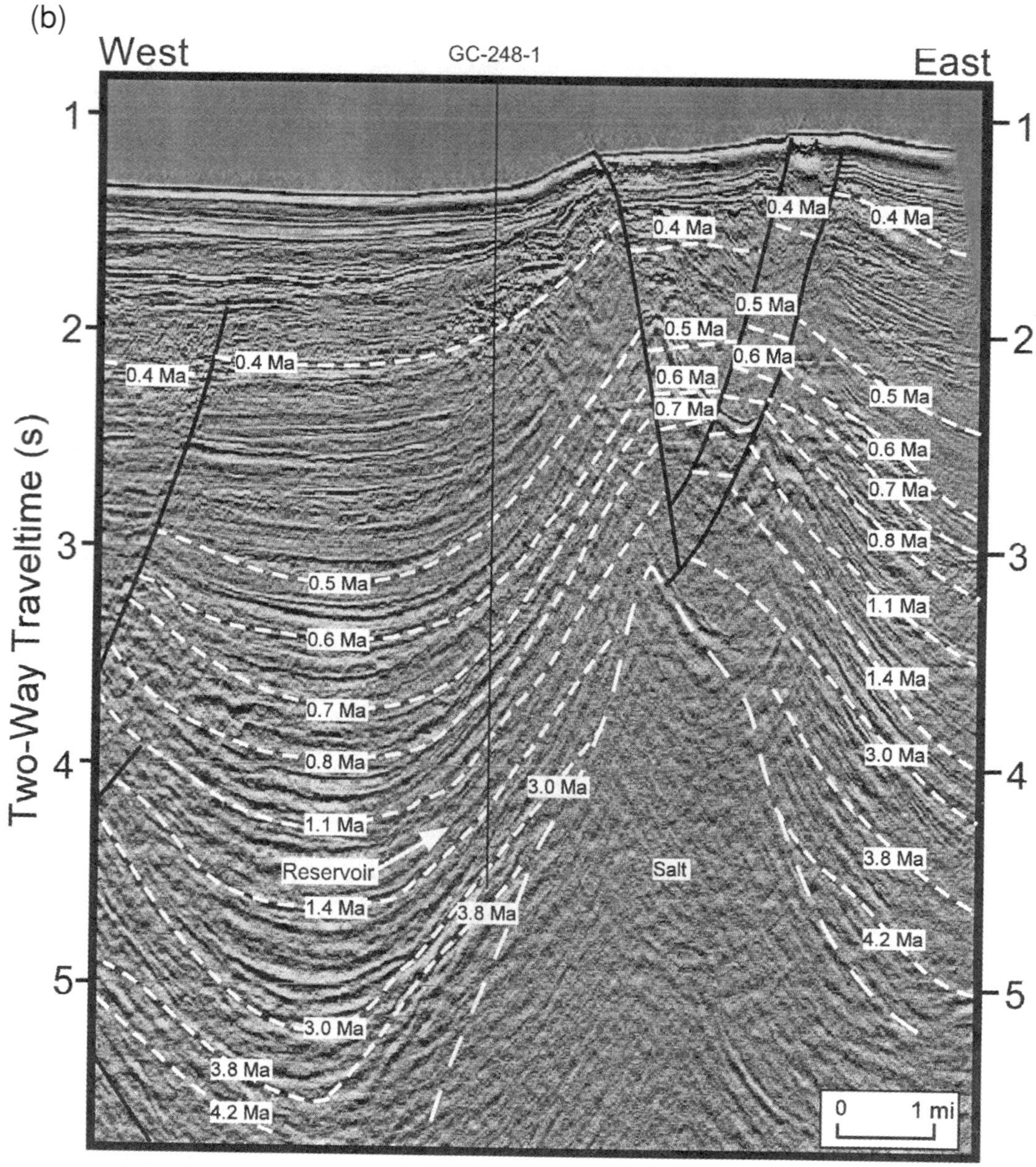


**Figure 10**—**Green Canyon 248 Discovery (Glider). (a) Location map of the Green Canyon 248 Discovery (Glider). Also shown is the location for (b). (b) Seismic profile across the Green Canyon 248 Discovery (Glider). Note that the well logs for the Glider discovery have not been released yet.**

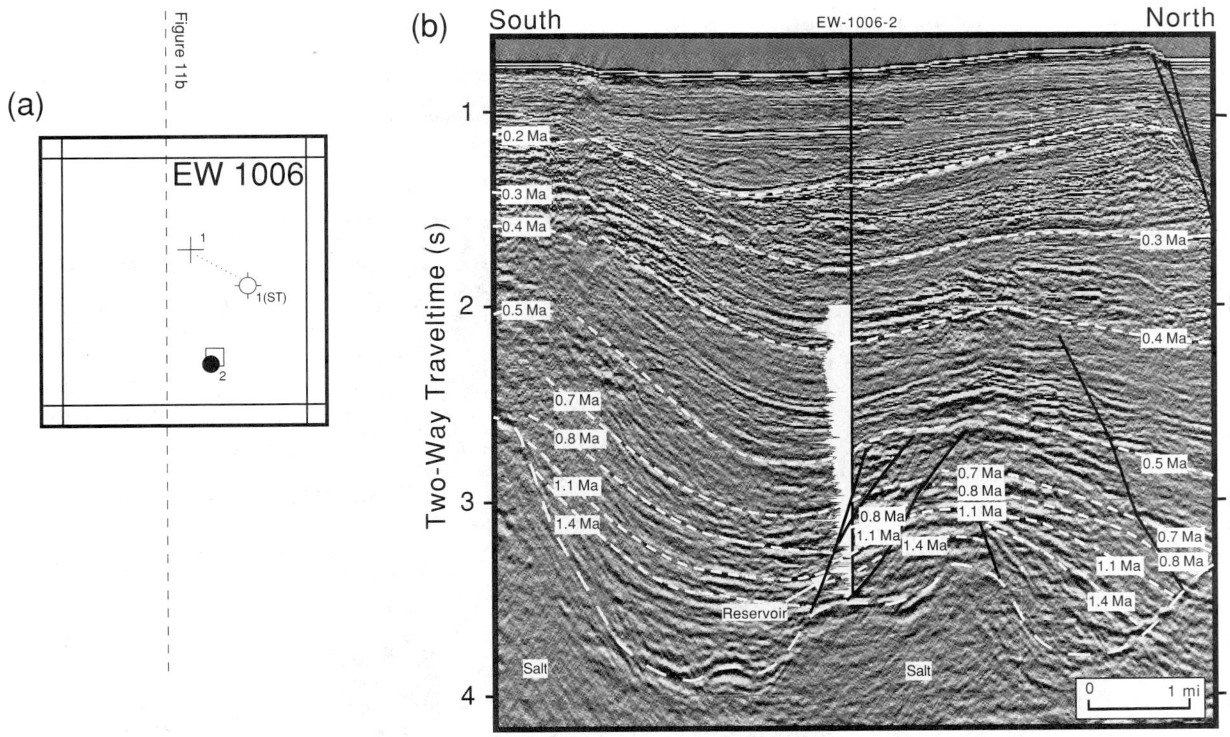


**Ewing Bank 1006**
**Manta Ray**

**Partners:** Walter Oil 100%
**Discovery Date:** January 1988
**Type of Production Facility:** Subsea tieback to Lobster (EW873) producing platform
**Water Depth:** 1882 feet
**Age of Reservoir:** late Pliocene
**Estimated Reserves:** 4–6 mmbo
**Reservoir Depth:** 11,310–11,350 Feet
**API Gravity:** 21
**Production:** waiting facilities hookup (1998)
**Source:** Press release, Mike Jobe, pers. comm., 1998

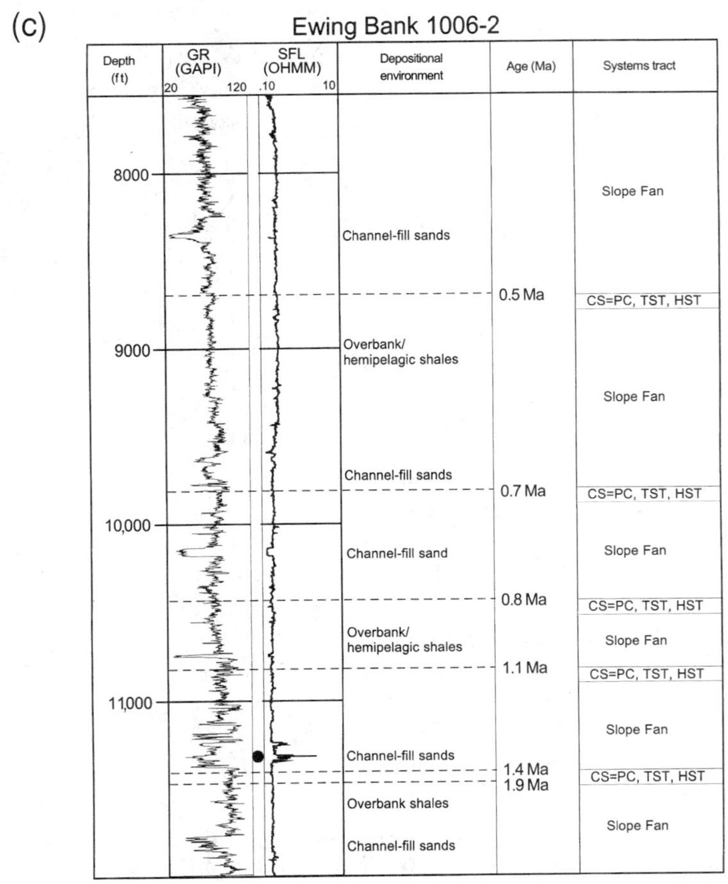


**Figure 11—Ewing Bank 1006 Discovery (Manta Ray). (a) Location map of the Ewing Bank 1006 Discovery (Manta Ray). Also shown are the locations for (b) and (c). (b) Seismic profile across the Ewing Bank 1006 Discovery (Manta Ray). (c) Gamma-ray and resistivity log for the EW 1006-1 well from the Manta Ray field. CS = condensed section, PC = prograding complex, TST = transgressive systems tract, HST = highstand systems tract.**

(a)

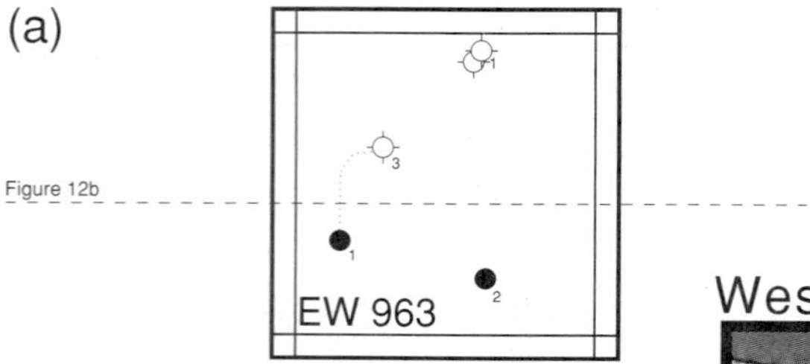


(b)

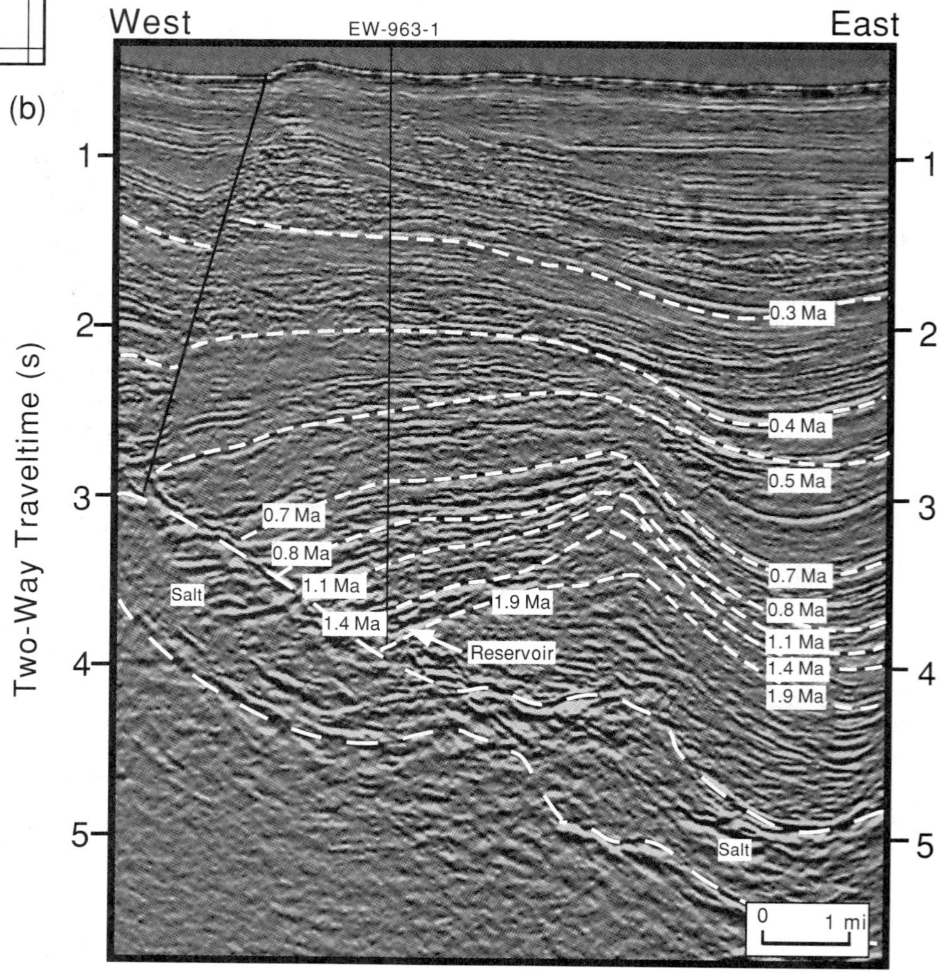


**Ewing Bank 963**
**Arnold**

**Partners:** Marathon 62.5%, Texaco 37.5%
**Discovery Date:** July 1996
**Production Began:** April 1998
**Production Facility:** subsea wells tied to Lobster (EW 873) producing platform
**Water Depth:** 1749 feet
**Estimated Reserves:** 20–25 mmbo
**Initial Tests:** 8700 bbls/day; 6.4 mcf/day
**Number of reservoirs completed:** 2
**Reservoir Depth:** Approx. 12,400 ft

**Gross Pay:** #1 well – 170′, #2 well – 84′, #3 well – 34′
**Depositional Environment:** channel-fill
**Number of wells:** 3
**Age:** late Pliocene
**HC Type:** Oil
**API Gravity:** 26
**Seals upper:** shale
**lateral:** shale
**Source of information:** press release

**Figure 12—Ewing Bank 963 Discovery (Arnold). (a) Location map of the Ewing Bank 963 Discovery (Arnold). Also shown is the location for (b). (b) Seismic profile across the Ewing Bank 963 Discovery (Arnold). Note that the well logs of the Arnold discovery have not been released yet.**

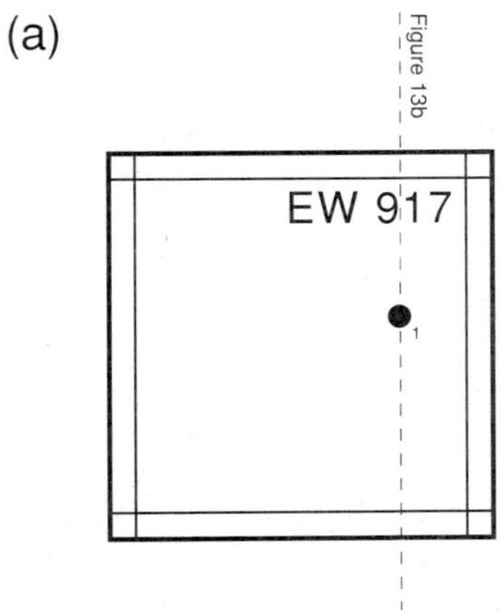


**Ewing Bank 917**
**Oyster**

**Partners:** Marathon 66.7%, Texaco 33.3%
**Discovery Date:** 1996
**Production Began:** January 1998
**Production Facility:** 1 subsea well tied to Lobster producing platform
**Water depth:** 1200 feet
**Reserves:** 10 mmbo
**Reservoir Depth:** 11,500 feet

**Number Wells:** 1 subsea
**Age:** late Pliocene
**HC Type:** oil
**API Gravity:** 25
**Trap:** undip pinchout
**Seals: Upper:** shale
**Lateral:** shale
**Source:** press release, 1997

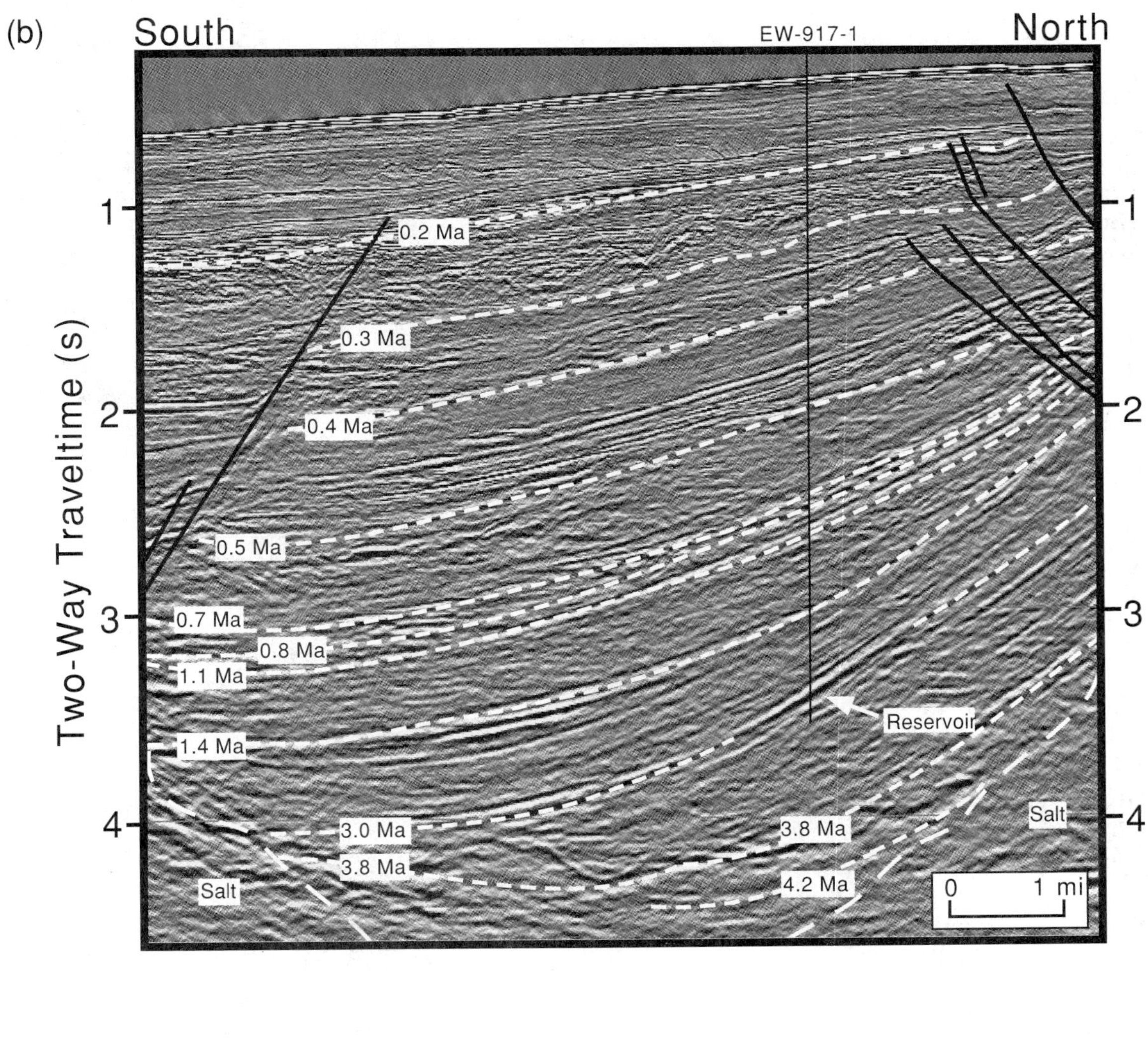


**Figure 13—Ewing Bank 917 Discovery (Oyster). (a) Location map of the Ewing Bank 917 Discovery (Oyster). Also shown is the location for (b). (b) Seismic profile across the Ewing Bank 917 Discovery (Oyster). Note that the well logs of the Oyster discovery have not been released yet.**

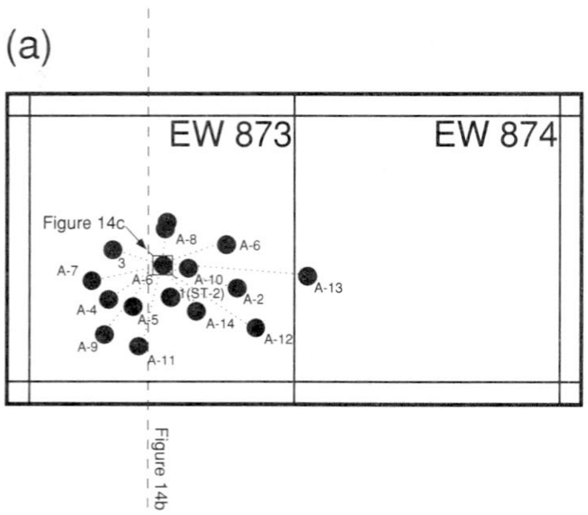


**Ewing Bank 873**
**Lobster**

**Partners:** Marathon 66.7%, Texaco 33.3%
**Discovery Date:** 1991
**Production began:** late 1995
**Type of production facility:** Fixed platform
**Water Depth:** 710 feet
**Production:** 46,500 BOPD, 31 MMCFGPD
**Number of reservoirs completed:** 12
**Depths:** 10,500–11,000 feet
**Gross thickness:** 100´ updip, 500´ in thickest part of reservoir
**Average Porosity:** 25–30%
**Average Permeability:** >1 darcy
**Depositional environment:** channels, sheet and amalgamated sheet sands
**Number of wells:** 16
**Age:** early Pliocene
**Seals upper:** shale
**lateral:** faults/shale
**Secondary recovery:** waterflood
**Oil API Gravity:** 18–26
**Sulfur:** 2.99-3.57 wt. %
**Initial GOR:** 500–1000
**Drive:** moderate aquifer drive
**Oil Column Thickness:** 3375 feet
**Status:** Producing, two additional development wells planned
**Source of information:** Edman and Burk, 1996; press release, 1997

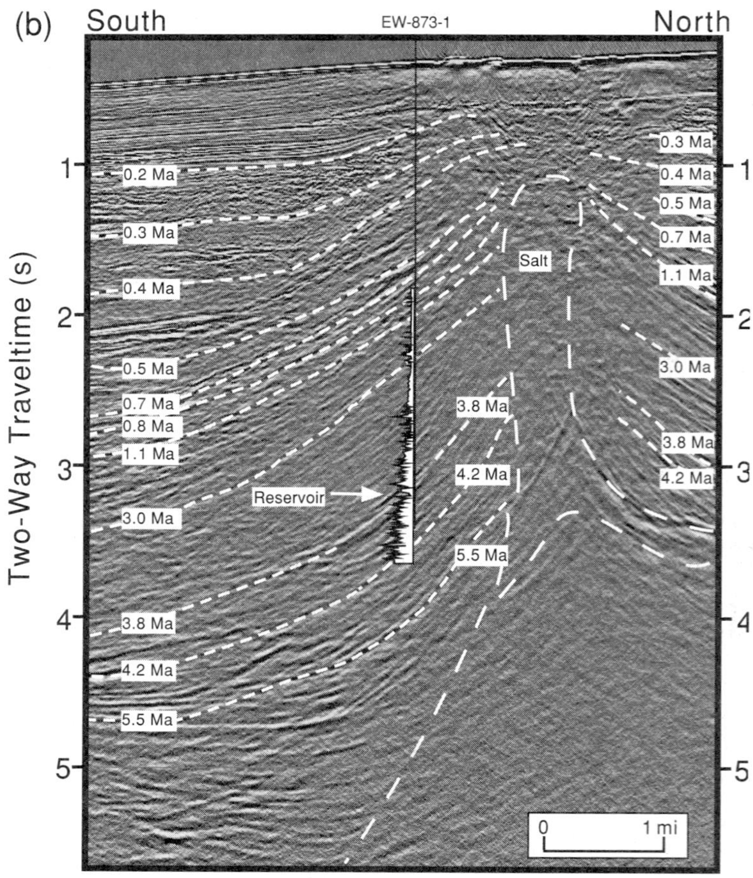


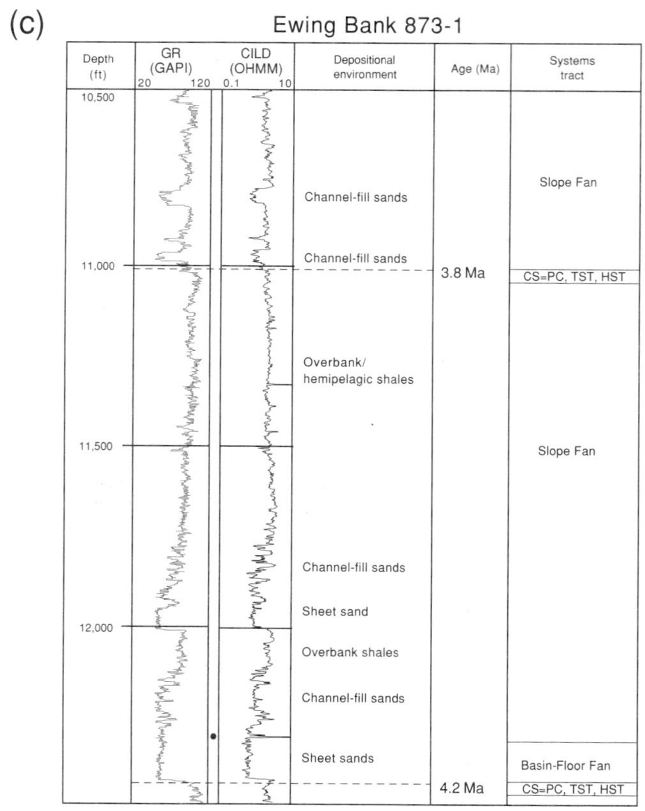


**Figure 14—Ewing Bank 873 field (Lobster). (a) Location map of the Ewing Bank 873 field (Lobster). Also shown are locations for (b) and (c). (b) Seismic profile across the Ewing Bank 873 field (Lobster). (c) Gamma-ray and resistivity log of the EW 873-1 well from the Lobster field. CS = condensed section, PC = prograding complex, TST = transgressive systems tract, HST = highstand systems tract.**

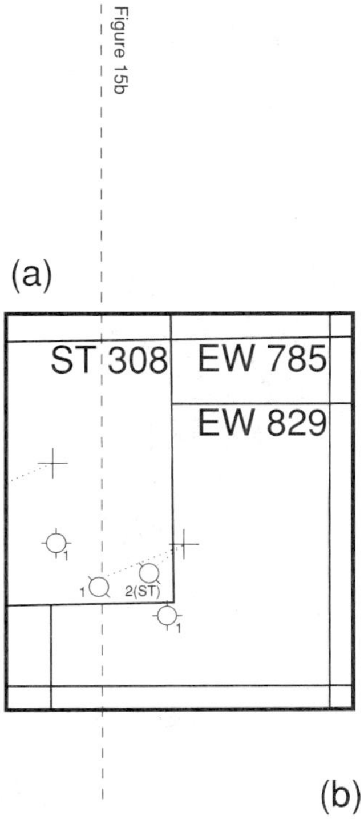


**South Timbalier 308/Ewing Bank 829**
**North Lobster**

**Operator:** Marathon 40%, Reading & Bates 40%, Amerada Hess 20%
**Discovery Date:** 1996
**Water Depth:** 550 feet
**Number of Wells:** 2
**Age:** late Miocene (?)
**Status:** Under evaluation

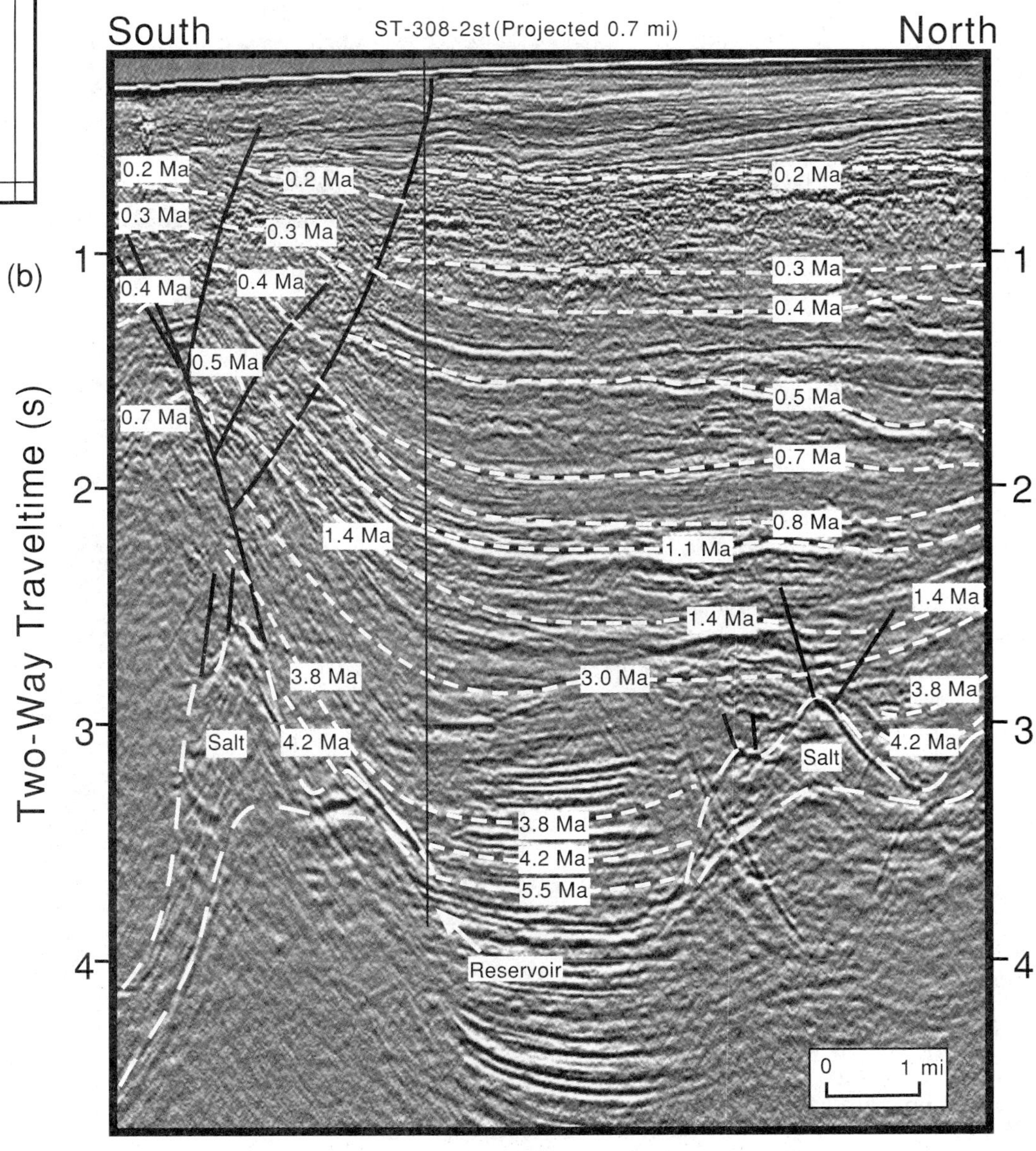


**Figure 15—South Timbalier 308 Discovery (North Lobster). (a) Location map of the South Timbalier 308 Discovery (North Lobster). Also shown is the location for (b). (b) Seismic profile across the South Timbalier 308 Discovery (North Lobster). Note that the well logs for the North Lobster discovery have not been released yet.**

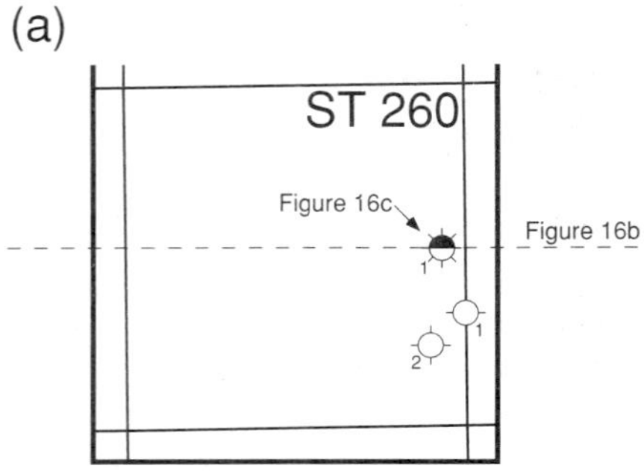


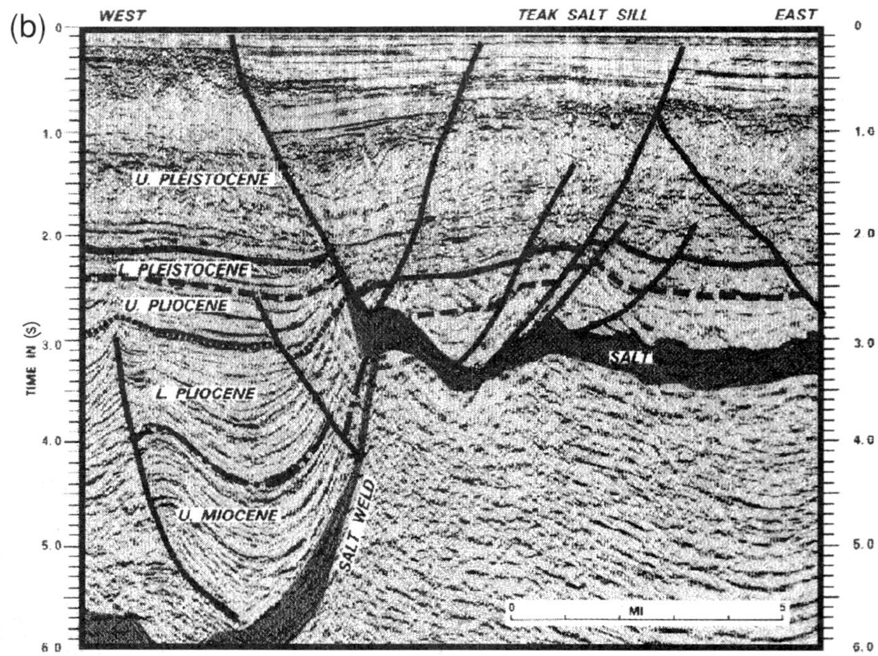


**South Timbalier 260**
**Teak**

**Note:** This block has both a nonproduced subsalt discovery (Teak) and a suprasalt gas field (operated by Walter Oil).
**Partners:** Phillips, Anadarko (subsalt); Walter Oil and Gas Co. 100% (suprasalt)
**Discovery Date:** December 1994
**Production Began:** January 1997 (suprasalt gas)
**Water Depth:** 2840 feet
**IP (Subsalt):** DST 3-channel-fill: 3742 BOPD 22/64" choke
**DST 1-thin bed:** 413 BOPD, 673 MCFD 16/64" choke
**DST 2-thin bed:** 335 BOPD, 10/64" choke
**Number of reservoirs completed:** 1 (suprasalt)
**Average Porosity:** 22 to 30% (subsalt)
**Average Permeability:** up to 2300 Md (subsalt)
**Depositional environment:** channel-fill, thin beds (levee)
**Number of wells:** 1
**Age:** middle Miocene (?) (subsalt); Pleistocene (suprasalt gas)
**Seals upper:** shales
**lateral:** shale
**API:** 27.3 to 31.3 degrees
**Status:** shallow gas producing
**Source of information:** Snyder and Nugent, 1995 (Subsalt)

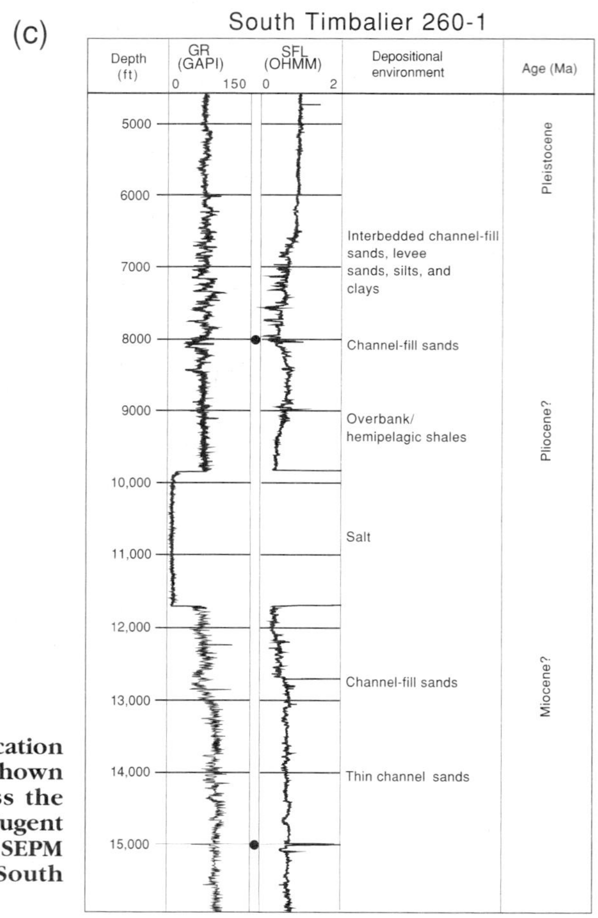


**Figure 16—South Timbalier 260 Discovery (Teak). (a) Location map of the South Timbalier 260 discovery (Teak). Also shown are locations for (b) and (c). (b) Seismic profile across the South Timbalier 260 discovery (Teak). After Snyder and Nugent (1995). Reprinted by permission of the Gulf Coast Section SEPM Foundation. (c) Gamma-ray and resistivity log of the South Timbalier 260-1 from the Teak discovery.**

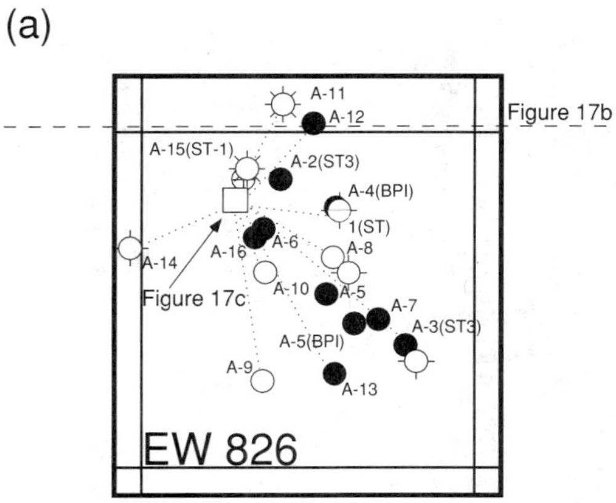


## Ewing Bank 826

**Partners:** BP 40%, Mobil 40%, Reading & Bates 20%
**Discovery Date:** May 1985
**Production began:** March 1989
**Type of production facility:** Fixed leg
**Water Depth:** 502 feet
**Number of reservoirs completed:** 4
**Gross pay thickness:** 120–157 feet
**Number of wells:** 6
**Age:** early Pleistocene
**Producing Interval:** 14,500–14,700 feet; 9900–10,100 feet
**Depositional environment:** channel-fill, overbank
**Drive mechanism:** Partial water, pressure depletion
**Average Porosity:** 29.0%
**Reservoir Temperature:** 181°F, 218°F
**Reservoir Pressure (BHP):** 5574, 6521
**Water Saturation:** 26.0, 34
**API Gravity:** 36.0
**Gas Gravity:** 0.63
**Gas-oil ratio:** 3777, 1714
**Yield:** 20.5
**Trap:** Sediments overlying dome
**Seals upper:** shales
**lateral:** shale
**Source of information:** MMS, 1997

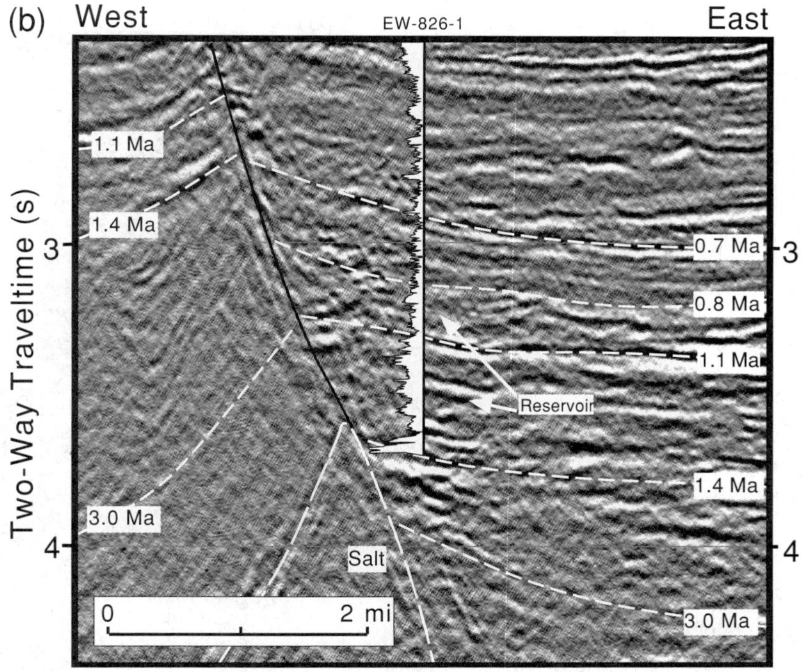


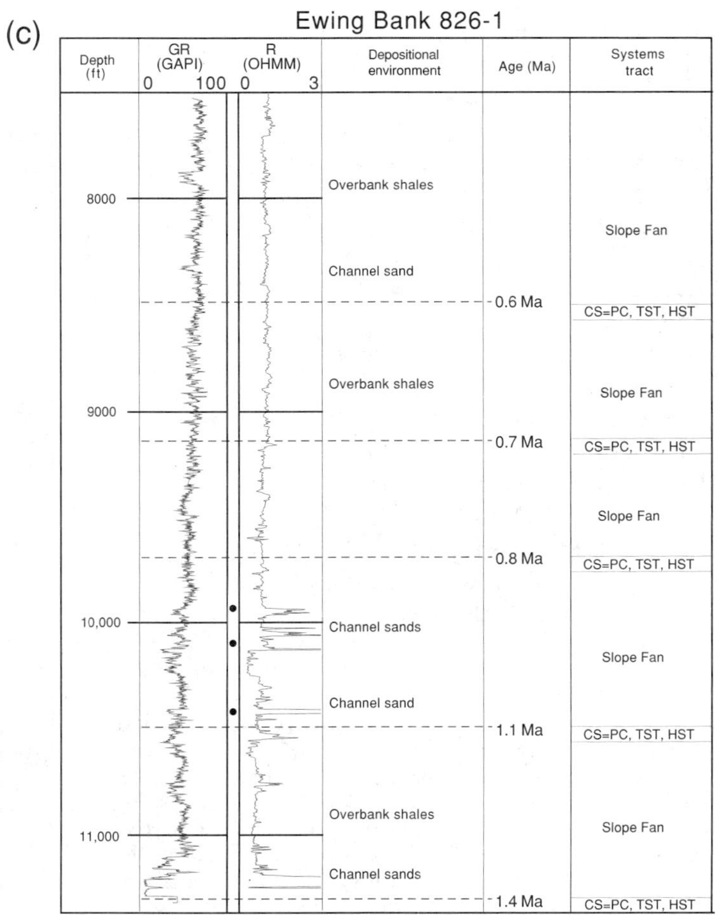


**Figure 17—Ewing Bank 826 field. (a) Location map of the Ewing Bank 826 field. Also shown are locations for (b) and (c). (b) Seismic profile across the Ewing Bank 826 field. (c) Gamma-ray and resistivity log of the EW 826-1 well. CS = condensed section, PC = prograding complex, TST = transgressive systems tract, HST = highstand systems tract.**

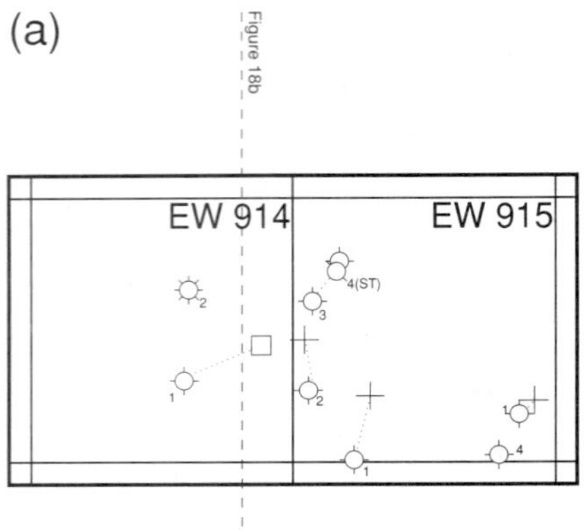


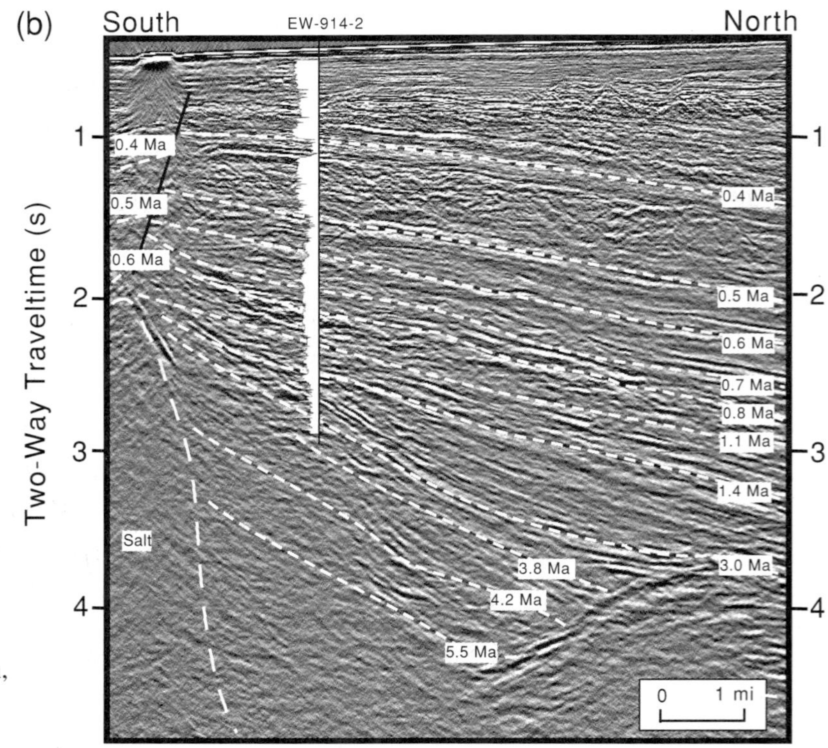


**Ewing Bank 914**
**Seattle Slew**

**Partners:** Tatham 100%
**Discovery Date:** 1991
**Production Began:** August 1993, shut in, then restarted
**Water Depth:** 920 feet
**Type of Production Facility:** Subsea tieback to EW 826
**Reserves:** 2.4 mmbo, 7 bcf
**IP:** 12mmcf, 2,500 bbls/day
**Number of Reservoirs:** 9, 1 produced to date
**Reservoir Depth:** 7800–9630 feet
**Drive mechanism:** Water
**Gross pay thickness:** 28 feet
**Average Porosity:** 27–29%
**Average Permeability:** 400 md
**Number of Wells:** 1
**Age:** late Pliocene–early Pleistocene
**Reservoir Temperature:** 175-190°F
**Reservoir Pressure (BHP):** 4078–4650
**Water Saturation:** 26–35.0
**API gravity:** 35.0 degrees
**Gas-oil ratio:** 2500–5521
**Trap:** Flank Trap Associated with Salt Diapir
**Status:** Production to begin late 1998
**Source:** MMS, 1997; John Pike, pers. comm., 1998

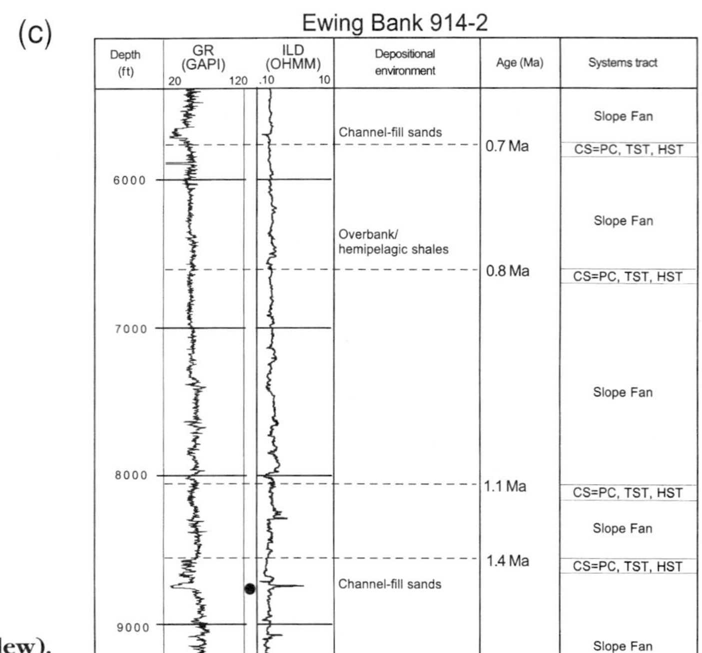


**Figure 18—Ewing Bank 914 Discovery (Seattle Slew). (a) Location map of the Ewing Bank 914 Discovery (Seattle Slew). Also shown is the location for (b). (b) Seismic profile across the Ewing Bank 914 Discovery (Seattle Slew). (c) Gamma-ray and resistivity log of the EW 914-2 well. CS = condensed section, PC = prograding complex, TST = transgressive systems tract, HST = highstand systems tract.**

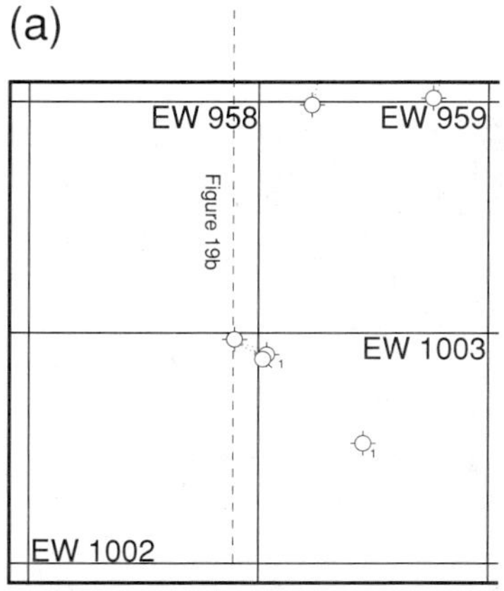


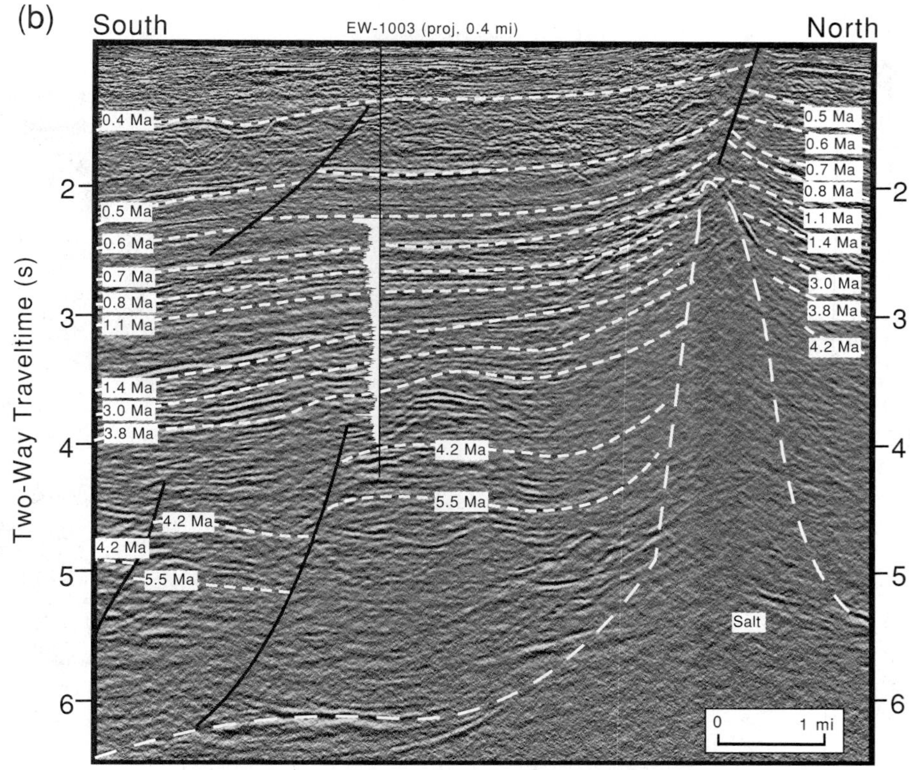


**Ewing Bank 958/1003**
**Sunday Silence**

**Partners:** Tatham 100%
**Discovery Date:** 1994
**Water depth:** 1490 feet
**IP:** 8,700 bbls/day, 24/64" choke
**Number of Reservoirs:** 14
**Reservoir Depth:** 10,400–13,600´
**Average Porosity:** >25%
**Average Permeability:** >100 md
**Number of Wells:** 2
**Age:** early Pliocene
**API Gravity:** 28–30 degrees
**GOR:** 800 ft$^3$/bbl
**Reservoir Pressure:** lower zones >10,000 psi
**Status:** Production to begin in 2000, 2001
**Source:** Press release, John Pike (pers. comm., 1998)

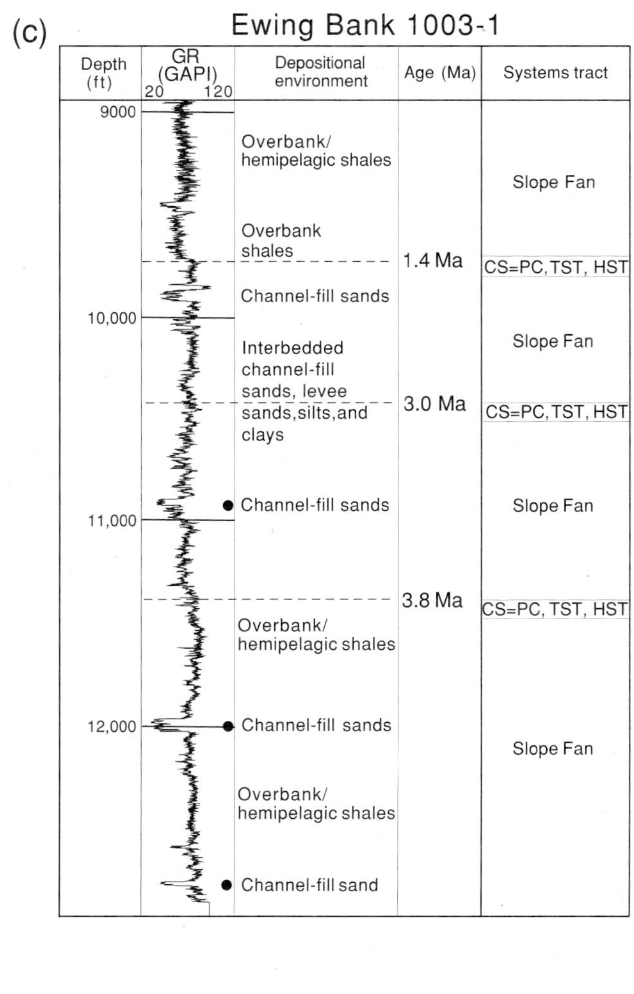


**Figure 19—Ewing Bank 1003 Discovery (Sunday Silence). (a) Location map of the Ewing Bank 1003 Discovery (Sunday Silence). Also shown are the locations for (b) and (c). (b) Seismic profile across the Ewing Bank 1003 Discovery (Sunday Silence). (c) Gamma-ray log from the EW 1003 well (Sunday Silence). CS = condensed section, PC = prograding complex, TST = transgressive systems tract, HST = highstand systems tract.**

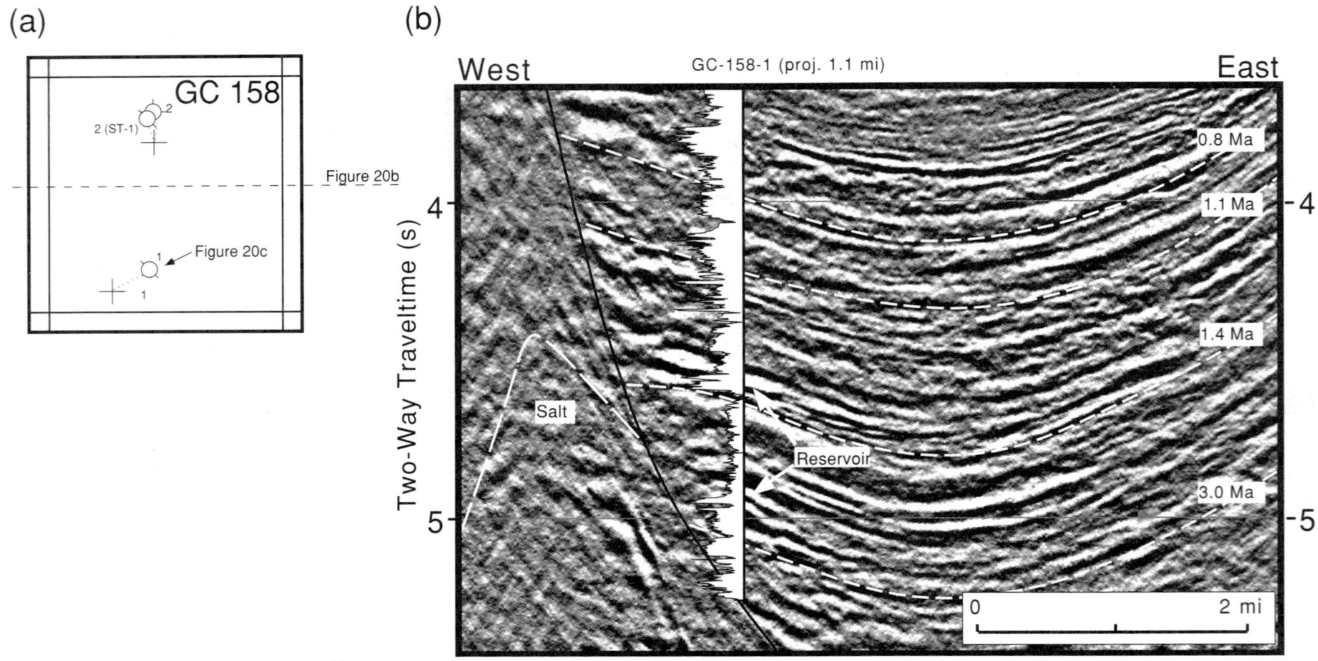


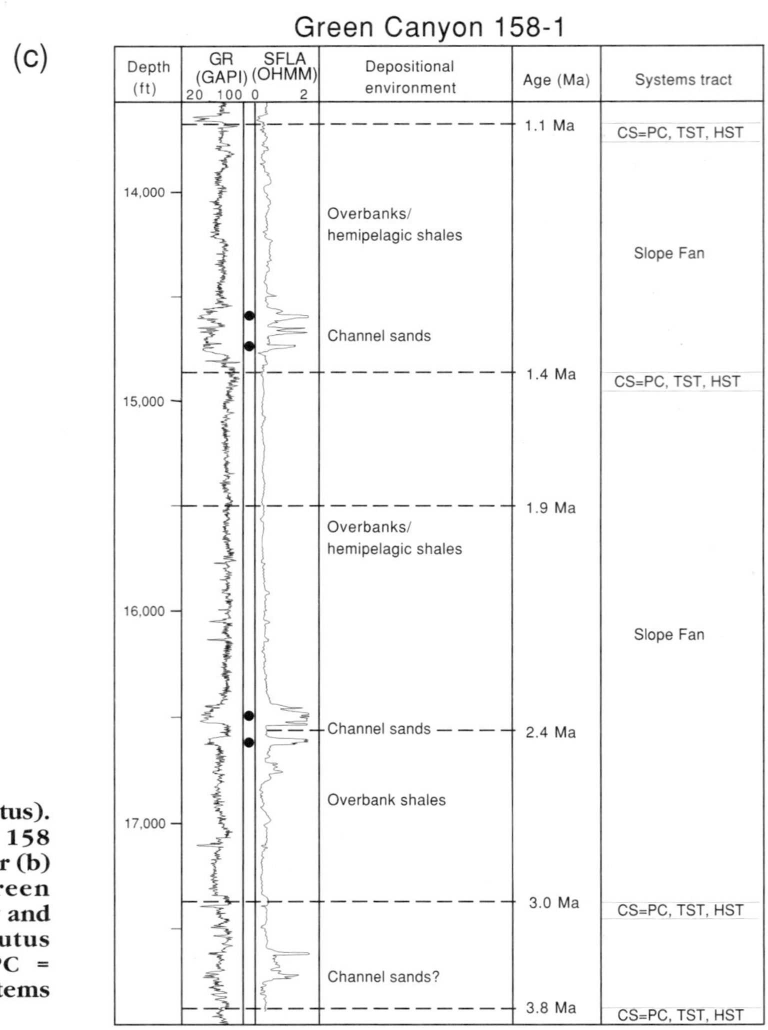


## Green Canyon 158, 202
## Brutus

**Partners:** Shell Offshore, Inc. 100%
**Discovery Date:** 1989
**Water Depth:** 2840 feet
**Number of reservoirs completed:** 2
**Depositional environment:** Amalgamated sheets sands, channel-fill
**Number of wells:** 3
**Age:** early Pleistocene
**Seals upper:** shales
**lateral:** shale
**Status:** Development plan underway

**Figure 20—Green Canyon 158 Discovery (Brutus). (a) Location map of the Green Canyon 158 Discovery (Brutus). Also shown are locations for (b) and (c). (b) Seismic profile across the Green Canyon 158 Discovery (Brutus). (c) Gamma-ray and resistivity log from the GC 158-1 well (Brutus Discovery). CS = condensed section, PC = prograding complex, TST = transgressive systems tract, HST = highstand systems tract.**

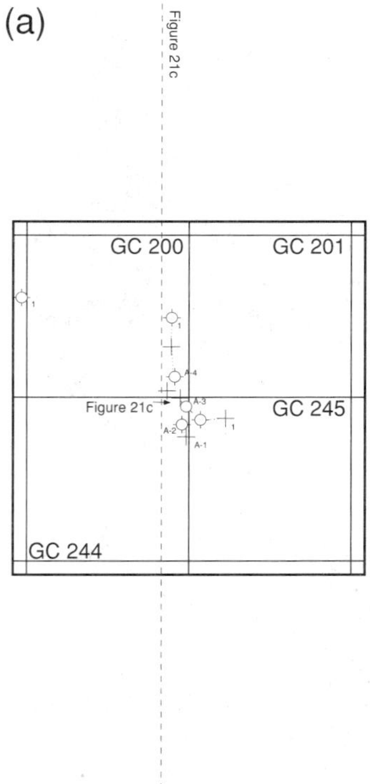


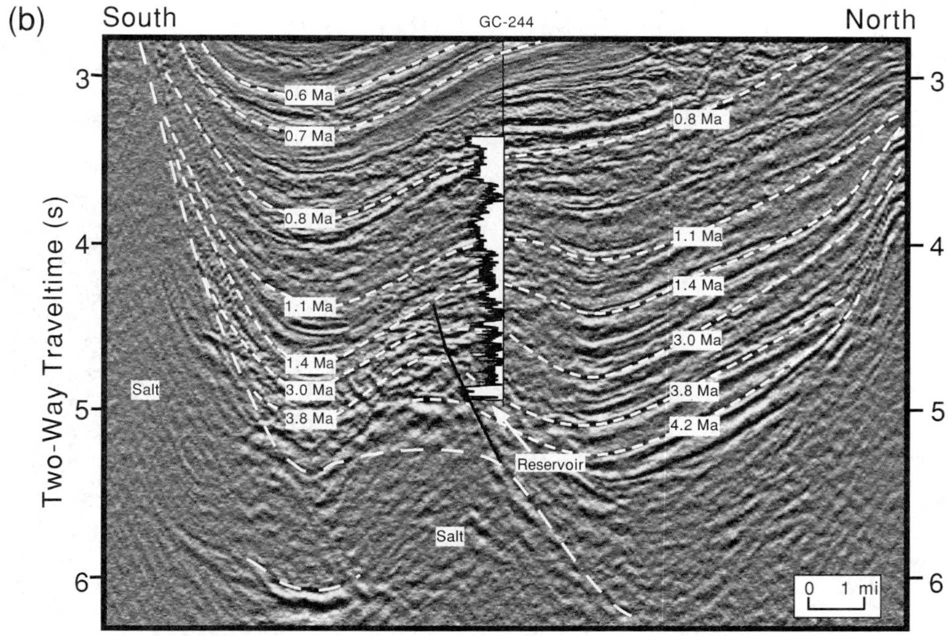


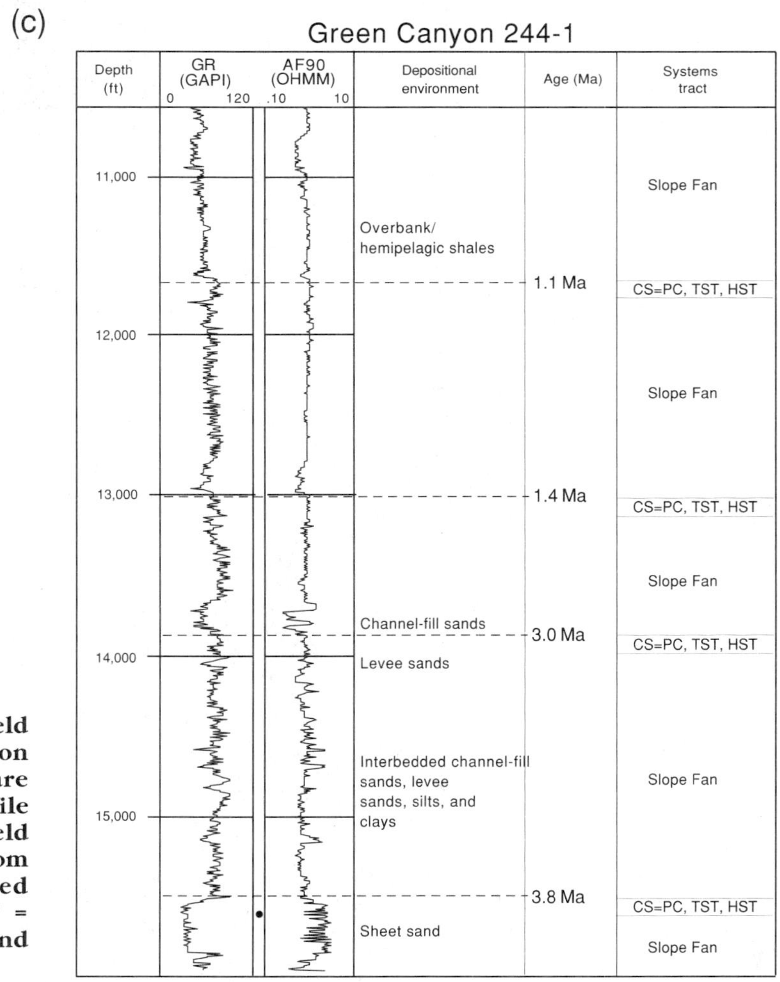


**Green Canyon 200, 201, 244, 245**
**Troika**

**Partners:** BP 33.33%, Shell 33.33% Marathon 33.33%
**Discovery Date:** June 1994
**Production began:** October 1997
**Type of production facility:** Subsea tieback to the Bullwinkle facility in GC 65/109
**Water Depth:** 2721 feet
**Reserves:** 200 million barrels
**Number of wells:** 5 to date
**Age:** early Pliocene (?)
**Seals: upper:** shale
**lateral:** shale/fault
**Source of information:** Haines, 1996

**Figure 21—Green Canyon 200, 244, 245 field (Troika). (a) Location map of the Green Canyon 200, 244, 245 field (Troika). Also shown are locations for (b) and (c). (b) Seismic profile across the Green Canyon 200, 244, 245 field (Troika). (c) Gamma-ray and resistivity log from the Troika GC 244-1 well. CS = condensed section, PC = prograding complex, TST = transgressive systems tract, HST = highstand systems tract.**

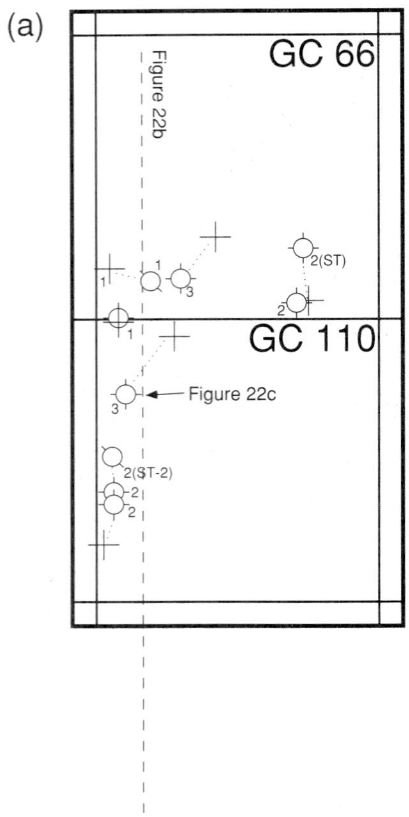


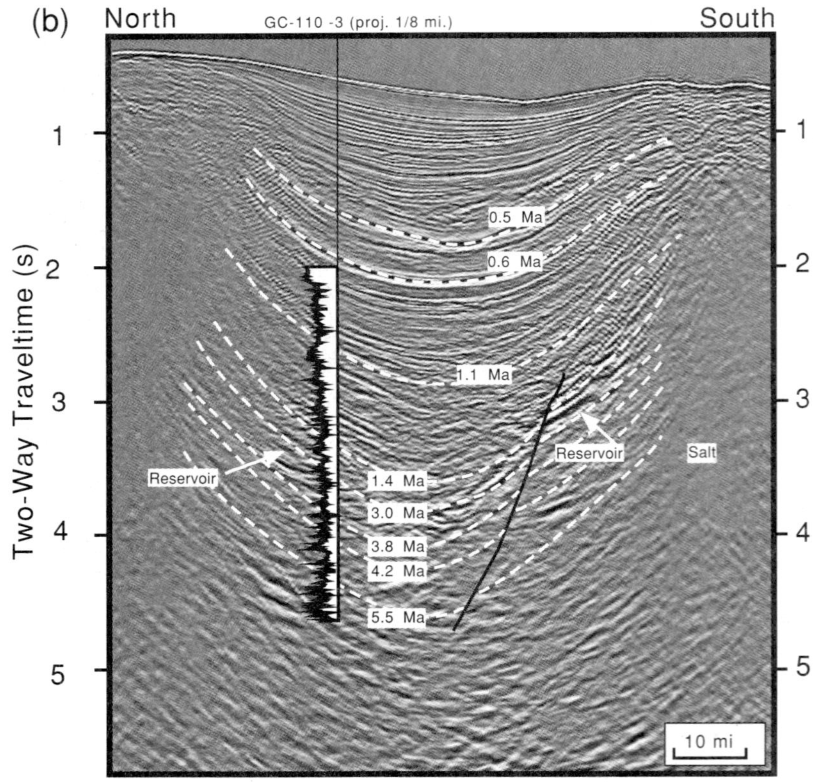


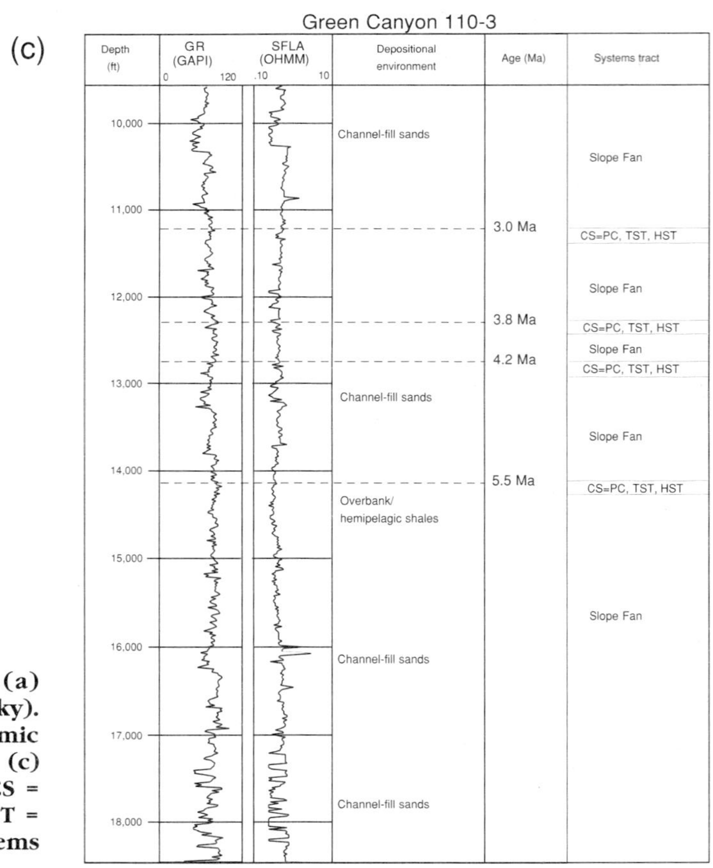


**Green Canyon 110**
**Rocky**

**Partners:** Shell Offshore, Inc. 100%
**Discovery Date:** October 1983
**Production began:** January 1995
**Type of production facility:** Subsea tieback to Bullwinkle platform (GC 65/109)
**Water Depth:** 1621 feet
**Number of wells:** 6
**Age:** late Pliocene
**Seals upper:** shale
**lateral:** shale
**Source of information:** Press Release. Note that the Shell GC 110-1 is the producing well.

**Figure 22—Green Canyon 110 field (Rocky). (a) Location map of the Green Canyon 110 field (Rocky). Also shown are locations for (b) and (c). (b) Seismic profile across the Green Canyon 110 field (Rocky). (c) Gamma-ray and resistivity log from GC 110-3. CS = condensed section, PC = prograding complex, TST = transgressive systems tract, HST = highstand systems tract.**

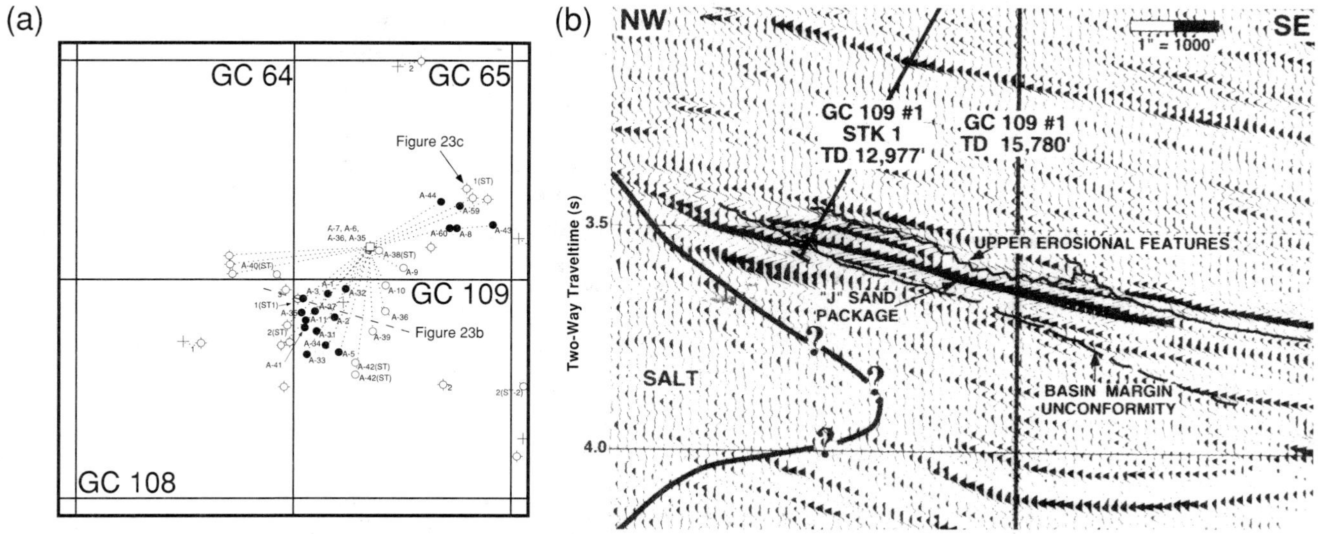


**Green Canyon 65**
**Bullwinkle**

**Operator:** Shell Offshore, Inc. 100%
**Discovery Date:** October 1983
**Type of production facility:** Fixed-Leg Platform
**Water Depth:** 1353 feet
**Reserves:** 100 Million BOE, 191 Bcf
**Reservoir Depth:** 10,455′
**Gross pay thickness:** 110 feet
**Depositional environment:** sheets, channels, levees
**Number of wells:** 20 producers, 5 injector wells
**Age:** Pliocene
**Producing Interval:** 11,200'–11,300', 12,200'-12,350'
**Drive mechanism:** Water drive
**Average Porosity:** 29.0%
**Reservoir Temperature:** 196°F, 214°F
**Reservoir Pressure (BHP):** 7543, 8626, 8705
**Water Saturation:** 34.0, 18, 20
**API Gravity:** 30.0
**Gas Gravity:** 0.60
**Gas-oil ratio:** 1163, 1278
**Yield:** 0.0, 14.3, 103
**Reservoir Depth:** 11,959 feet
**Gross pay thickness:** 103
**Average Porosity:** 31.0%
**Reservoir Depth:** 12,069 feet
**Gross pay thickness:** 25 feet
**Number of reservoirs completed:** 4 primary
**Trap:** Flank Trap Associated with Salt Diapirs
**Seals upper:** shales
**lateral:** shales
**Size acres:** 105 acre spacing
**Status:** Producing since 1989
**Source of information:** Holman and Robertson, 1994; MMS, 1997

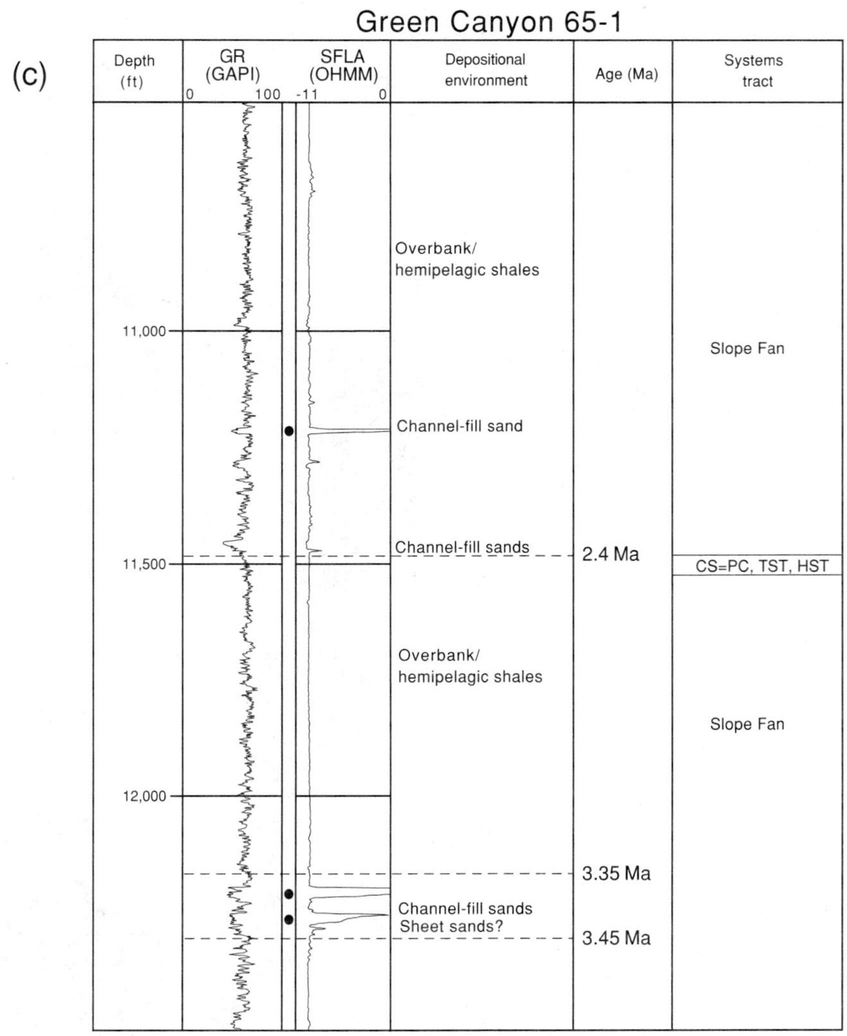


**Figure 23—Green Canyon 65, 109 field (Bullwinkle). (a) Location map of the Green Canyon 65, 109 field (Bullwinkle). Also shown are locations for (b) and (c). (b) Seismic profile across the Green Canyon 65, 109 field (Bullwinkle). Reprinted by permission of the Gulf Coast Section Society of Economic Paleontologists and Mineralogists Foundation. (c) Gamma-ray and resistivity log of the Bullwinkle field. CS = condensed section, PC = prograding complex, TST = transgressive systems tract, HST = highstand systems tract.**

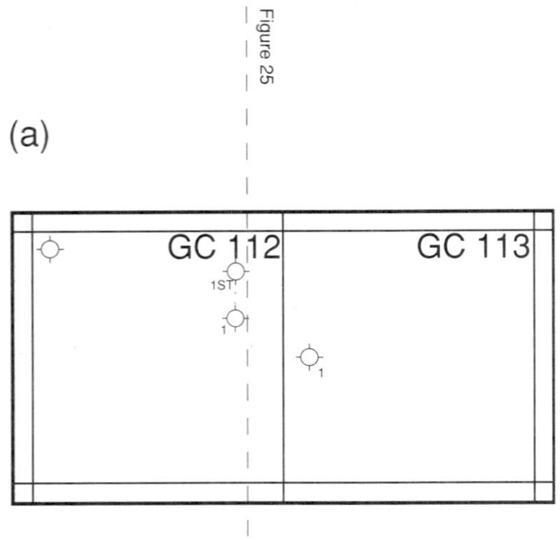


**Green Canyon 112, 113**
**Stellaria**

**Partners:** Marathon 65%, Shell 35%
**Discovery Date:** 1997
**Water depth:** 2045 feet
**Net Pay:** #1-70' in two zones, #1ST-67', 113#1-120'
**Number of Wells:** 2
**Age:** early Pliocene (?)
**TD:** 20,422'
**Status:** Under evaluation

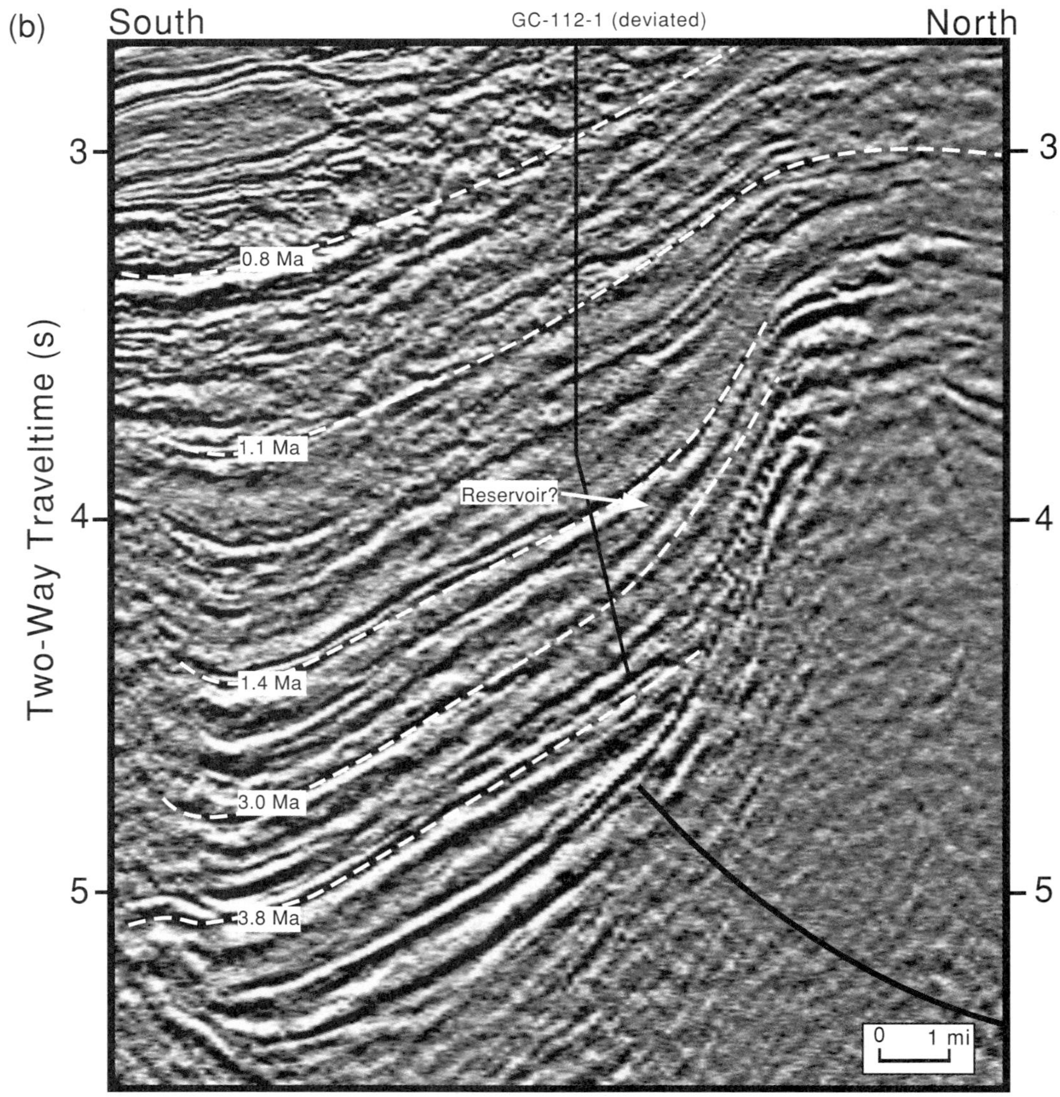


**Figure 24—Green Canyon 112/113 Discovery (Stellaria). (a) Location map of the Green Canyon 112 Discovery (Stellaria). Also shown is the location for (b). (b) Seismic profile across the Discovery (Stellaria). Note that the well logs for the Stellaria discovery have not been released.**

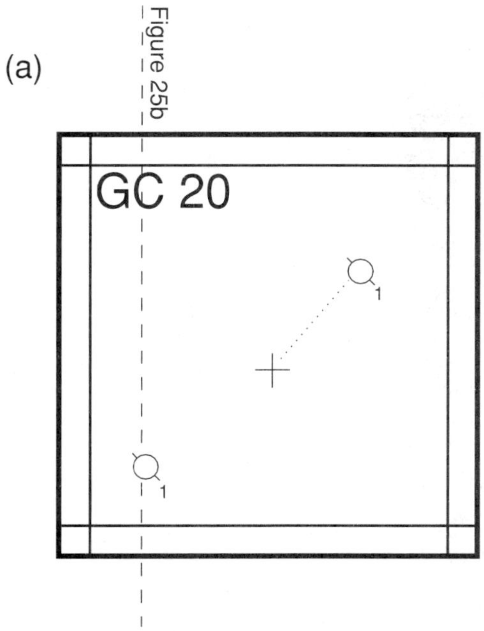


**Green Canyon 20**

**Partner:** Reading & Bates, 100%
**Discovery Date:** March 1997
**Type of Production Facility:** Subsea well tied to Boxer (GC 19) platform
**Water Depth:** 815 feet
**Depth to Reservoir:** 18,000 ft
**Number of Wells:** 1
**Status:** Production to begin in February 1999
**Source:** Jim Fox, pers. comm., 1998

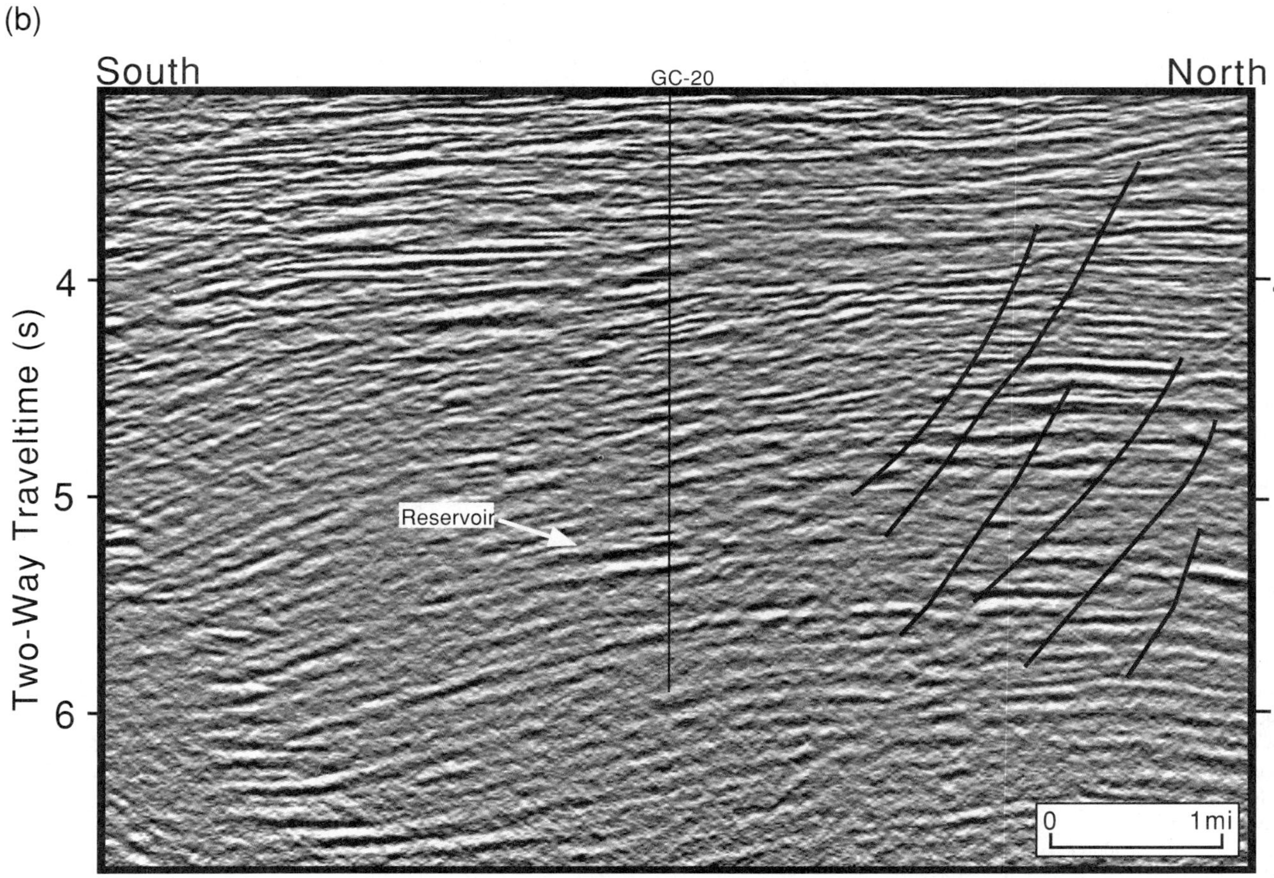


**Figure 25—Green Canyon 20 field. (a) Location map of the Green Canyon 20 field. Also shown is location for (b). (b) Seismic profile across the Green Canyon 20 field. Note that the logs for the GC 20 discovery have not been released yet.**

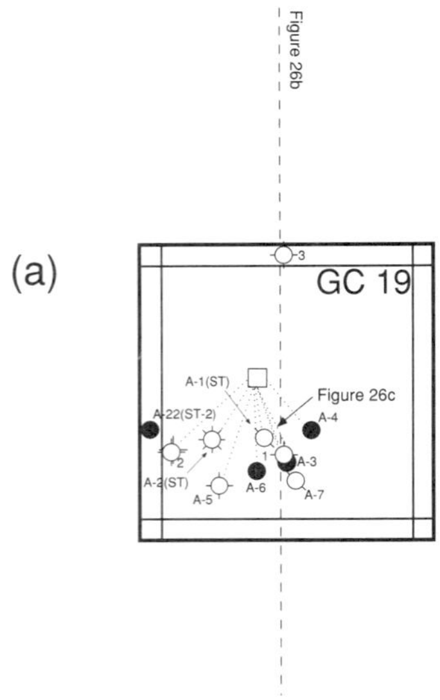


Green Canyon 19
Boxer

**Partners:** Shell Offshore, Inc. 100%
**Discovery Date:** Spring 1980
**Production Began:** November 1988
**Type of Production Facility:** Fixed Platform
**Water Depth:** 757 feet
**Initial Rates:** Variable–Depends on Reservoir Thickness/Quality. 4–12 mmcf/d Gas, and 700–2,000 bpd Oil.
**Number of Reservoirs Completed:** 8 to date.
**Reservoir Depth:** Gas: 5785-gas; Oil: 10,079 feet, 13, 275 feet
**Gross pay thickness:** Variable, 20–70 feet
**Average Porosity:** 28–32%
**Depositional Environment:** Channelized Turbidites
**Number of wells:** 9 (2 exploration tests, 7 development wells, 3–4 Additional Wells Anticipated During 1998)
**Age:** late Pliocene to Pleistocene
**Hydrocarbon Type:** Gas in Shallow $D_4$ Sand (5,200´); Oil and Gas in Deep $I_6J$, $J_2J_3$ Sands (11,000´–12,000´).
**Trap:** Faulted, NE-SW trending, deep seated, salt-cored anticline
**Seals upper:** shale
**lateral:** shale/fault
**Drive mechanism:** Partial to strong water drive
**Reservoir Temperature:** Shallow $D_4$ Sand at 130°F. Deep $I_6$-J Sands 170–180°F.
**Reservoir Pressure (BHP):** Shallow $D_4$ Sand at 3600 psi. Deep $I_6$-J Sands at 8800–9800 psi.
**Water Saturation:** Variable–Most Reservoirs in 10% to 40% Range
**API Gravity:** 26–31
**Gas Gravity:** 0.60
**Gas-oil ratio:** 1100–1300
**Cumulative Oil-Cond Production:** 6.5 MMB (Through 10/97)
**Cumulative Gas Production:** 45 BCF (Through 10/97)
**Expected Ultimate Recovery:** Approximately 8 MMBO and 65 BCF (20 MMBE)
**Secondary Recovery:** None
**Status:** Produced, abandoned, production reestablished in 1995
**Source:** MMS, 1997; Tom Wilson, pers. comm., 1998

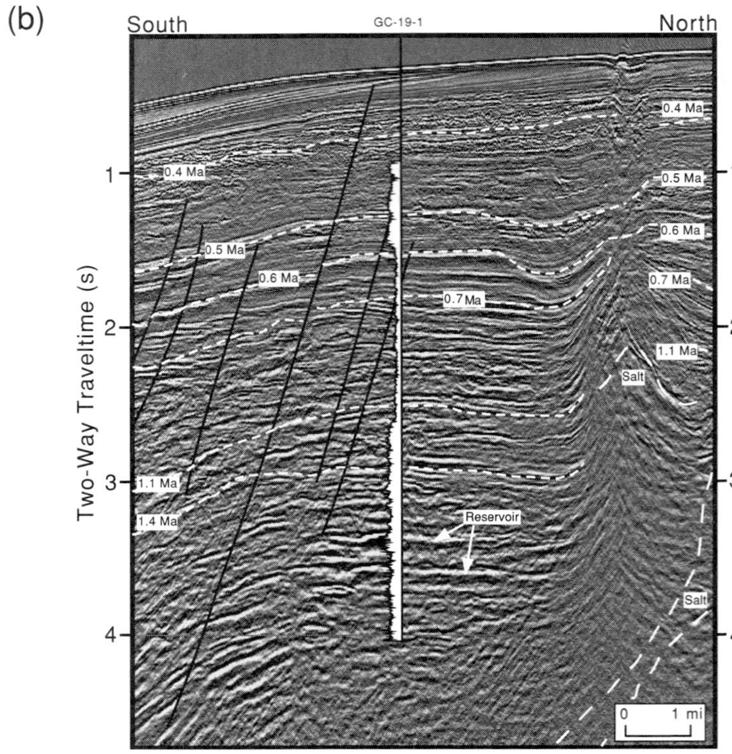


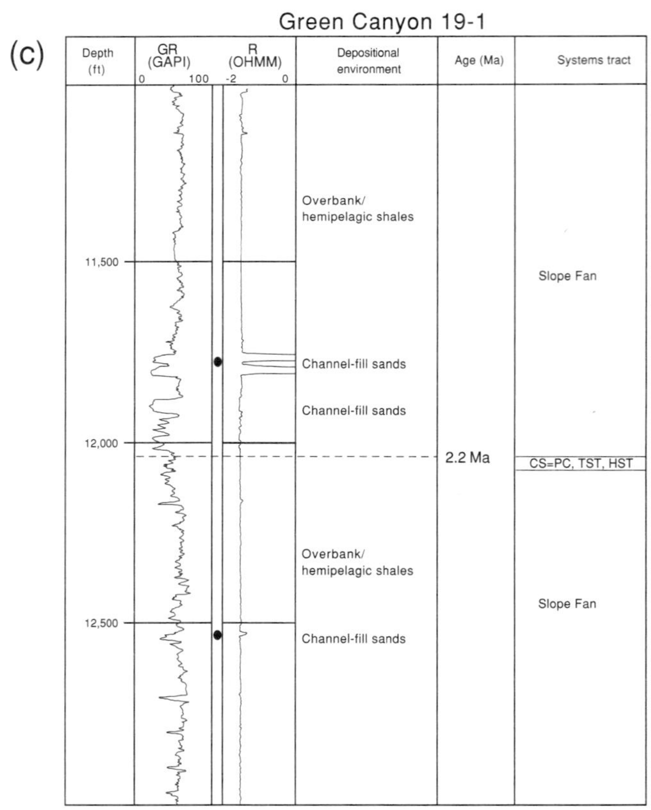


**Figure 26—Green Canyon 19 field (Boxer). (a) Location map of the Green Canyon 19 field (Boxer). Also shown are locations for (b) and (c). (b) Seismic profile across the Green Canyon 19 field (Boxer). (c) Gamma-ray and resistivity log from the GC 19-1 well. CS = condensed section, PC = prograding complex, TST = transgressive systems tract, HST = highstand systems tract.**

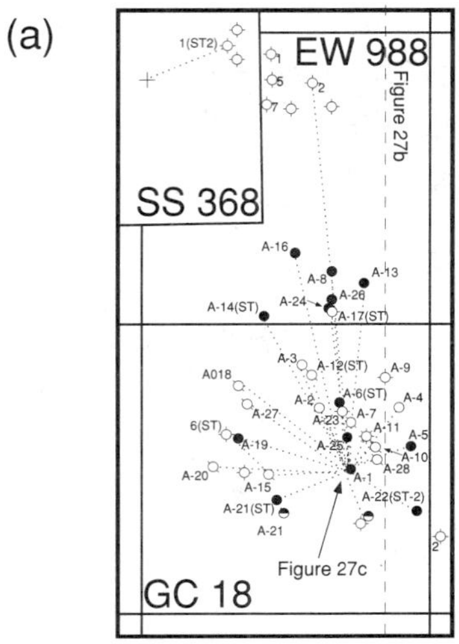


**Green Canyon 18/Ewing Bank 988**

**Partners:** Mobil Oil Co. 55%, BHP 25%, Burlington 15%, Kerr-McGee 5%
**Discovery Date:** Spring 1982
**Production began:** May 1987
**Type of production facility:** Fixed platform
**Water Depth:** 760 feet
**Reserves:** 60 million barrels (MMS)
**IP:** 800-3000 bbls/day, variability depends upon which reservoir
**Number of reservoirs completed:** 16 (to date)
**Depths:** shallowest 6700 feet, deepest 13,850 feet
**Average Porosity:** 28-36%, mean is 31%
**Average Permeability:** 400–500 md, highest is 1500 md (air permeability)
**Depositional environment:** channelized (upper reservoirs), sheet (lower reservoirs)
**Number of wells:** 29 (to date)
**Age:** middle-late Pliocene
**Producing Interval:** middle Pleistocene (GC18), top Pliocene (GC 18), middle Pliocene (EW 988)
**Seals upper:** shale
**lateral:** shale
**Size acres:** 2000
**Secondary recovery:** None yet
**API:** 28 degrees
**Sulfur:** 2.6%
**Initial GOR:** 1000–1500 $ft^3$/bbl, average is 1200
**Viscosity:** 0.53 to 0.97 centipoise
**Initial Pressure:** Shallow reservoir 8000 psi, deeper 11,400 psi
**Oil Column Thickness:** Shallow reservoirs 700 feet; deeper reservoirs 1200 feet
**Status:** Producing
**Source of information:** David Walz, personal communication, 1996; MMS, 1997

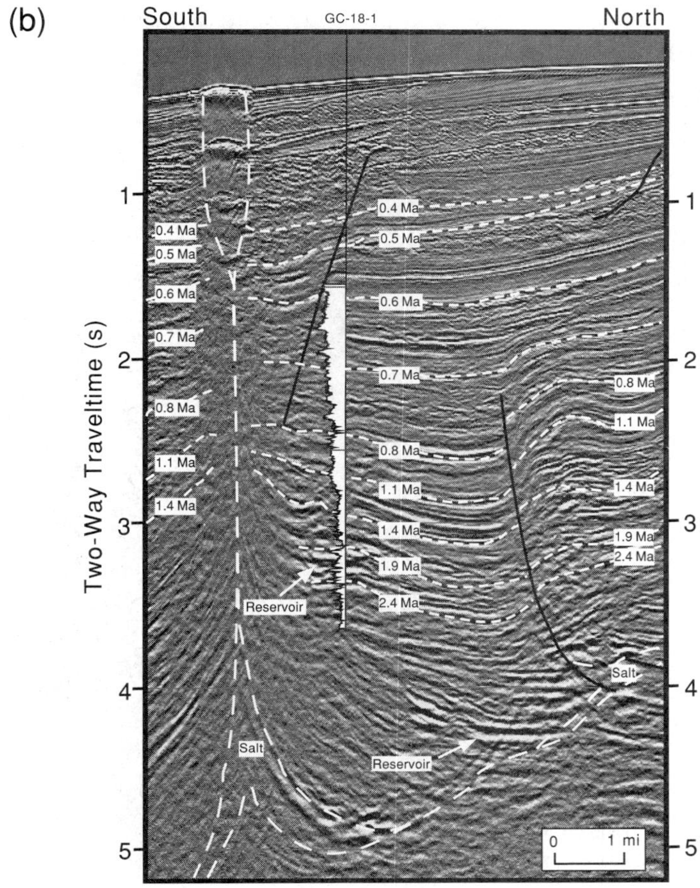


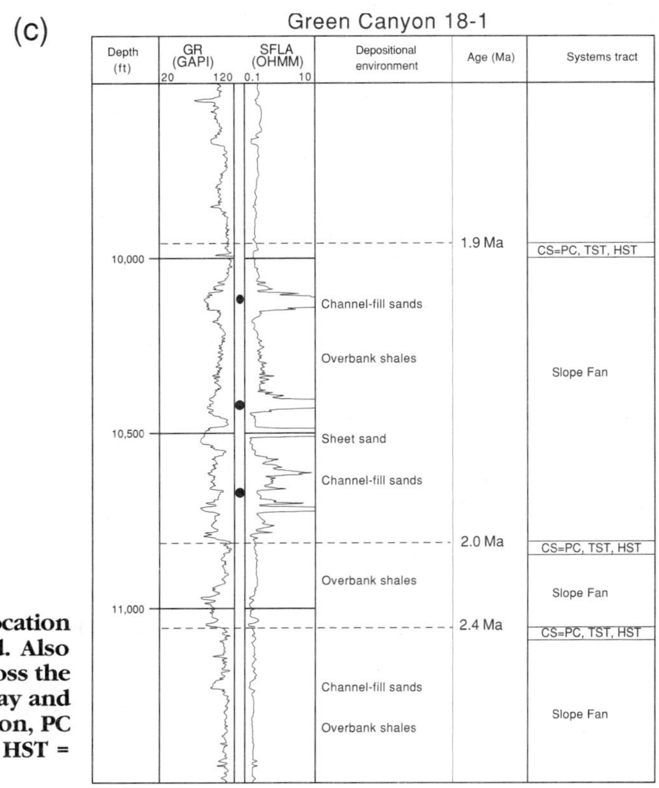


**Figure 27—Green Canyon 18/Ewing Bank 988 field. (a) Location map of the Green Canyon 18 field/Ewing Bank 988 field. Also shown are locations for (b) and (c). (b) Seismic profile across the Green Canyon 18 field/Ewing Bank 988 field. (c) Gamma-ray and resistivity log from the GC 18-1 well. CS = condensed section, PC = prograding complex, TST = transgressive systems tract, HST = highstand systems tract.**

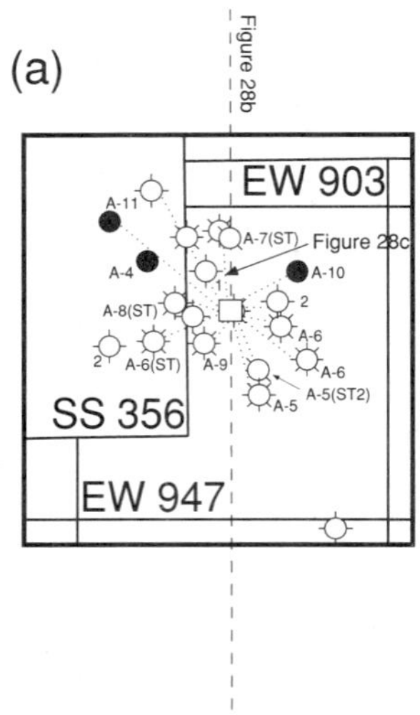


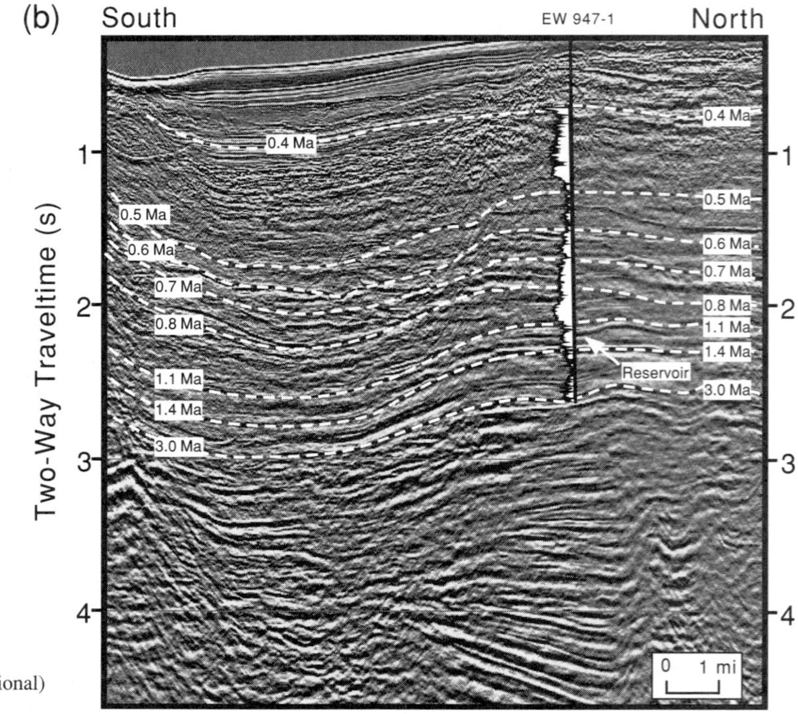


**Ewing Bank 947**

Note: 2 separate sands divide the field E-W

**Partners:** Newfield 60%, ELF Aquitane 12%, Continental Land and fur 11% Colton 10%, Del Mar 10%, Fidelity 6%

**Discovery Date:** January 1984

**Production began:** late 1991

**Type of production facility:** (18 slot, 4 pile conventional) Fixed platform

**Water Depth:** 460 feet

**IP Rate: 7-**15 Mmcf/d

**Reservoir Depth:** 7000', 8200'

**Number of Reservoirs Completed:** 2 to date

**Depths:** shallowest – 6900', deepest – 8400'

**Gross Pay Thickness:** 22 feet

**Gross Production:** 50Mmcf/day

**Cumulative oil-cond production (7/31/97):** 1934.5 MBL

**Cumulative gas production (7/31/97):** 58.1 Bcf, 1.9 MMBC

**Gas Recovery Efficiency:** 70%

**Daily Recovery:** 40 Mmcf/d

**Average Porosity:** 28-34%, mean is 31%

**Average Permeability:** 100-300 md, highest is 2000 md (air permeability)

**Depositional environment:** Channel-fill, levee and overbank

**Number of wells:** Primary development – 9 wells (A1-A9)
Secondary – 2 wells
Tertiary – 2 wells

**Age:** Pleistocene (Trim B)

**HC Type:** Gas Condensate, Oil

**Trap:** Sediments Overlying Dome; Combination stratigraphic (updip pinchout) and structural

**Seals upper:** shale

**Aerial Closure:** 9,821 acres

**Drive mechanism:** Partial Water

**Reservoir Temperature:** 140°F

**Reservoir Pressure (BHP):** 5300, 5950

**Water Saturation:** 49%, 41%

**API gravity:** 60, 32

**Gas Gravity:** 0.59

**Gas-oil ratio:** 10588

**Yield:** 25

**Status:** Producing

**Source of information:** MMS, 1997, Todd Stone, pers. comm., 1998

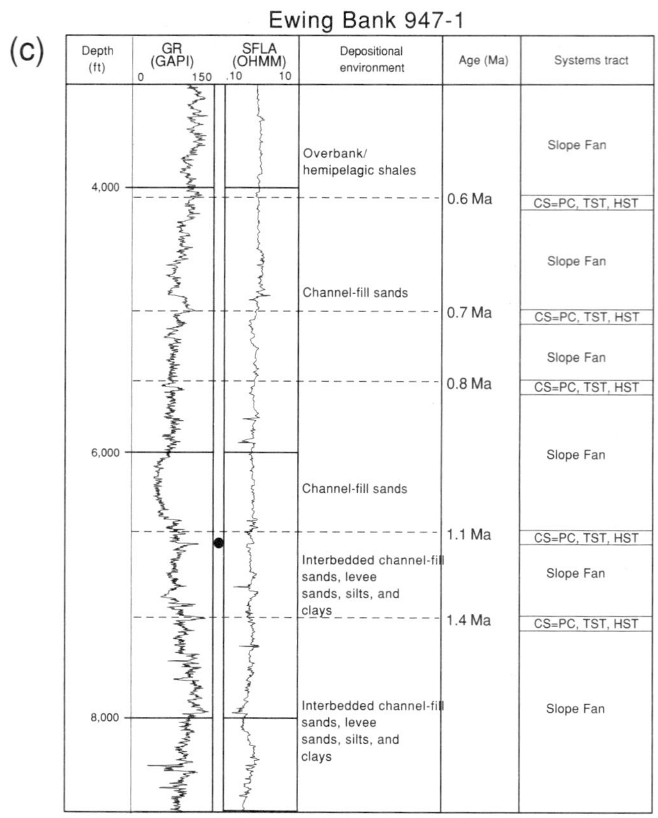


**Figure 28—Ewing Bank 947 field. (a) Location map of the Ewing Bank 947 field. Also shown are locations for (b) and (c). (b) Seismic profile across the Ewing Bank 947 field. (c) Gamma-ray and resistivity log from the EW 947-1 well. CS = condensed section, PC = prograding complex, TST = transgressive systems tract, HST = highstand systems tract.**

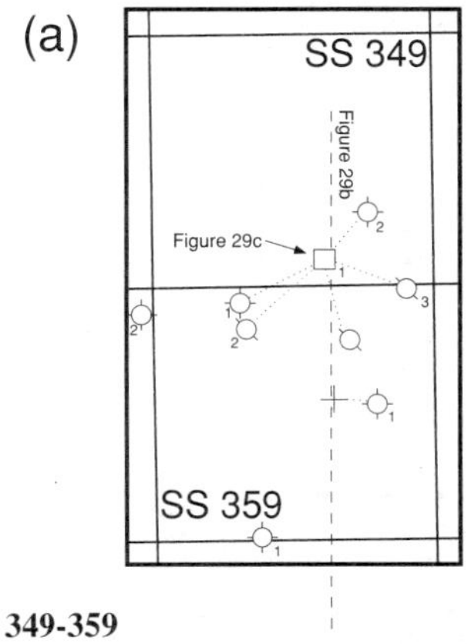


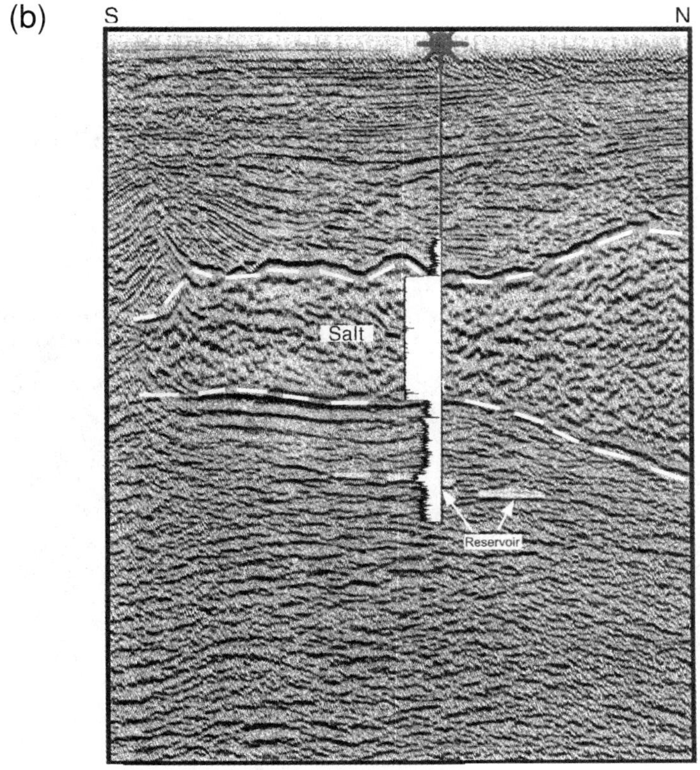


**Ship Shoal 349-359**
**Mahogany**

**Partners:** Phillips 37.5%, Anadarko 37.5%, Amoco 25%
**Discovery Date:** September 1993
**Production began:** December 31, 1996
**Type of production facility:** Fixed Platform Structure
**Water depth:** 375 ft
**Reserves:** 100 MMBOE
**Estimated IP Rate:** 22,000 bbls/day
**Actual IP Rate:** 18,000 BOPD, 30 MMCFD, July 1997 (Camp and McGuire, Oct. 1997)
**Number of reservoirs completed:** 3 pays, O, P, Q sands; P sand (best sand) is deepest
**Depths:** 13,640 to 15,200 ft
**Producing Zones Depth Range:** 13,640–14,990 feet TVD
**Gross thickness:** 180 ft (P sand in discovery well A-1)
**Conventional Core Measurements for P sand:**
- **Average Porosity:** 29%
- **Porosity Range:** 20-36%
- **Average Permeability:** 1500 md
- **Permeability Range:** 0.5–7460 md

**Average Porosity:** 25% (18–33%)
**Average Permeability:** 338 md (up to 6.8 Darcies)
**Test rates:** Up to 3700 bbls/day. 559,000 $Ft^3$ gas/day 14/64" choke basal sand–P sands best tests 7200 bbls/day, 7 million cubic feet/day
**Production Rates:**
- **O sand:** 4366 bbls/day, 5.3 mmcfpd
- **P sand:** 10,000 bbls/day, 7 mmcfpd
- **Q sand:** 5000 bbls/day, 5 mmcfpd

**Depositional environment:** Channel-fill and levees, low resistivity play (levee/overbank thin beds) (from log intepretation and core)
**Number of wells:** 7 to date, 5 wells completed, 2 expected by 1998
**Age:** middle Miocene (7)
**Seals upper:** shale
**lateral:** shale (possibly salt)
**Secondary recovery:** N/A
**Initial Pressure:** 7200 PSI
**Oil Column Thickness:** Minimum of 200 ft
**Source of information:** Harrison et al., 1995; Clint Moore, pers. communication, 1996; Camp et al., 1997

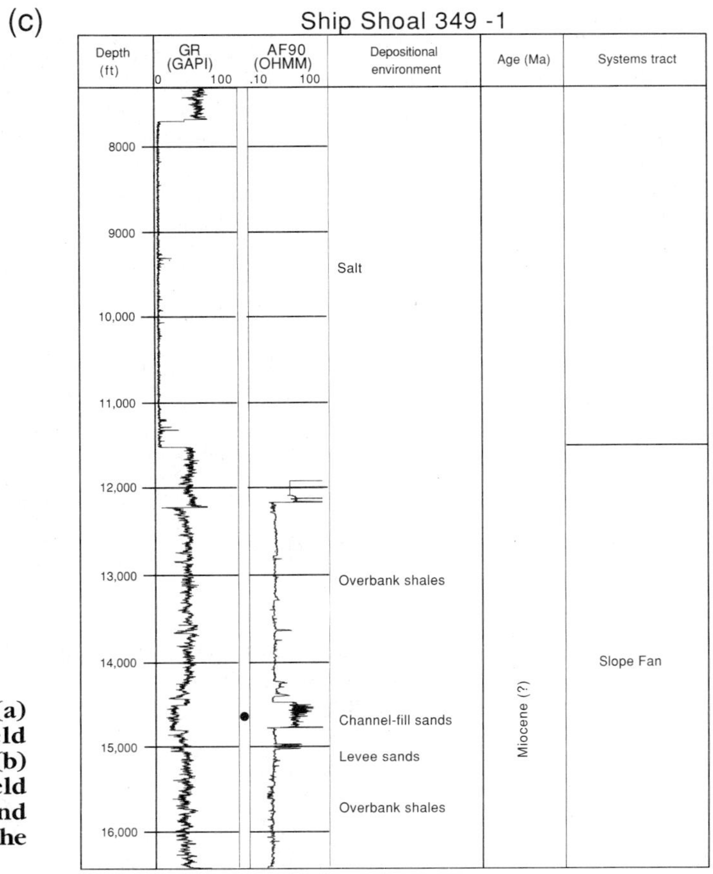


**Figure 29—Ship Shoal 349, 359 field (Mahogany). (a) Location map of the Ship Shoal 349, 359 field (Mahogany). Also shown are locations for (b) and (c). (b) Seismic profile across the Ship Shoal 349, 359 field (Mahogany). Reprinted with permission of Diamond Geophysical. (c) Gamma-ray and resistivity log from the SS 349-1 well, Mahogany field.**

(a)

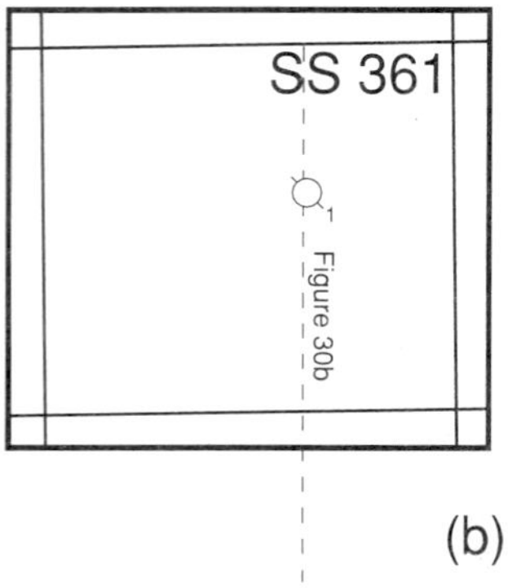


(b)

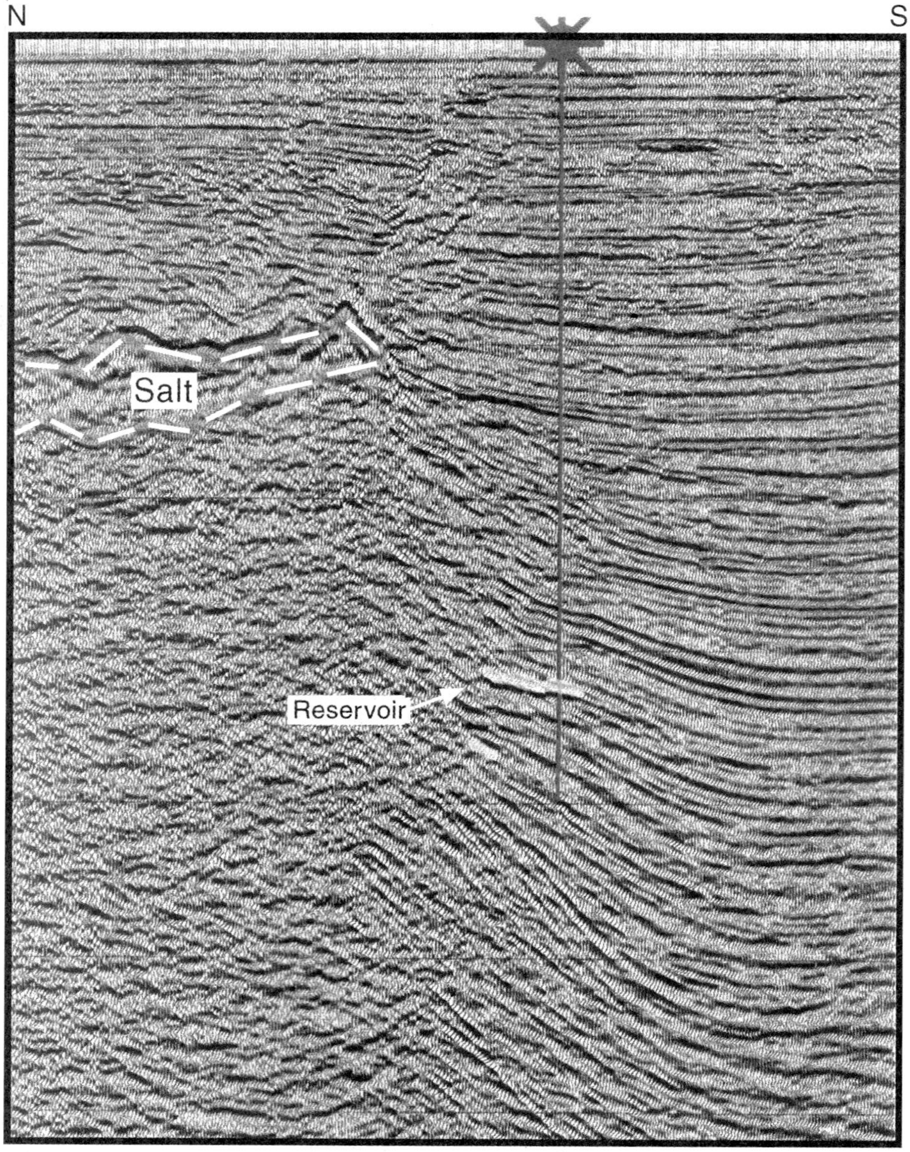


**Ship Shoal 361**
**Agate**

**Partners:** Anadarko 50%, Phillips 50%
**Discovery Date:** July 1996
**First Production:** February 1998
**Type of Production Facility:** Subsea tieback to Mahogany platform
**Water Depth:** 400 feet
**IP:** 2 zones (1) 4126 bbls/day, 24 Mcfpd; (2) 2738 bbls/day, 14.4 mcfpd with 17/64" choke, 6773 psi
**Number of reservoirs completed:** 2
**Number of wells:** 1
**Age:** middle Miocene (?)
**Seals upper:** shale
**lateral:** shale
**Source of information:** Press releases, Wayne Camp, pers. comm., 1997

**Figure 30—Ship Shoal 361 field (Agate). (a) Location map of the Ship Shoal 361 field (Agate). Also shown is the location for (b). (b) Seismic profile across the Ship Shoal 361 field (Agate). Reprinted with permission of Diamond Geophysical. Note that the well logs for the Agate discovery have not been released yet.**

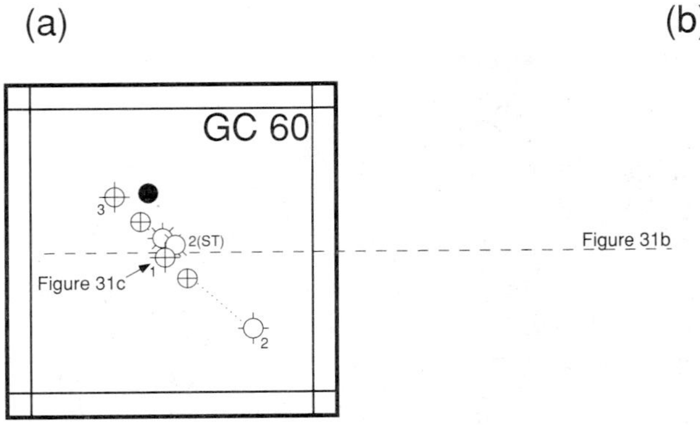


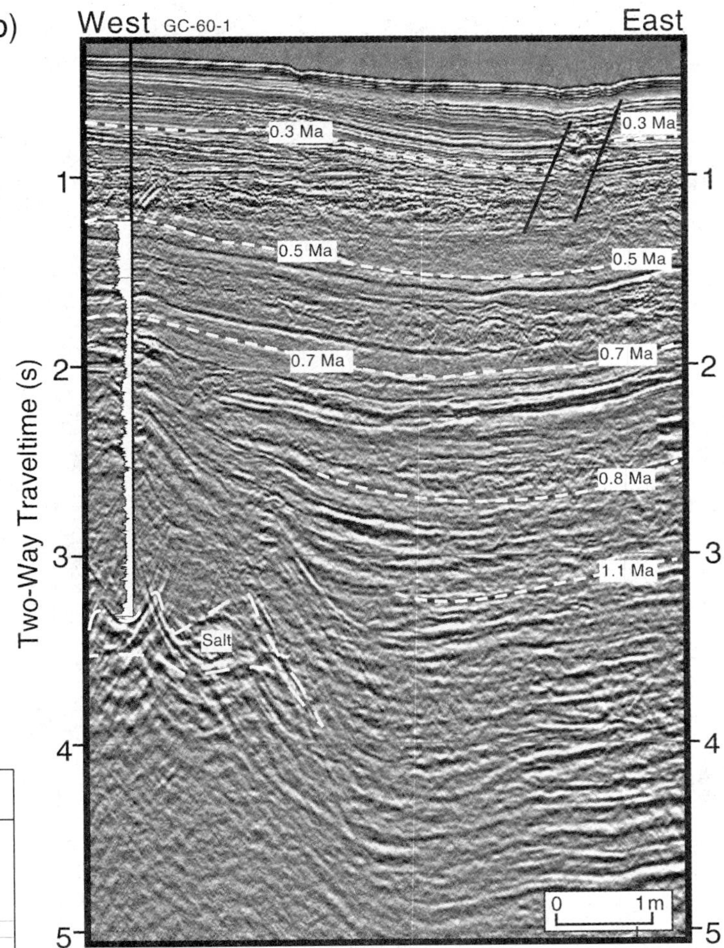


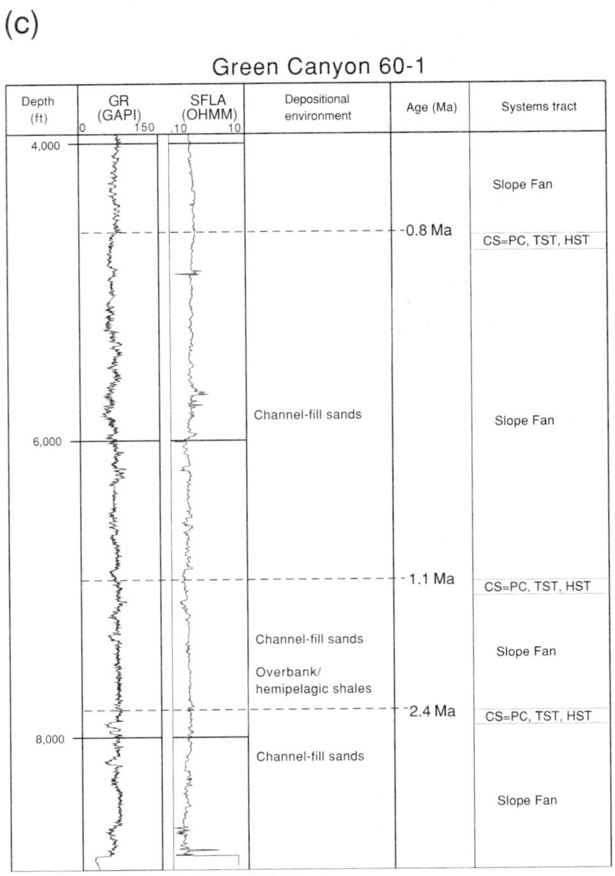


**Green Canyon 60**

**Partners:** Mobil Oil Co. 100%
**Discovery Date:** 1984
**Production Began:** September 1996
**Type of Production Facility:** Subsea tieback to GC18 producing platform
**Water Depth:** 950 feet
**Age:** early Pleistocene
**Status:** Under evaluation
**Source:** Press release

**Figure 31—Green Canyon 60 field. (a) Location map of the Green Canyon 60 field. Also shown are locations for (b) and (c). (b) Seismic profile across the Green Canyon 60 field. (c) Gamma-ray and resistivity log from the GC 60-1 well. CS = condensed section, PC = prograding complex, TST = transgressive systems tract, HST = highstand systems tract.**

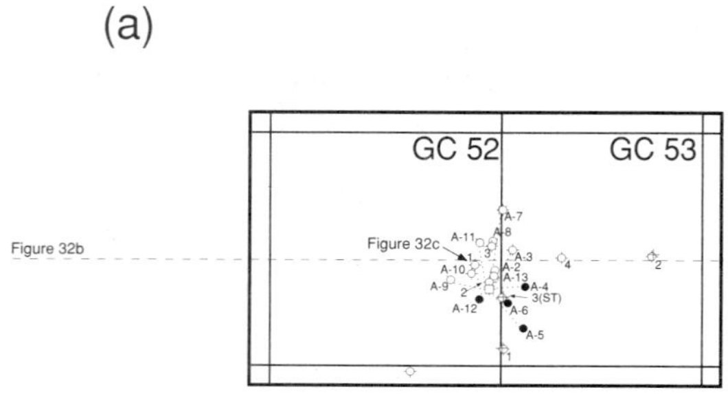


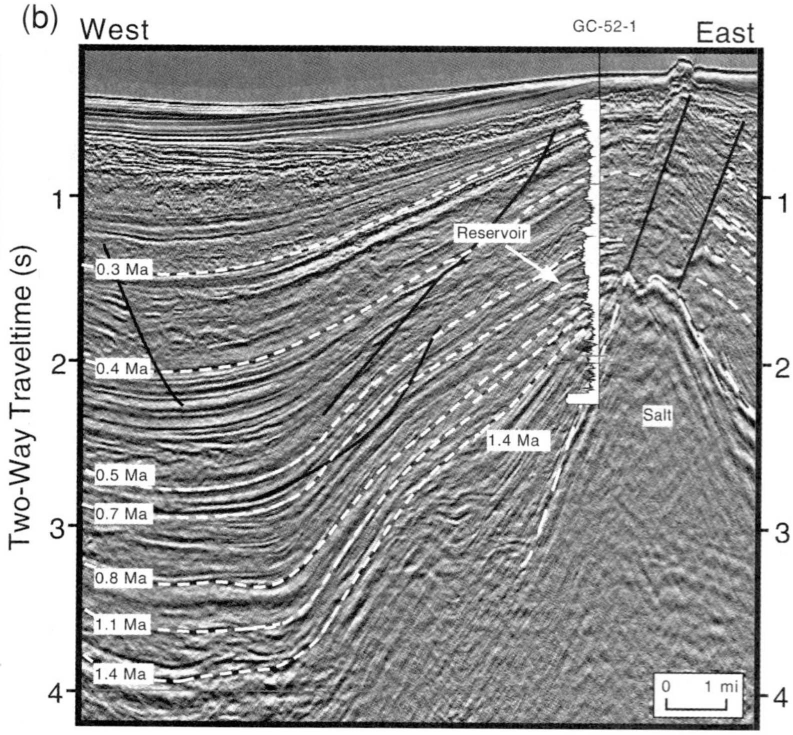


## Green Canyon 52-53 Marquette

**Partners:** Conoco 50%, Occidental 37.5%, Texaco 12.5%
**Discovery Date:** July 1984
**Production began:** late 1989/early 1990
**Type of production facility:** Fixed leg, attached to a catwalk to central production facility
**Water Depth:** 628 feet, 590 feet
**Gross pay thickness:** 60 feet
**Depositional environment:** channel-fill and thin-bed levee sands
**Number of wells:** Approx. 17
**Age:** Pleistocene
**Producing Interval:** 8250'–8500'
**Drive Mechanism:** Water Drive
**Average Porosity:** 35.0
**Reservoir Temperature:** 117°F
**Reservoir Pressure (BHP):** 2381
**Water Saturation:** 29.0
**API Gravity:** 35.0
**Gas Gravity:** 0.60
**Gas-oil ratio:** 1363
**Yield:** 36.3
**Seals upper:** shale
**lateral:** shale
**Status:** Producing since 1990
**Source of information:** W. Schneider, pers. comm., 1996; MMS, 1997

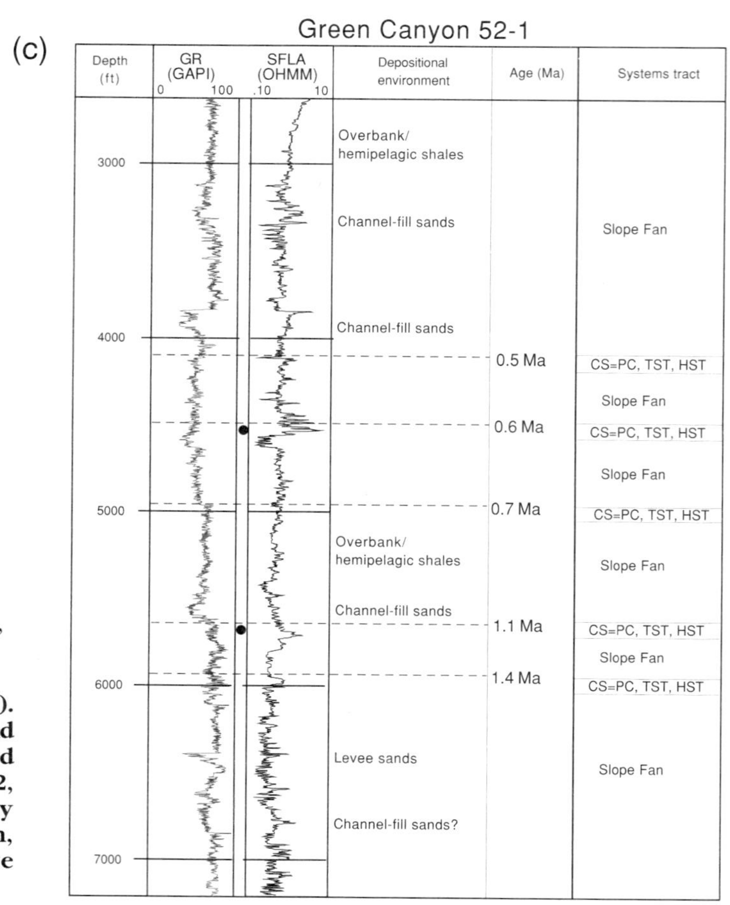


**Figure 32—Green Canyon 52, 53 field (Marquette). (a) Location map of the Green Canyon 52, 53 field (Marquette). Also shown are locations for (b) and (c). (b) Seismic profile across the Green Canyon 52, 53 field (Marquette). (c) Gamma-ray and resistivity log from the GC 52-1 well. CS = condensed section, PC = prograding complex, TST = transgressive systems tract, HST = highstand systems tract.**

(a)

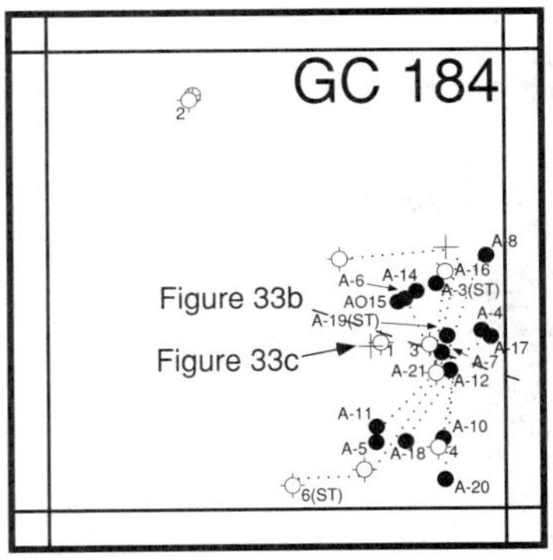


**Green Canyon 184**
**Jolliett**

**Partners:** Conoco 33.3%, Texaco 33.3%, Occidental 33.3%
**Discovery Date:** August 1981
**Production began:** November 1989
**Type of production facility:** TLP
**Water Depth:** 1716 ft
**IP:** Did not test first well
**Number of reservoirs completed:** 21 zones (104 with faults) most of reserves from the west 1400 acre fault block
**Depths:** 7178 ft
**Gross pay thickness:** 277 ft
**Depositional environment:** channel-fill; thin-bedded levee sands
**Number of wells:** 20 (5 sidetracks as workovers)
**Age:** early to late Pleistocene
**Drive mechanism:** Partial Water
**Reservoir Temperature:** 149°F
**Reservoir Pressure (BHP):** 4537
**Water Saturation:** 23.0
**Gas Gravity:** 0.60
**Gas-oil ratio:** 3529
**Yield:** 36.3
**Seals upper:** shale
**lateral:** shale
**Size acres:** 1000, 8 fault blocks
**Secondary recovery:** No waterflood, depletion reservoirs
**API:** 20–60, bulk is 25–35
**Initial GOR:** 600 to 10,000
**Initial Pressure:** 3900 PSI to 6500 (above thrust fault), 8000 PSI below fault
**Reservoir Temperature:** 106° (shallowest), 130° (deepest)
**Sources of information:** Schneider and Clifton, 1995; W. Schneider, pers comm., 1996; MMS, 1997

(b)

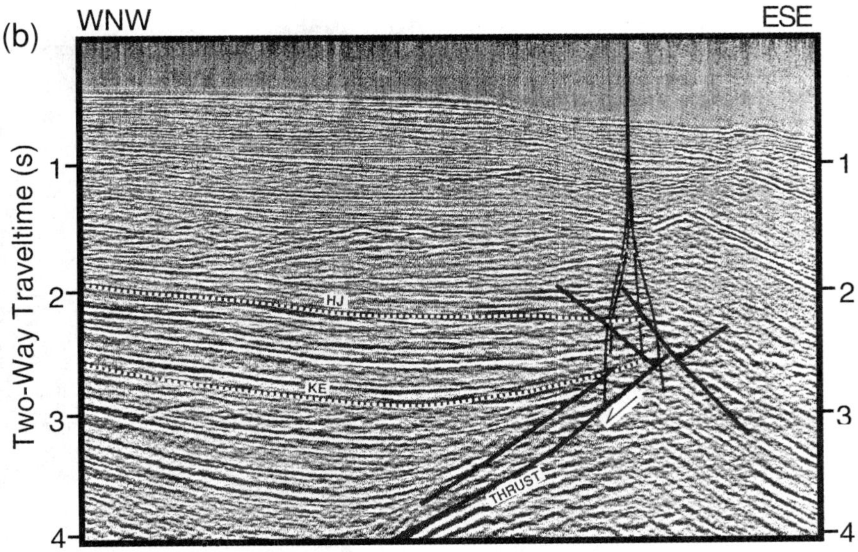


(c)

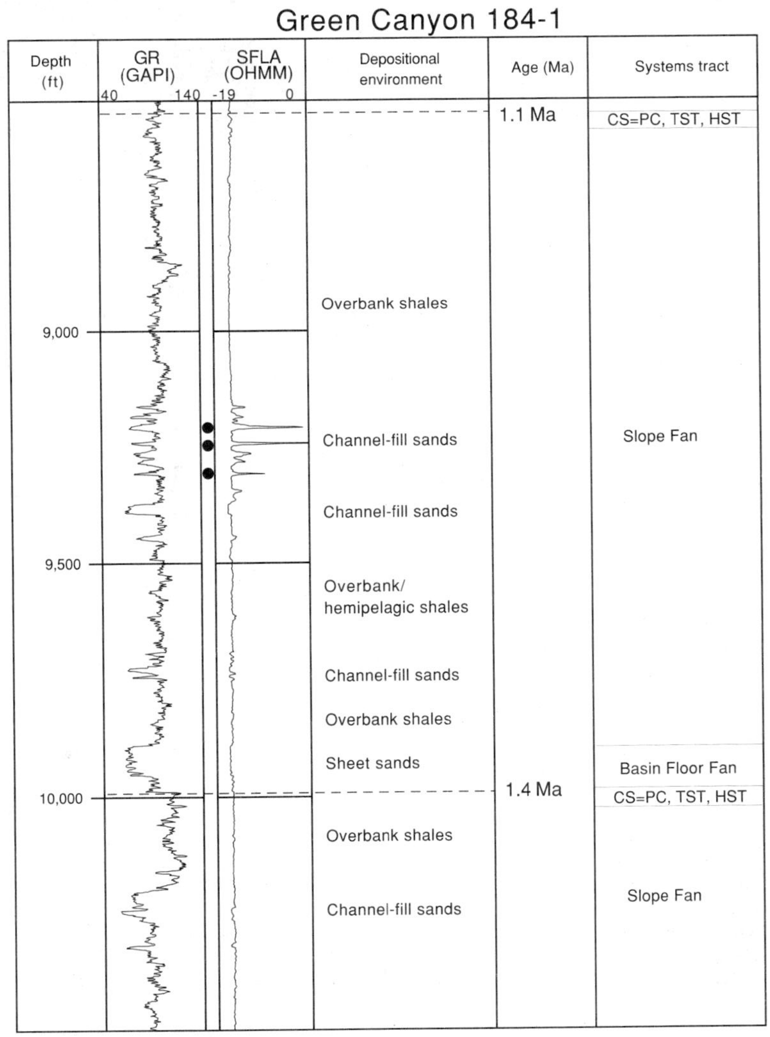


**Figure 33—Green Canyon 184 field (Jolliet). (a) Location map of the Green Canyon 184 field (Jolliet). Also shown are locations for (b) and (c). (b) Seismic profile across the Green Canyon 184 field (Jolliet). After Schneider and Clifton (1995). Reprinted with permission of SEPM. (c) Gamma-ray and resistivity log from the GC-184-1 well. CS = condensed section, PC = prograding complex, TST = transgressive systems tract, HST = highstand systems tract.**

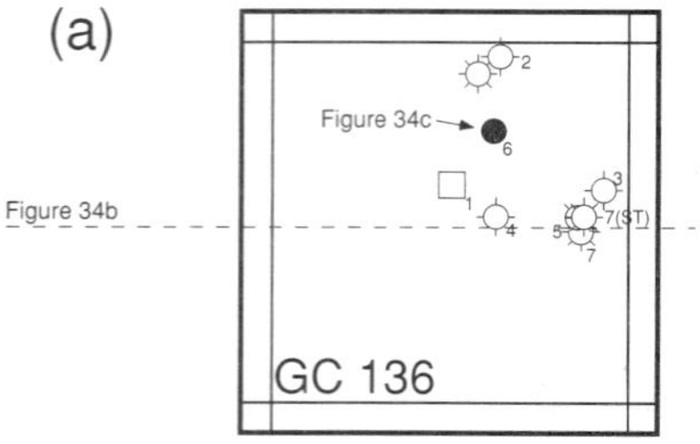


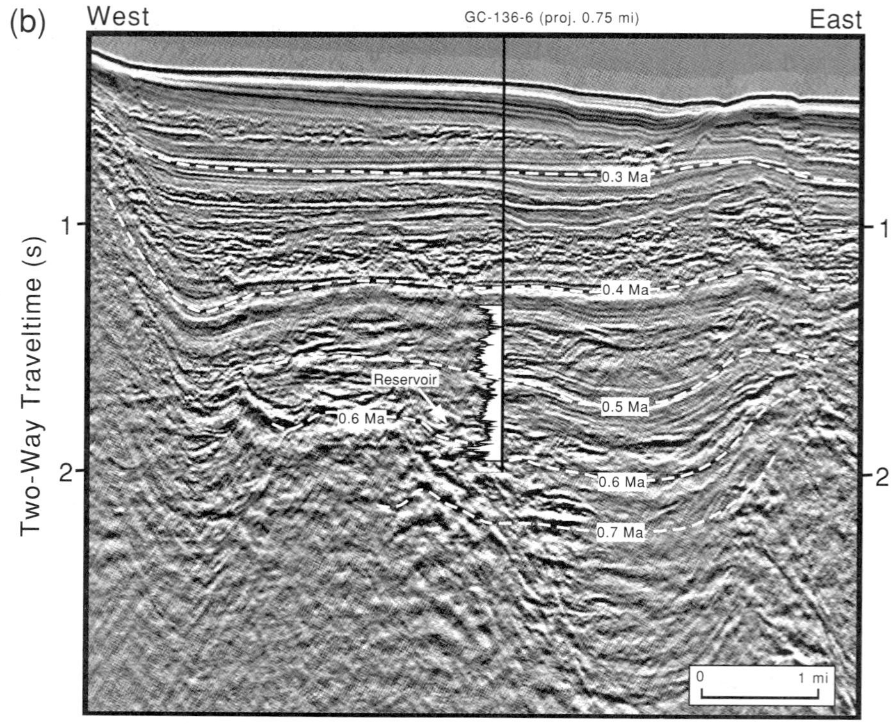


**Green Canyon 136**
**Shasta**

**Partners:** Texaco 50%, Mariner 25%, Samedan 25%
**Discovery Data:** Original wells 1981
**Discovery Year:** 1994 (136#6) & 1995 (136#7ST)
**Production Began:** 3/94 (136#6) & 12/95 (136#7ST)
**Production Facility:** 2 subsea wells tied to Kilauea (GC6) producing platform
**Water Depth:** 850'–1000'
**Reserves:** 38BCF (136#6), 27BCF (136#7ST) estimated
**IP Rate:** 28,023–31,000 mcf/d
**Number or Reservoirs Completed:** 2
**Reservoir Depth:** 5200'–8000'
**Gross Pay:** 62´ (136#6), 57' (136#7ST)
**Gross Thickness:** 130–150 feet
**Average Porosity:** 27%
**Average Permeability:** 120–1700 md.
**Reserve Temperature:** 119–127°F
**Current Pressure:** 2600–3180 psia
**Water Saturation:** .33–.36
**Depositional Environment:** Slope fan channel & overbank sands
**Number of Wells:** 2 subsea
**Age:** Pleistocene; lower 'Trim A' sand
**Producing Interval:** Mid-upper Pleistocene
**Play:** Flexure trend
**HC Type:** Gas (136#6), gas/condensate (136#7ST)
**Drive Mechanism:** Depletion (136#6), partial water drive (136#7ST)
**Aerial Closure:** 443 ac. (136#6), 276 ac. (136#7ST)
**API Gravity:** 59.9 (136#7ST)
**Gas Gravity:** .556–.65
**Cumulative Oil/Condensate:** 171 MBO (1/97-136#7ST)
**Cumulative Gas Production:** 8.8–12.1 BCF
**Gas Reserves:** 17.4–33.8 BCF
**Gas Recovery Efficiency:** 63.5–65%
**Trap:** Channel & overbank sands truncated against shale filled channels
**Seals: upper:** shale (136#6)
**lateral:** shale (136#7ST)
**Secondary Recovery:** N/A
**Oil Type:** Condensate 59.9 (136#7ST)
**Initial BHP:** 4341–4500 psia
**Status:** Producing

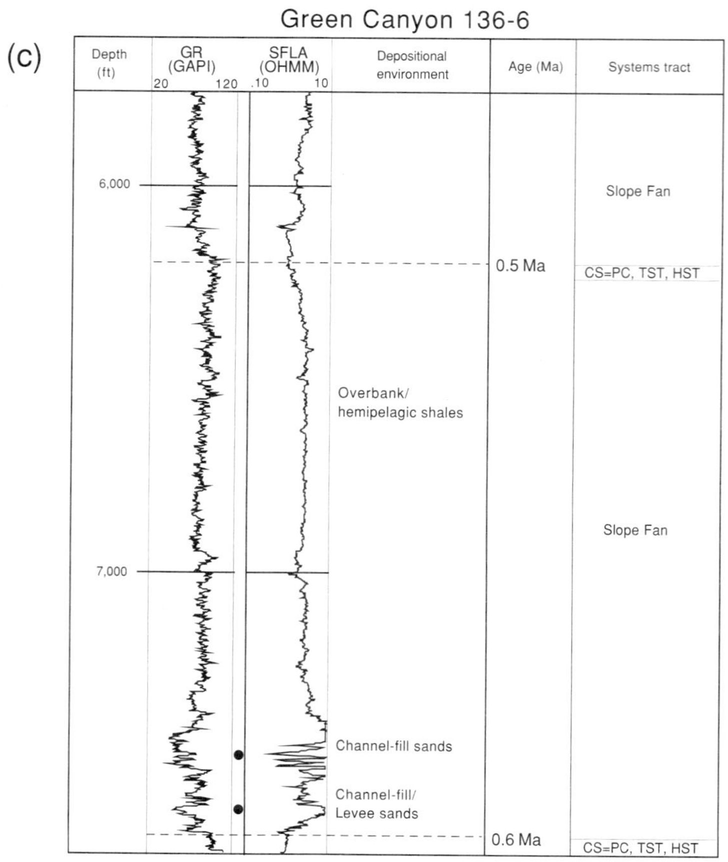


**Figure 34—Green Canyon 136 field (Shasta). (a) Location map of the Green Canyon 136 field (Shasta). Also shown are locations for (b) and (c). (b) Seismic profile across the Green Canyon 136 field (Shasta). (c) Gamma-ray and resistivity log from the GC 136-7 well. CS = condensed section, PC = prograding complex, TST = transgressive systems tract, HST = highstand systems tract.**

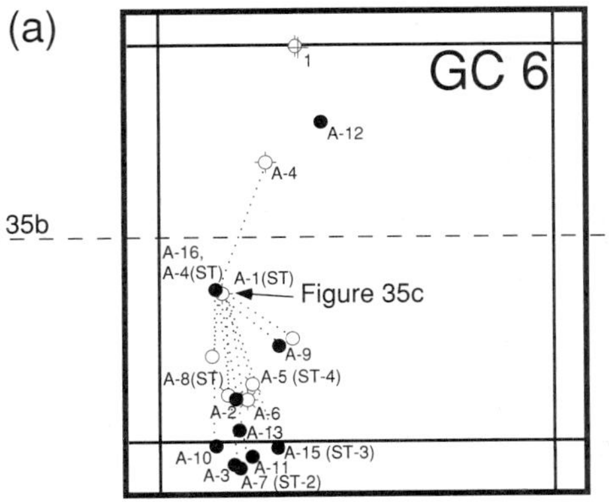


**Green Canyon 6**
**Kilauea**

**Partners:** Texaco 50%, Spirit Energy 76 (Unocal) 50%
**Discovery Date:** 1985
**Production Began:** 1990
**Type of Production Facility:** 16 slot fixed platform
**Water Depth:** 622 feet
**Oil Reserves:** 10 mbo
**Gas Reserves:** 60 BCF
**Number of Reservoirs Completed:** 16 different reservoirs with production
**Reservoir Depth:** 6950'–7400'
**Depositional Environment:** Shelf-edge delta (LST), Channel-fill
**Number of Wells:** 16 (12 active)
**Age:** Pleistocene
**Producing Interval:** 3500–7800′ (16 individual sand layers)
**Average Porosity:** 25–32%
**Average Permeability:** 200 md–1.7 d
**HC Type:** oil, gas
**API Gravity:** 27.8–31.4
**Drive Mechanism:** Partial to strong water drive
**Trap:** Radial faulting
**Seals upper:** shale
**lateral:** shale
**Secondary Rec.:** N/A
**Status:** Producing

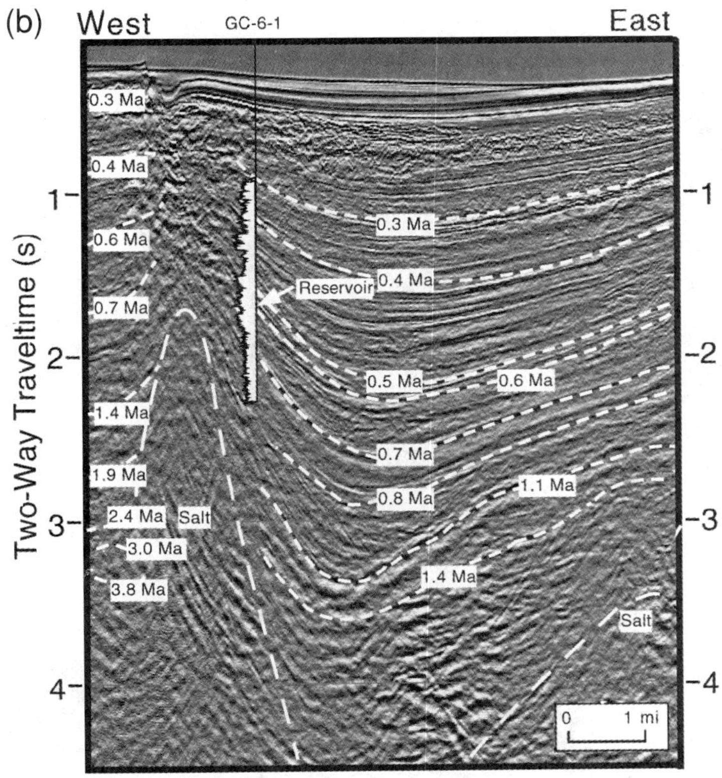


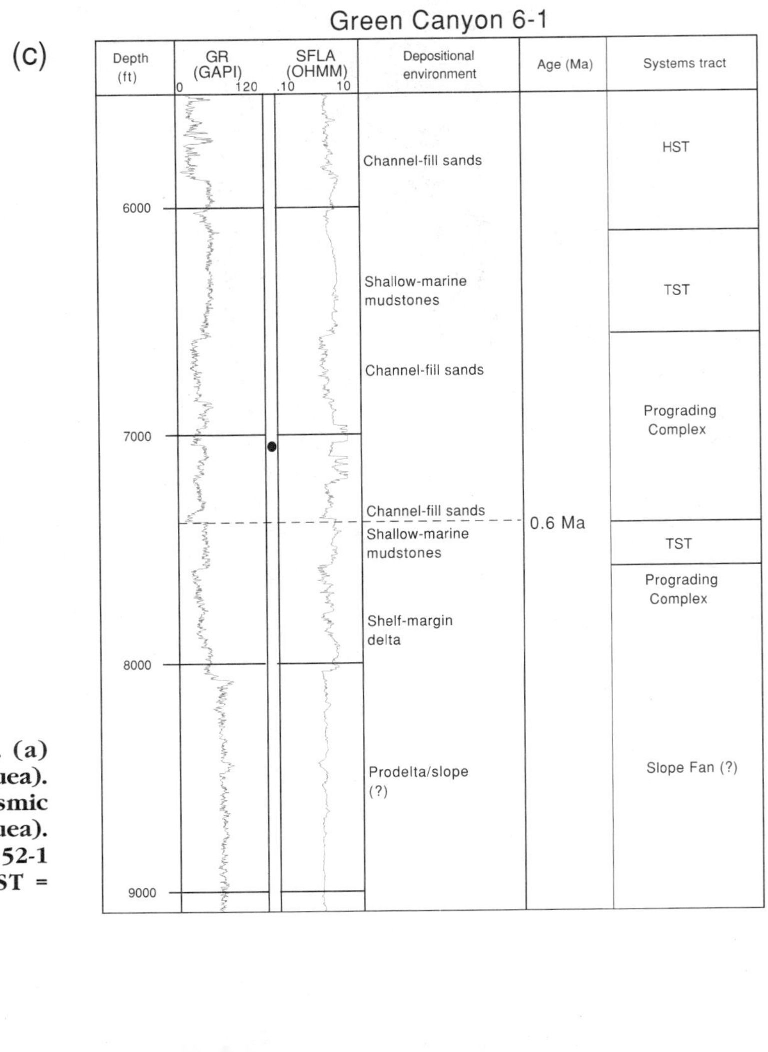


**Figure 35—Green Canyon 6 field (Kilauea). (a) Location map of the Green Canyon 6 field (Kilauea). Also shown are locations for (b) and (c). (b) Seismic profile across the Green Canyon 6 field (Kilauea). (c) Gamma-ray and resistivity log from the GC 52-1 well. TST = transgressive systems tract, HST = highstand systems tract.**

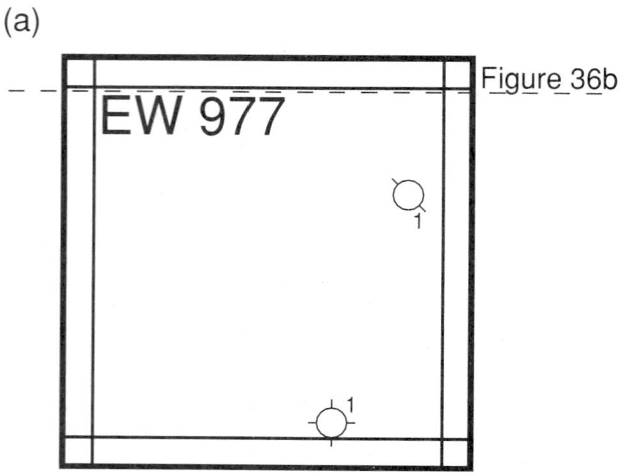


**Ewing Bank 977**
**Operator:** Walter Oil 100%
**Discovery Date:** 1996
**Water Depth:** ft
**Type of Production Facility:**
Subsea tieback to Eugene Island 390 platform
**Number of Wells:** 1
**Age:** early Pleistocene
**HC Type:** gas
**Status:** producing

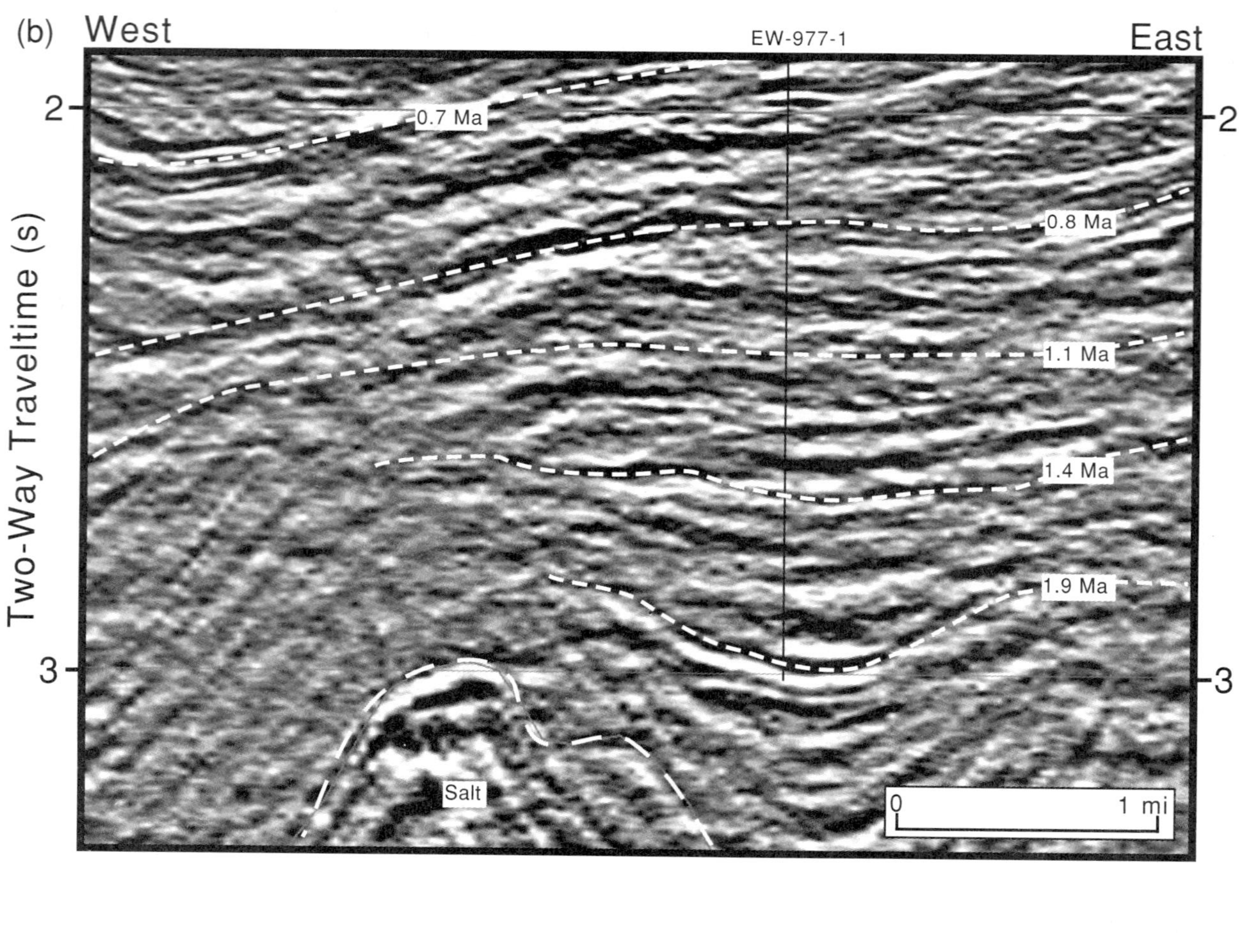


**Figure 36—Ewing Bank 977 Field. (a) Location map of the Ewing Bank 977 field. Also shown is location for (b). (b) Seismic profile across the Ewing Bank 977 Field. Note that the well logs for the EW 977 field have not been released yet.**

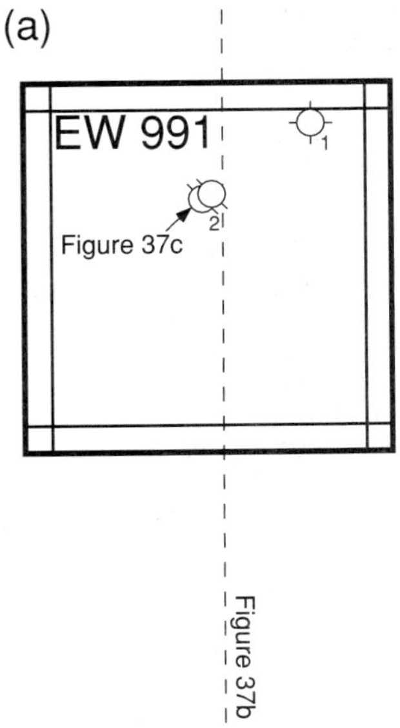


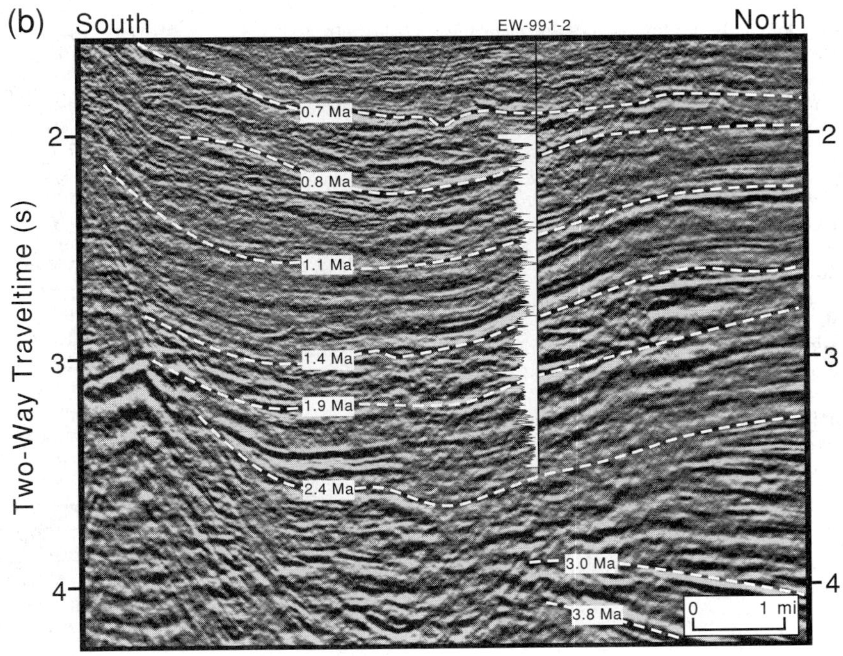


**Ewing Bank 991**
**Operator:** Walter Oil 100%
**Discovery Date:** 1995
**Water Depth:** 750 ft
**Type of Production Facility:** Subsea tieback to EW947 fixed platform
**Number of Wells:** 1 (twin of Exxon #2 well)
**Age:** late Pliocene
**HC Type:** Oil
**Status:** Producing

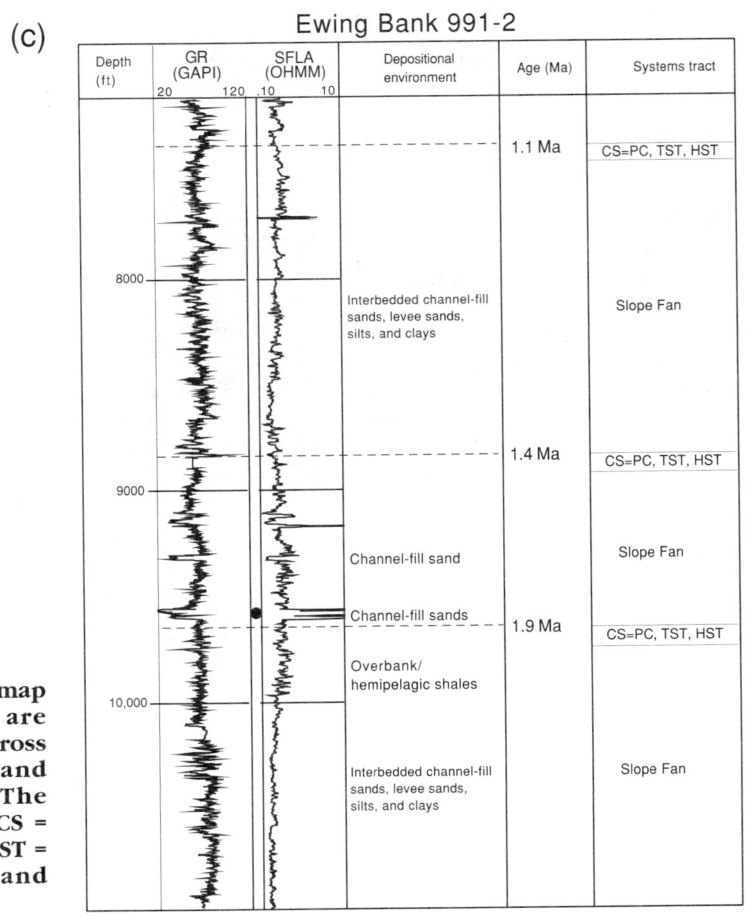


**Figure 37—Ewing Bank 991 Field. (a) Location map of the Ewing Bank 991 Field. Also shown are locations for (b) and (c). (b) Seismic profile across the Ewing Bank 991 Field. (c) Gamma-ray and resistivity log from the EW 991-2 well. The discovery well was a twin of the 991-2 well. CS = condensed section, PC = prograding complex, TST = transgressive systems tract, HST = highstand systems tract.**

## REFERENCES CITED

Beaumont, E. A., and N. H. Foster, compilers, 1990–1992,: AAPG treatise of petroleum geology: Atlas of Oil and Gas Fields, v. 1-8.

Camp, W. K., and D. McGuire, 1997, Mahogany field, a subsalt legend: a tale of technology, timing, and tenacity, offshore Gulf of Mexico: Houston Geological Society Bulletin, v. 40, no. 2, p. 14–16.

Chapin, M. A., P. Davies, J. L. Gibson, and H. S. Pettingill, 1994, Reservoir architecture of turbidite sheet sandstones in laterally extensive outcrops, Ross formation, western Ireland, *in* P. Weimer, A. H. Bouma, and B. F. Perkins, eds., Submarine fans and turbidite systems: GCS-SEPM Foundation 15th Annual Research Conference, p. 53–68.

Cook, D., and P. D'Onfro, 1991, Jolliet field thrust fault structure and stratigraphy, Green Canyon Block 184, offshore Louisiana: Gulf Coast Association of Geological Society Transactions, v. 41, p. 100–121.

Edman, J. D., and M. K. Burk, 1996, An integrated reservoir compartmentalization at Ewing Bank 873: AAPG/EAGE Research Conference, 4 p.

Haines, L., 1996, Going deep: Oil and Gas Investor, v. 16, no. 9, p. 24–32.

Harrison, H. L., D. Moore, and P. Hodkins, 1995, The Mahogany subsalt discovery: a unique hydrocarbon play, offshore Louisiana, *in* C. J. Travis, H. Harrison, M. R. Hudec, B. C. Vendeville, F. J. Peel, and B. E. Perkins, eds., Salt, sediment and hydrocarbons: GCS-SEPM Foundation 16th Annual Research Conference, p. 95–97.

Holman, W. E., and S. S. Robertson, 1994, Field development, depositional model and production performance of the turbidite "J" sands at the prospect Bullwinkle, Green Canyon 65 fields, outer shelf Gulf of Mexico, *in* P. Weimer, A. H. Bouma, and B. F. Perkins, eds., Submarine fans and turbidite systems: GCS-SEPM 15th Annual Research Conference, p. 139–150.

McBride, B. C., M. G. Rowan, and P. Weimer, 1998, The evolution of allochthonous salt systems, northern Green Canyon and Ewing Bank (offshore Louisiana), northern Gulf of Mexico; AAPG Bulletin, v. 82, no. 5B, p. 1013–1036.

MMS (Minerals Management Service), 1997, Gulf of Mexico OCS region: Internet site www.gomr.mms.gov

Rafalowski, J. W., B. W. Regel, D. J. Jordan, and D. O. Lucidi, 1994, Green Canyon block 205 lithofacies, seismic facies, and reservoir architecture, *in* P. Weimer, A. H. Bouma, and B. F. Perkins, eds., Submarine fans and turbidite systems: GCS-SEPM Foundation 15th Annual Research Conference, p. 293–306.

Schneider, W., and H. E. Clifton, Jr., 1995, Green Canyon 184 upper slope turbidites, *in* R. Winn, Jr., ed., Turbidites and associated facies: SEPM Core Workshop 20, p. 55–73.

Shew, R. D., D. R. Rollins, G. M. Tiller, and C. J. Hackbarth, 1994, Thin-bedded turbidite deposits in the Gulf of Mexico: characterization and modeling using subsurface data, outcrops, and modern analogs, *in* P. Weimer, A. H. Bouma, and B. F. Perkins, eds., Submarine fans and turbidite systems: GCS-SEPM 15th Annual Research Conference, p. 327–334.

Shew, R. D., 1997, Deepwater Gulf of Mexico core workshop course notes: GCAGS 47th Annual Meeting.

Snyder, F. C., and J. Nugent, 1995, Teak—testing a subsalt hydrocarbon trap geometry, South Timbalier block 260, Gulf of Mexico, *in* C. J. Travis, H. Harrison, M. R. Hudec, B. C. Vendeville, F. J. Peel, and B. E. Perkins, eds., Salt, sediment and hydrocarbons: GCS-SEPM Foundation 16th Annual Research Conference, p. 257–267.

Weimer, P., P. Varnai, F. M. Budhijanto, Z. M. Acosta, R. E. Martinez, A. F. Navarro, M. G. Rowan, B. C. McBride, T. Villamil, C. Arango, J. R. Crews, and A. J. Pulham, 1998, Sequence stratigraphy of Pliocene and Pleistocene turbidite systems, northern Green Canyon and Ewing Bank (offshore Louisiana), northern Gulf of Mexico; AAPG Bulletin, v. 82, no. 5B, p. 918–960.

## ABOUT THE AUTHORS

### Paul Weimer

Paul Weimer is an associate professor and director of the Energy and Minerals Applied Research Center at the University of Colorado at Boulder. His research interests are in sequence stratigraphy and basin analysis, three-dimensional seismic interpretation, and reservoir geology.

### Jennifer R. Crews

Jennifer R. Crews received her Bachelor's degree in geology from the University of Colorado at Boulder in 1995; she will receive her M.S. degree during the summer of 1998. Her thesis topic is high-resolution biostratigraphic analysis of Green Canyon and Ewing Bank areas, northern Gulf of Mexico. Upon graduation, she will be employed by Texaco Exploration and Production, Inc., New Orleans. Research interests include applied biostratigraphic methods and sequence analysis.

**Peter Varnai**

Peter Varnai received his B.S. degree (1985) in geology from Eötvös Loránd University, Budapest, Hungary; an M.S. degree (1991) in geophysics from Miskolc University, Miskolc, Hungary; and a Ph.D. (1996) in geology from the University of Colorado, Boulder, Colorado. He worked for the Hungarian National Oil Company in Budapest as an exploration geologist for six years. He was responsible for seismic interpretation, stratigraphic interpretation of well logs, construction of stratigraphic and tectonic models, and basin analysis in the Pannonian basin. During his graduate studies in Colorado, he participated in the Gulf of Mexico industry consortium. In 1996 he joined Exxon Exploration Company as a senior geologist.

**Ryan Crow**

Ryan Crow is studying geology at the University of Colorado-Boulder. He anticipates receiving his B.A. in 1999.

# Sequence Stratigraphy of Pliocene and Pleistocene Turbidite Systems, Northern Green Canyon and Ewing Bank (Offshore Louisiana), Northern Gulf of Mexico[1]

**Paul Weimer,[2] Peter Varnai,[3] Fadjar M. Budhijanto,[2] Zurilma M. Acosta,[4] Rafael E. Martinez,[4] Alonso F. Navarro,[5] Mark G. Rowan,[2] Barry C. McBride,[6] Tomas Villamil,[7] Claudia Arango,[8] Jennifer R. Crews,[2] and Andrew J. Pulham[2]**

## ABSTRACT

Neogene turbidite systems are major reservoirs in the northern deep Gulf of Mexico. Few publications have described the stratigraphic variations or the three-dimensional geometries of these turbidite systems in detail; hence, an understanding of the stratigraphic characteristics of the producing sands is important for deep-water exploration in the Gulf of Mexico and similar basins worldwide.

This study focuses on the northern Green Canyon and central Ewing Bank protraction (lease) areas, where the Pliocene–Pleistocene turbidite systems were mapped using an integrated exploration database. Interpretation of 10,000 km of two-dimensional seismic, 185 well logs, and biostratigraphy from 180 wells allowed us to define the regional sequence stratigraphic framework for this area and potential areas for future exploration. A complex Pliocene–Pleistocene geologic evolution of the area is indicated by the seismic and geologic facies, depositional rates, nature of turbidite systems, and sand content.

Significant sand deposits (basin-floor fans) were deposited in most sequences and directly overlie sequence boundaries. Salt tectonics and faulting greatly influenced the loci of these fans' deposition. Large, thick fans fill entire salt-withdrawal minibasins at the base of the Pliocene sequences. In the Pleistocene sequences, where the rate of salt withdrawal was less, smaller and thinner fans were deposited downdip of faults and adjacent to shallow salt bodies.

Channel systems, interbedded with overbank shales, constitute most of the sediments in the sequences. The older sequences contain more channels and sandier channel fills than the younger sequences. Analysis of all sequences indicates a complex depositional history where significant sands were deposited where abrupt decreases in bathymetric gradient are associated with salt tectonics or faulting.

## INTRODUCTION

Neogene turbidite systems of the northern Gulf of Mexico form the primary reservoirs in the deepwater and subsalt plays. Complex salt tectonics has



[1]Manuscript received December 12, 1996; revised manuscript received January 16, 1998; final acceptance January 20, 1998.

[2]Energy and Minerals Applied Research Center, Department of Geological Sciences, University of Colorado, Boulder, Colorado 80309-0399.

[3]Energy and Minerals Applied Research Center, Department of Geological Sciences, University of Colorado, Boulder, Colorado 80309-0399. Present address: Exxon Exploration Company, P.O. Box 9778, Houston, Texas 77210-4778.

[4]PDVSA Exploracion y Produccion, Gerencia de Planificacionde Exploracion, Edificio La Estancia, Oficina 1359, Chuac, Caracas, Venezuela.

[5]PDSVA Exploracion y Production, Piso 14, Ofic. 14115-C, Avenida La Estancia, Chuao-Caracas 1010A Venezuela.

[6]Energy and Minerals Applied Research Cent., Department of Geological Sciences, University of Colorado, Boulder, Colorado 80309-0399. Present Address: HS Resources Inc., 1999 Broadway, Suite 3600, Denver, Colorado 80202.

[7]Energy and Minerals Applied Research Center, Department of Geological Sciences, University of Colorado, Boulder, Colorado 80309-0399. Present address: Conoco, P.O. Box 2197, Houston, Texas 77252-2197.

[8]Energy and Minerals Applied Research Center, Department of Geological Sciences, University of Colorado, Boulder, Colorado 80309-0399. Present address: Marathon Oil Company, P.O. Box 3128, Houston, Texas 77252-3128.

We thank the following companies and their representatives for their support and input as part of the Gulf of Mexico research consortium at the University of Colorado: Agip, Amoco, Anadarko, BP Exploration, BHP, Burlington Resources, Chevron USA, CNG, CXY, Conoco, Enterprise, Exxon, Marathon, Mobil, Occidental, PanCanadian, Pennzoil, Petrobras, Phillips, Shell, Texaco, Total, Union Pacific, and Unocal. Their support has been crucial for the success of this research. We extend special thanks to Halliburton Geophysical (now Western Geophysical), specifically to Bob Graebner, John Anderson, and Gary White, for the seismic data used in this project. We are indebted to PaleoData Inc. and Art Waterman for use of their biostratigraphic data. Digital logs were kindly provided by Marathon (Tricia Santogrossi). Software donations from GeoQuest, CogniSeis Development (Pat Poe, Roy Kligfield, Robert Ratliff), Landmark Graphics (Thom Fisher), Platte River Associates (Jay and China Leonard), and Zeh Graphics (Jerry Martin) are essential to the success of our study. We are grateful for the reviews by Laurie Lamar and M. S. Wacker, and AAPG reviewers Kevin Biddle, Robert Mitchum, Tor Nilsen, and George Ramsayer, all of whom improved the manuscript considerably. Acknowledgment is made to the donors of the Petroleum Research Fund, administered by the American Chemical Society, for partial support of this work.

created a series of salt highs and intraslope basins throughout the Neogene. These producing turbidite systems were deposited at bathyal water depths in a series of salt-withdrawal minibasins. Seismic and geologic facies change abruptly across the area in all sequences, reflecting the lateral and vertical changes in lithofacies from sand-rich to shale-rich systems.

Recent significant Neogene turbidite discoveries yielded good production rates [e.g., Auger field (McGee et al., 1994), and Mars field (Haines, 1996)]; therefore, stratigraphic prediction of these reservoirs is of utmost importance to companies exploring for the deep-water or subsalt plays. Most papers published on the petroleum-producing Neogene turbidite systems in the northern Gulf of Mexico have focused on the regional setting (e.g., Mitchum et al., 1993; Yeilding and Apps, 1994) or on specific fields (see many references in Weimer et al., 1994; Weimer et al., 1998a, this issue). Although the northern Gulf of Mexico currently is the major exploration province in the United States, no significant regional studies have been published that systematically describe the distribution and occurrence of sand-prone facies and how the turbidite systems stack in one subregional area.

In this paper, we summarize the sequence stratigraphy of the Pliocene–Pleistocene turbidite systems in northern Green Canyon, Ewing Bank, and the southern parts of Grand Isle, South Timbalier, Ship Shoal, and Eugene Island protraction (lease) areas (Figure 1). The study area is located in water depths between 200 and 4500 ft (65 to 1400 m) and reflects the transition from shelf to slope water depths (Figure 1). The study area includes 35 fields or discoveries, primarily in turbidite-related reservoirs (Figure 2) (Weimer et al., 1998a, this issue). Three subsalt discoveries have been made in the area, with several more prospects expected to be tested within the next year. Good potential for future exploration exists both in subsalt opportunities and in the deeper water portions of the area.

The objectives of this paper are to (1) describe the general sequence stratigraphic setting of the Pliocene–Pleistocene sediments in the area, with special emphasis on the sand-rich turbidite-related facies; (2) characterize the turbidite elements with two-dimensional (2-D) seismic data, wireline logs, and benthic biofacies; (3) systematically describe two representative sequences using a series of maps, seismic profiles, and log cross sections; (4) discuss the relative controls on deposition of various elements; and (5) propose a generalized depositional model of the turbidite systems in the area.

## DATABASE

We used an integrated data set from three primary sources (Figure 2). First, we interpreted 6300 mi (10,200 km) of 48- and 96-fold 2-D seismic data. These data consist of 4300 mi (6960 km) of dip direction (north-south) and 2000 mi (3240 km) of strike direction (east-west). Line spacing is 1 mi (1.62 km) in the dip direction and 2–3 mi (3.2–4.8 km) in the strike direction. These seismic data were collected between 1984 and 1986 as part of a 3-D data set, and were reprocessed as 2-D lines in 1988–1989. Second, we used digital logs from 185 of the approximately 480 wells in the area. Synthetic seismograms and time-based logs were generated for all wells. Third, the biostratigraphic database consists of reports from foraminifera from 180 wells, calcareous nannoplankton from 83 wells, and high-resolution foraminifera from 60 wells. This biostratigraphic data set has been used for dating the sequences, correlating abundance curves to seismic profiles to identify condensed sections, and interpreting the benthic biofacies of each sequence (Villamil et al., 1998, this issue).

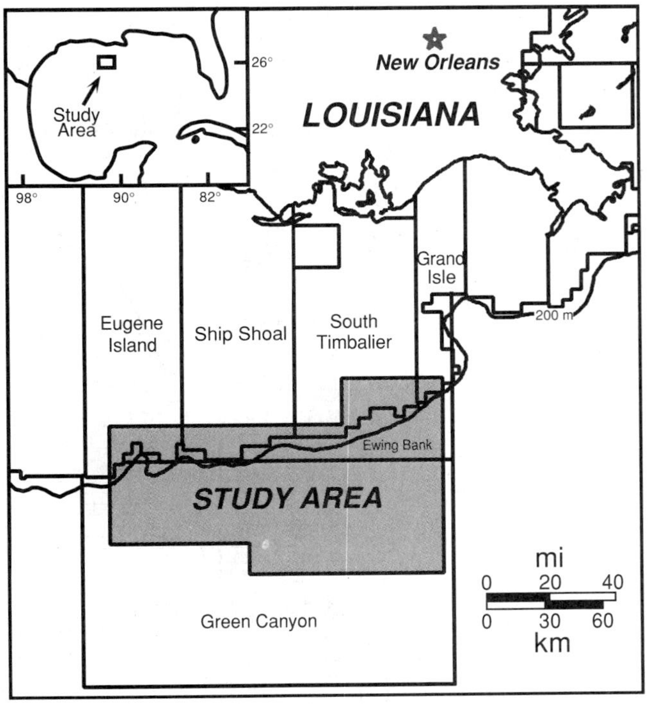


Figure 1—Location of study area in northern Green Canyon, Ewing Bank, and southernmost portions of Eugene Island, Ship Shoal, South Timbalier, and southeastern Grand Island (offshore Louisiana), northern Gulf of Mexico.

## STRUCTURAL SETTING

A major theme in the Gulf of Mexico geological literature is the complex relationship between tectonics and sedimentation in the northern continental

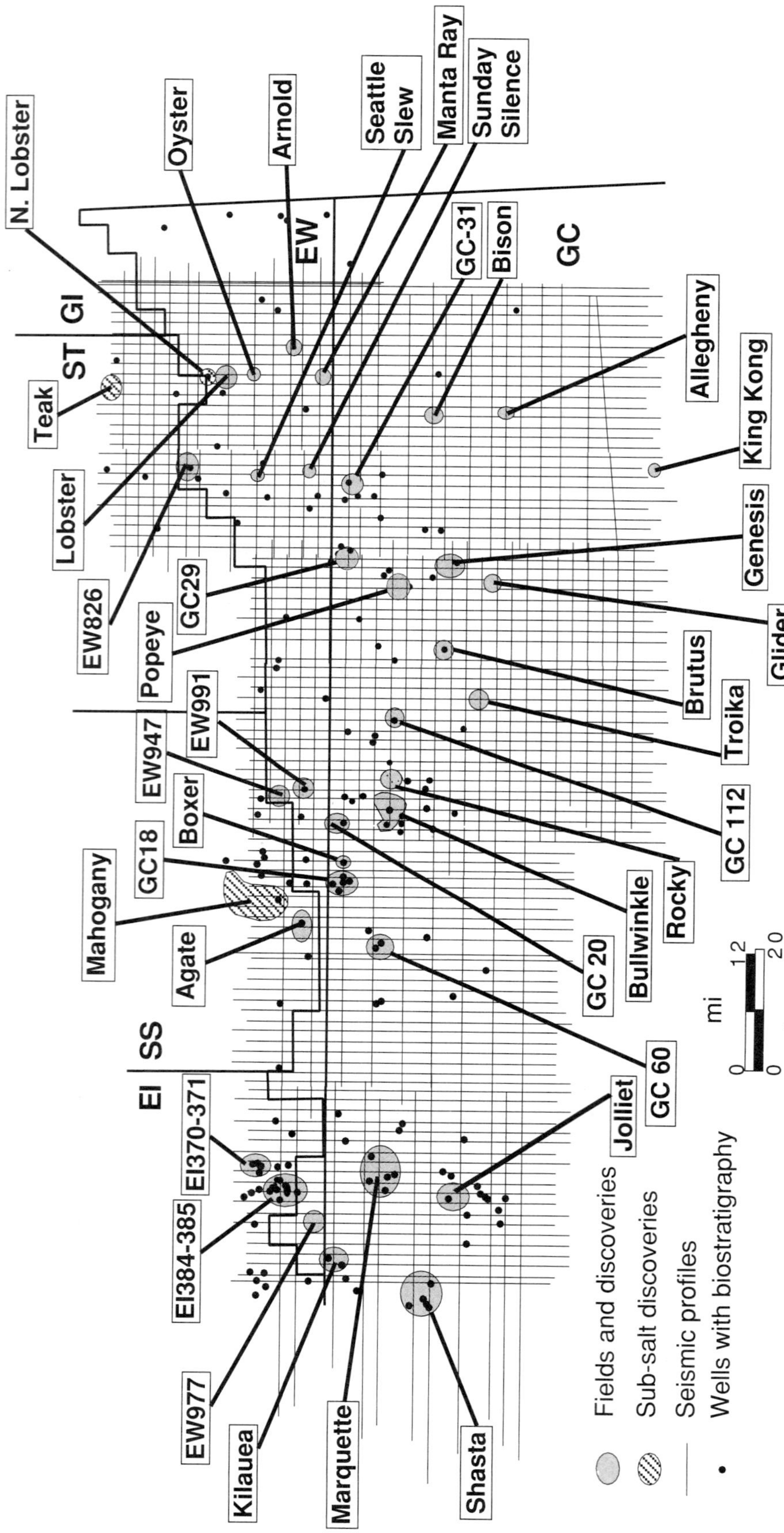


Figure 2—Database used for this study. Seismic profiles are shown by solid lines, wells with biostratigraphy are shown by circles, and fields and discoveries also are shown.

slope. Any discussion of the stratigraphy also must address the geometry and evolution of salt bodies and associated faults. The structural geology of the region is described briefly here; for further details, see Rowan (1995) and McBride et al. (1998, this issue).

## Salt Geometry

A structure contour map of the salt system in northern Green Canyon/Ewing Bank is shown in Figure 3. The top of salt has been mapped where identifiable; where the salt has been evacuated, the resultant salt weld has been mapped. The most prominent feature of the salt system is its large relief, with the tops of the shallowest diapirs at less than 1.0 s two-way traveltime and the deepest salt welds at more than 7.5 s two-way traveltime (Figures 3, 4). The difference does not represent the thickness of salt in diapirs that are rising from a common deep source level; instead, there is significant topographic relief on both the weld and the base salt, such that there are several levels of allochthonous salt (Figure 4; also see regional profiles in McBride et al., 1998, this issue; Weimer et al., 1998, this issue).

Most lows in the salt weld (deeper than 5.0 s two-way traveltime) are elliptical, generally with a north-south elongation (Figure 3). The remainder display relatively minor structural relief over broad areas, similar to the bases of shallow salt sheets present in the northern part of the area (Figure 4) (also see figures in McBride, 1998, this issue; McBride et al., 1998, this issue; Weimer et al., 1998b, this issue). Sometimes different salt/weld levels are separated by prominent overhangs; in other cases, different levels are connected by ramps in the weld surface that have various orientations. These ramps are interpreted to be primary features because they are locally observed to truncate subsalt reflections that are roughly horizontal. Thus, the geometry of the salt weld system is interpreted to represent the original base-salt geometry. The pattern is a complex one in which different salt bodies have evolved and amalgamated through time to create a composite salt system with multiple levels. Sometimes the levels are connected; sometimes the levels are separated by overhangs. Subsequent modification of the base salt and equivalent welds due to deeper deformation is thought to have been negligible.

Diapirs generally are found either along the frontal or lateral edges of shallow sheets or over the ridges and saddles between separate elliptical lows. As a consequence, these diapirs usually do not have deep roots. The salt bodies have variable geometries. In plan view, they range from circular to highly irregular (Figure 3); in profile view, they may have gently sloping, vertical, or overhanging flanks with smooth to uneven tops.

A transition zone trending approximately east-west through the center of the study area separates two regions with varying salt geometries (Figure 3). To the north, the system is dominated by the southern portions of shallow sheets that extend to the north onto the shelf. These northern areas generally have thin remnant salt between scattered small diapirs with no deep roots, and the diapirs along the southern edges of the sheets commonly have been modified by contractional squeezing. To the south, salt bodies usually are thicker and of larger areal extent, with roots that connect them to the underlying ridges and saddles in the base-salt surface.

## Origin and Evolution

Interpreted salt geometries and the results of sequential structural restoration (Rowan, 1995; McBride et al., 1998, this issue; Rowan and Weimer, 1998, this issue) are compatible with two end-member models of allochthonous salt systems. The first model describes the formation and evolution of salt sheets that are sourced from counterregional (north-dipping) feeder stocks (Schuster, 1995). These sheets may accommodate significant downslope translation of the overburden along families of listric growth faults (roho systems), or may be evacuated to form large, basinward-dipping growth wedges and associated southward-leaning diapirs linked by counterregional faults (stepped counterregional systems). Examples of both are observed in the study area (McBride et al., 1998, this issue; Rowan and Weimer, 1998, this issue).

The second model describes the growth and evolution of salt-stock canopies that are sourced from vertical feeders (Rowan, 1995). These canopies eventually collapse and are loaded such that the original salt stocks are replaced by elliptical minibasins, with salt displaced into diapirs located over saddles in the base salt. Several old salt-stock canopies, consisting of up to five amalgamated stocks, have been identified in the study area (Rowan and Weimer, 1998, this issue).

The different types of salt systems impact the stratigraphic evolution of minibasins in several ways. First, these salt systems determine the shifting location and shape of minibasins through time. Second, each type of salt system is characterized by distinctive styles of salt bodies, folds, and fault arrays. These salt systems define the local sea-floor paleobathymetry during each depositional cycle, and also dictate subsequent basin modification (thereby producing structural traps). Third, the evolution of each type of system determines whether there is adequate underlying salt to provide accommodation for vertical stacking of ponded sands. Further details of the

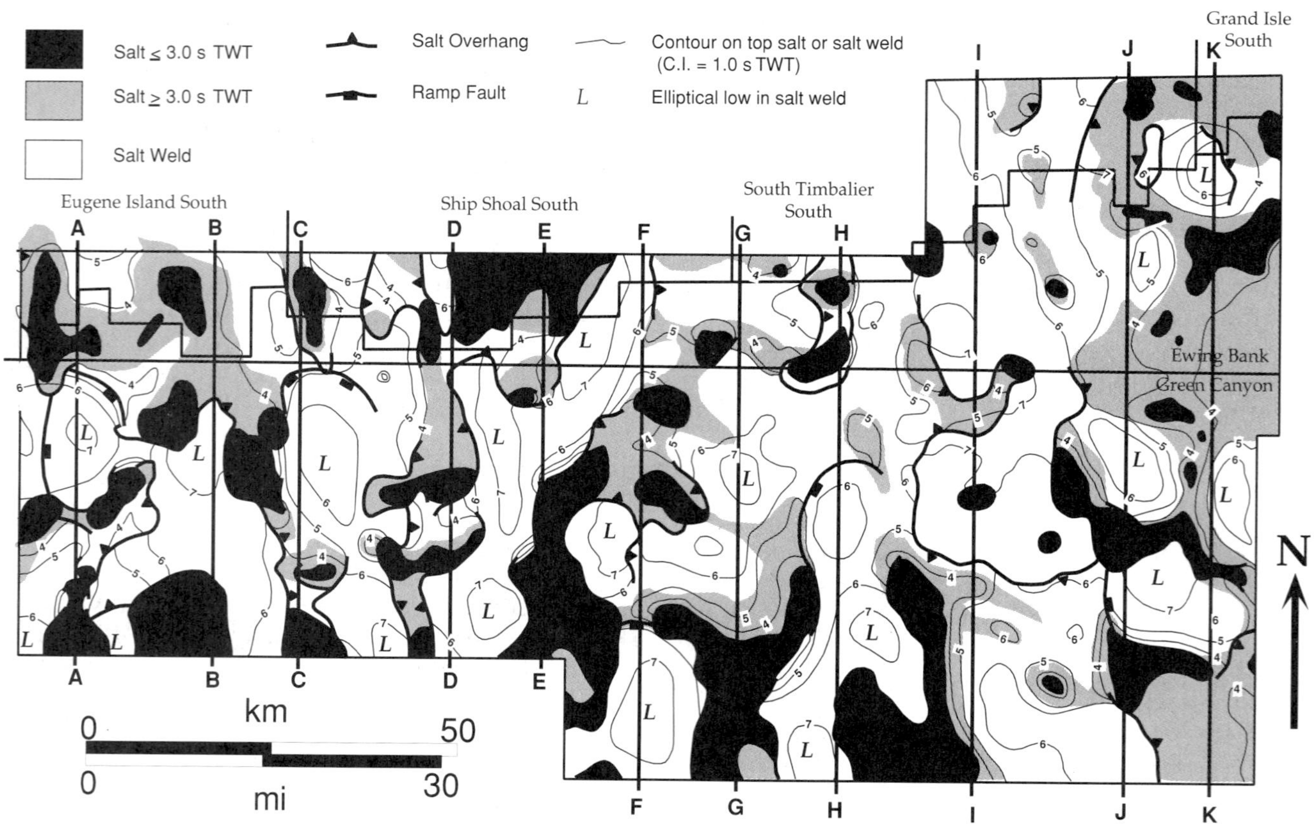


**Figure 3—Structure contour map (in time) of the top salt or equivalent salt weld. The salt system is characterized by large relief, with different levels of salt either separated by overhangs or connected by ramps in the salt weld. Top salt contours shallower than 3.0 s are not shown. Salt interpretation is by Mark Rowan. Location of Figure 4 (profile I) as shown.**

interactions between salt tectonics and sedimentation within the study area can be found in Rowan and Weimer (1998, this issue).

## SEQUENCE STRATIGRAPHY

### Methodology and Interpretation Techniques

For the regional interpretation we used basic sequence-stratigraphic interpretation techniques, with special emphasis on the Gulf Basin stratigraphy (Mitchum et al., 1993). Specifically, syndepositional tectonics affects the stratigraphic fill of the slope region. Movement along growth faults and the development of salt withdrawal minibasins generate enormous thickness changes in equivalent sediments and affect the external shape of the stratigraphic packages (Figure 4) (McBride et al., 1998, this issue; Rowan and Weimer, 1998, this issue).

The regional interpretation for this study included four steps: (1) establishing the chronostratigraphic framework by identifying and mapping the sequence boundaries, (2) mapping the seismic facies within each sequence, (3) interpreting their depositional environments, and (4) determining the geologic evolution of the area.

All three data sets were integrated to create a detailed chronostratigraphic framework. First, seismic horizons were identified by sequence analysis. Sequence boundaries in bathyal settings were best identified by the onlap terminations onto high-amplitude reflections. The sequence boundaries were tied to available biostratigraphic data to correlate them between the minibasins. In the wells, we integrated high-resolution abundance curves, calcareous nannoplankton datums, and lithologic information to help identify and confirm condensed depositional intervals. Villamil et al. (1998, this issue) present details on the biostratigraphy of the study area. All of these data were combined with the relative coastal onlap curve to date the condensed sections and sequence boundaries (see figure 3 in Villamil et al., 1998, this issue).

In the second phase, the seismic facies interpretation technique outlined by Ramsayer (1979) was used to map three seismic facies parameters in each sequence: reflection configuration, amplitude, and continuity (Table 1). Seismic facies maps were created by outlining areas with similar parameters.

For the third phase, the depositional environments were interpreted based on integrating seismic facies maps, lithologic and paleoecologic information from wells, and isopach maps. The resulting geologic facies maps show the distribution of different turbidite systems and elements.

Finally, in the fourth phase of interpretation, a series of interpreted depocenter maps was prepared to summarize the depositional history through more than 5.5 m.y.

## Turbidite Elements

Twelve depositional sequences were correlated, mapped, and described in the study area: 5.5, 4.2, 3.8, 3.0, 2.6., 2.4., 1.9, 1.4, 1.1, 0.8, 0.7, and 0.5 Ma (Figures 4, 5). Each sequence consists primarily of sediments deposited in turbidite systems. For this report, a turbidite system is defined as a body of genetically related mass-flow and turbidity-current facies and facies associations that were deposited in virtual stratigraphic continuity (after Mutti and Normark, 1987, 1991). This generic definition includes a wide spectrum of depositional processes and environments that is especially useful when using different data sets of lower resolution.

In their seminal papers, Mutti and Normark (1987, 1991) recognized five elements common to both modern and ancient turbidite systems: major erosional features, channels, overbank deposits, depositional lobes, and channel-lobe transition deposits. Four of these elements (erosional features, lobes, channels, and overbank shales) are recognized in the Pliocene–Pleistocene succession of the study area. These four elements form the main facies mapped in the area, and are briefly reviewed here. The fifth element (channel-lobe transition deposits) is not resolvable on conventional multifold seismic data; this element is best identified in outcrops and in modern fans with side scan sonar, and possibly on horizon slices in three-dimensional (3-D) seismic data. Two additional elements recognized in this study are slides and unchannelized mud-rich upper slope facies. Table 2 summarizes the main characteristics of these elements.

### *Major Erosional Features*

Major erosional features include four types: (1) shelf-edge failures, (2) slope failures, (3) failures within the basinal turbidites, and (4) canyons (Mutti and Normark, 1991). The fill of these erosional features is highly variable. In this study, all four of these features are recognized. Slides mapped in this area are interpreted as both local slope and shelf-edge failures. These slides, however, were apparently not transported for long distances (a few kilometers). Large submarine canyons were mapped in the younger sequences (1.1–0.5 Ma) (Weimer and Dixon, 1994).

**Table 1. Seismic and Geologic Facies Codes and Abbreviations for the Study Area**

| | |
|---|---|
| **Seismic Facies** | |
| Internal Reflection Configuration | |
| Sp = subparallel | |
| P = parallel | |
| Dv = divergent | |
| Hu = hummocky | |
| Ch = chaotic | |
| Mo = mounded | |
| Sp-ch = subparallel-chaotic | |
| Hu (underlined) = hummocky over parallel | |
| Amplitude (A) modifiers | |
| l = low | |
| f = fair | |
| m = moderate | |
| h = high | |
| m/h = moderate/high | |
| Continuity (C) modifiers | |
| p = poor | |
| f = fair | |
| m = moderate | |
| g = good | |
| f/g = fair over good | |
| **Geologic Facies** | |
| ovbnk = overbank | |
| cls = channel systems; overbank strata with more channels | |
| bff = basin-floor fan | |
| sl = slide | |
| d ovbnk = slides deforming overbank | |
| cl, pel = channel systems and hemipelagic shales | |
| ovbnk (underlined), bff = overbank overlying basin-floor fan | |
| d = deformed (modifier) | |

### *Depositional Lobes (Basin-Floor Fans)*

Generally, depositional lobes are represented by roughly tabular, nonchannelized sand bodies that have individual thicknesses of generally 10–50 ft (3–15 m) (Mutti and Normark, 1991). Individual sandstone lobes in ancient systems commonly form vertical successions up to several hundred meters thick. The sand layers alternate with variable thicknesses of mudstone and thinner bedded and finer grained sandstones. The sandstone lobes

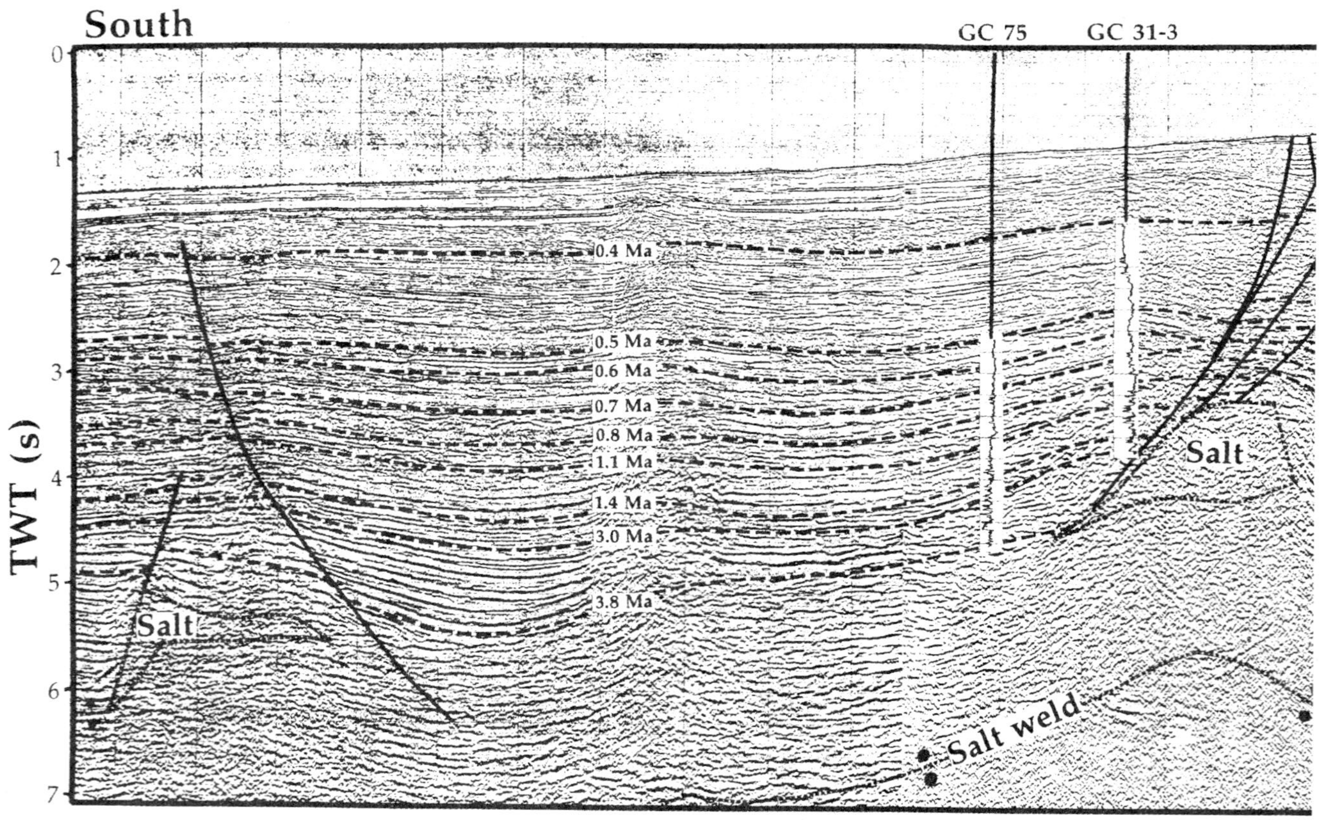


**Figure 4—Regional north-south seismic profile across the eastern part of the study area showing Pliocene–Pleistocene sequences (3.8–0.4 Ma). To the north, the section contains a deep, low relief, salt weld/base salt surface that dips gently to the south and extends across the length of the profile. Several salt rollers are above this surface and are associated with listric, normal faults. It is unclear whether the salt stock, in the middle of the profile, is pinched off at depth. A shallow salt weld/base appears to have flowed around the salt stock and contains several salt rollers and thicker salt bodies. The shallow salt weld/base salt surface also has very low relief. See Figure 3 for location of profile (profile I).**

can be basinwide features with abrupt onlap terminations against the basin margins, as well as smaller lobes that gradually thin and shale out into peripheral thinner bedded lobe-fringe deposits. These sandstone bodies tend to consist of the most complete Bouma sequences, forming "classical" turbidites.

In the study area, amalgamated sheetlike and sand-rich turbidites (depositional lobes) were deposited at the base of the early Pliocene–early Pleistocene sequences. On the seismic profiles, these deposits are represented by one or two parallel reflections having good lateral continuity and onlapping onto the underlying sequence boundary (Table 2). Well penetrations of basin-floor fans show 50–150-ft- (15–50-m-) thick sand bodies with blocky gamma-ray log patterns (Figure 5, EW 868, 13,850–13,950 ft); however, most of these features are not penetrated by wells, so the sand-rich lithologies are not always confirmed. Thus, the term "basin-floor fan" is used here as a seismic facies description. Where this facies is penetrated and shown on well logs, we use a more geometric definition of the sand body (see the following discussion). Most of the basin-floor fans that were deposited in the older sequences (early Pliocene) fill the entire minibasins, and the fans of the younger sequences (late Pliocene–early Pleistocene) filled only parts of minibasins.

### *Channels*

Channels are defined as relatively long-lived erosional features associated with sediment transport pathways (Mutti and Normark, 1991). As channels cease to transport sediments, they fill. Three main types of channel-fill deposits are commonly observed in ancient turbidite systems: (1) erosional channel-fill deposits, where basal coarse-grained sediments are overlain by fine-grained channel abandonment deposits, (2) depositional channel-fill deposits,

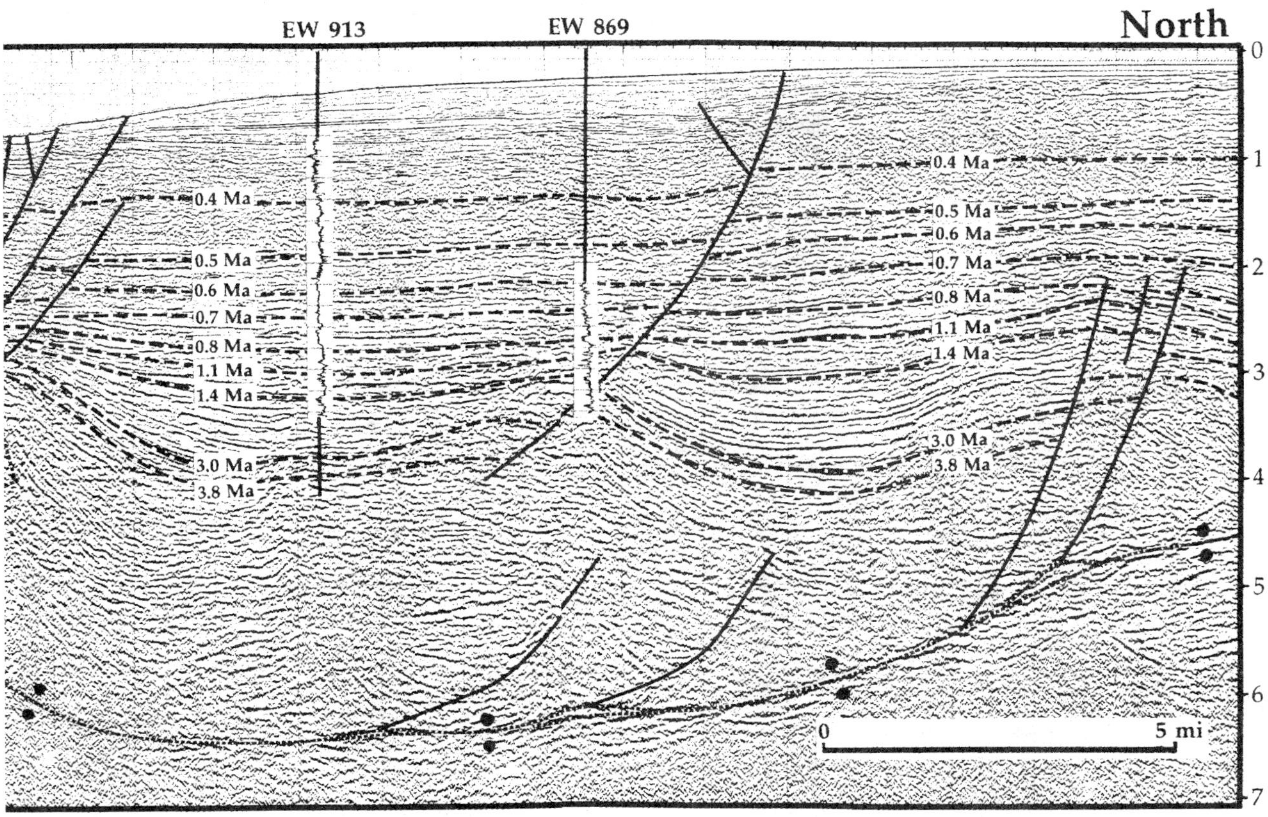


Figure 4—Continued.

which form similar sand layers to the depositional lobes, and (3) mixed channel-fill deposits, which are a transition between the two end members (Mutti and Normark, 1991).

In this study, mostly depositional channel fills were identified; erosional channel fills were only occasionally observed and can be confidently interpreted only from 3-D seismic data (Varnai, 1998, this issue). The well penetrations show that channel-fill sands either have blocky gamma-ray log shape or they form sand packages of individual thin sand layers with crescentic or bell (fining-upward) overall log shape. Because of the low resolution of the 2-D seismic data, individual channels are not mappable using the 2-D seismic data. Using 3-D seismic data in the southeastern part of the area, Varnai (1996, 1998, this issue) identified and mapped both erosional and depositional channel-fill deposits.

### *Overbank Deposits*

Overbank deposits are laterally extensive, fine-grained, thin-bedded turbidites adjacent to the channels in a turbidite system (Mutti and Normark, 1991). Overbank deposits are divided into two main groups: (1) channel-related overbank deposits (levees) and (2) overbank wedges. The first group includes a variety of fine-grained, thin-bedded sandstone layers along the edges of channels and within interchannel areas. Overbank wedges are mainly muddy turbidite sediments with basinwide extent. Channel-fill sands and associated thin-bedded sandstones of channel-related overbank deposition form a volumetrically minor component of overbank wedges.

"Overbank sediments" is used as a generic term in this study, defined as all fine-grained "extra-channel" sediments, including overbank wedges, channel-related overbank deposits (levees), and minor channels interbedded with hemipelagic shales. The well penetrations show primarily shale with varying amounts of thin interbedded sand layers, resulting in a "nervous" log pattern (Mitchum et al., 1993) (Figure 5, EW 868, 10,000–11,400 ft). These minor channels were identified only in wells and are too small to be resolved on the 2-D seismic data. In addition, hemipelagic shales, identified by local biostratigraphic abundance peaks, are present in all sequences; their relative volume is difficult to evaluate without cores. The amount of interbedded hemipelagic shale is interpreted to

**Table 2. Seismic Facies of Turbidite Elements***

| Turbidite Element | Reflection Configuration | Amplitude | Continuity | Lithology | Thickness | Areal Extent |
|---|---|---|---|---|---|---|
| Turbidite lobe or basin-floor fan | Parallel (mounded) | High | Good | 15–50-m-thick sand with blocky GR log pattern, occasionally forming up to 115-m-thick sand packages | 75–300 m | 20–210 $km^2$ |
| Channels | Subparallel | Moderate (high) | Fair (moderate) | 20–30-m-thick sand with blocky GR pattern, or 5–10-m-thick sands forming 100–150-m-thick packages of crescentic, fining-up GR log pattern | Several hundred meters for stacked systems | Not mappable using two-dimensional seismic data |
| Overbank deposits | Subparallel | Fair-moderate | Fair-moderate | Shale with interbedded 1–10-m-thick sands (nervous log) | Several hundred meters | Widespread, entire area |
| Hemipelagic | Parallel | High | Good | Shale, faunal abundance peaks | 15–30-m | Widespread, several minibasins |
| Slides** | Subparallel-hummocky to chaotic | Poor-moderate | Poor | Varying, depends on the original strata | Several hundred meters | Varying, limited around salt, and faults (30–100 $km^2$ to widespread 100–400 $km^2$ |

* Thickness and lateral extent were calculated from dimensions of corresponding seismic reflections.
**Slides interpreted around shallow salt and faults may be the result of poorer seismic imaging.

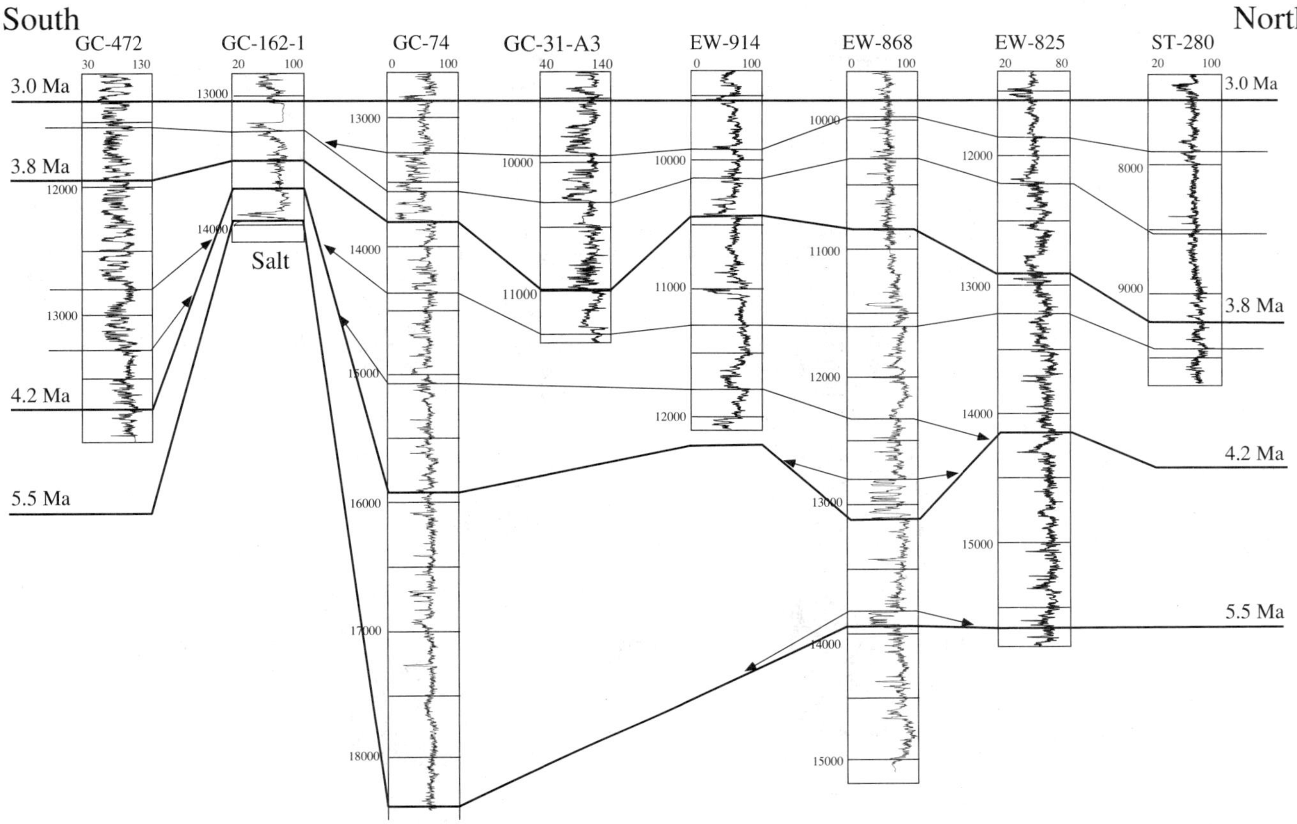


**Figure 5—Regional north-south gamma-ray wireline cross section across northeastern part of the area for sequences 5.5–3.0 Ma. See Figure 7 for location of cross section. Depths in feet.**

increase in thinner sequences, especially during the 3.0–1.4 Ma depositional interval.

## SEQUENCE STRATIGRAPHIC INTERPRETATION

To properly illustrate the key aspects of each depositional sequence, a series of maps and cross sections were constructed for making geologic interpretations. Isopach maps were constructed based on a velocity gradient ($Y = 25.1 + 2843x + 202x^2$) that was generated from ten wells across the area. Biofacies maps were constructed from the benthic foraminifera in the wells for an entire depositional sequence (e.g., 1.4–1.1 Ma). Because the thickness of these sequences can range anywhere from 50 to 3000 ft (15–915 m), these maps are considered to be sequence-averaged maps. Additional maps that reflect the paleoecology at specific instants of time are included in Villamil et al. (1998, this issue).

The creation of meaningful seismic facies maps is a challenge in this kind of active tectonic setting where there are abrupt lateral and vertical facies changes across the area within a single sequence. Several facies consist of two or three vertical subfacies. These complex facies are marked by underlined facies attributes in Table 1 and figures. Although these maps are more qualitative than seismic attribute maps generated from the workstation, the maps do reveal important regional patterns that are internally consistent throughout the data set.

Wireline-log cross sections illustrate the regional lithologic and biostratigraphic markers that have been tied to the seismic data. These show important stratigraphic relationships and regional variabilities, in both dip and strike orientations. Interpreted absolute age dates are shown and have been determined from extinction datums and seismic correlations in some places. For proprietary reasons, we cannot list the biostratigraphic extinctions or abundance or diversity curves, although we used these to interpret and generate the cross sections.

For each sequence, several representative seismic profiles with time-based logs are shown to illustrate key seismic facies and stratal relationships. From the integration of all of these data, geologic facies maps were generated. Finally, interpreted

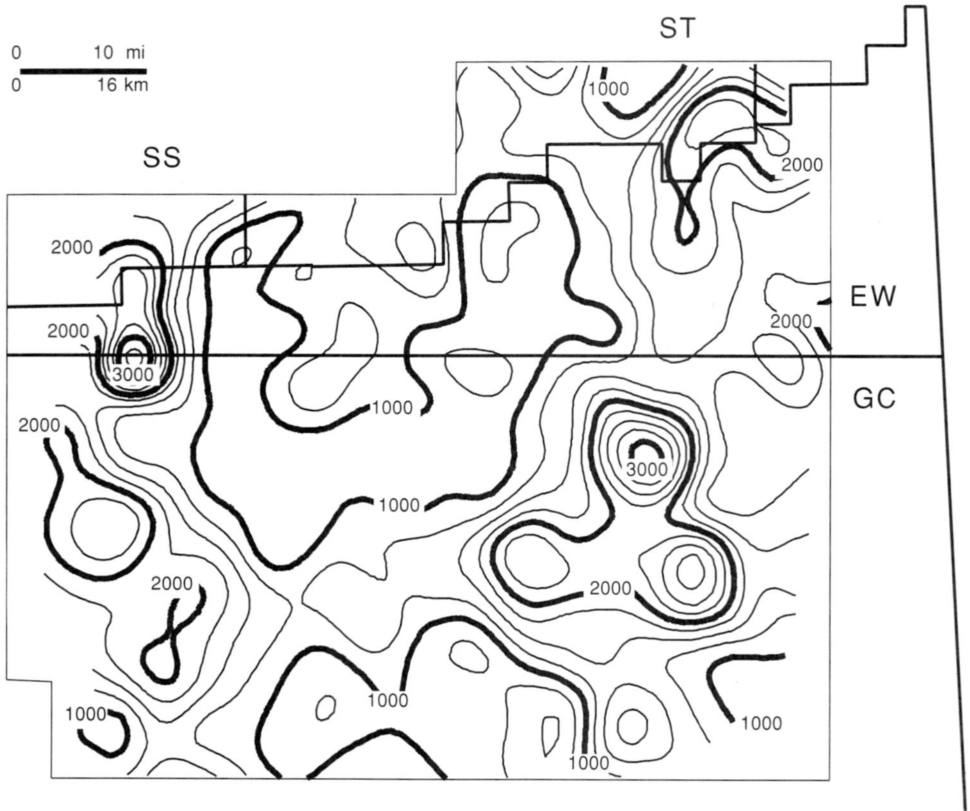


**Figure 6—Isopach map of the 3.8–3.0 Ma sequence. Contour interval is 250 ft.**

depocenter maps were created based on our best estimate of what the regional transport direction was for the major sediment pathways through the area for each sequence. These maps were generated from the regional interpreted geologic facies, specifically the distribution of basin-floor fan facies (i.e., amalgamated sheet and sheet sands) and the interpreted distribution of channel-levee systems in the basin.

The depositional environments of turbidite elements are interpreted from wireline log patterns (e.g., Mitchum et al., 1993; Chapin et al., 1994; Yeilding and Apps, 1994). Based on these wireline log classifications, three different kinds of sands are interpreted in this study: (1) sheet and amalgamated sheet sands (basin-floor fan) having blocky gamma-ray log patterns and good lateral continuity on the seismic profiles, (2) channel-fill sand bodies that are either thick sand layers with blocky gamma-ray log patterns (similar to the sheet sands), or multiple thinner layers forming sand packages with "crescentic" or increasing-upward overall gamma-ray log patterns, indicating fining or thinning upward (these sands have poor to fair reflection continuity on the seismic profiles), and (3) overbank sands that are thin layers with a "nervous" gamma-ray log pattern (Mitchum et al., 1993) (Figure 5, EW-868, 10,100–10,500 ft).

Because of space limitations, we summarize the general sequence stratigraphy of only two representative sequences, the 3.8–3.0 Ma and the 1.4–1.1 Ma. Details of specific areas are included in Acosta (1994), Martinez (1994), Navarro (1994), Budhijanto (1995), and Varnai (1996).

## 3.8–3.0 Ma Sequence (Lower–Upper Pliocene)

This sequence is best imaged in the eastern portion of the study area, where there is better definition of the older sequences and more well control in the area to tie to the ages of the sequences. To the west, these sequences are deeper and difficult to date because of a lack of well penetrations. These sequences are deeper in the section and have less resolution on the seismic profiles. Fields or discoveries that have reservoirs in this interval include the GC 254 (Allegheny) discovery, Lobster-EW 873, and Bison GC 166 (Figure 2).

Isopach maps indicate that the sequence ranges from 500 to more than 3250 ft (150–1000 m) in thickness (Figure 6). A benthic biofacies map indicates that the area extended from ecozones 5, 5.5, and 6 (lower bathyal) (see figure 13A, Villamil et al., 1998, this issue). In the north, a few areas were

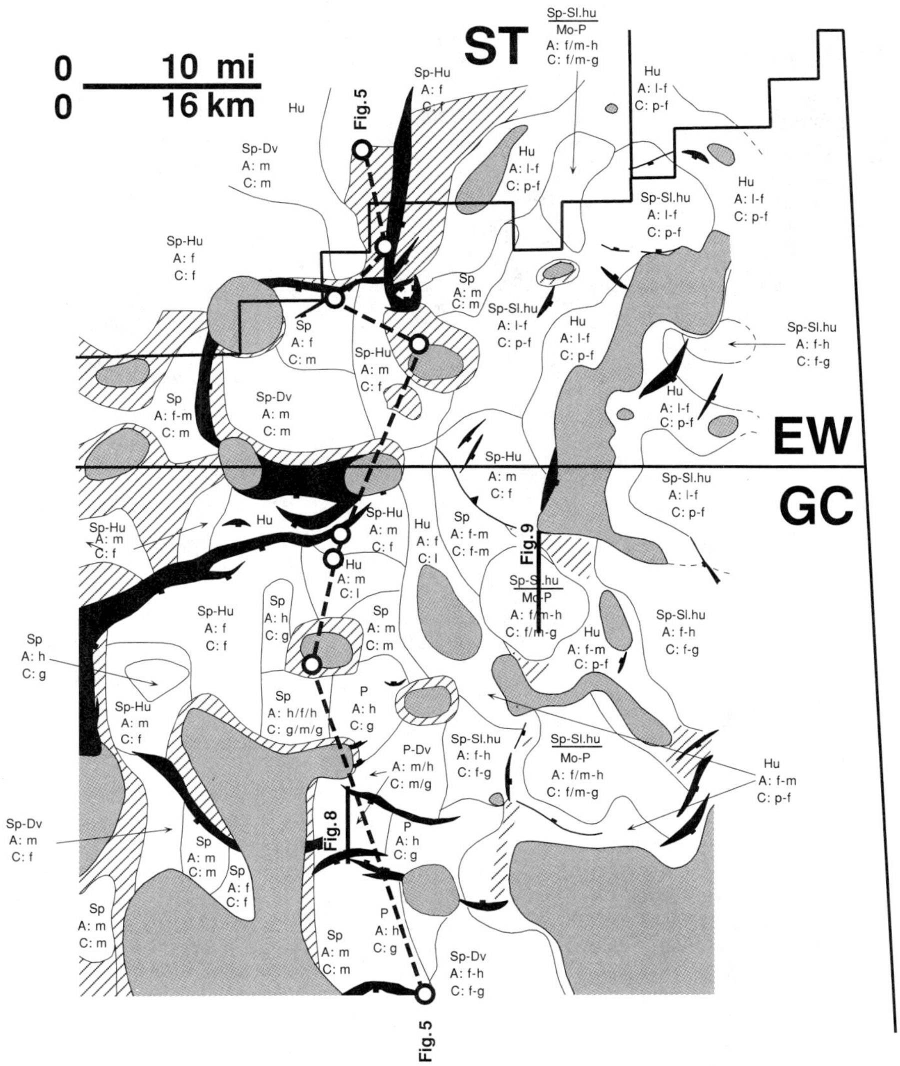


**Figure 7—Seismic facies map of the 3.8–3.0 Ma sequence. See Table 1 for seismic facies coding. Location of Figures 8 and 9 as shown.**

identified where local highs associated with salt could be interpreted from paleoecologic zones of lower value.

Seismic facies maps and profiles indicate dominantly subparallel to slightly divergent reflections in the area, with variable amplitude and continuity (Figures 4, 7–9). In the southern portion of the area, a distinct vertical change in the seismic facies is present in some of the minibasins (Figures 8, 9). At the base, laterally continuous high-amplitude reflections onlap against the flanks of the minibasin. Overlying the minibasin are more subparallel reflections having moderate continuity and amplitude. Overlying that are more laterally continuous subparallel reflections. These upward changes in seismic facies are interpreted to represent changes in the sedimentary processes in the depositional systems, from which we interpret an upward change in grain size.

Wireline log cross sections indicate that the southeastern portion of the study area is more sand rich, whereas the northern portion of the area is more shale dominated (Figure 5). Thick, well-developed sands are present in GC 74, GC 167, and GC 472 (Figures 5, 9). GC 167 has up to 50% sand in this sequence. Gamma-ray curves range from blocky to thinning or fining upward in the area. Wells with shale with some thin beds are EW 868 and ST 280 (Figure 5).

The seismic and lithofacies patterns in the southern portion of the area are interpreted to reflect an upward change in the environment of deposition of the turbidite system (Figure 10). A basin-floor fan (amalgamated sheet sand) is interpreted at the base, overlain by more shale-dominated channel-levee systems, and then upward to more hemipelagic sediments near the top with possibly interbedded, unchannelized mud-rich facies.

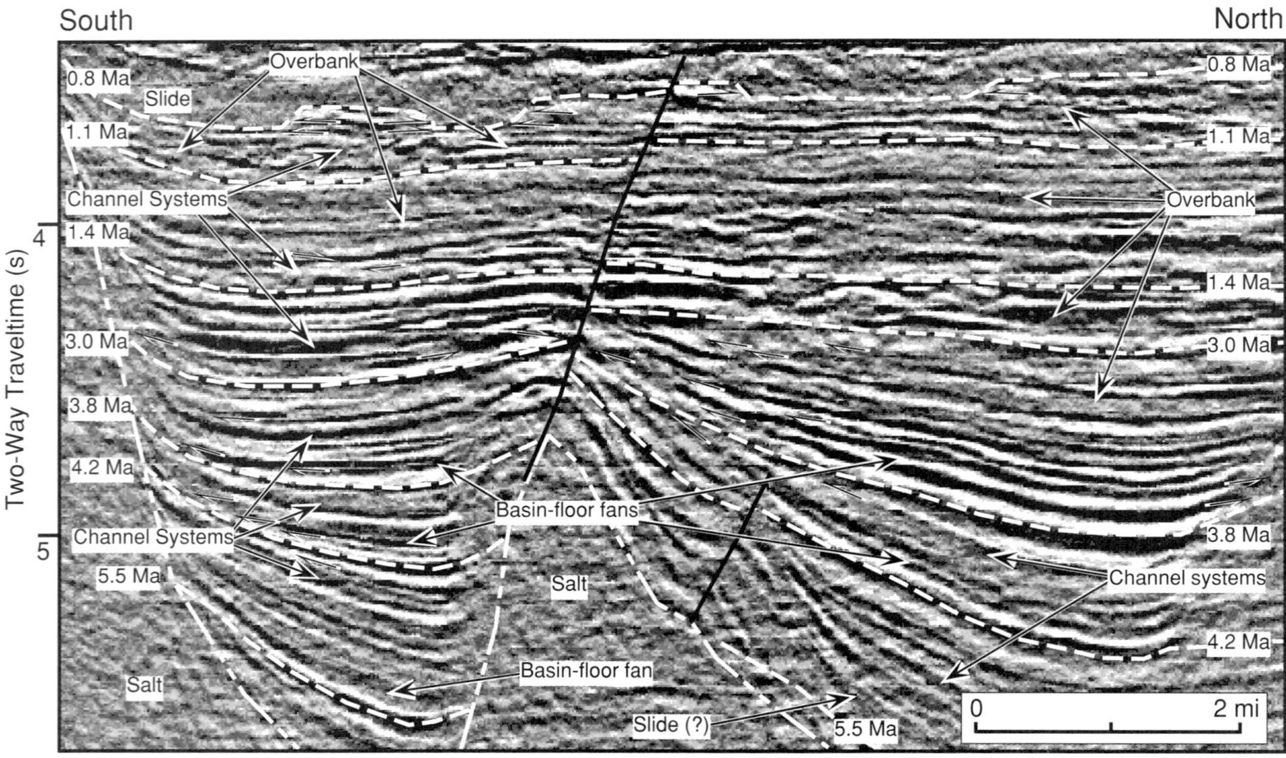


**Figure 8—Seismic profile showing the 5.5–0.8 Ma sequences. Syndepositional salt movement is reflected by divergent reflections and external wedge shape in the three oldest sequences, reflections onlapping onto the 4.2 and 3.8 Ma sequence boundaries, and erosion below the 3.8 Ma sequence boundary. Onlapping high-amplitude continuous reflections at the base of the four older sequences were interpreted as basin-floor fans (amalgamated sheet sands). Moderate-amplitude discontinuous reflections in the upper parts of all sequences are interpreted to be channel-levee systems; the individual higher amplitude reflections were interpreted as channel fill. Lower amplitude reflections with better continuity in the younger sequences (1.4–0.8 Ma) are interpreted as overbank strata. Slides overlie the 0.8 Ma sequence boundary. See Figure 7 for location of profile.**

The interpreted depocenter map in Figure 11 shows important regional depositional relationships. Four discrete basin-floor fans are stacked in the southern part of the area in a series of salt-withdrawal minibasins (see also McBride et al., 1998, this issue; Rowan and Weimer, 1998, this issue). To the north, only channel-levee systems are interpreted to be present (indicated by arrows in Figure 11). Based on the wireline log sections (Figure 5) and the regional distribution of the seismic facies (Figures 4, 7–9), a distinct pattern can be seen where the more sand-rich facies bypasses the upper slope and is being deposited in a series of salt-withdrawal minibasins. The channel-levee systems develop throughout the entire sequence. Thus, we interpret the basin-floor fans to be the first sediment delivered to this basin. After their deposition, channel-levee systems begin to form in response to an overall increase in the mud content in the turbidite flows.

## 1.4–1.1 Ma Sequence (Lower Pleistocene)

This sequence ranges in thickness from less than 750 ft (230 m) to more than 2500 ft (765 m) (Figure 12). Benthic biofacies indicate primarily lower bathyal water depths for the sequence (ecozones 5, 5.5), with some upper bathyal forms to the northwest (Figure 13). For the first time, ecozone 3 (outer neritic) is present in the northwestern portion of the area, reflecting the progradation of deltaic sediments into the area. Fields or discoveries that have reservoirs from this interval include the GC 184 (Jolliet) with several different intervals, EW 826, Popeye-GC 72/116, Brutus-GC 158, and GC 29 (Figure 2).

Extensive variability is found in the seismic and lithofacies in the area. Seismic facies consist primarily of subparallel to hummocky reflections with highly variable continuity and amplitude (Figures 4, 14–16). In fifteen discrete areas, facies consisting of high-amplitude parallel reflections

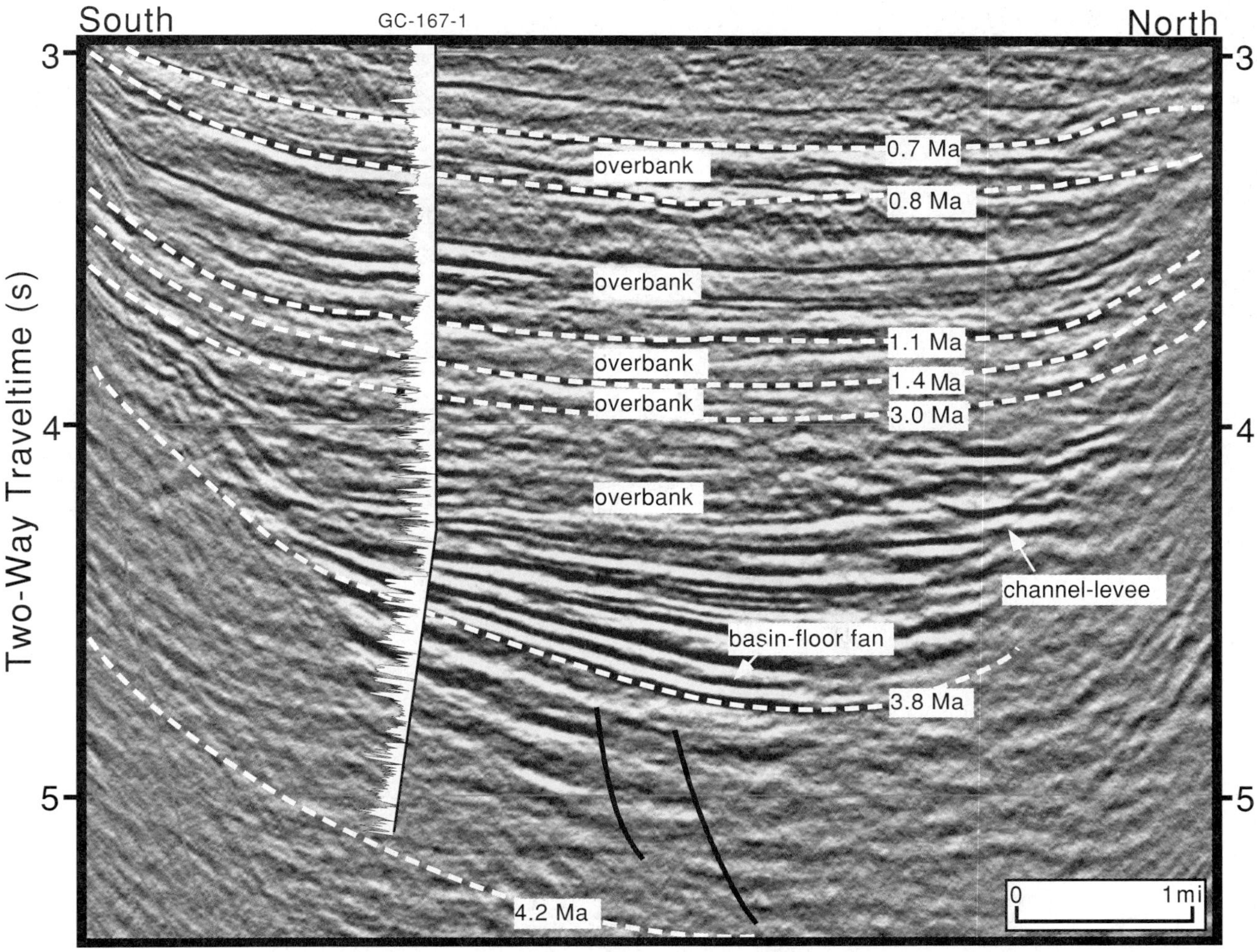


**Figure 9—Seismic profile showing the 4.2–0.7 Ma sequences. High-amplitude, mounded to parallel reflections at the base of the 4.2 and 3.8 Ma sequences are interpreted as basin-floor fans. Time-based gamma-ray log for GC 167 shows a blocky pattern, overlain by crescentic to spiky patterns (channel-levee systems). Sequences younger than 3.0 Ma are characterized by subparallel to slight hummocky reflections with poor to fair continuity; gamma-ray log indicates interbedded shale and sand with increasing shale upward. This interval is interpreted as dominantly overbank deposits. See Figure 7 for location of profile.**

with good continuity overlie the sequence boundary (Figures 14–16). These features have been penetrated in six different wells. For these sand bodies, gamma-ray curves exhibit a blocky shape averaging 100 ft (30 m) in thickness (SS355, Figure 15, GC-23, Figure 17). This seismic facies is interpreted as basin-floor fans (Figures 18, 19).

Overlying the basin-floor fans, and more laterally widespread in the area, are a series of subparallel to hummocky reflections with variable amplitude and continuity. This facies consists of sand bodies that are 10–50 ft (3–15 m) thick, which occur as upward thinning/fining patterns (Figure 17, GC 23, 9200–9600 ft; GC 199, 10,850–11,050 ft), interbedded with more shale-rich facies. Based on the seismic facies and lithofacies, most of the area appears to consist of a series of channel-levee systems flowing though the minibasins (Figures 18, 19).

Another significant facies consists of subparallel reflections with low amplitude and moderate to good continuity (Figures 4, 8, 9, 14–16). Wireline logs indicate shale for this facies (e.g., Figure 17, GC 199, 10,100–10,800 ft; GC 67, 7550–8000 ft). The facies generally can be traced laterally to a channel-levee facies, and therefore is interpreted to be overbank facies. This is also the oldest sequence where extensive slides are interpreted based on the hummocky, mounded, or chaotic reflections with poor continuity.

**Figure 10—Geologic facies map of the 3.8 to 3.0 Ma sequence. See Table 1 for geologic facies coding.**

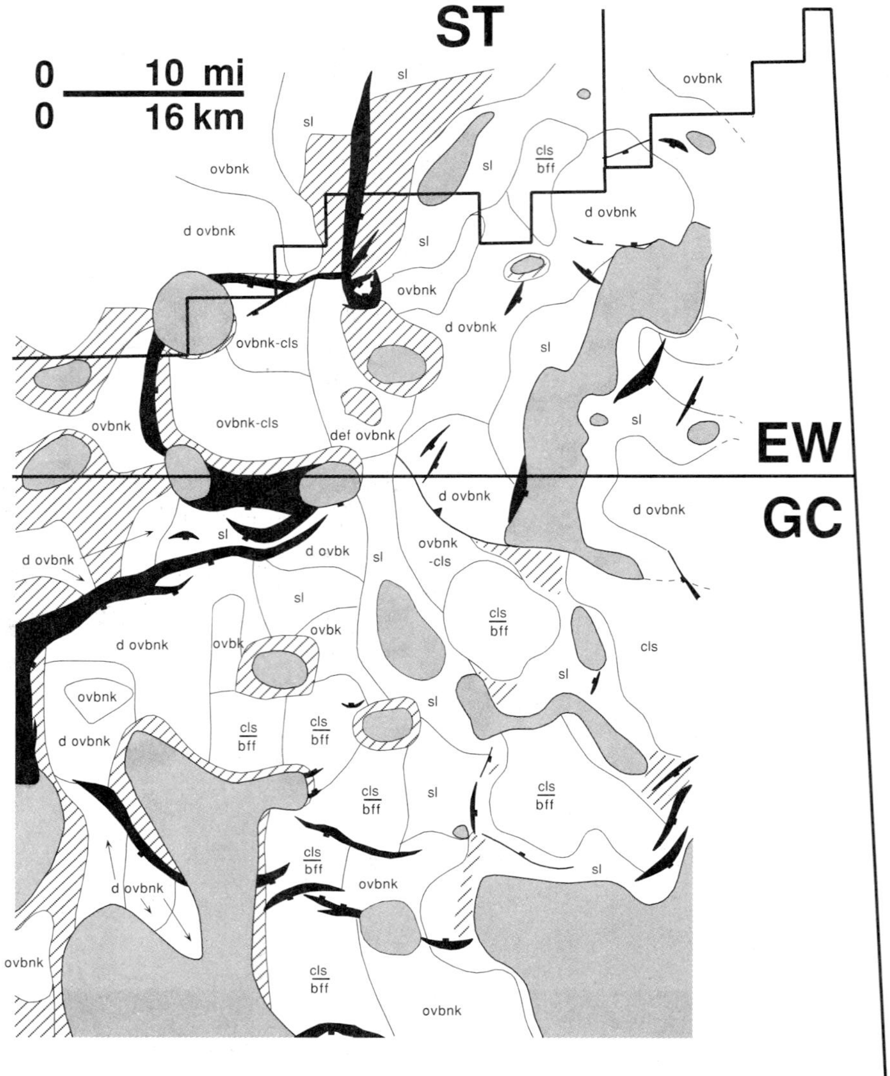


This sequence is unique because it has the most basin-floor fans interpreted in any Pliocene–Pleistocene sequence (Figures 18–20; Table 3). This may be an artifact of the data, because the distribution of the lower Pliocene sequences to the west are not as well constrained as this lower Pleistocene sequence. Where the basin-floor fans have been identified in this sequence, they tend to line up in a discrete north-south trend, indicating probable sand transport axes into the basin. In some places the features lap out against salt, and in other cases the features are juxtaposed next to faults. Similar to the Pliocene sequences, most sands occur at the base of the sequence; the sequence becomes more mud dominated upward. The sediment transport axes also reflect the location of the main channel-levee systems that can be mapped (Figure 19).

## DISCUSSION

The regional depocenter maps for the remaining ten sequences are shown in Figure 20. Each map shows the interpreted axes of sediment transport interpreted from the location of basin-floor fans and the overlying channel-levee systems. Because the seismic facies are complex, different workers might draw the location of the transport axes slightly differently. The five main depositional facies are discussed here, with emphasis on temporal and spatial variations and possible regional controls on their deposition.

### Basin-Floor Fans

Sheet sands (depositional lobes or basin-floor fans) represent only a minor portion of the

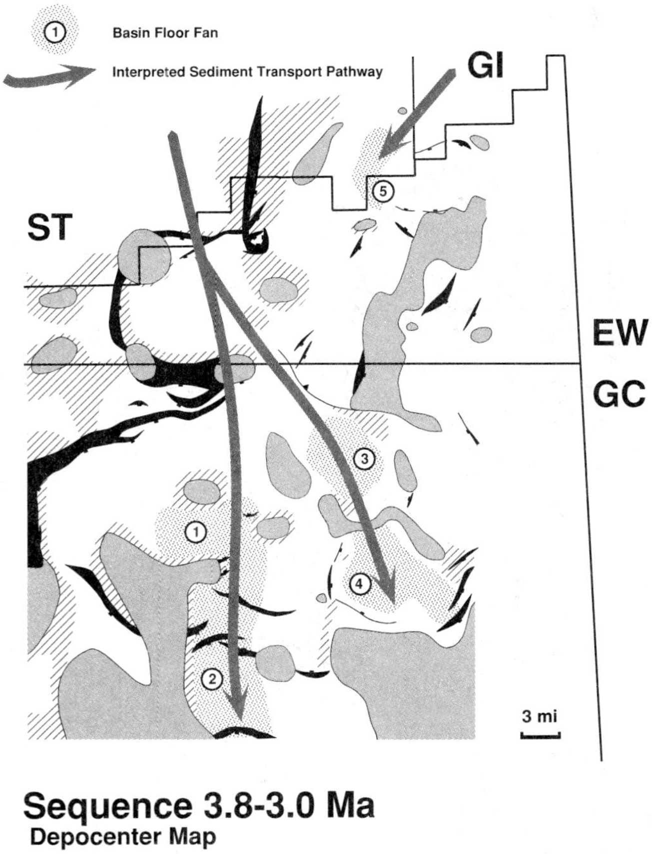


**Figure 11—Regional depocenter map of the 3.8–3.0-Ma sequence showing interpreted sediment pathways. Basin-floor fans (amalgamated sheet and sheet sands) are numbered and summarized in Table 3.**

sequences in terms of thickness and areal distribution; they consist of relatively thin [300–500 ft (100–150 m)] units at the base of sequences and are concentrated in only a few minibasins. The seismic stratigraphic expression of the basin-floor fans illustrated here is fairly representative of all of the basin-floor fans in the area. The key features of each basin-floor fan are summarized in Table 3. Of the 52 basin-floor fans in the area, 19 have been penetrated by wells. The remaining 33 features were not penetrated by wells because they are not in a petroleum trap. The producing fields in the study area all have amplitude anomalies associated with the reservoirs (Weimer et al., 1998a, this issue). More drilling of these features is expected during the next few years.

Most of the basin-floor fans occur in paleoecology zones 5 and 5.5 (lower bathyal) (Table 3; see also figure 5 in Villamil et al., 1988, this issue). Only one basin-floor fan (sequence 1.4–1.1 Ma) was deposited in ecozone 4 (upper bathyal). Nine basin-floor fans deposited in the 5.5–3.0 Ma sequences occurred in paleoecology zone 6 (lowermost bathyal) (Table 3). The sediments that correspond to the seismic amplitude associated with individual basin-floor fans range in maximum thickness from 100 to 600 ft (30–180 m), depending upon setting and penetration. Many of the thicker sands associated with any of the amplitudes are thinner than the actual amplitude, suggesting that the amplitude is made up of a complex layering of sands with different geometries. The areal extent of the basin-floor fans ranges from 4 to 68 mi$^2$ (10–176 km$^2$), based on seismic amplitude.

Two different types of basin-floor fans were recognized in the area. In the older sequences (5.5–3.0 Ma), most basin-floor fans are concentrated in the southeastern part of the area (Figures 11, 20A, B). The area, thickness, and sand content of these basin-floor fans gradually increase upward. The 3.8 Ma sequence has the most basin-floor fans, and they are interpreted to be the sandiest (Table 3; Figure 11). These basin-floor fans were deposited in lower bathyal water depths (ecozone 6) and fill entire minibasins or the entire downdip side of a minibasin, onlapping in the updip direction (e.g., fan 2 in the 3.8 Ma sequence, Figure 11). These basin-floor fans are interpreted to be the result of sediment ponding by local bathymetric lows. Syndepositional salt withdrawal resulted in bathymetric lows across entire minibasins, with highs on the downdip flanks. The highs acted as barriers for the turbidity currents, trapping the sandiest materials in the minibasins updip from them.

The second type of basin-floor fan was deposited in the 3.0–0.7 Ma sequences. These fans do not fill entire minibasins, unlike the older ones, so their areal extent is less (Table 3). These kinds of fans were deposited in the downthrown side of growth faults and adjacent to shallow salt bodies in lower bathyal water depths (ecozones 5.5, 5). The tectonic elements helped to create localized bathymetric lows on the sea floor, and the lobes were deposited in these restricted areas. Kneller and McCaffrey (1995) pointed out that highs on the sea floor (e.g., salt highs) may focus the turbidity currents into narrow pathways (constrained flow). As the turbidity currents enter a minibasin downdip of the salt bodies, the flow becomes divergent and the velocity decreases (unconstrained flow). The decrease in current velocity also may have contributed to the deposition of the sand-rich fans. These fans are interpreted to be sand-rich parts of the sequences, but sand content and thickness are generally less than the salt withdrawal controlled fans of the older sequences.

The differences in basin-floor fan types suggest that the active large salt withdrawal basins were more effective in trapping coarse-grained sediments than the fault-related basins. Possibly, the

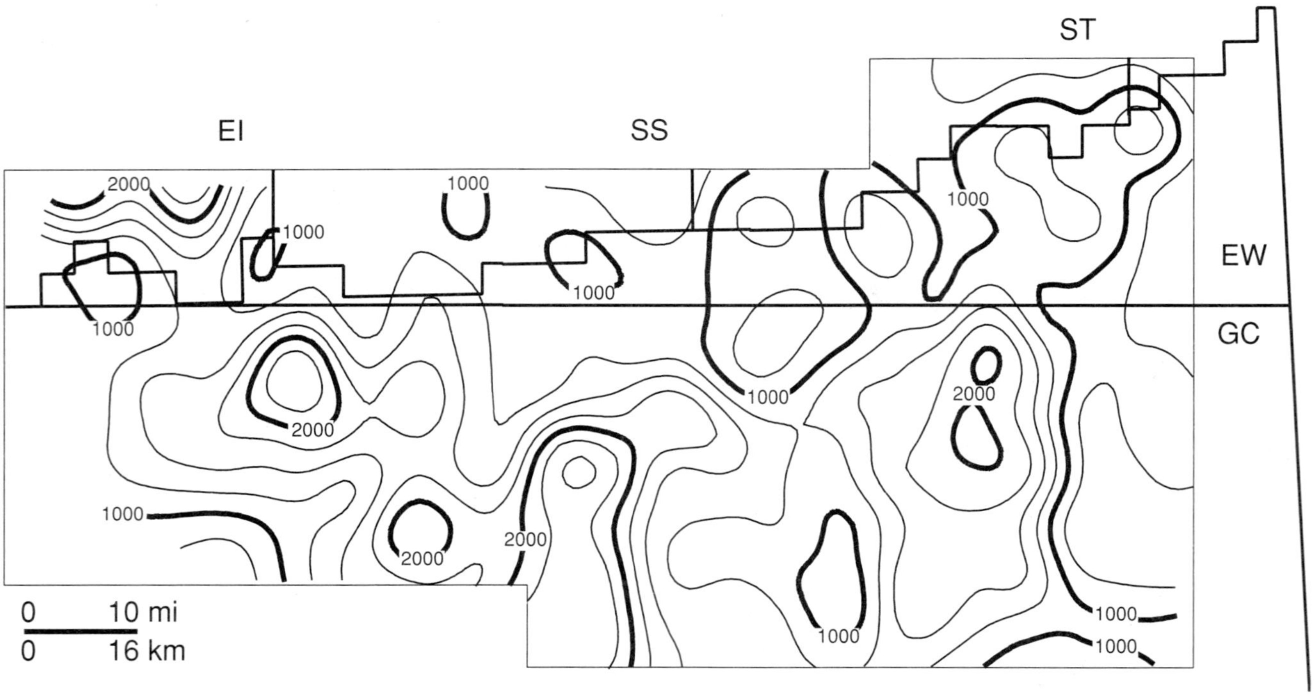


**Figure 12—Isopach map for the 1.4–1.1 Ma sequence. Contour interval is 250 ft.**

abrupt changes in bathymetric relief that caused sand deposition from turbidity currents were greater in these salt-withdrawal basins.

No basin-floor fans were interpreted in the 0.7 Ma and younger sequences. The lack of fans might be caused by increases in clay in turbidity currents delivered to the basin, decreased tectonic activity, high depositional rates, increase in regional slope gradient (upper slope), and possibly by fewer bathymetric lows on the slope. These lows were quickly filled by shaly overbank sediments, and the sandier parts of turbidity currents bypassed to the south into deep water basins.

## Channel-Levee Systems

Channel-levee systems were mapped in each sequence. Channel-levee systems have been characterized in the geologic literature as having characteristic "gull-wing patterns" on seismic reflection data. Although characteristic of large modern submarine fans, few of the gull-wing patterns were observed in the study area. Instead, the typical seismic patterns consisted of subparallel-hummocky reflections with moderate amplitude and fair continuity. The well penetrations show either sand bodies [50–100 ft (15–30 m)] with blocky gamma-ray log patterns, or thinner sands with an overall crescentic, fining-up pattern; both patterns are interpreted as channel-fill sands. These sands may coalesce into 200–300-ft-(60–90-m-) thick intervals interpreted as stacked channel systems. Wireline logs also indicate that these channels are interbedded with varying amounts of overbank shales.

The seismic facies shows that the high-amplitude reflections that correspond to the channels usually are interbedded in the lower amplitude, more laterally continuous reflections of the overbank facies. Due to the wide spacing of the 2-D seismic data, individual channel systems are not mappable. Varnai (1996, 1998, this issue) demonstrated that the greater data density of the 3-D seismic data, however, allows for the mapping of the individual channels in the 3.8–1.4 Ma sequences in the southeastern part of the area.

The seismic facies and well penetrations indicate that the number and sand content of the channel systems gradually decreased through time throughout the area. More channel systems with thicker channel-fill sands were deposited in the Pliocene sequences than in the younger sequences. The exception is the 0.5 Ma sequence, in which several intervals were built dominantly of channel systems. Excluding the 0.5 Ma sequence, the decrease in channel systems, similar to the

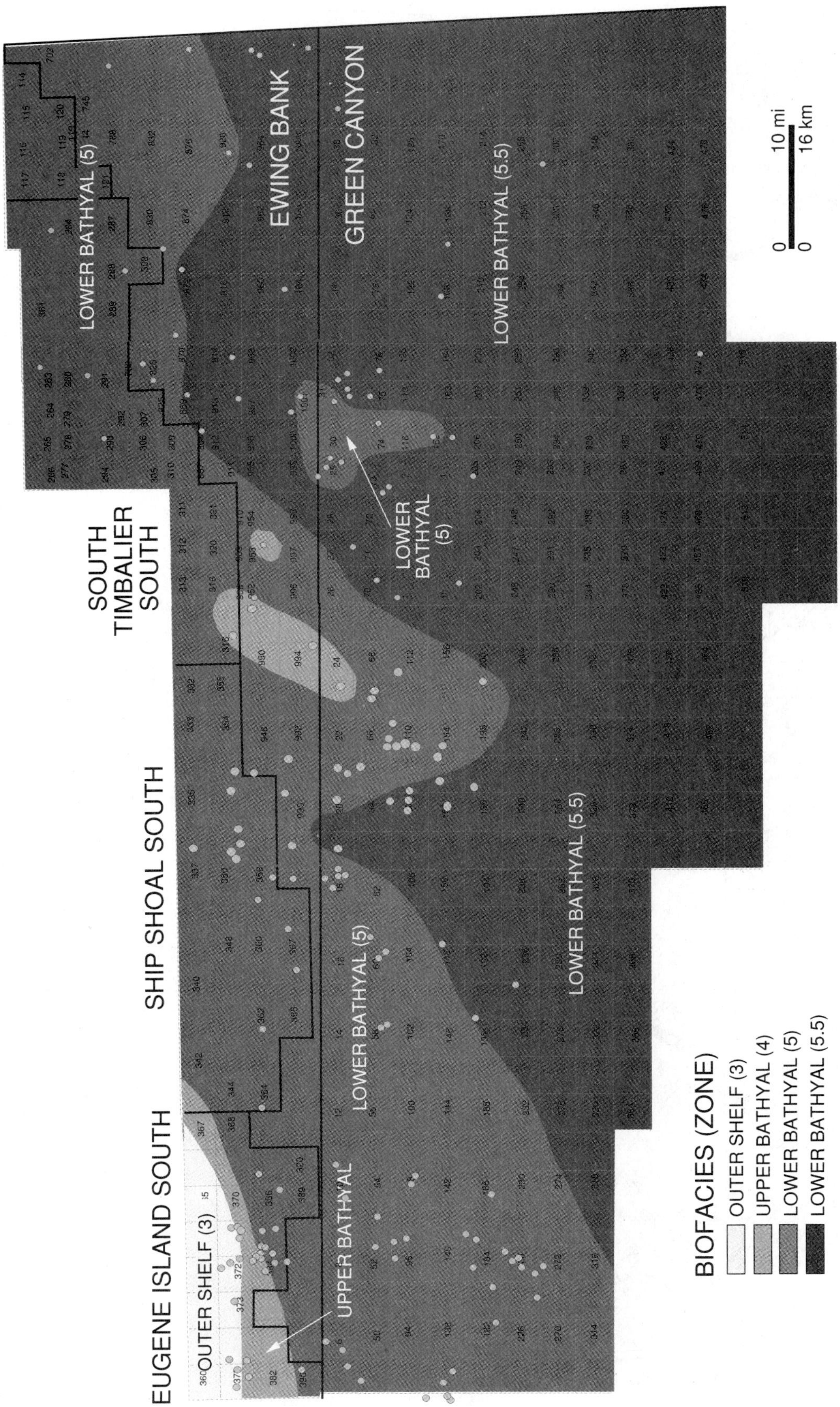


**Figure 13—Benthic biofacies map for the 1.4 1.1-Ma sequence. Biofacies zones (3–5.5) classification is shown in figure 5 of Villamil et al. (1998, this issue).**

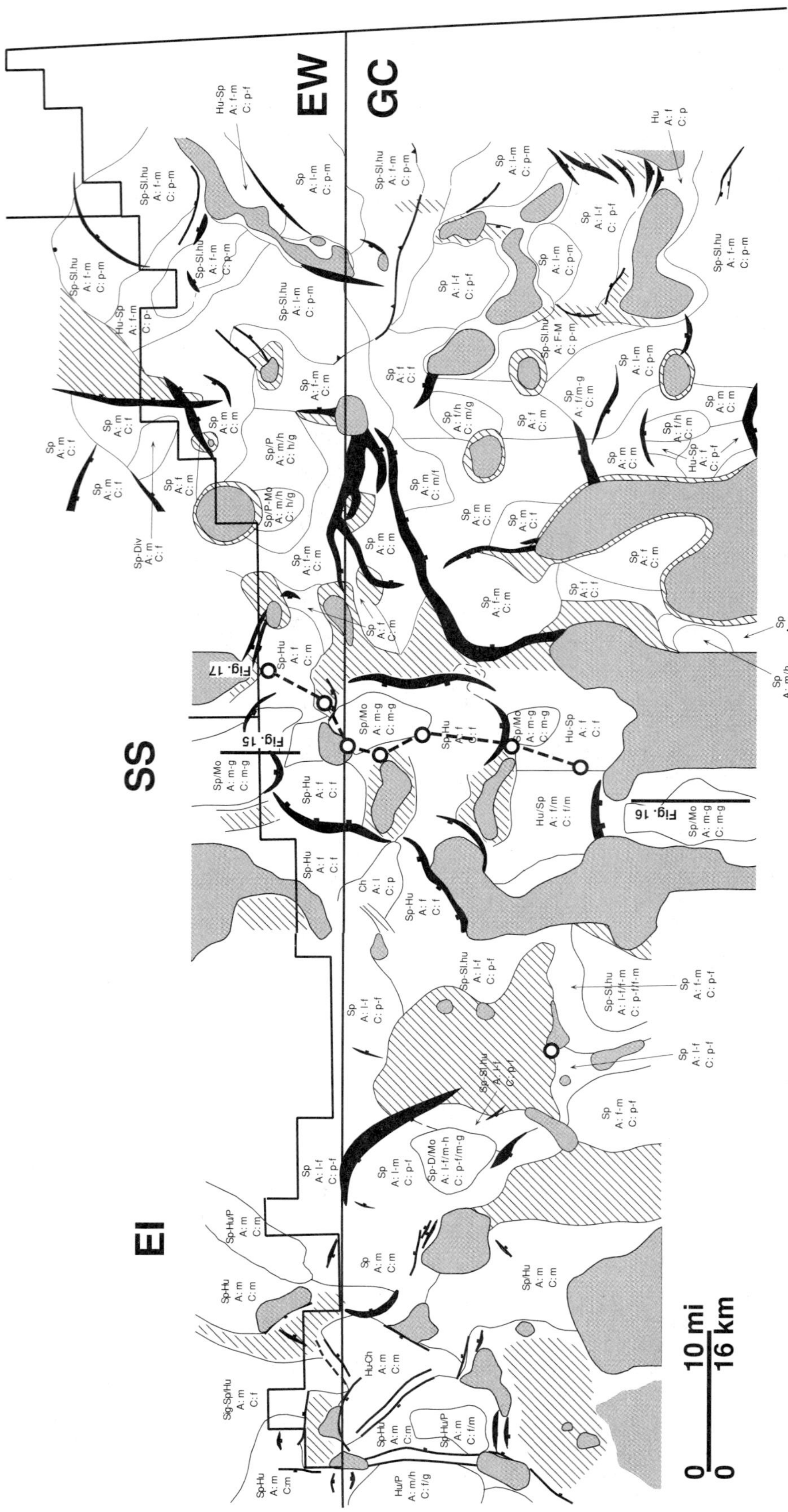


**Figure 14—Seismic facies map for the 1.4–1.1 Ma sequence. See Table 1 for coding. Location of Figures 15–17 as shown.**

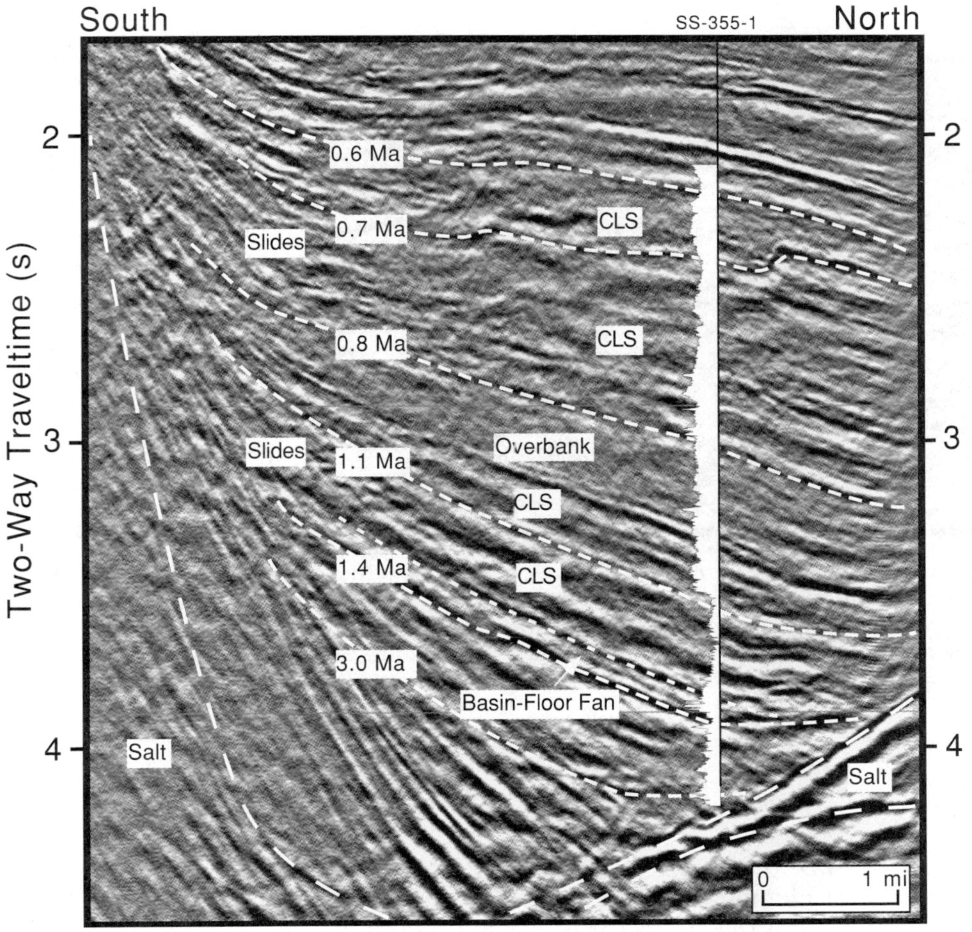


**Figure 15—Dip-oriented seismic profile showing the 1.4–0.6 Ma sequences and time-based gamma-ray log (Ship Shoal 355). High-amplitude, parallel reflection overlies the 1.4 Ma sequence boundary and pinches out to the south. Gamma-ray log indicates good sand development; the feature is interpreted as a basin-floor fan. Discrete slides are common in all sequences and are characterized by hummocky, mounded and chaotic seismic reflections. Channel-levee systems are prominent and defined by subparallel to hummocky reflections with poor to fair continuity and fair amplitude. See Figure 14 for location profile.**

changes in the basin-floor fans, probably is related to the generally shalier sediments of the younger sequences because most of the sands bypassed to the area farther downslope. Additionally, the large unchannelized flows of mud-rich sediments (discussed in a following section) may have become more dominant in the shalier Pleistocene sequences.

## Overbank

Two different facies, overbank sediments and nonchannelized mud-rich sediments, are grouped under the "overbank deposits" term. Both facies are characterized by subparallel reflections with fair to moderate amplitude and moderate continuity. These facies were penetrated by numerous wells and are dominantly shales with thin interbedded sands (spiky gamma-ray log pattern). A few channel systems, however, also were deposited in these facies. The channels are represented by thicker [up to 50 ft (15 m)] sands with a blocky gamma-ray log pattern. These channels are too small to be identified using the 2-D seismic data, and thus are not mapped separately from the overbank deposits. The overbank facies is interpreted as mostly shales interbedded with thin silty sands.

The two overbank facies are differentiated based on their stratigraphic setting and interpreted transport mechanisms. The overbank sediments associated with channels are inferred to be the fine-grained component of turbidity currents, which were transported through channels, overflowed the levees, and then spread out across the area. These overbank sediments are dominant in the older sequences.

In the 0.8 Ma and younger sequences, increasing amounts of nonchannelized, mud-rich sediments are present. As discussed, however, the number of channel systems apparently decreased in these young sequences. Due to this decrease in channel systems, these overbank deposits are interpreted as large unchannelized flows of mud-rich sediments transported into the area. These flows originated in the outer shelf and upper slope south of the Mississippi Delta and probably are related to large upper slope failures and slides.

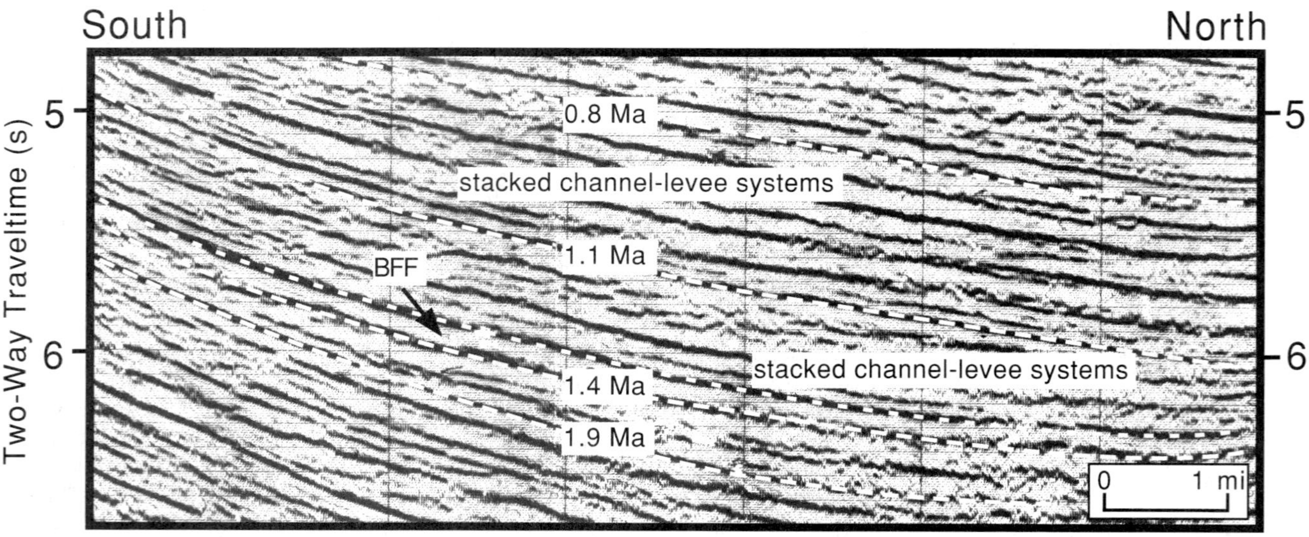


**Figure 16—Dip-oriented seismic profile showing the 1.9–0.8 Ma sequences. Most of the sequences consist of subparallel to slightly hummocky reflections with moderate amplitude and continuity. These are interpreted as a series of stacked channel-levee systems. A laterally continuous, high amplitude reflection overlying the 1.4 Ma sequence boundary is interpreted as a basin-floor fan (BFF). The GC 506 well, 2 miles south of the study area, corroborates the interpretation. See Figure 14 for location of profile.**

### Hemipelagic Shale

The timing of deposition of the turbidite systems was determined by the relative changes of sea level and the location of the Mississippi Delta during the Pliocene–Pleistocene. Turbidite systems are interpreted to be deposited only during major relative lowstands in sea level in the lowstand systems tracts (basin-floor fans and slope fans). Prograding complexes are generally thin in the midslope region. In the transgressive and highstand systems tracts, most sediments were trapped in shelf areas; thus hemipelagic shales were deposited in the study area. These shales are high-amplitude, laterally continuous reflections underlying the sequence boundaries, and they drape large areas. Well penetrations show that shales with increased faunal abundance values (condensed sections) correspond to these reflections (Villamil et al., 1998, this issue); therefore, these reflections are not interpreted as basin-floor fans.

Hemipelagic shales also were interbedded with the overbank shales, especially in the thinner sequences that have low depositional rates. The hemipelagic shales were most dominant during the 3.0–1.4 Ma interval, where they are interbedded with channels and overbanks. Increased faunal abundance values indicate that these shales were also deposited during the 5.5, 4.2, and 3.8 Ma sequences, when depositional rates were lower.

### Slides

Slides are characterized by hummocky to chaotic reflections with poor to fair reflection continuity and generally lower amplitudes. Well penetrations of the slides show dominantly shales with few thin [3–15 ft (1–5 m)] interbedded sands; most of the slides are interpreted to have been deformed overbank sediments. The decrease in reflection continuity probably indicates increased deformation, ranging from slight deformation (deformed overbank) to highly reworked sediments.

Three major types of slides have been identified. First, in the older sequences (5.5–1.1 Ma), slides are associated with shallow salt bodies and growth faults. The slides range in size between 30 and 100 $km^2$ and may have deformed the entire sequence in some minibasins. These slides are interpreted as the result of syndepositional tectonic movements and are less common in the younger sequences as a result of reduced tectonic activity. However, the lower amplitude and continuity might also be the result of poorer seismic imaging.

Second, in the younger sequences (0.8 Ma to present), slides affected larger areas (100–400 $km^2$) and were not restricted to the vicinity of salt or faults. These slides usually deform only parts of the sequence (most commonly the basal parts), but in the northeastern area, the 0.5 Ma sequence

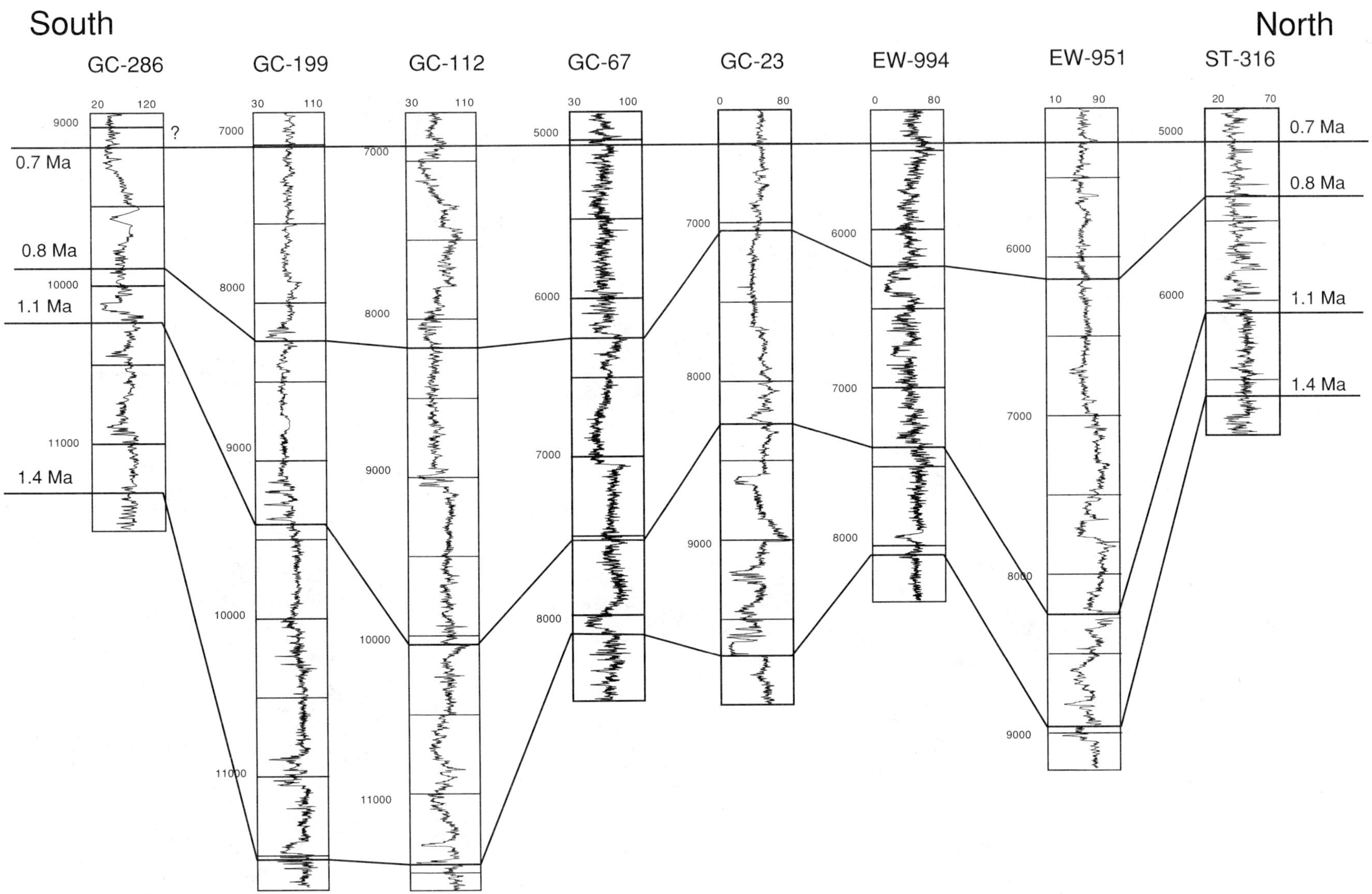


**Figure 17—Regional north-south gamma-ray wireline-log cross section across the central part of the area for sequences 1.4 to 0.7 Ma. See Figure 14 for location of cross section. Depths in feet.**

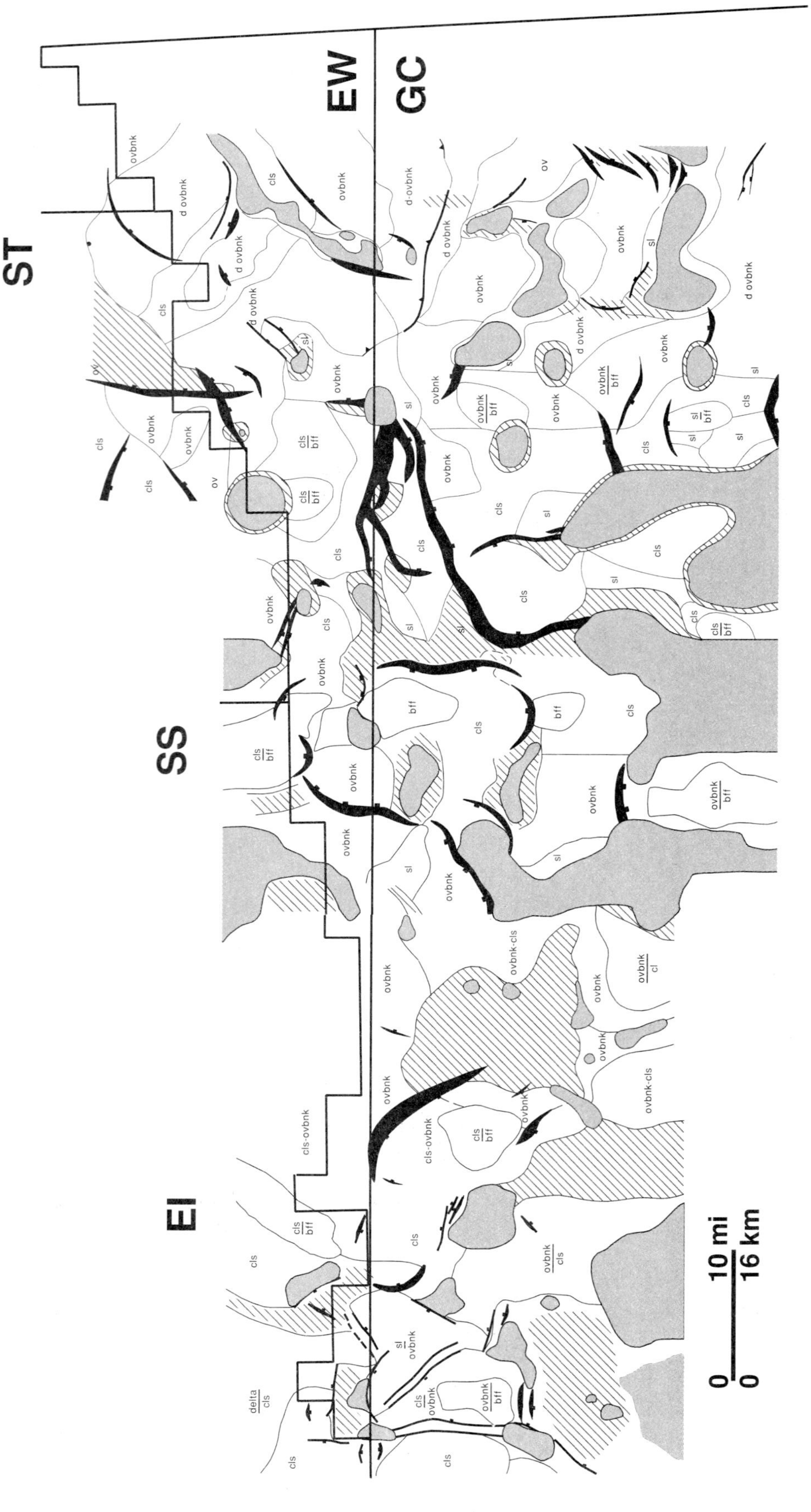


**Figure 18. Geologic facies maps for the 1.4–1.1 Ma sequence. See Table 1 for geologic facies coding.**

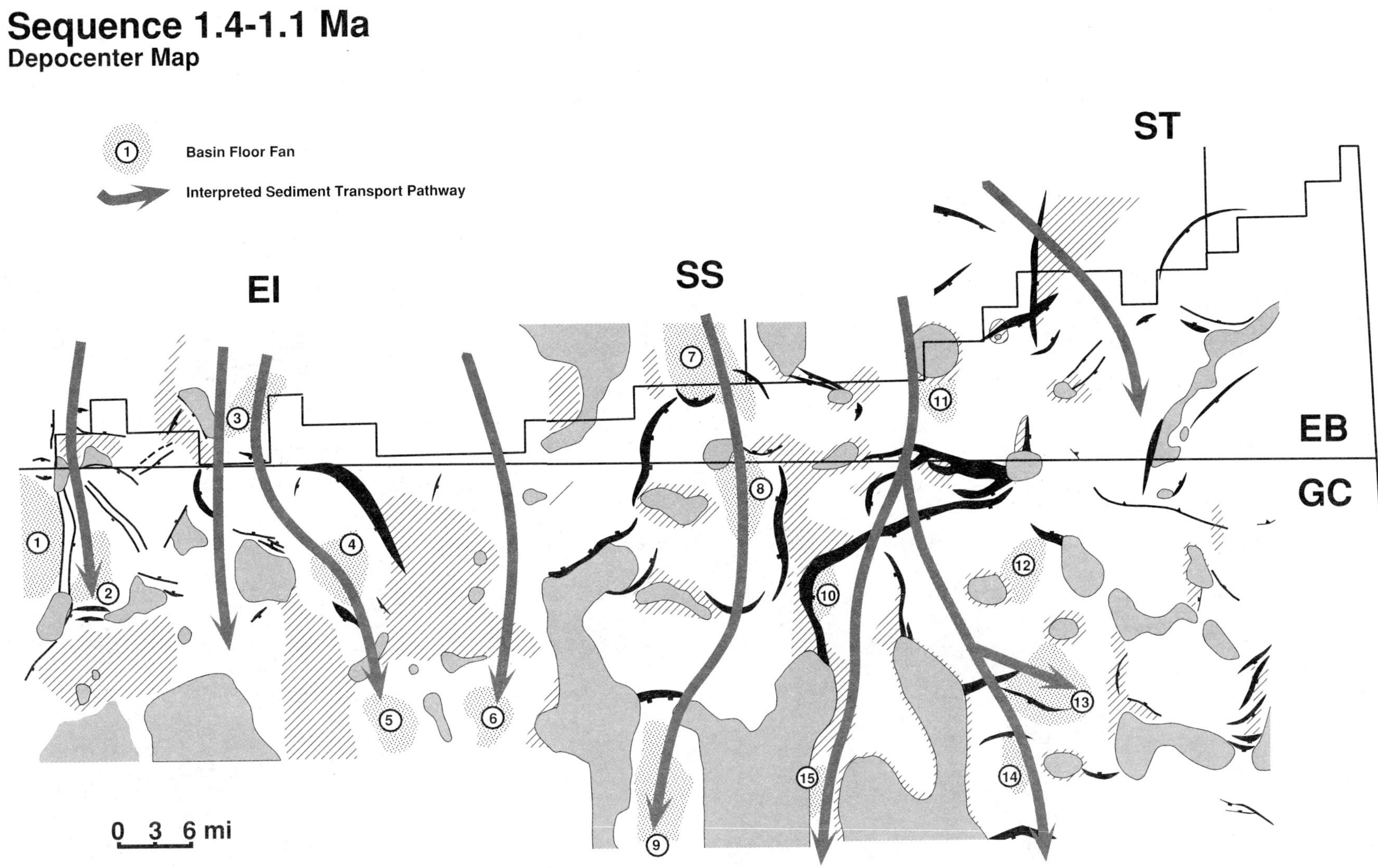


Figure 19—Regional depocenter map of the 1.4–1.1 Ma sequence showing interpreted sediment pathways. Basin-floor fans (amalgamated sheet and sheet sands) are numbered and summarized in Table 3.

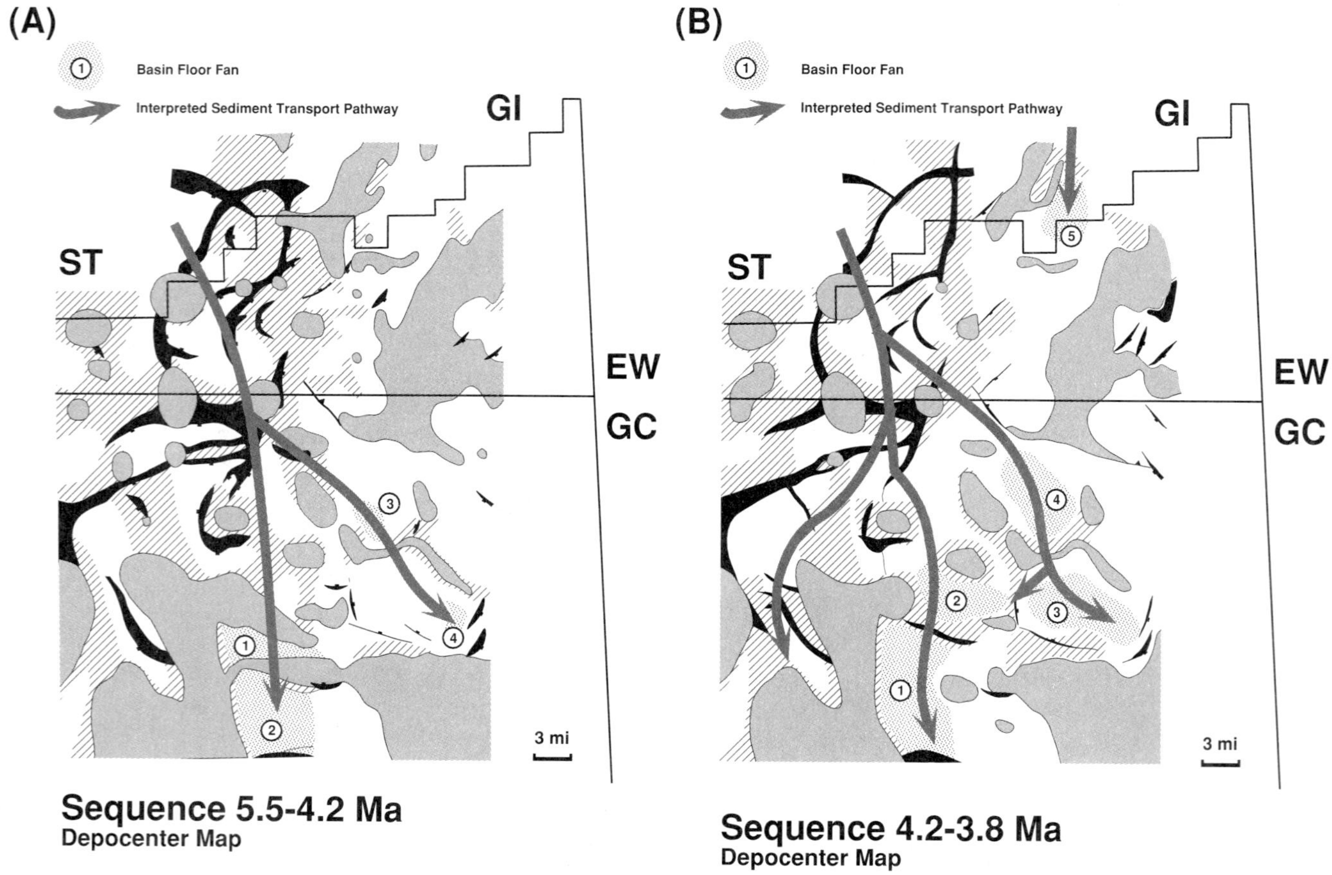


**Figure 20—Regional depocenter maps for the 5.5–0.5 Ma sequences. Distribution of basin-floor fan features for each sequence is shown in dotted pattern; axes of interpreted sediment transport direction are shown by arrows. Present distributions of salt and faults are shown; qualitative restorations of the sea-floor topography are shown in Rowan and Weimer (1998, this issue). (A) 5.5–4.2 Ma, (B) 4.2–3.8 Ma, (C) 3.0–2.6 Ma, (D) 2.6–2.4 Ma, (E) 2.4–1.9 Ma, (F) 1.9–1.4 Ma, (G) 1.1–0.8 Ma, (H) 0.8–0.7 Ma, (I) 0.7–0.5 Ma sequences (merged), (J) 0.5 to 0.4 Ma. Note that the 3.8–3.0 Ma sequence is shown in Figure 11, and the 1.4–1.1 Ma sequence is shown in Figure 19.**

is completely deformed by slides. These large slides may have deformed the older sequences too, but if so, they are not mappable on the seismic data, possibly due to poorer seismic imaging. This second type of slide is possibly the result of localized overpressuring of shales related to high rates of sedimentation. Other possible factors include the increasing shale content of sediments, changes in slope gradient, changes in clay mineralogy, and changes in rates of deposition. The amount of erosion at their bases increases upward from the 0.8-Ma sequence [320 ft (100 m)] to the 0.4 Ma sequence boundary [450–500 ft (140–150 m)].

Third, in the younger sequences the slides are thicker than in the older sequences, and caused more erosion at the sequence base. The third kind of slide forms in the upper slope/outer shelf and is associated with overpressuring in the prodelta slope.

## Erosional Features

Erosional features associated with these turbidite systems also show great variability. Most of the turbidite systems in the Pliocene sequences appear to have been deposited conformably over sequence boundaries, with only few erosional features present within the sequences. Erosion occurred primarily at the top of uplifted areas associated with syndepositional salt tectonics. Seismic resolution is lower in the Pliocene sequences than in the younger ones, hence the amount of erosion may be greater in the older sequences than what can be recognized.

In the Pleistocene sequences, erosion became more common. The first significant erosion developed at the 1.1 Ma sequence boundary, indicating a possible submarine canyon. In the overlying sequences, areally widespread slides became more common, with erosion at their bases. Submarine

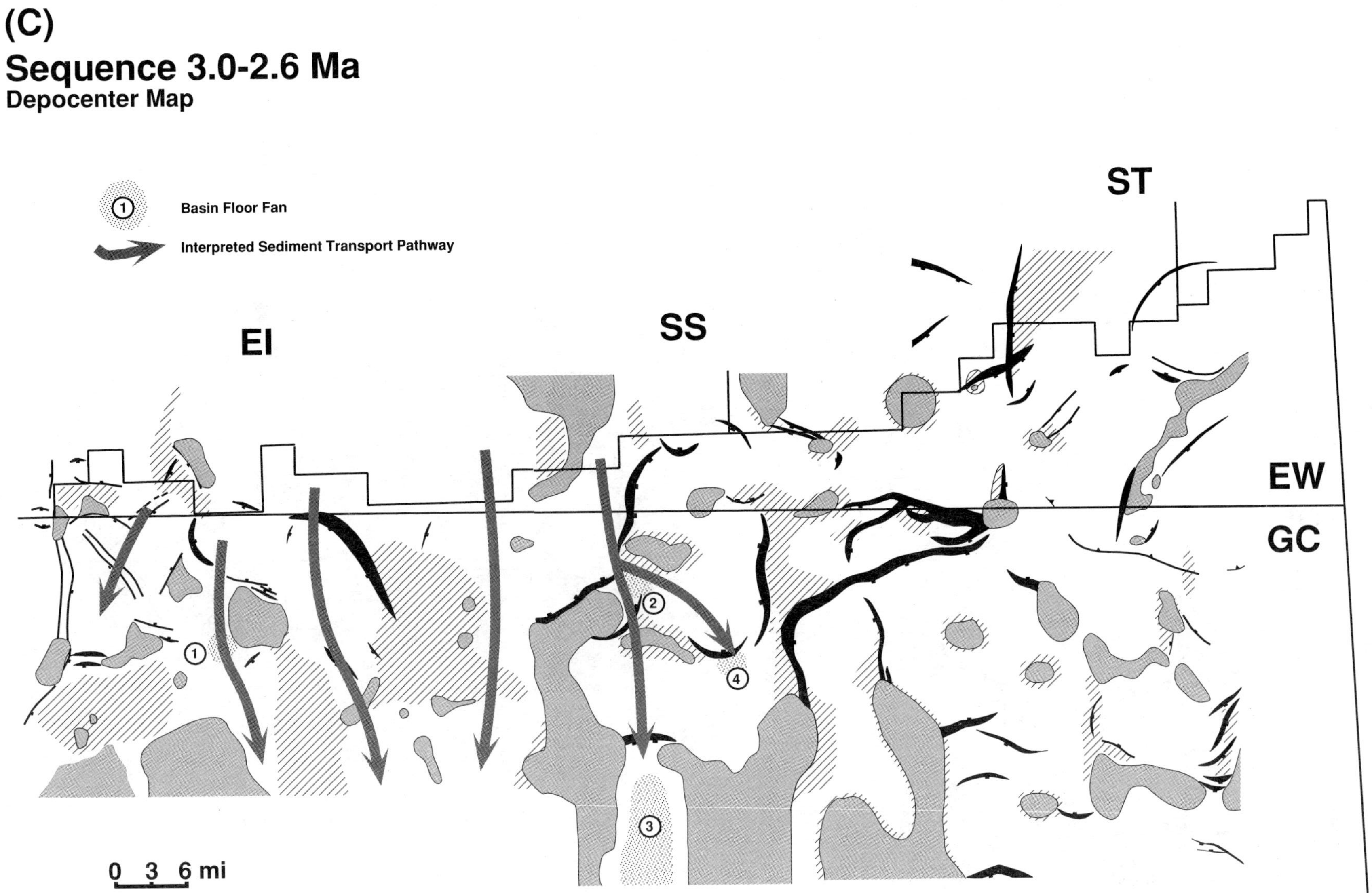


Figure 20—Continued.

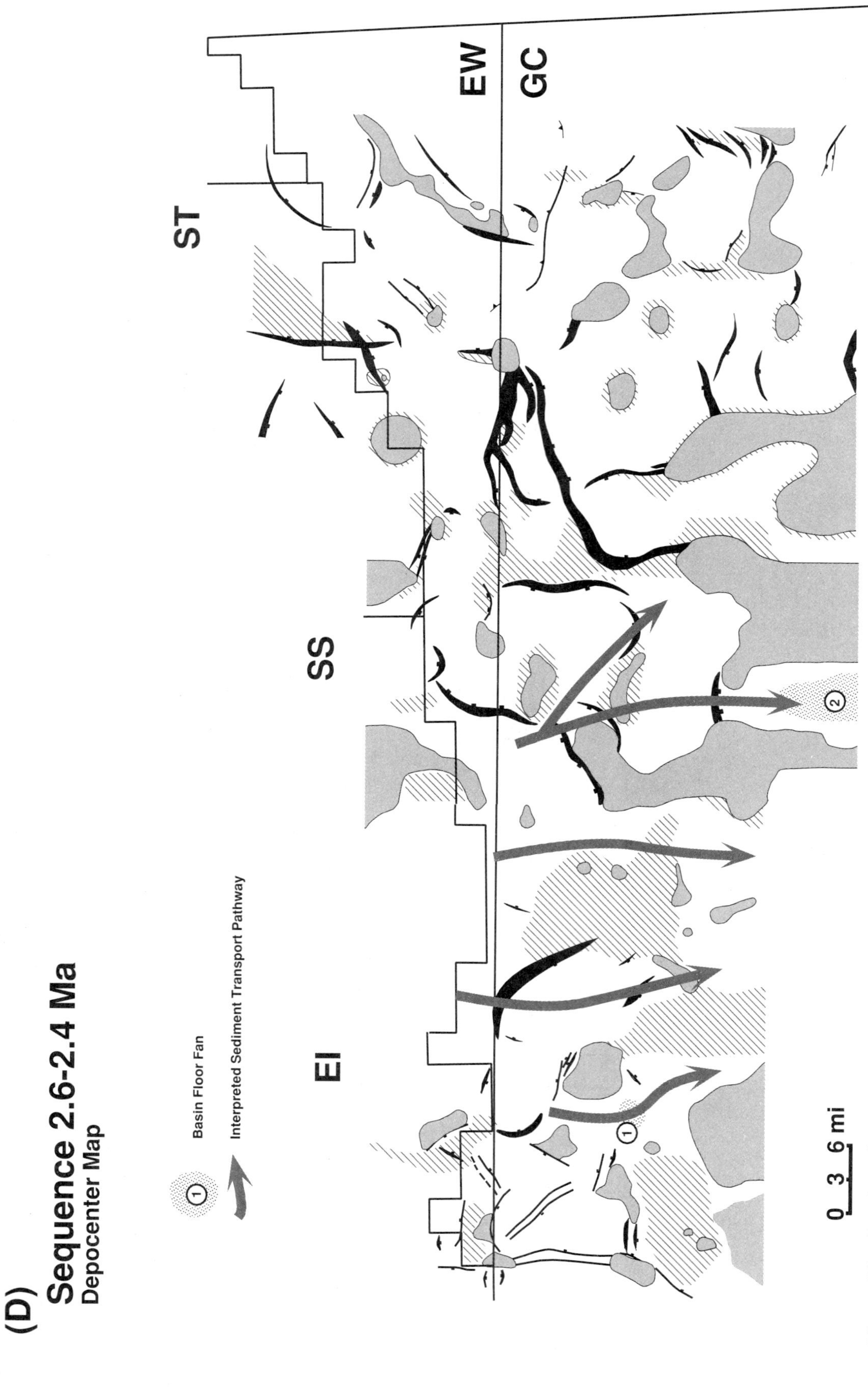


Figure 20—Continued.

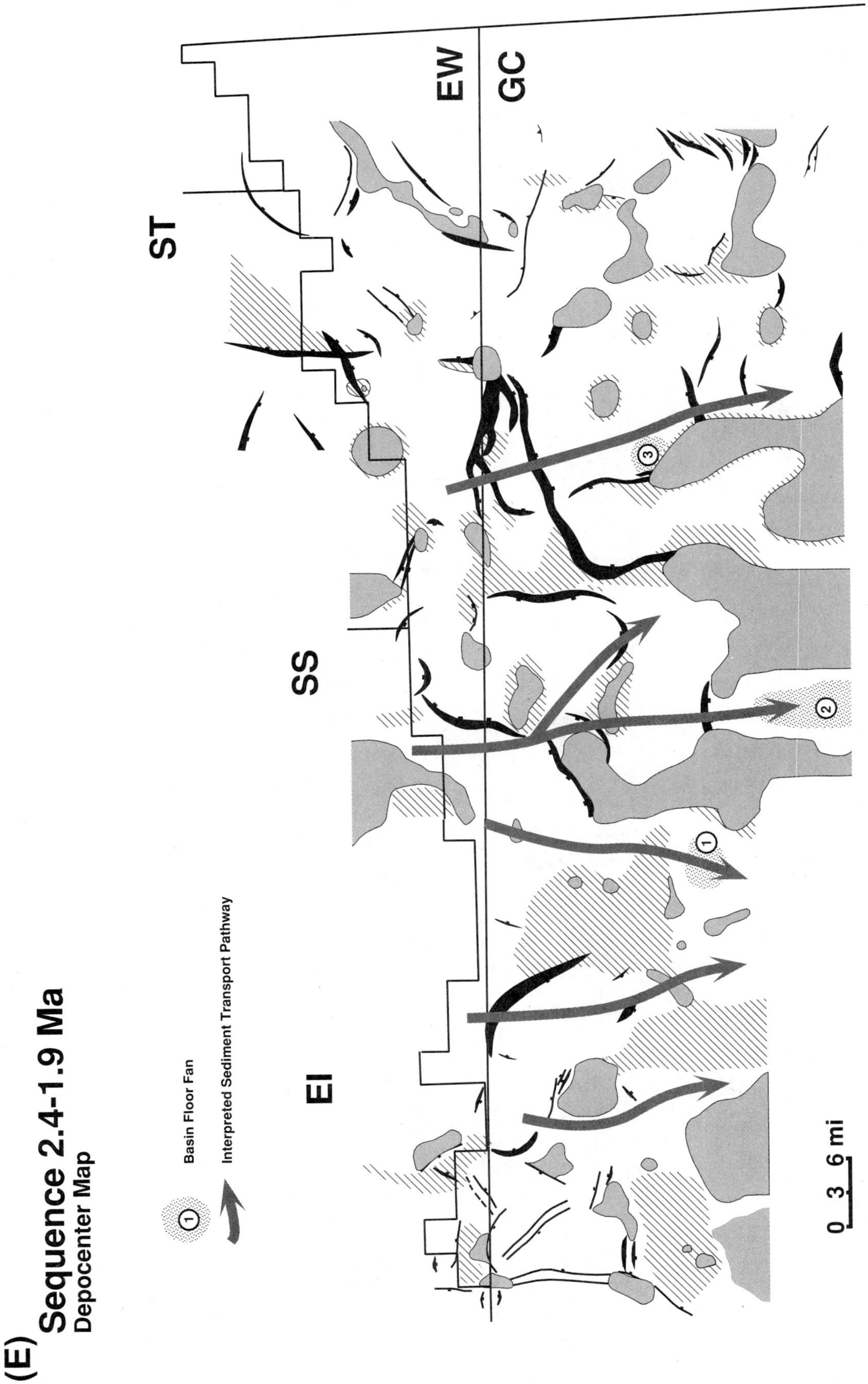


Figure 20—Continued.

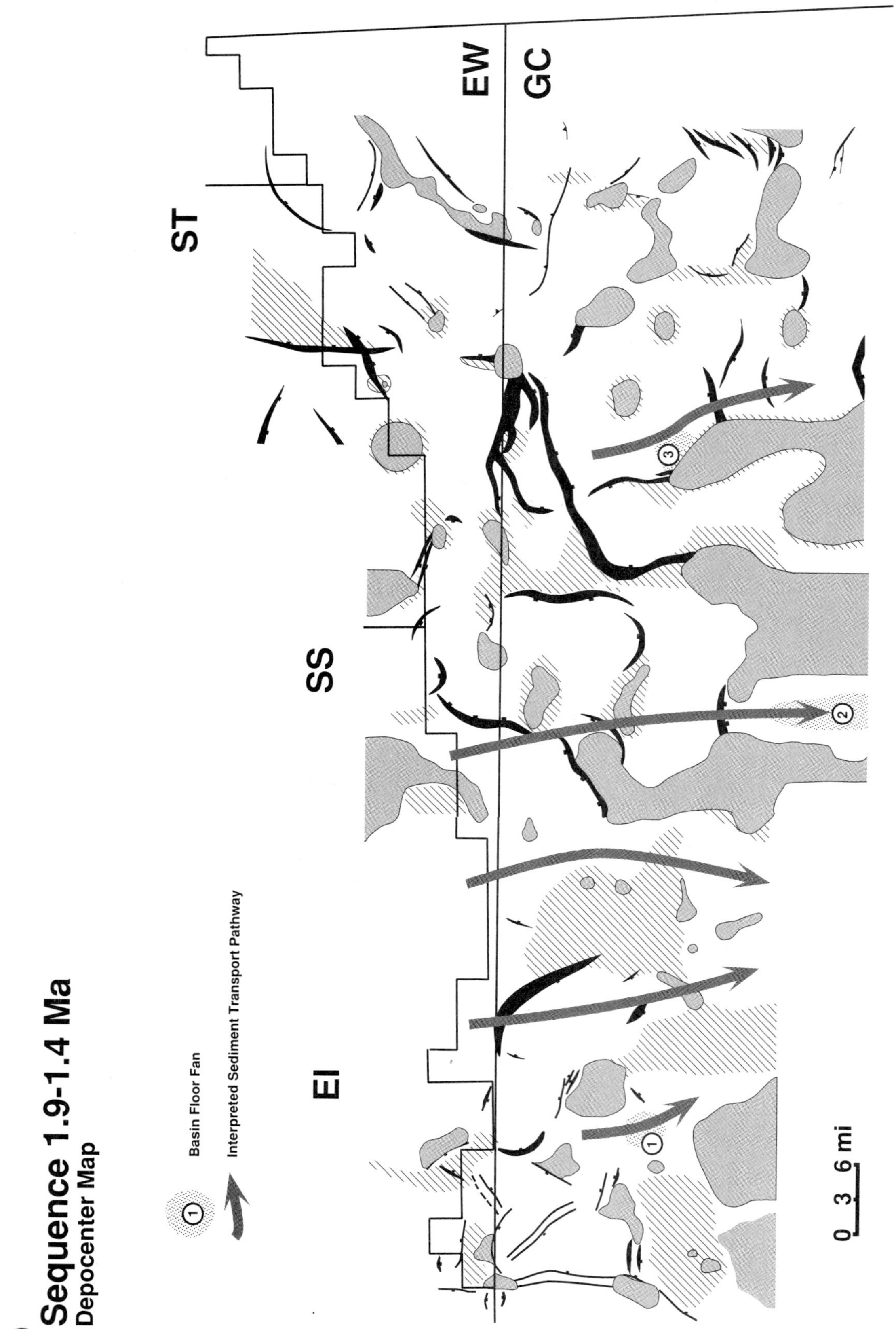


Figure 20—Continued.

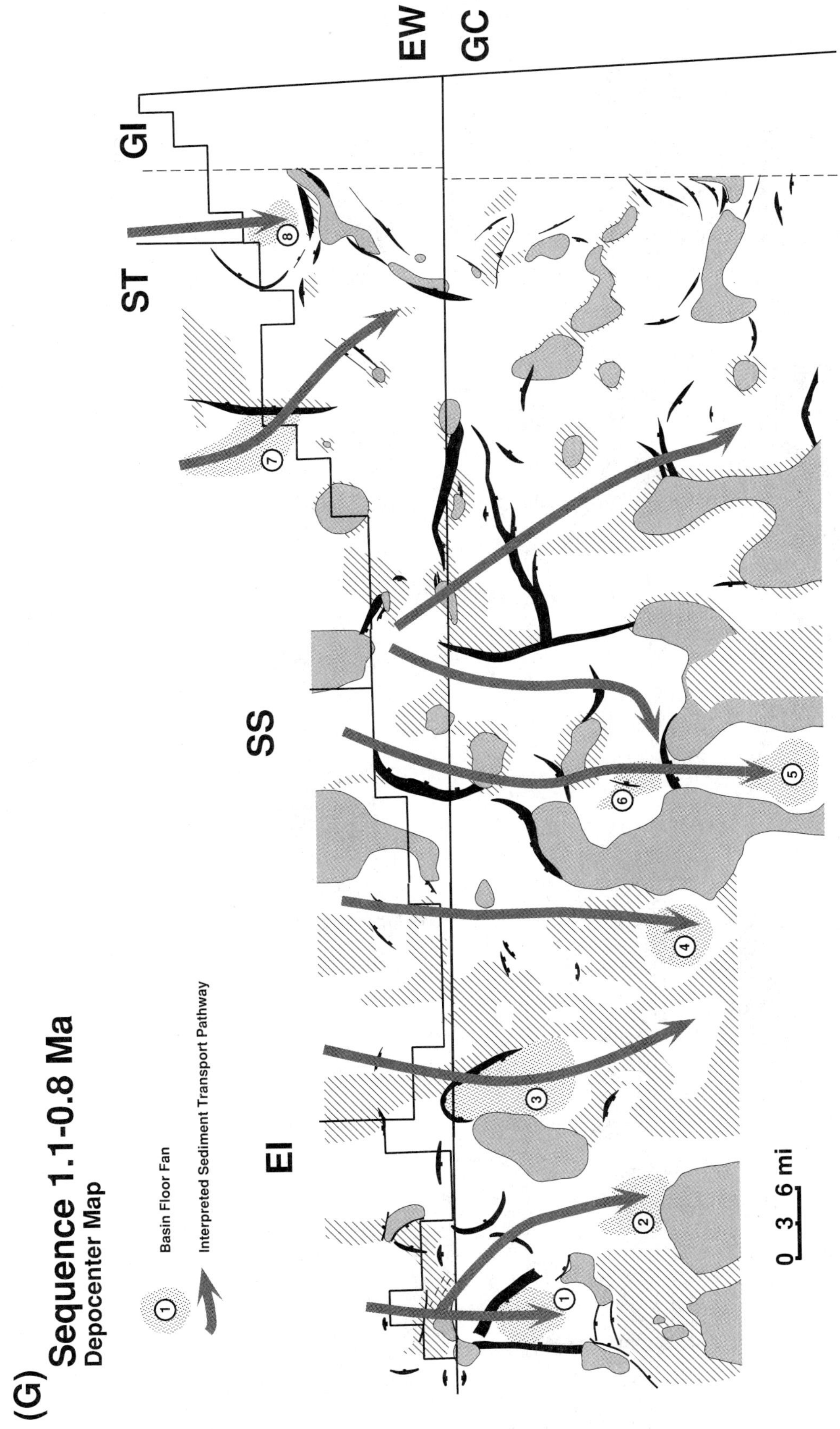


Figure 20—Continued.

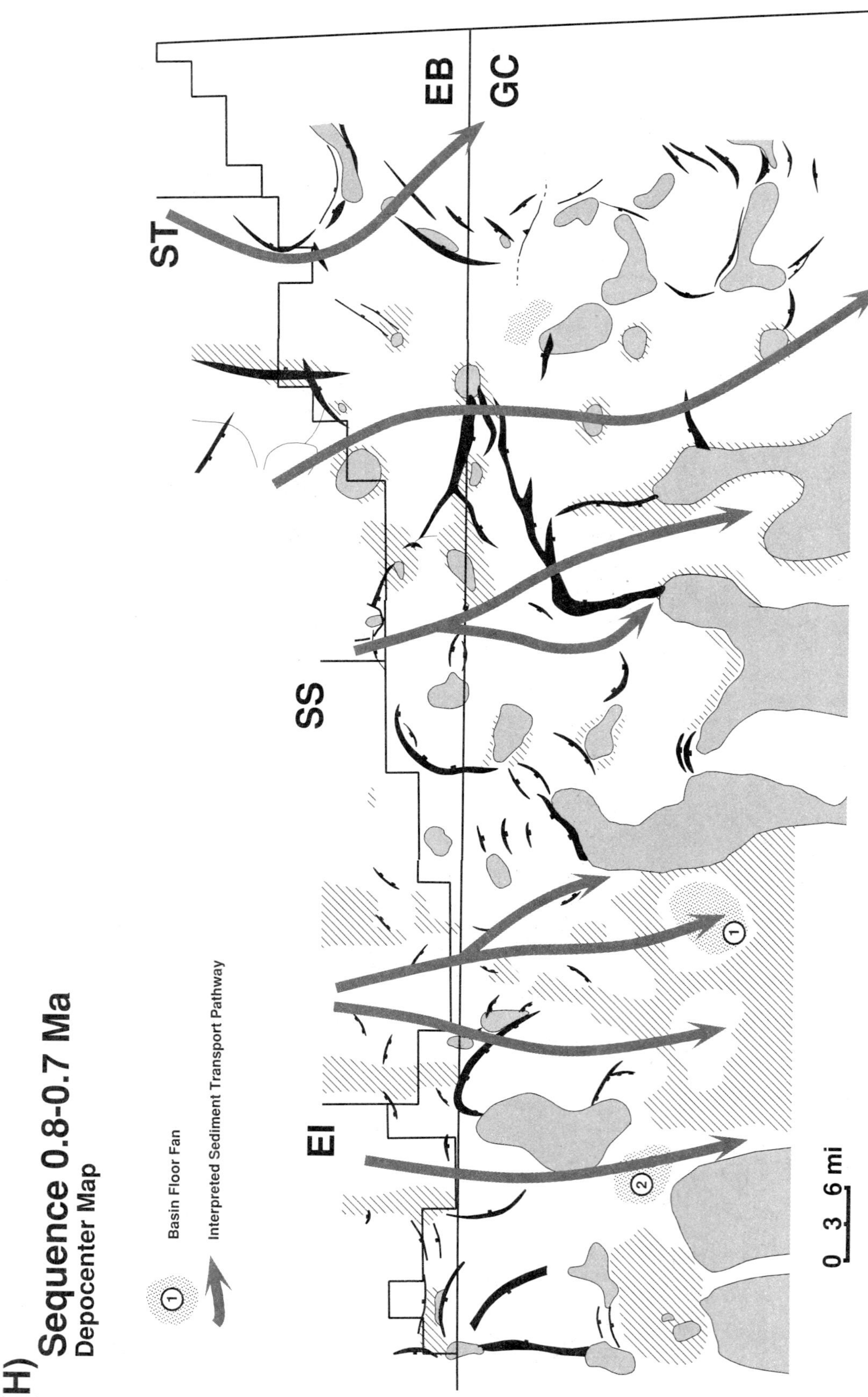


Figure 20—Continued.

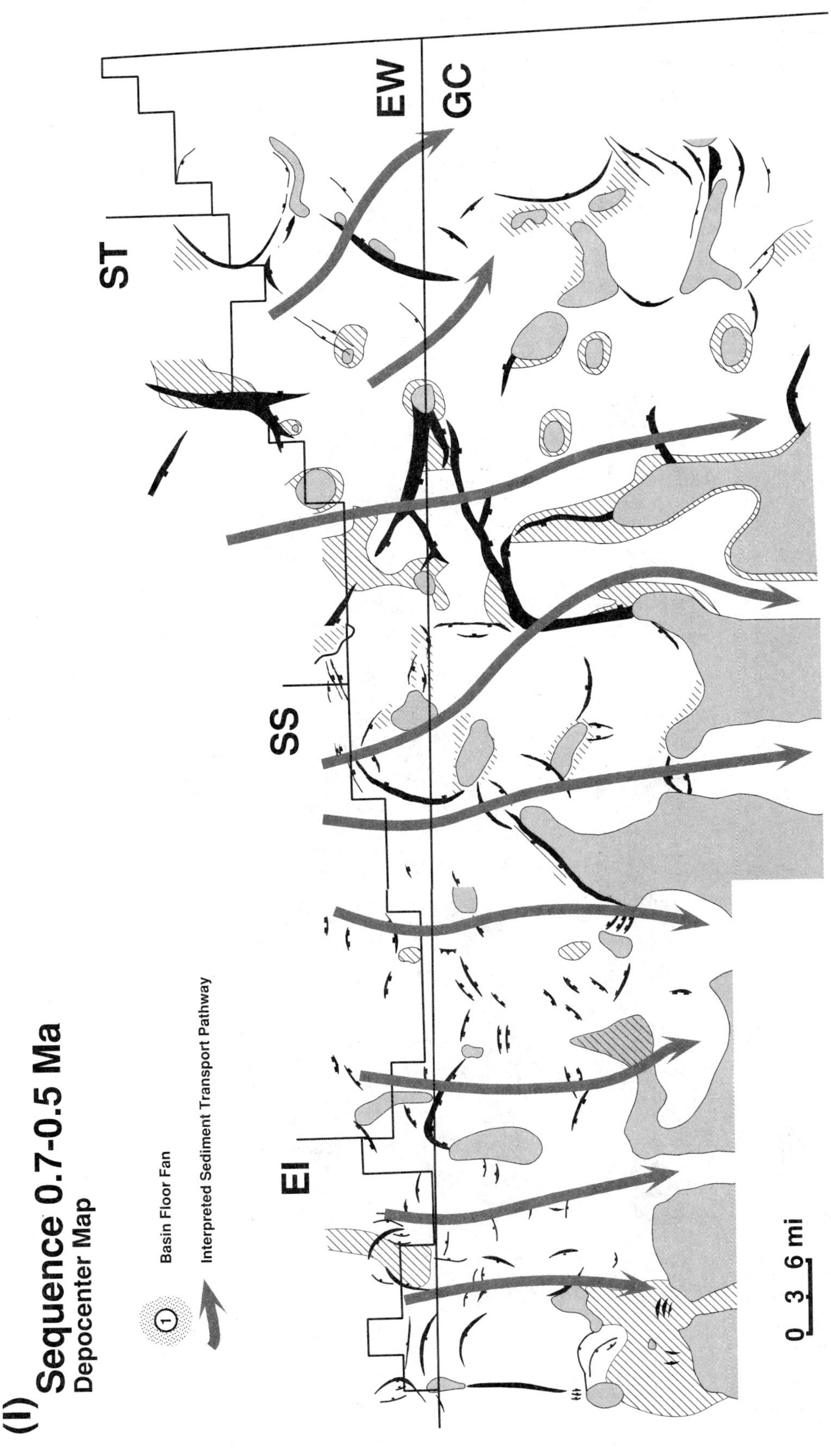


Figure 20—Continued.

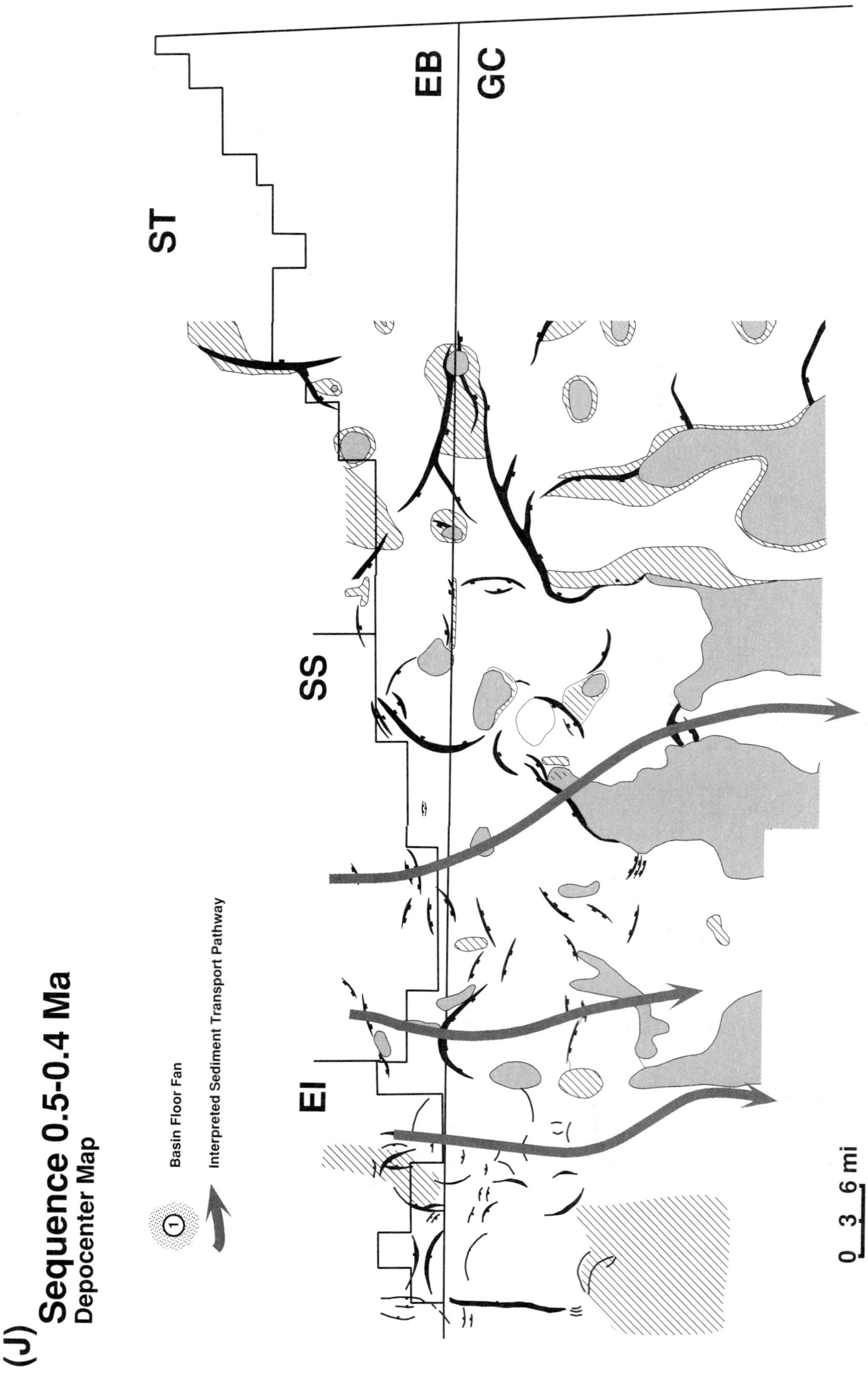


Figure 20—Continued.

canyons developed on the 0.6, 0.5, and 0.4 Ma sequence boundaries (Weimer and Dixon, 1994), and on the modern sea floor. The areal extent of these canyons and the thickness of the eroded sediments become larger in the younger canyons.

Based on these changes in erosion and sliding, the Pliocene–Pleistocene interval can be divided into three broad intervals. During the 5.5–1.4 Ma interval, salt- and fault-related slides are relatively common, showing the effects of syndepositional tectonics or the result of poorer seismic imaging. In the 1.4–0.7 Ma interval, one possible submarine canyon and a few erosional slides were mapped. The number of salt-related slides decreased, indicating less syndepositional salt activity. In the 0.7-Ma and younger sequences, erosion became common (canyons, erosive slides). The increased volumes of sediments associated with the onset of North American glaciation (about 0.8 Ma) were transported, in part, through these canyons farther down to the basin (Pulham, 1993) and deposited in the Mississippi Fan (Weimer and Dixon, 1994). The rest of the sediments were deposited on the slope.

### Shallow-Marine Sediments

Shallow- to marginal-marine sediments were first deposited in the study area during the early Pleistocene (about 0.8 Ma), as evidenced by the paleoecology, wireline-log patterns, and well-defined clinoforms on the seismic (see figure 7 in McBride et al., 1998, this issue). These sediments reflect the progradation of the Mississippi Delta and associated environments into the study area. During the Pliocene, the shelf edge sat about 50–75 mi (80–120 km) north of the study area in south central Eugene Island (Yeilding and Apps, 1994).

## DEPOSITIONAL MODELS FOR PLIOCENE–PLEISTOCENE TURBIDITE SYSTEMS

### General Considerations

In constructing a depositional model for the transport and deposition of sediments in these intraslope basins and for the formation of reservoir-size sand bodies, several observations must be kept in mind. First, enormous volumes of sediment were transported and deposited in these intraslope basins; presumably, large volumes of sediment also flowed through the basins to the south and were deposited in the lower slope and unconfined abyssal plain. Second, based on the approximately 185 well log penetrations that we studied, although sand was deposited in nearly all depositional sequences, the amount ranges both within and between sequences (e.g., Figures 5, 17). Sands generally occur in the lower portions of each sequence in these bathyal facies. Third, burial history curves from the wells indicate that the rates of deposition varied between sequences and across the area within sequences (see figures 7 and 8 in Villamil et al., 1998, this issue). The variation in rates of deposition was controlled by the location of the coeval shelf-margin depocenters. Fourth, most of the sands were deposited during periods of lower sea level (lowstand systems tracts), suggesting that high rates of deposition probably only occurred during 50–60% of the time for any sequence. Lower rates of deposition characterize the rest of the sequence, as reflected by the condensed sections equivalent to the prograding complex, transgressive and highstand systems tracts (see figures 10 and 11 in Villamil et al., 1998, this issue). Fifth, as the margin prograded to the south throughout the Pliocene–Pleistocene, the resulting paleoecology changed from lower to upper bathyal, or in some places to neritic (see summary paleoecology maps (Figure 13; see also figure 13 in Villamil et al., 1998, this issue). Associated with these changes is a change in the regional slope gradient, thus affecting the grain size and location of deposited sediments. Finally, in nearly all sequences, a distinct vertical change in the seismic and interpreted lithofacies occurs, reflecting a transition from more sand-rich to more shale-rich sediments. A similar vertical change in lithofacies and interpreted depositional setting has been described in turbidite systems in many sedimentary basins with different tectonic settings (e.g., Mitchum, 1985; Mutti, 1985; Mitchum et al., 1993; Richards et al., 1998), indicating that the change is associated with decrease in the grain size of the gravity flows.

Limitations of our data set must be kept in mind when developing a depositional model. The first and most important limitation is the areal limits of the study area. The study area does not include a complete shelf-to-lower slope transect for any of the sequences. We cannot see the entire depositional systems as they evolved, which makes it difficult to ascertain and interpret some of the important regional relationships. We do not know the nature of the Pliocene–early Pleistocene deltas and related shorelines, or the sediment transport into the deeper basin. Also, because of the changing position of the shelf depocenter, we cannot see the entire slope depocenter from one end to the other. Specifically, as the Mississippi Delta switched to

**Table 3. Summary of Basin-Floor Fans by Sequence***

| Number (ft) | Paleoecologic Zone | Maximum Thickness of bff Reflections (ft) | Area (mi²) | Influence on Deposition | Well Penetration? | Log Shape/ Thickness (ft) |
|---|---|---|---|---|---|---|
| Sequence 5.5 to 4.2 Ma | | | | | | |
| 1 | 6 | 540 | 15 | Salt | | |
| 2 | 6 | 310 | 38 | Salt | | |
| 3 | 6 | 380 | 6 | Salt | | |
| 4 | 6 | 540 | 4 | Salt | GC 301 | Blocky (200) |
| Sequence 4.2 to 3.8 Ma | | | | | | |
| 1 | 6 | 610 | 53 | Salt | | |
| 2 | 6 | 410 | 28 | Salt | | |
| 3 | 6 | 490 | 34 | Salt | GC 301 | Blocky (200) |
| 4 | 6 | 370 | 29 | Salt | | |
| 5 | 5.5 | 250 | 16 | Salt | | |
| Sequence 3.8 to 3.0 Ma | | | | | | |
| 1 | 6 | 600 | 32 | Salt | | |
| 2 | 6 | 400 | 68 | Salt | GC 472 | Crescentic (200) |
| 3 | 6 | 480 | 30 | Salt | GC 254 | Blocky, upward fining (100) |
| 4 | 5/5.5 | 520 | 44 | Salt | | |
| 5 | 5 | 750 | 16 | Salt | | |
| Sequence 3.0 to 2.6 Ma | | | | | | |
| 1 | 5.5 | 390 | 5 | Salt | | |
| 2 | 5.5 | 200 | 5.4 | Salt | GC 65 | Blocky (150) |
| 3 | 5.5 | 350 | 34 | Fault/salt | GC 506 | Blocky (100) |
| 4 | 5.5 | 700 | 5.2 | Salt | | |
| Sequence 2.6 to 2.4 Ma | | | | | | |
| 1 | 5.5 | 400 | 6 | Salt | | |
| 2 | 5.5 | 600 | 26 | Fault/salt | GC 506 | Blocky (50) |
| Sequence 2.4 to 1.9 Ma | | | | | | |
| 1 | 5.5 | 540 | 14.5 | Salt | | |
| 2 | 5.5 | 360 | 31 | Fault/salt | GC 506 | Blocky (75) |
| 3 | 5.5 | 70 | 10 | Salt | GC 205 | Blocky (150) |

**Table 3. Continued**

| Number (ft) | Paleoecologic Zone | Maximum Thickness of BFF Reflections (ft) | Area (mi²) | Influence on Deposition | Well Penetration? | Log Shape/ Thickness |
|---|---|---|---|---|---|---|
| Sequence 1.9 to 1.4 Ma | | | | | | |
| 1 | 5.5 | 390 | 9 | Salt | | |
| 2 | 5.5 | 410 | 24 | Fault/salt | GC 506 | Fining up (50) |
| 3 | 5.5 | 110 | 10 | Salt | GC 205 | Fining up (100) |
| Sequence 1.4 to 1.1 Ma | | | | | | |
| 1 | 5.5 | 350 | 26 | Salt | GC 5 | Fining up (75) |
| 2 | 5.5 | 400 | 14 | salt | | |
| 3 | 5 | 450 | 16 | | | |
| 4 | 5/5.5 | 370 | 20 | Fault | GC 145 | Blocky (100) |
| 5 | 5.5 | 530 | 14 | Salt | | |
| 6 | 5.5 | 490 | 13 | Salt | | |
| 7 | 5 | 570 | 39 | Fault | SS 355 | Blocky (100) |
| 8 | 4 | 350 | 21 | Fault | GC 23 | Blocky (100) |
| 9 | 5.5 | 380 | 33 | Fault/salt | | GC 506 Upward Fining (75) |
| 10 | 5.5 | 250 | 9 | Fault | GC 158 | Blocky/ Upward Fining (200) |
| 11 | 5 | 370 | 10 | Salt | | |
| 12 | 5.5 | 450 | 14 | Salt | | |
| 13 | 5.5 | 350 | 32 | Fault/salt | | |
| 14 | 5.5 | 230 | 9 | Fault | | |
| 15 | 5.5 | 350 | 7 | Salt | | |
| Sequence 1.1 to 0.8 Ma | | | | | | |
| 1 | 5 | 470 | 21 | Fault | | |
| 2 | 5.5 | 220 | 20 | Salt | | |
| 3 | 5 | 650 | 50 | Fault | GC 145 | Blocky (50) |
| 4 | 5.5 | 360 | 24 | Salt | | |
| 5 | 5.5 | 700 | 25 | Fault/salt | GC 506 | Blocky (50) |
| 6 | 5.5 | 250 | 12 | Salt/fault | | |
| 7 | 5 | 400 | 33 | Salt | | |
| 8 | 5 | 650 | 11 | Fault/salt | | |
| Sequence 0.8 to 0.7 Ma) | | | | | | |
| 1 | 5.5 | 280 | 26 | Salt | | |
| 2 | 5.5 | 380 | 20 | Salt | | |
| 3 | 5.5 | 420 | 14 | Fault | | |

* See Figures 11, 19, and 20.

(A)

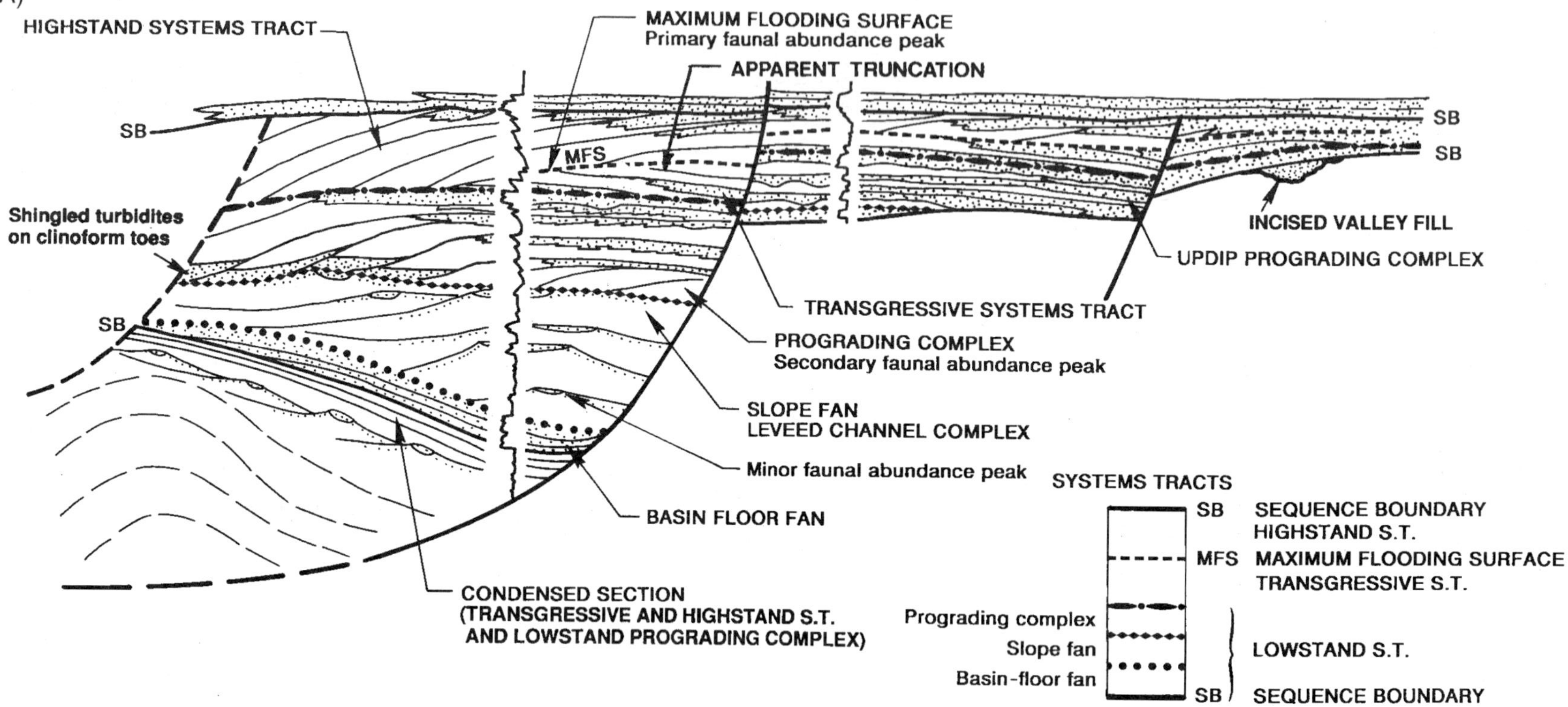


Figure 21—Depositional models for the northern Gulf of Mexico: (A) significant sands deposited on downthrown side of major expansion fault (after Mitchum et al., 1993) and (B) this study showing the relationship of filling of minibasins formed from salt withdrawal. Basin-floor fans (bff) are deposited in adjacent minibasins in sequence 1, and are overlain by channel-levee systems (cls) and condensed section (cs). In sequence 2, one basin-floor fan is deposited in the minibasin to the right; channel-levee systems are then deposited in both minibasins. In sequence 3, only channel-levee systems (bypass phase) are deposited as major sands bypass farther downslope.

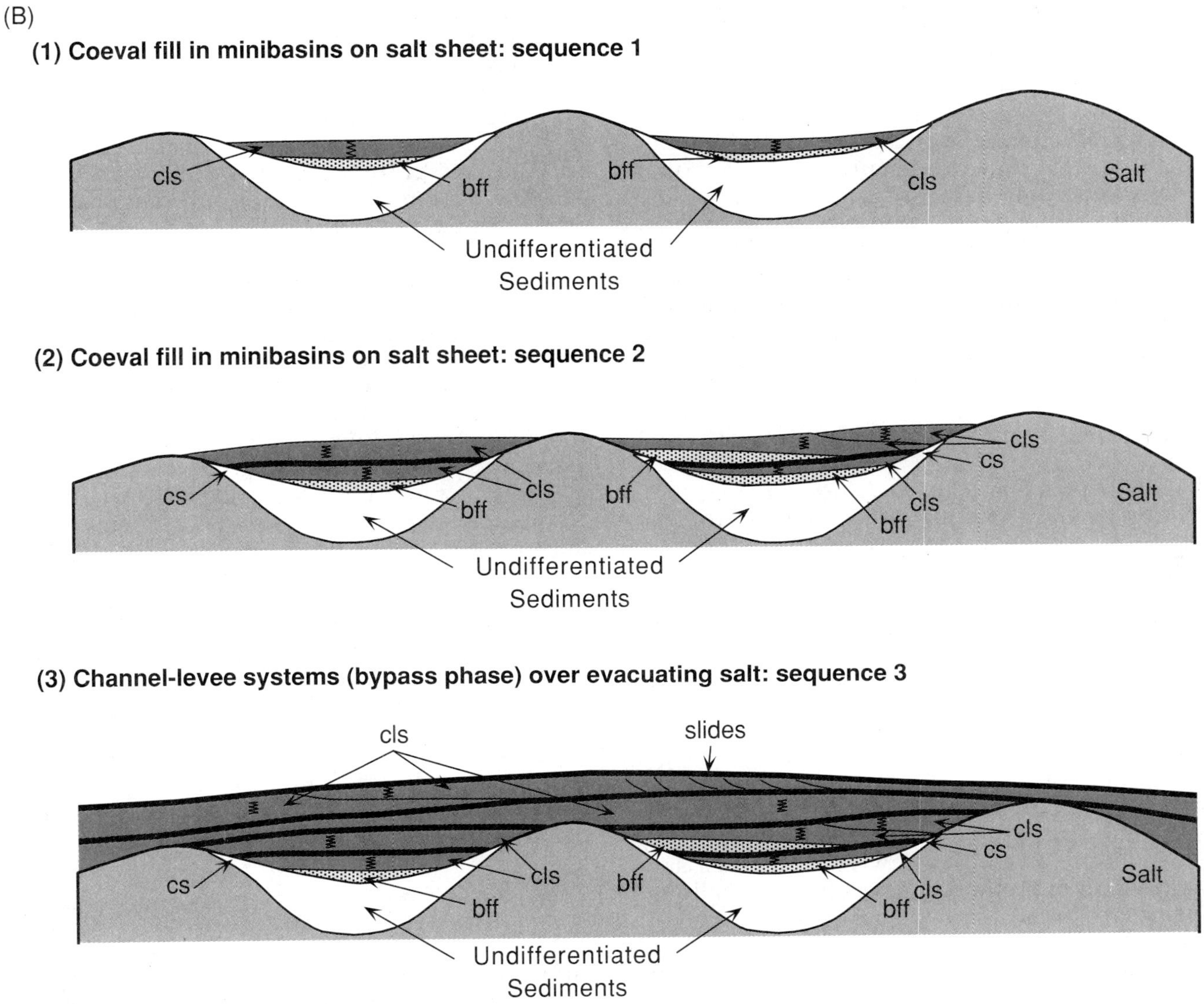


**Figure 21—Continued.**

the west around 3.0 Ma, the regional distribution of sediment changed, and sediments become considerably thinner to the east (eastern Green Canyon), Mississippi Canyon. With the study area extended to the west and east, the complete Pliocene–Pleistocene depocenter would be included.

## Previous Models

A variety of stratigraphic models have been proposed for the distribution of lithofacies in the Gulf of Mexico intraslope turbidite systems. All of these models are viable and reflect the portion of the northern Gulf of Mexico they were drawn from in terms of the style and amount of tectonics and slope gradients (e.g., amount and style of salt deformation, faulting, nature of depocenter, paleobathymetry). One model emphasized how thick bathyal turbidite systems (lowstand systems tract) preferentially stack on the downthrown side of major normal (expansion) faults (Pacht et al., 1992; Mitchum et al., 1993; Figure 21A). In their model, continued movement on the fault is interpreted to create necessary gradient and space in bathyal water depths for the sands to be deposited.

Another model describes the sediment filling of a succession of intraslope basins during the Neogene (Satterfield and Behrens, 1990; Winker, 1996; Prather et al., 1998). Upslope basins fill with sediment first, and once the basin is completely filled, new sediment bypasses the basin to downdip basins. This model was important because it showed that the intraslope basins fill at different times, and the sediment fill becomes younger downslope. In this two-phase stratigraphic fill, sand-rich turbidite facies tend to stack and onlap (pond) onto the flanks of salt highs. Once the basin is filled, a "bypass" situation develops where sediment is transported downslope to the next minibasin, and it fills with sediments. The early bypass facies that develops in the first basin consists of a shale-dominated channel-levee system, channel- fills, and shallow erosional features.

## Integrated Depositional Model

The depocenter maps in Figures 11, 14a, and 20 show the interpreted trend of sediment transport throughout the Pliocene–Pleistocene in the study area. This series of maps illustrates how the slope depositional systems evolved through time. We can immediately observe that there are different bathymetric and structural controls on the distribution of the interpreted sand-rich facies (basin-floor fans). Table 3 lists end members that control of the distribution of these features—fault controlled, salt controlled, or a combination of both.

In some places, sands are clearly deposited on the downthrown sides of faults, similar to what Mitchum et al. (1993) showed (see Figure 19, fans 7, 9; Figure 21A). In a few places, thick sands develop in multiple sequences in some fault-related basins (Figure 20D–F, fan 2).

The ponding-to-bypass model (Figure 21B) discussed by Satterfield and Behrens (1990), Winker (1996), and Prather et al. (1998, this issue) is recognized in a few sequences. In the 1.4–1.1 Ma sequence, three basin-floor fans are aligned north to south in the central part of the basin (Figure 19). The northernmost basin-floor fan was deposited on the downthrown side of the fault (fan 7, Figure 19). The next basin-floor fan was deposited within a minibasin created by a thrust fault in the back limb of a fold (fan 8, Figure 19). The third basin-floor fan was deposited in the Fuji minibasin (GC 506). Thick sands [50–80 ft (15–25 m)] are present in the three wells that penetrated these features. In addition, several basin-floor fans were deposited in the southeastern portion of the area in the lower Pliocene sequences (i.e., 5.5, 4.2, 3.8 Ma; Figures 11, 20A, B). Each of these basin-floor fans is deposited in the same minibasins.

In both of these examples, the relative order of deposition of basin-floor fans in the same sequence is beyond resolution of the biostratigraphy (Figure 21B). Thus, it is possible that these basin-floor fans were deposited in a ponded-to-bypass mode within one sequence or, more importantly, there may be a lateral shift in the deposition in sand bodies to the east or west (lower Pliocene sequences); however, there is a distinct vertical change in the seismic facies and interpreted grain sizes of sediments within a sequence from sand-rich to more mud-rich turbidite systems (Figure 21B). These changes reflect the change from a "fill" mode to a "bypass" mode at a sequence level, in different kinds of basins. Thus, we interpret the ponding-to-bypass facies to occur at two levels. In a simplified regional sense, this process occurs though several depositional sequences (Prather et al., 1998, this issue). The process also occurs at the minibasin level within one sequence. Careful correlation of sequence boundaries and integration of the biostratigraphy from the wells that penetrate these basins indicate coeval filling of the adjacent minibasins in a downslope direction. In adjoining minibasins, the ponded-to-bypass mode of sand deposition between individual minibasins cannot be seen.

In summary, we interpret several ways in which sand is deposited in these intraslope minibasins. The critical factor appears to have been an abrupt decrease in local gradient that caused sands to be deposited from the gravity flows. The abrupt changes in sea-floor gradient were caused primarily by salt evacuation or fault movement.

## CONCLUSIONS

The seismic and geologic facies of twelve depositional sequences (5.5, 4.2, 3.8, 3.0, 2.6, 2.4, 1.9, 1.4, 1.1, 0.8, 0.7, and 0.5 Ma) were mapped in the area. Benthic biofacies indicate deposition of sediment occurred primarily in lower bathyal water depths during the Pliocene–Pleistocene. These sequences are interpreted as deep-water turbidite systems deposited in intraslope minibasins. Turbidite systems were deposited primarily during relative sea level lowstands (basin-floor fans were overlain by overbank and channel-levee deposits of slope fans). During transgressive and highstand deposition, sediment was trapped in the shelfal areas, and condensed sections developed on the slope. The repetition of these systems tracts resulted in cyclic deposition.

Deposition and size of the basin-floor fans were controlled by the sand content of the turbidite flows and by salt tectonics or faults. Areally extensive fans fill the basal parts of the southeastern minibasins in the 5.5–3.8-Ma sequences. Salt withdrawal created bathymetric lows here, which caused deposition of sand-rich turbidite systems.

In the northern and western parts of the area, salt withdrawal was less active, and no large fans were deposited.

Decreased salt withdrawal, and possibly more mud-rich turbidite systems, resulted in smaller basin-floor fans around growth faults and shallow salt bodies in the younger sequences (3.0–0.7 Ma). The areal extent and number of these fans decreased as the margin prograded to the south, and possibly shale-rich sediments were transported to the area. No fans were deposited in the 0.7 Ma and younger sequences, even though active salt withdrawal created deep bathymetric lows. Most of the sands are interpreted to have been transported through the area to deeper basins.

Channel systems interbedded with overbank shales constitute most of the sediments in each sequence. More channels apparently were deposited in the older sequences (and the channel fills are more sand rich) than in the younger sequences. The individual channel systems are too small to be mapped using the regional two-dimensional seismic data.

An integrated depositional model includes the idea that sand occurrences were controlled by abrupt decreases in the bathymetric gradient in the minibasins. These gradient decreases were controlled by faults or salt withdrawal. Thus, the many depositional models proposed for intraslope sedimentation are accurate, but local factors can complicate the depositional patterns.

## REFERENCES CITED


Acosta, Z. M., 1994, Sequence stratigraphy of Plio-Pleistocene sediments in north-central Green Canyon and western Ewing Bank, northern Gulf of Mexico: Master's thesis, University of Colorado, Boulder, Colorado, 187 p.

Budhijanto, F. M., 1995, Sequence stratigraphy of Plio-Pleistocene sediments of the northeastern Green Canyon and eastern Ewing Bank areas, northern Gulf of Mexico: unpublished Master's thesis, University of Colorado, Boulder, Colorado, 154 p.

Chapin, M. A., P. Davies, J. L. Gibson, and H. S. Pettingill, 1994, Reservoir architecture of turbidite sheet sandstones in laterally extensive outcrops, Ross formation, western Ireland, *in* P. Weimer, A. H. Bouma, and B. F. Perkins, eds., Submarine fans and turbidite systems: GCS-SEPM Foundation 15th Annual Research Conference, p. 53–68.

Haines, L., 1996, Going deep: Oil and Gas Investor, September, v. 16, no. 9, p. 24–32.

Kneller, B., and B. McCaffrey, 1995, Modeling the effects of salt-induced topography on deposition from turbidity currents, *in* C. J. Travis, H. Harrison, M. R. Hudec, B. C. Vendeville, F. J. Peel and B. F. Perkins, eds., Salt, sediment and hydrocarbons: GCS-SEPM Foundation 16th Annual Research Conference, p. 137–149.

Martinez, R. E., 1994, Sequence stratigraphy of late Pliocene and Pleistocene sediments of northwestern Green Canyon and Ewing Bank, northern Gulf of Mexico: Master's thesis, University of Colorado, Boulder, Colorado, 138 p.

McBride, B. C., 1998, The evolution of allochthonous salt along a megaregional profile across the northern Gulf of Mexico: AAPG Bulletin, v. 82, no. 5B, p. 1037–1054.

McBride, B. C., M. G. Rowan, and P. Weimer, 1998, The evolution of allochthonous salt systems, northern Green Canyon and Ewing Bank (offshore Louisiana), northern Gulf of Mexico: AAPG Bulletin, v. 82,p. 1013–1036.

McGee, D. T., P. W. Bilinski, P. S. Gary, D. S. Pfeiffer, and J. L. Sheiman, 1994, Geologic models and reservoir geometries, Auger field, deep-water Gulf of Mexico, *in* P. Weimer, A. H. Bouma, and B. F. Perkins, eds., Submarine fans and turbidite systems: GCS-SEPM 15th Annual Research Conference, p. 245–256.

Mitchum, R. M., Jr., 1985, Seismic stratigraphic recognition of submarine fans, *in* O. R. Berg and D. G. Woolverton, eds., Seismic stratigraphy II: AAPG Memoir 39, p. 117–136.

Mitchum, R. M., Jr., J. B. Sangree, P. R. Vail, and W. W. Wornardt, 1993, Recognizing sequences and systems tracts from well logs, seismic data, and biostratigraphy: examples from the late Cenozoic, *in* P. Weimer and H. W. Posamentier, eds., Siliciclastic sequence stratigraphy: AAPG Memoir 58, p. 163–199.

Mutti, E., 1985, Turbidite systems and their relation to depositional sequences, *in* G. G. Zuffa, ed., Provenance of arenites: NATO-ASI series, Dordrecht, Reidel, p. 65–93.

Mutti, E., and W. R. Normark, 1987, Comparing examples of modern and ancient turbidite systems: problems and concepts, *in* J. K. Leggett and G. G. Zuffa, eds., Marine clastic sedimentology: concept and case studies: London, Graham and Trotman, p. 1–38.

Mutti, E., and W. R. Normark, 1991, An integrated approach to the study of turbidite systems, *in* P. Weimer and M. H. Link, eds., Seismic facies and sedimentary processes of submarine fans and turbidite systems: New York, Springer-Verlag, p. 75–106.

Navarro, A. F., 1994, Sequence stratigraphy of Pleistocene sediments of northwestern Green Canyon area, northern Gulf of Mexico: Master's thesis, University of Colorado, Boulder, Colorado, 108 p.

Pacht, J. A., B. E. Bowen, J. H. Beard, B. L. Shaffer, and W. R. Pottorf, 1992, Systems tracts, seismic facies, and attribute analysis within a sequence stratigraphic framework—example from the offshore Louisiana Gulf Coast, *in* E. G. Rhodes and T. F. Moslow, eds., Marine clastic reservoirs: New York, Springer-Verlag, p. 27–38.

Prather, B. E., J. R. Booth, G. S. Steffens, and P. A. Craig, 1998, Classification, lithologic calibration, and stratigraphic succession of seismic facies in intraslope basins, deep-water Gulf of Mexico: AAPG Bulletin, v. 82, no. 5A, p. 701–728.

Pulham, A. J., 1993, Variations in slope deposition, Pliocene-Pleistocene, offshore Louisiana, northeast Gulf of Mexico, *in* P. Weimer and H. W. Posamentier, eds., Siliciclastic sequence stratigraphy: AAPG Memoir 58, p. 199–234.

Ramsayer, G. R., 1979, Seismic stratigraphy, a fundamental exploration tool: Offshore Technology Conference Proceedings, Paper 3568, v. 3, p. 1859–1867.

Richards, M. T., and M. Bowman, in press, Submarine fans II: variability in reservoir architecture and log character: Marine and Petroleum Geology.

Rowan, M. G., 1995, Structural styles and evolution of allochthonous salt, central Louisiana outer shelf and upper slope, *in* M. P. A. Jackson, D. G. Roberts, and S. Snelson, eds., Salt tectonics: a global perspective: AAPG Memoir 65, p. 199–228.

Rowan M. G., and P. Weimer, 1998, Salt-sediment interaction, northern Green Canyon and Ewing Bank, offshore Louisiana, northern Gulf of Mexico: AAPG Bulletin, v. 82, p. 1055–1082.

Satterfield, W. M., and W. E. Behrens, 1990, A late Quaternary canyon/channel system, northwestern Gulf of Mexico continental slope: Marine Geology, v. 92, p. 51–67.

Schuster, D. C., 1995, Deformation of allochthonous salt and evolution of related salt/structural systems, eastern Louisiana

Gulf Coast, *in* M. P. A. Jackson, D. G. Roberts, and S. Snelson, eds., Salt tectonics: a global perspective: AAPG Memoir 65, p. 177–198.

Varnai, P., 1996, Integrated 2-D and 3-D sequence stratigraphic analysis of Plio-Pleistocene turbidite systems, northern Green Canyon and central Ewing Bank areas, northern Gulf of Mexico: Ph.D. dissertation, University of Colorado, Boulder, Colorado, 508 p.

Varnai, P., 1998, Three-dimensional seismic stratigraphic expression of Pliocene-Pleistocene turbidite systems, northern Green Canyon, offshore Louisiana, northern Gulf of Mexico: AAPG Bulletin, v. 82, no. 5B, p. 986–1012.

Villamil, T., C. Arango, P. Weimer, A. Waterman, M. G. Rowan, P. Varnai, A. J. Pulham, and J. R. Crews, 1998, Biostratigraphic techniques for analyzing benthic biofacies, stratigraphic condensation, and key surface identification, Pliocene and Pleistocene sediments, northern Green Canyon and Ewing Bank (offshore Louisiana), northern Gulf of Mexico: AAPG Bulletin, v. 82, no. 5B, p. 961–985.

Weimer, P., and B. T. Dixon, 1994, Regional sequence stratigraphic setting of the Mississippi Fan complex, northern Gulf of Mexico: implications for evolution of the northern Gulf Basin margin, *in* P. Weimer, A. H. Bouma, and B. F. Perkins, eds., Submarine fans and turbidite systems: GCS-SEPM Foundation 15th Annual Research Conference, p. 373–381.

Weimer, P., A. H. Bouma, and B. F. Perkins, eds., 1994, Submarine fans and turbidite systems: GCS-SEPM Foundation 15th Annual Research Conference, 440 p.

Weimer, P., J. R. Crews, R. Crow, and P. Varnai, 1998a, Atlas of petroleum fields and discoveries, northern Green Canyon, Ewing Bank, and southern Ship Shoal and South Timbalier areas (offshore Louisiana), northern Gulf of Mexico: AAPG Bulletin, v. 82, no. 5B, p. 878–917.

Weimer, P., M. G. Rowan, B. C. McBride, and R. M. Kligfield, 1998b, Evaluating the petroleum systems of the northern deep Gulf of Mexico through integrated basin analysis: an overview: AAPG Bulletin, v. 82, no. 5B, p. 865–877.

Winker, C. D., 1996, High-resolution seismic stratigraphy of a late Pleistocene submarine fan ponded by salt-withdrawal mini-basins on the Gulf of Mexico continental slope: Offshore Technology Conference Proceedings, OTC 8024, p. 619–628.

Yeilding, C. A., and G. M. Apps, 1994, Spatial and temporal variations in the facies ssociations of depositional sequences on the slope: examples from the Miocene–Pleistocene of the Gulf of Mexico, *in* P. Weimer, A. H. Bouma, and B. F. Perkins, eds., Submarine fans and turbidite systems: GCS-SEPM Foundation 15th Annual Research Conference, p. 425–437.

## ABOUT THE AUTHORS

### Paul Weimer

Paul Weimer is an associate professor and director of the Energy and Minerals Applied Research Center at the University of Colorado at Boulder. His research interests are in sequence stratigraphy and basin analysis, three-dimensional seismic interpretation, and reservoir geology.

### Peter Varnai

Peter Varnai received his B.S. degree (1985) in geology from Eötvös Loránd University, Budapest, Hungary; an M.S. degree (1991) in geophysics from Miskolc University, Miskolc, Hungary; and a Ph.D. (1996) in geology from the University of Colorado, Boulder, Colorado. He worked for the Hungarian National Oil Company in Budapest as an exploration geologist for six years. He was responsible for seismic interpretation, stratigraphic interpretation of well logs, construction of stratigraphic and tectonic models, and basin analysis in the Pannonian basin. During his graduate studies in Colorado, he participated in the Gulf of Mexico industry consortium. In 1996 he joined Exxon Exploration Company as a senior geologist.

### Fadjar M. Budhijanto

Fadjar M. Budhijanto worked for five years as a senior geologist with Mobil Oil Indonesia Inc., and for five years with EMARC Research Center at the University of Colorado at Boulder. His main research interest is the sequence stratigraphy of turbidite reservoirs. He has worked on the deep-water Gulf of Mexico Basin and the Mahakam Delta of Indonesia. He has worked as a consultant in Boulder, Colorado, since 1995.

### Zurilma Acosta

Zurilma Acosta received her B.S. degree in geological engineering from the Universidad de Oriente in Venezuela in 1978, and her M.S. degree in geology from the University of Colorado at Boulder in Boulder, Colorado, in 1994. She has worked for Petroleos de Venezuela for 19 years as a production geologist in Maturin subbasin reservoirs. Her current responsibilities are the geological development of the Cerro Negro area in the Orinoco Oil Belt.

**Rafael Martinez**

Rafael Martinez received his B.A. degree in geology from the Universidad Central de Venezuela in 1980 and his M.S. degree in geological sciences from Colorado University in 1994. In 1980, he joined Lagoven in Venezuela and currently works on exploration projects in Western and Eastern basins. He is presently team leader of the North of Monagas Group in the Eastern Venezuela basin.

**Alonso Navarro**

Alonso Navarro graduated from the Universidad Central de Venezuela as a geophysicist in 1988; he joined the exploration department of Maraven, an affiliate of Petroleos de Venezuela, working on seismic interpretation and seismic velocity analysis in the Andean mountain front until 1992. He obtained an M.S. degree in geology from the University of Colorado in 1994. He returned to Venezuela and continued working as a seismic interpreter in Zulia Oriental, a frontier area in the east side of the Maracaibo basin.

**Mark G. Rowan**

Mark Rowan is a graduate of Caltech (1976), University of California, Berkeley (1982), and the University of Colorado (1991), where he is currently an assistant research professor in structural geology. He has also worked in the petroleum industry, both as an exploration geologist for Sohio and as a consultant with Geo-Logic Systems and Alastair Beach Associates. His research interests focus on the three-dimensional geometry and evolution of both extensional and contractional structures, with a recent emphasis on salt tectonics.

**Barry C. McBride**

Barry McBride is a geologist at HS Resources Inc. He received his Ph.D. in geology at the University of Colorado, his M.S. degree in geology from Western Michigan University, and his B.S. degree in geology from Furman University. From 1988 to 1993 he worked for Mobil Exploration and Producing U.S. Inc. and Mobil Global Interpretation Support as an exploration geologist. His research interests focus on complex structures, petroleum exploration, and petroleum systems.

**Tomas Villamil**

Tomas Villamil is an upstream exploration geologist in Conoco's Advance Exploration Organization. His expertise is in stratigraphy and tectonostratigraphy of northern South America and the Gulf of Mexico. He graduated from the National University of Colombia (Bogotá, 1988), and then received a Ph.D from the University of Colorado at Boulder (1994) and a postdoctorate degree from the same school. He is a member of AAPG and SEPM.

**Claudia Arango**

Claudia Arango is a geologist in Marathon Oil Company's Gulf of Mexico Exploration group. Her expertise is in stratigraphy and biostratigraphy of Colombia and the Gulf of Mexico. After graduating from the National University of Colombia (Bogotá, 1992), she received an M.S. degree from the University of Colorado at Boulder (1995). She is a member of AAPG.

### Jennifer R. Crews

Jennifer R. Crews received her bachelor's degree in geology from the University of Colorado at Boulder in 1995; she will receive her M.S. degree during the summer of 1998. Her thesis topic is "High-resolution biostratigraphic analysis of Green Canyon and Ewing Bank areas, northern Gulf of Mexico." Upon graduation, she will be employed by Texaco Exploration and Production, Inc., New Orleans. Research interests include applied biostratigraphic methods and sequence analysis.

### Andrew J. Pulham

Andy Pulham has a B.Sc. in physical geography and geology from Liverpool University, United Kingdom and a Ph.D. in geology from the University of Wales, United Kingdom. After graduating, he spent 12 years as an international petroleum sedimentolgist for British Petroleum. He has worked in the North Sea and South America and spent over three years in Houston working the deep-water Gulf of Mexico. Currently, Pulham is leading a Reservoir Geology Research Consortium within the Energy and Minerals Applied Research Center in Boulder, Colorado. This is an industry supported project and is a collaborative venture with S. George Pemberton and his Ichnology Research Group at the University of Alberta.

# Biostratigraphic Techniques for Analyzing Benthic Biofacies, Stratigraphic Condensation, and Key Surface Identification, Pliocene and Pleistocene Sediments, Northern Green Canyon and Ewing Bank (Offshore Louisiana), Northern Gulf of Mexico[1]

**Tomas Villamil,[2] Claudia Arango,[3] Paul Weimer,[4] Art Waterman,[5] Mark G. Rowan,[4] Peter Varnai,[6] Andrew J. Pulham,[4] and Jennifer R. Crews[4]**

## ABSTRACT

The northern deep Gulf of Mexico is a geologically complex province consisting of Neogene intraslope minibasins created by sediment loading onto and evacuation of allochthonous salt. Sedimentary fill in the minibasins consists of bathyal turbidite systems with highly variable facies distribution. In this paper, we present three biostratigraphic techniques developed to increase our understanding of the geologic evolution of the northern Green Canyon and Ewing Bank lease areas.

The first two techniques address the nature of stratigraphic condensation in minibasins and the relative areal extent of the condensed sections. The third technique uses benthic foraminifers to improve the stratigraphic knowledge of the area. (1) The first technique is concerned with sediment accumulation plots from wells helping to identify possible areas of condensation based on variations in the rate of accumulation. By displaying several wells together, the areal extent of the condensed section can be identified, and the causes for the condensation can be evaluated. (2) The second technique addresses condensation within one minibasin during a 2.5-m.y. interval. Relative abundance values are plotted from a series of wells at selected time surfaces and tied to seismic horizons that are correlated throughout the basin. This allows plotting a series of maps that show how stratigraphic condensation changes throughout a basin. (3) The third technique develops benthic biofacies maps that more accurately reflect the highly irregular paleobathymetry of the slope at the time of deposition. These maps were created by plotting the benthic biofacies of the sediments recovered from a well at the stratigraphic extinction of 29 selected planktonic foraminifer and calcareous nannoplankton species. The maps are integrated with lithostratigraphic and structural palinspastic reconstructions. The resulting maps are interpreted to more accurately reflect benthic biofacies of the continental slope at specific geological times during the Pliocene and Pleistocene.

## INTRODUCTION

Biostratigraphy is an important aspect of basin analysis, especially when integrating seismic



[1]Manuscript received January 6, 1997; revised manuscript received November 10, 1997; final acceptance November 19, 1997.

[2]Energy and Minerals Applied Research Center, Department of Geological Sciences, University of Colorado, Boulder, Colorado 80309-0250. Present address: Advance Exploration Organization, Conoco, Inc., P.O. Box 2197, Houston, Texas 77252-2197.

[3]Energy and Minerals Applied Research Center, Department of Geological Sciences, University of Colorado, Boulder, Colorado 80309-0250. Present address: Marathon Oil Company, P.O. Box 3128, Houston, Texas 77252-3128.

[4]Energy and Minerals Applied Research Center, Department of Geological Sciences, University of Colorado, Boulder, Colorado 80309-0250.

[5]Paleo-Data, Inc., 6619 Fleur de Lis Dr., New Orleans, Louisiana 70124.

[6]Energy and Minerals Applied Research Center, Department of Geological Sciences, University of Colorado, Boulder, Colorado 80309-0250. Present address: Exxon Exploration Company, P.O. Box 4480, Houston, Texas 77210.

We thank the following companies and their many representatives for their support and input: Agip, Anadarko, Amoco, BHP, BP Exploration, Burlington Resources, Chevron USA, CNG, Conoco, CXY Energy, Enterprise, Exxon, Marathon, Mobil, Occidental, Pan Canadian, Pennzoil, Petrobras, Phillips, Shell, Texaco, Total, Union Pacific, Unocal, and the donors of the Petroleum Research fund, administered by the American Chemical Society. Their support has been crucial for the success of this research. We extend special thanks to Halliburton Geophysical (now Western Geophysical), specifically to Bob Graebner, John Anderson, and Gary White, for the seismic data used in this project. We are indebted to Paleo-Data, Inc. for use of their biostratigraphic data. Digital logs were kindly provided by Marathon (Tricia Santogrossi). Software donations from GeoQuest, CogniSeis Development (Pat Poe, Roy Kligfield, Robert Ratliff), Landmark Graphics (Thom Fisher), and Zeh Graphics (Jerry Martin) were also essential to the success of our study. Thanks to Ryan Crow for his help with drafting some of the figures. We thank Conoco for defraying costs of color figures. We thank Laurie Lamar, M. S. Wacker, and AAPG reviewers Scott W. Starratt, Finn Surlyk, Kevin Biddle, and an anonymous reviewer for their valuable comments that improved the quality of this paper.

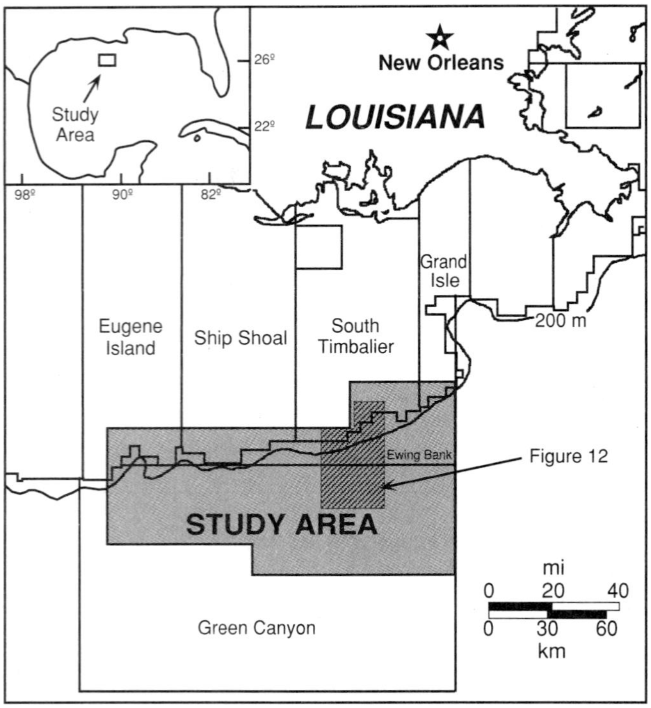


**Figure 1—Map of the study area, northern Green Canyon and Ewing Bank, northern Gulf of Mexico. Location of minibasin presented in Figure 12 is shown.**

reflection data and wireline logs for establishing the chronostratigraphic framework of a sedimentary basin. Biostratigraphy, however, is underused in the petroleum industry. Oil companies tend to outsource most biostratigraphic analyses, and fewer people have time to integrate and extract all the information biostratigraphy provides in the accelerated world of lease sales or acreage evaluation.

The Neogene biostratigraphy of the northern Gulf of Mexico is one of the best subsurface data sets in the world. The biostratigraphic zonations are based primarily on planktonic foraminifers and calcareous nannoplankton. These zonations give some of the best stratigraphic resolution in any sedimentary basin. Important stratigraphic concepts developed within the last decade include the recognition of a hierarchy in the development of localized vs. more regional condensed sections and their causes (cf. Vail and Wornardt, 1991; Pulham et al., 1993); the occurrence of condensed sections and planktonic microfossil last appearance data (LAD) within systems tracts; high vs. low frequency sequences; and the effects of eustatic fluctuations on benthic biofacies patterns (Armentrout, 1991).

This paper builds upon and refines these earlier biostratigraphic concepts for the Gulf of Mexico to develop a more accurate application of biostratigraphy toward integrated basin analysis. Our challenge was to fully use the biostratigraphic data available, so that we could better understand the geologic evolution of our study area in northern Green Canyon and Ewing Bank areas (Figure 1), and complement the other aspects of basin analysis (see Weimer et al., 1998, this issue; McBride et al., 1998, this issue; Rowan and Weimer, 1998, this issue). Although some of the results are specific to this basin, the general techniques and approaches are applicable to other basins in which high-resolution biostratigraphic information is available.

The objectives of this paper are to (1) illustrate the use of sediment accumulation plots for identifying different key surfaces and systems tracts; (2) evaluate how stratigraphic condensation occurs within one minibasin and how sites of condensed sedimentation shift through time; (3) demonstrate a technique of estimating biofacies of a sedimentary basin with significant variations in slope bathymetry during deposition; and (4) develop a modified model for benthic biofacies for northern Gulf of Mexico.

## DATA SET

The biostratigraphic data used for this study come from analyses of the cuttings from exploration wells that generally penetrated 3000–6500 m (10,000- 20,000 ft) of partially consolidated, middle Miocene–Pleistocene sediments. In establishing the regional sequence stratigraphic framework, we analyzed and synthesized planktonic and benthic foraminifera reports from 180 wells, calcareous nannoplankton reports from 83 wells, and high-resolution foraminifer analyses from 60 wells (Figure 2).

In addition to traditional first downhole occurrence biostratigraphy, the database also contains information on nannofossil abundance. Nannofossil abundance data can be used with traditional biostratigraphy in the interpretation of the paleoenvironment, in depicting regional stratigraphic trends, and in generating the chronostratigraphy of a region. The data used here are internally consistent because the same group of paleontologists analyzed all samples. The biostratigraphic zonation that was used for this paper is from Paleo-Data, Inc. (1993) (Figure 3). This biostratigraphic data set has been essential for dating the sequences, correlating abundance curves to seismic profiles, identifying condensed sections, and establishing benthic biofacies for each sequence.

## REGIONAL SETTING

The regional stratigraphic setting of the study area is summarized by Weimer et al. (1998, this

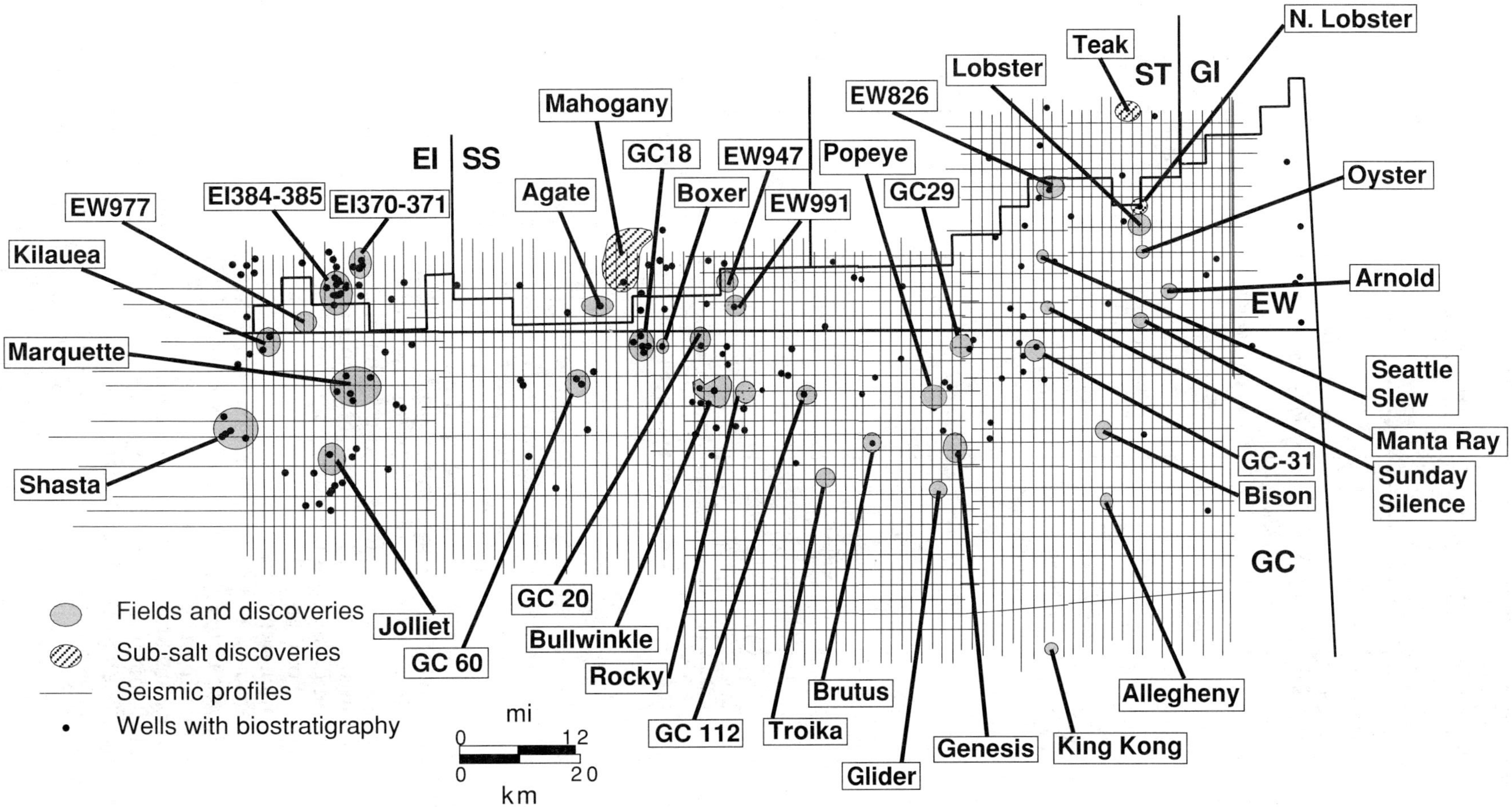


Figure 2—Map showing the location of the seismic and well database used for this study.

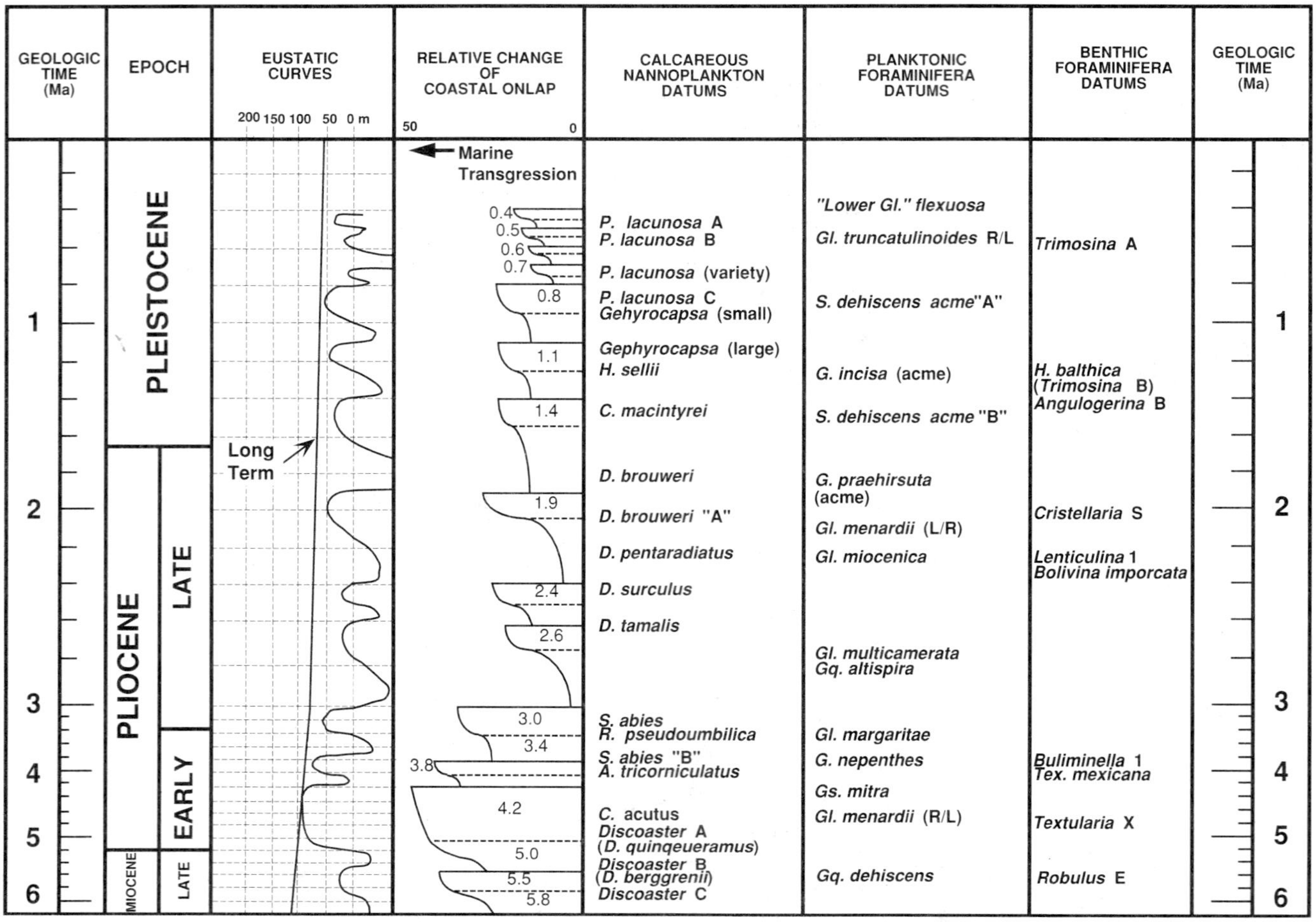


**Figure 3—Onlap chart and biozonation for the Neogene sediments, northern Gulf of Mexico. After Paleo-Data, Inc. (1993).**

issue), and the structural setting by Rowan (1995) and McBride et al. (1998, this issue). These are briefly reviewed here to establish the geologic setting for biostratigraphic analyses.

About one-half of the drainage area of North America delivered sediment into the northern Gulf of Mexico during the Pliocene and Pleistocene (Pulham, 1993). Up to 7600 m (25,000 ft) of sediments accumulated in northern Green Canyon and Ewing Bank areas during the past 5.5 m.y., indicating high rates of deposition. Deposition occurred primarily in turbidite systems on the lower to upper slope for most of the Pliocene and Pleistocene. Neritic to nonmarine settings were present only in the upper Pleistocene sequences in the northwestern part of the area. There were significant shifts in the shallow-marine depocenters that fed the slope turbidite systems during the Pliocene and Pleistocene, thus affecting local rates of deposition, as well as the kinds of sediments delivered to the basin.

Sediments were deposited on an unstable substrate affected by extensive allochthonous salt bodies and related faults. The presence of salt has tremendous effect on the distribution of sediments and on the biofacies patterns. The timing of sedimentation in the slope was largely controlled by eustatic variations during the Pliocene and Pleistocene. The shifts in the water mass and circulation patterns associated with eustatic variations had profound effects on the foraminiferal distribution patterns. All of these interdependent factors are reflected in the biostratigraphy of the area.

## INTERPRETING BIOSTRATIGRAPHIC DATA AND SEQUENCE STRATIGRAPHY

Biostratigraphic data are an important component of the sequence stratigraphic interpretation in deep-water settings, where sediments were deposited in partially isolated intraslope minibasins. Planktonic foraminifera and calcareous nannoplankton last appearance data (LAD) and high-resolution faunal abundance curves were used for identifying condensed sections and sequence boundaries, and for correlating minibasins. Biofacies analyses were used combined with facies interpretations derived from seismic and well log studies.

Condensed sections have two discrete signatures on wireline logs. First, a slight increase in radioactivity ("hot" shales) observed on the gamma-ray logs commonly corresponds to increased fossil content on faunal abundance curves (Figure 4). These faunal abundance peaks in shales are considered to represent condensed sections dominated by clay with little or no sand and silt. The slight increase in radioactivity is probably caused by low depositional rates that allow concentration of trace and rare elements relative to underlying and overlying stratigraphy. In addition, local sea bottom conditions could produce a reducing environment favoring concentration of trace amounts of radioactive elements.

Second, hemipelagic carbonate-rich zones are identified by decreased gamma-ray response and muted SP response that indicate little or no effective or connected porosity and limited permeability. These lithologies have super abundant levels of foraminiferal tests and nannofossils in a clay matrix (Pulham et al., 1993). In these planktonic foraminiferal oozes, the density and sonic log responses typically differ from the overlying and underlying sediments. Transit time can be either considerably larger or smaller than surrounding shale and sand layers. Large impedance contrasts can commonly give rise to the high-amplitude and continuous seismic reflection associated with these types of condensed sections. Also, the first downhole occurrences of many index fossils correspond to these zones. These deep-water hemipelagic carbonates are interpreted to reflect extremely low rates of siliciclastic sedimentation, during which pelagic sedimentation and benthic accumulation of fauna occurs undiluted by siliciclastic gravity flows and suspended clays and silts. Foraminiferal and nannofossil oozes are the result of such time periods and are the most useful lithologies for regional and subregional correlations.

In this deep-water setting, regionally significant condensed sections developed during deposition of the transgressive and highstand systems tracts (including the maximum flooding surfaces), and during major delta avulsions when little sediment was delivered to any particular region of the slope. Sequence boundaries overlie condensed sections as correlative non-erosive surfaces separating sequences in the middle to lower slope (Figure 4).

The abundance curves were correlated with the relative onlap curve using LAD (Figure 3). This correlation shows that abundance peaks almost always correspond to the condensed sections directly below the sequence boundaries. Groups of calcareous nannoplankton LAD usually are associated with these condensed sections (abundance peaks); hence, groups of LAD generally indicate the position of the condensed sections immediately below the sequence boundaries.

LAD and true extinctions tend to occur within the condensed section in deep-water settings. True extinctions that occurred during times of transgression and highstand are recorded as LAD in condensed sections. The combined duration of the transgressive and highstand systems tracts may be

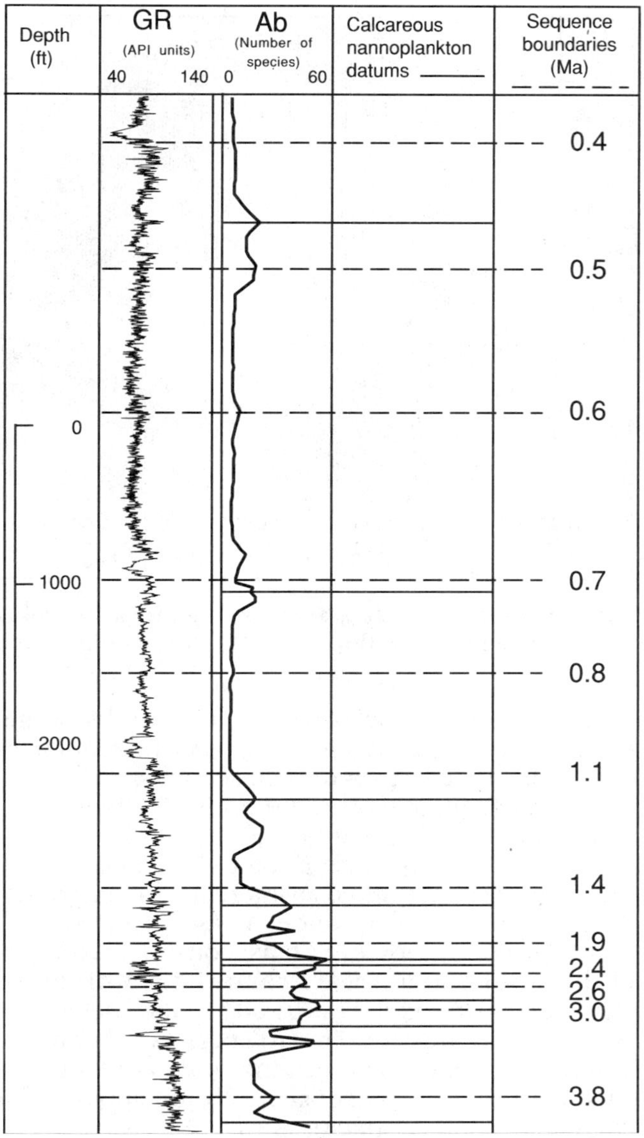


**Figure 4—Gamma ray (GR), calcareous nannoplankton abundance curve (Ab), calcareous nannofossil last appearance data (LAD) (solid lines), and interpreted sequence boundaries (dashed lines) from a representative well in the study area.**

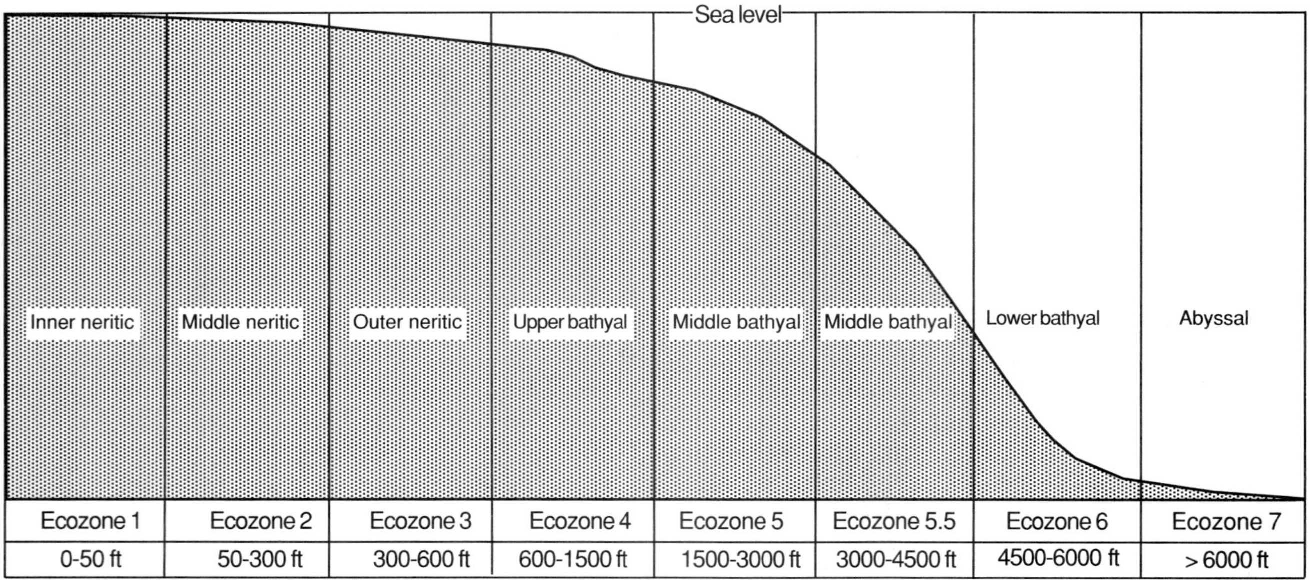


**Figure 5—Schematic diagram showing the paleoecologic zonation of benthic foraminiferal assemblages along a depositional profile in the Gulf of Mexico (modified from Paleo-Data, Inc., 1993).**

longer than the duration of the lowstand systems tract, thus increasing even further the geological time represented in condensed sections.

Several sequences contain multiple abundance peaks in the biostratigraphic data (Figure 4; 1.4–1.1- and 0.7–0.6-Ma sequences). Some abundance peaks, therefore, are interpreted as internal condensed sections within one sequence. These internal condensed sections, however, have considerably lower faunal abundance values than the major, more regionally distributed condensed sections below the sequence boundaries. Due to sparse well control, the internal condensed sections could not be correlated in the area. They are generally not apparent on two-dimensional seismic profiles.

Internal condensed sections are present in most sequences and occur locally on the top of sheet sands and within the levee or overbank sediments of the slope fans. Given the high rates of deposition within any one sequence, these local condensed sections develop after minor avulsions within individual channel systems and after major avulsions (deposition switched to another minibasin). These avulsions are the result of autocyclic processes. Condensed sections developed with limited areal extent in the abandoned areas. Also, in the younger sequences (<0.7 Ma), condensed sections are interpreted locally to separate the slope fan facies from the prograding complex, reflecting a slight hiatus between the sedimentation on the slope and the overlying prograding shallow marine sediments.

Biostratigraphic data also help in the interpretation of paleowater depths. Biofacies zones are based on specific benthic foraminiferal assemblages. These zones indirectly suggest water depth of the sea floor at time of deposition (Figure 5). This information is essential for accurate depositional environment interpretation.

## SEDIMENT ACCUMULATION PLOTS AND SEQUENCE STRATIGRAPHY

### Sediment Accumulation Plots

Sediment accumulation plots were used to help interpret the location of stratigraphic condensation, key surfaces, diffuse boundaries between systems tracts, and to evaluate the regional significance of the condensed sections (Figures 6–8). A sediment accumulation plot is constructed by plotting all LAD from a well so that absolute geologic time is on the vertical axis and stratigraphic thickness is on the horizontal axis (Figures 7, 8) (see papers in Mann and Lane, 1995). After the data are in computers, sediment accumulation plots can be generated quickly and can be integrated quickly into routine exploration work.

On the sediment accumulation plot, the line of correlation (LOC) is derived from crossplotting the theoretical position of microfossil extinctions or LAD in the "absolute" time domain vs. the stratigraphic location of extinctions. The slope of the different segments of the LOC reflects the history of sediment accumulation not decompacted. Line segments with steep slopes represent relatively low sedimentation rates with small thickness accumulating during long periods of time. Gently sloping

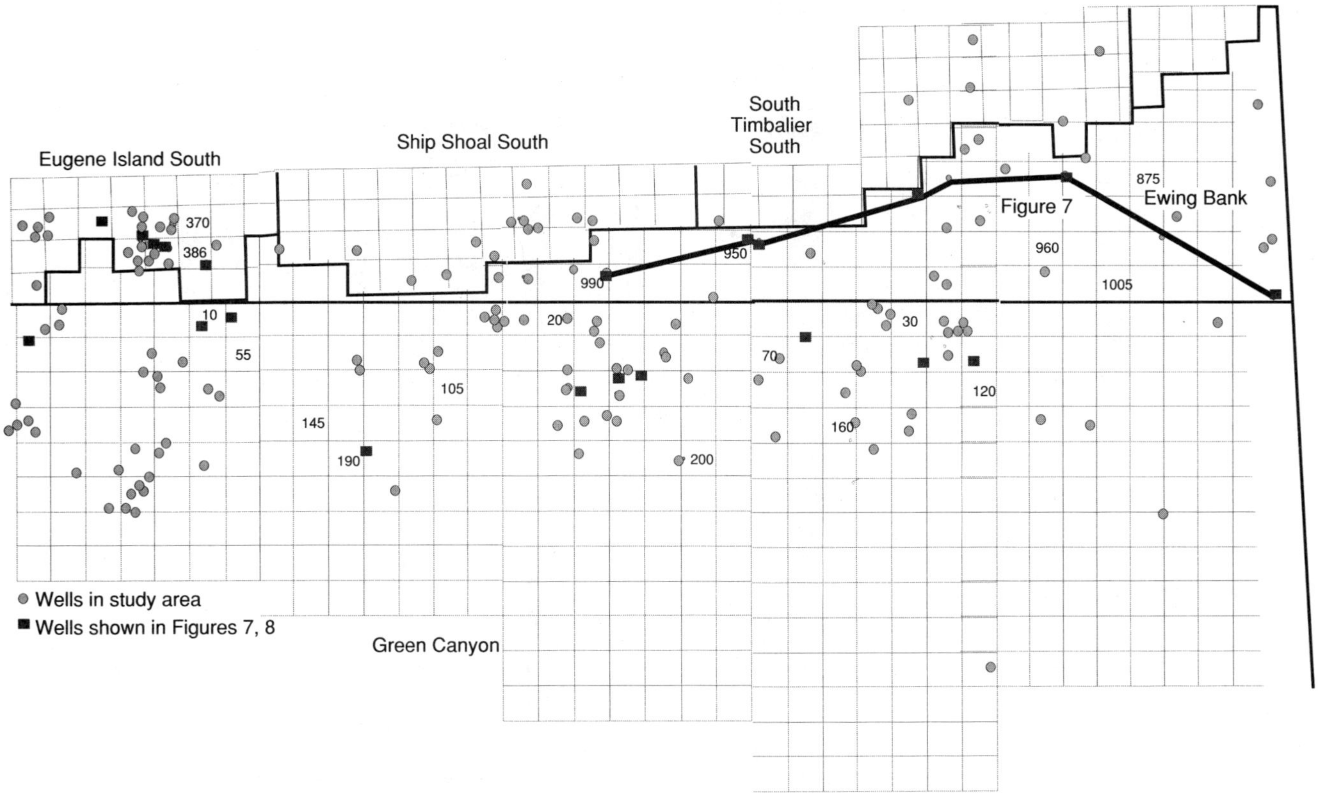


**Figure 6—Location map of wells used for sediment accumulation plots shown in Figures 7 and 8. Lease block numbers also are shown.**

segments of the line represent periods of high accumulation rates, i.e., thick packages deposited during short periods of time.

## Local vs. Widespread Sequence Stratigraphic Features

To evaluate the relationship between sediment accumulation plots and sequence stratigraphy, seventeen plots are shown from the study area for the Pliocene and Pleistocene (Figures 6–8). This method allows for sequence stratigraphic interpretations at the regional, three-dimensional scale because several sedimentation history plots can be analyzed simultaneously in the same chart.

Sediment accumulation plots from seven wells in the northeastern part of the area indicate two different condensed sections (Figure 7). Maximum condensation occurred from approximately 2.8 to 1.4 Ma. A more localized condensed section, delineated by the near vertical line (~1.2–0.87 Ma), is present in the EW 952 and 991 wells, with the coeval units in nearby wells characterized by relatively high sedimentation rates. Plots from additional wells show that the older condensed section is of longer duration, having started at approximately 3.5 Ma (Figure 8). This condensed section gradually disappeared in the west at approximately 2.0 Ma; condensation ended to the east at about 1.45 Ma. Plots of the GC 49, GC 110, and GC 190 wells also show a condensed section from approximately 1.2 to 0.87 Ma (Figure 8).

This major condensed section spans four depositional sequences (3.0, 2.6, 2.4, 1.9 Ma; Figures 3, 8). The origin of this condensed section is caused by major allocyclic changes associated with a regional transgression and shifting of the deltaic depocenter that fed the area (Pulham, 1993; Weimer et al., 1998, this issue). Data points from 33 plots and wells shown in Figures 6–8 were used to create a map that shows the general architecture of the two most widely-spread condensed intervals (Figure 9). The condensed section from approximately 3.5 to 1.45 Ma is better developed in eastern portions of the area, but loses its character westward where active accumulation was taking place. The condensed section from approximately 1.2 to 0.87 Ma occurs in the western regions of the basin, and loses its character eastward where more active sedimentation was taking place. This regional change in condensation through time is

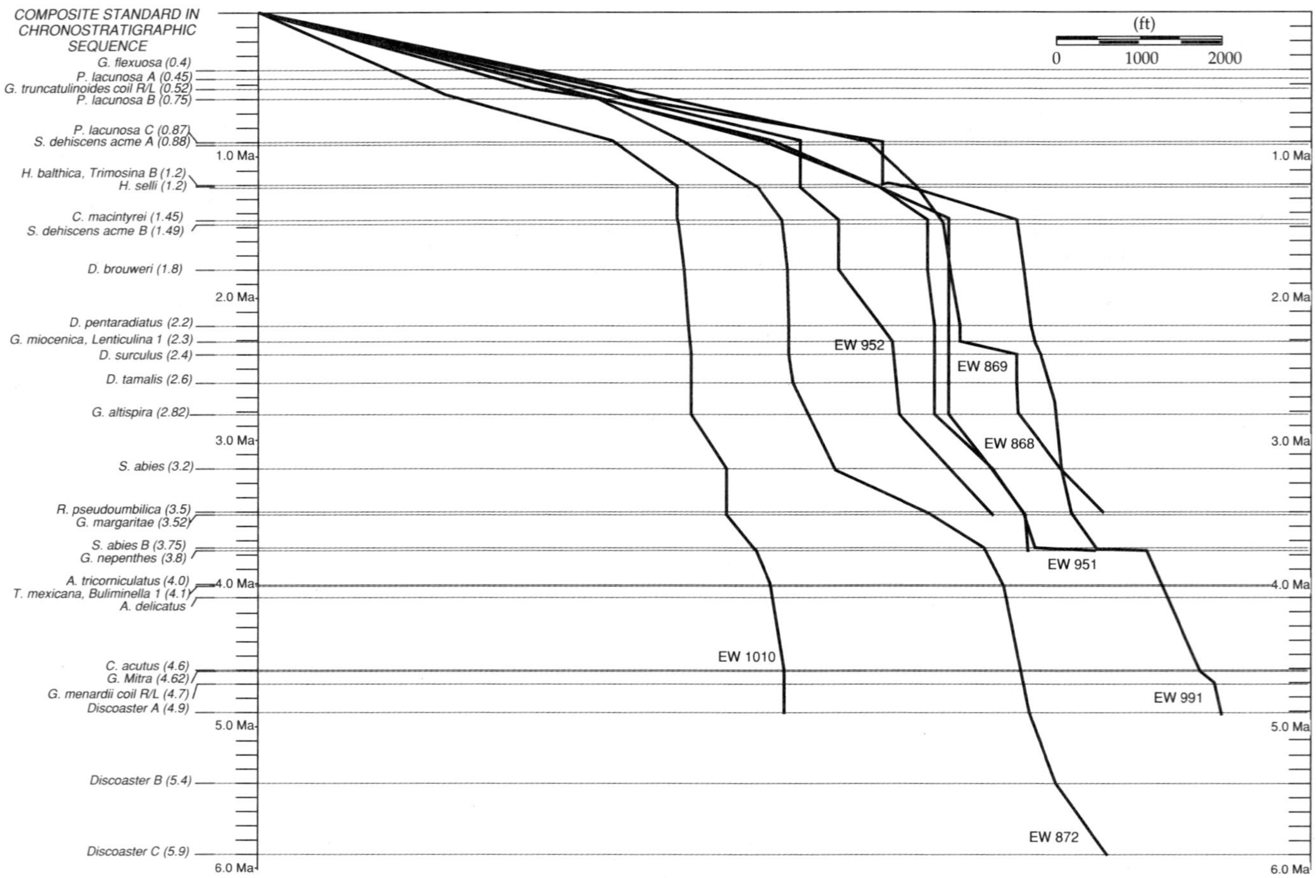


**Figure 7—Sediment accumulation plots for Ewing Bank wells EW 868, EW 869, EW 872, EW 951, EW 952, EW 991, and EW 1010. Vertical axis is geological time and was derived by using graphical correlation (Mann and Lane, 1995). Horizontal axis represents the thickness of sediments in the well in ft. Condensation occurred between 3.5 and 1.45 Ma throughout the area, and from 1.2 to 0.87 Ma in the western region. Note the abrupt increase in rates of sedimentation in EW 869 at about 2.4 Ma. See Figure 6 for location of wells.**

interpreted as reflecting avulsion of the shallow marine sediment source from west to east. In addition, the compilation of sediment accumulation plots in Figure 8 also shows a major increase in sedimentation at about 0.87 Ma, caused by the influx of prograding shallow marine sediments north of the area. Wells located in distal regions are more condensed (steeper slope) than proximal locations. Wells located in distal regions are more condensed (steeper slope) than proximal locations (Figure 8).

## Sequence Stratigraphy and Graphic Correlation: Conceptual Model

Reviewing the regional accumulation plots from different minibasins with different sedimentation histories, we observed a relationship between key sequence stratigraphic surfaces and systems tracts and variations in sedimentation rate (Figures 7, 8). A conceptual model shown in Figure 10 illustrates the relationship between sedimentation rate and sequence stratigraphic divisions for bathyal sedimentation. In basinal regions, such as the study area throughout the Pliocene and Pleistocene, most of the active deposition occurred during times of lower sea level. This relationship is reflected in the sediment accumulation curves as relatively gently sloping segments (Figure 10).

The lowstand systems tract has been divided into three components: basin-floor fan, slope fan, and prograding complex (Vail, 1987; Mitchum et al., 1993). The possible occurrence of basin-floor fans can be evaluated by analyzing several sediment accumulation curves. Our results indicate that the basin-floor fan has the highest rate of deposition, and can be identified as a gentle slope in the line of sediment accumulation. As shown in Figure 8, several basin-floor fans developed in different regions at different times (see Weimer et al., 1998,

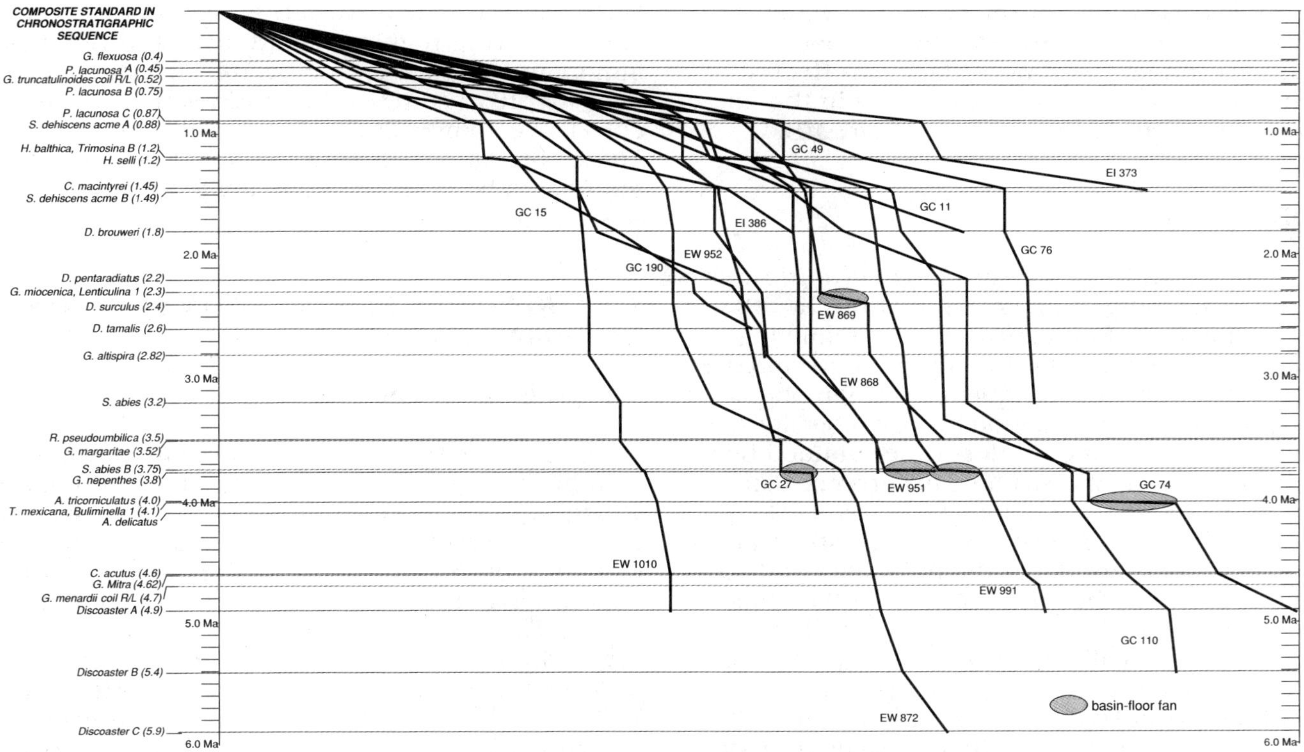


**Figure 8—Compilation of sediment accumulation plots for 17 wells in the area. Note the several marked changes in the slope of the line of correlation in some regions of the basin at different times marking high and low rates of deposition. These abrupt increases are interpreted as representing basin-floor fans, and are corroborated by wells (see Weimer et al., 1998, for corresponding maps). The slight offset in the position of basin-floor fans along a sequence boundary is an artifact of the graphic correlation technique. Note the major increase in sedimentation rates at about 0.87 Ma. See Figure 6 for location of wells.**

this issue, for the complete distribution of basin-floor fans in the study area). Sediment accumulation plots can, therefore, be used to help in identifying basin-floor fans.

The slope fan and the prograding complex cannot be differentiated clearly and consistently from one another based on changes in the slope of the accumulation curve, probably because they are characterized by similar sedimentation rates, the lack of prograding complexes (i.e., deltas) in sequences older than 0.7 Ma, or because differentiating these two systems is beyond the resolution of present-day biostratigraphy, that is, there are not enough points on the line of correlation (Figure 10). The sediment accumulation lines for these two divisions of the lowstand systems tract are relatively gently sloping, but are steeper than the basin-floor fan (Figure 10, upper portion). The model presented here is simplified, but fits the data and examples outlined. This model, however, does not consider minor variations in stratigraphy nor the notion that boundaries between components of the lowstand systems tracts are diffuse.

The change between deposition of the lowstand systems tract and the overlying condensed transgressive and highstand systems tracts is dramatic, and can be easily identified with sedimentation rate curves (Figure 10). In distal regions of the basin, such as the study area, throughout the Pliocene and Pleistocene, transgressive systems tracts are characterized by sediment starvation because most of the sediments are trapped in proximal areas. The maximum flooding surface is recorded in deep water as a condensed section. Therefore, the transgressive and the highstand systems tracts are represented by a single, condensed section. Sedimentation rate is characterized by a segment with a steep slope (Figure 10).

The easiest surfaces to identify using sediment accumulation curves are the sequence boundaries and the maximum flooding surfaces (Figures 8, 10). The sequence boundary marks the base of the lowstand systems tract and is marked by an abrupt change from low to high sedimentation rates. The sediment accumulation line shows a marked change of slope from steep to gentle, representing the sequence boundary. The change happens regardless of the type of sediments and systems tract overlying the sequence boundary. The maximum flooding surface is reflected by a change from gentle to steep slopes in the line of sediment accumulation history (Figure 10).

### Discussion

The model presented here complements sequence stratigraphic interpretations; it can also be applied as a first approach to wireline log data when seismic data are not available for independent interpretations. Sequence stratigraphic interpretations of sedimentation history plots must consider that the shape of plots is determined by the correlation of first downhole occurrence data (last appearances or extinctions), but these LAD do not necessarily coincide with sequence stratigraphic surfaces. This introduces the possibility of discrepancies between the "real" position of the sequence stratigraphic surface and the position identified by analyzing the sedimentation history plot. In practice, however, even though the position of a key surface potentially may be offset from its real depth, sediment history curves change so dramatically that surfaces identified using sediment accumulation plots are generally valid.

The shape of the sedimentation plot in relationship with sequence stratigraphic surfaces varies dramatically across the basin. The model shown in the upper portion of the Figure 10 applies only for distal regions of the basin, i.e., the position of the hypothetical well and surrounding areas. As shown in the lower portion of Figure 10, lateral changes in thickness of systems tracts directly affect the sedimentation history plot. The slopes of line segments for deep-water deposition are not applicable to the proximal setting where the lowstand systems tract is characterized by lower thicknesses caused by sediment bypass or erosion. The highstand systems tract is relatively thick. Note that a thick deep-water stratigraphic package deposited during a lowstand in sea level has a similar slope to a thick stratigraphic package deposited in shallower water during a highstand in sea level.

There is widespread areal distribution of true extinctions during times of highstand in sea level because they are preserved in proximal regions within the TST (transgressive systems tracts) and HST (highstand systems tracts), and within the condensed section in distal regions (Figure 11). In contrast, the distribution of real extinctions during times of lowstand is restricted to the regions where LST (lowstand systems tracts) deposits accumulated. In proximal areas, apparent tops can be found associated with the termination of the HST immediately below the sequence boundary (Figure 11). Differentiating the distribution of taxa that went extinct during highstand and lowstand times is important for sequence stratigraphy because it can help identify sequence boundaries that become correlative conformities within similar rock types. These correlative conformities cannot be easily differentiated based on wireline log petrophysics or seismic data.

## STRATIGRAPHIC CONDENSATION BASED ON ABUNDANCE MAPS

### Mapping Abundance Values

An important concept that is now used routinely in biostratigraphic studies is the recognition of stratigraphic condensation from the biostratigraphic reports from wells; these condensed sections are important for interpreting local and regional stratigraphic patterns. Peaks in nannofossil abundance have been interpreted as representing condensation events and used for regional chronostratigraphic correlations (Loutit et al., 1988). The value of these correlations, however, depends on the nature of these condensation events. If condensation events are regional in nature and are related to allocyclic processes, then abundance peaks can be used as correlation horizons. Biostratigraphic data in the study area show that many abundance peaks are localized areally and respond to autocyclic condensation events, such as lateral shifts in sedimentation, development or destruction of minibasins, and the history of paleotopography of the sea bottom. The goal was to develop a technique to evaluate the evidence for local condensed sections using nannofossil abundance data and to discuss some consequent pitfalls associated with correlating based uniquely on abundance peaks.

One intraslope minibasin in eastern Ewing bank and northern Green Canyon was selected as a case study to evaluate these concepts because of the large number of wells, the relatively simple structure of the basin with relatively minor faults and salt bodies, adequate seismic-to-well ties, and easily correlated seismic reflections (Figure 12).

A series of maps were created in the following manner. (1) Abundance values from wells (e.g., Figure 4) were plotted onto seismic data using time-based logs. (2) Abundance values were then transferred into plane views of seismic horizons (shotpoint maps). These maps show the abundance values plotted at several mapped seismic reflections between 3.4 and 0.5 Ma (Figure 12). Analysis of the maps reveals the development of isolated condensation (areas with light, high abundance, colors surrounded by dark, low abundance

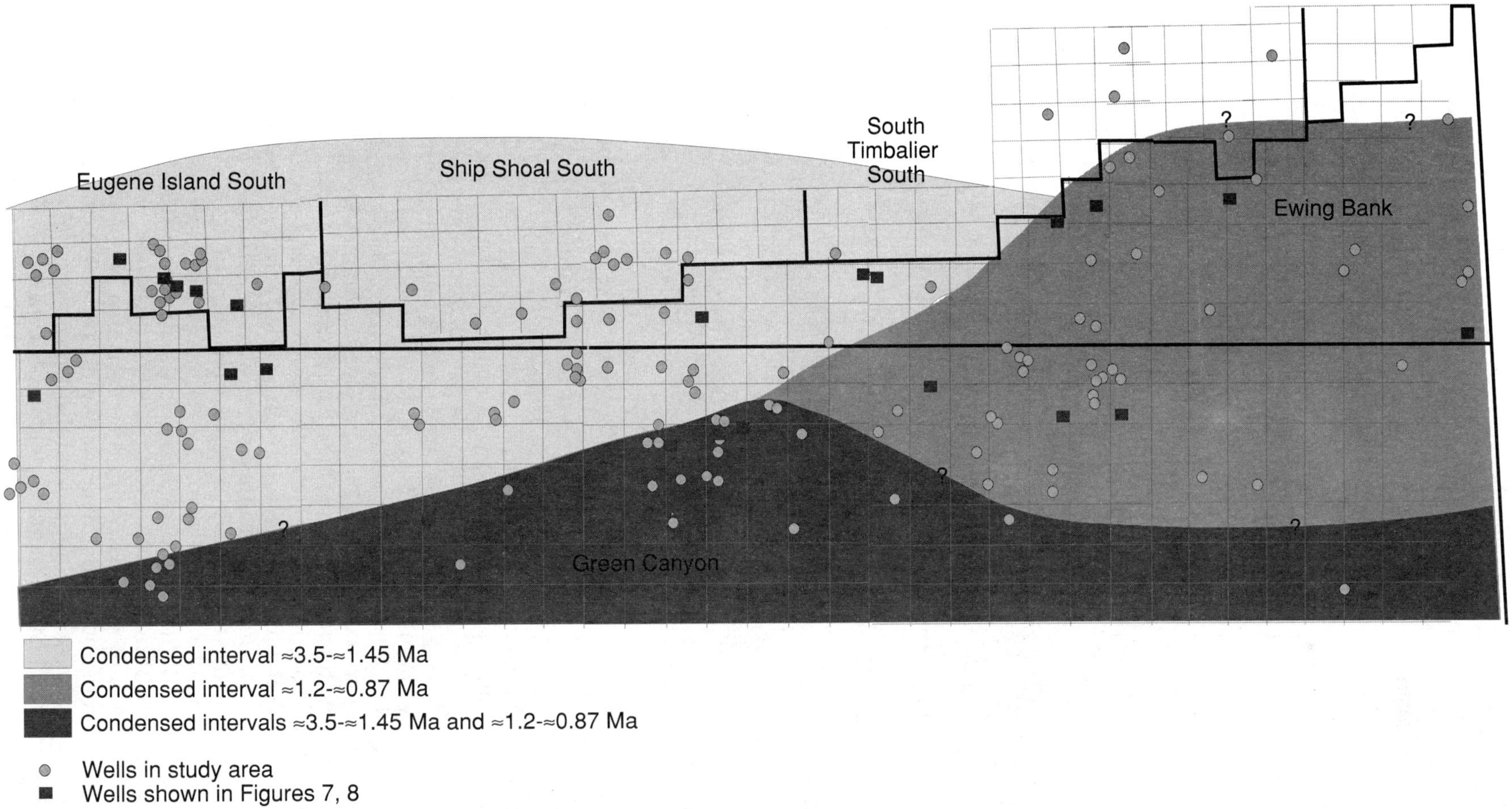


**Figure 9—Map showing the distribution of the two regional condensed intervals in the study area. The condensed interval lasting from 3.45 to 1.45 Ma is better developed in the eastern region of the area and loses its character westward where active sedimentation was taking place. The condensed section from 1.2 to 0.87 Ma is best developed in the western region of the basin and loses its character to the east. This regional change in condensation is caused by the shifting of the shallow marine depocenter (i.e., Mississippi Delta) from west to east. The large dark area to the south is where both condensed sections overlap.**

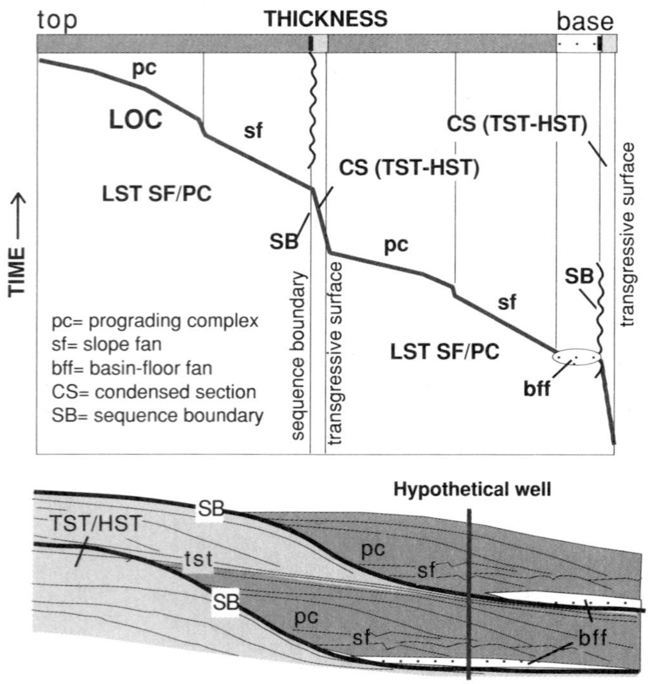


Figure 10—Conceptual model for interpreting sequence stratigraphy from sediment accumulation plots. The model is derived from the analysis of 33 wells. The lower portion of the figure presents a simple sequence stratigraphic model. The upper portion of the figure has time in the vertical axis and thickness in the horizontal axis. The slope of the line of correlation (LOC) is representative of sediment supply; gentle and steep slopes represent fast and low sedimentation rates, respectively. Systems tracts occur in direct relationship with sediment supply; the model links both.

values) within portions of the basin. The development of regional condensed sections can also be observed in maps that show the entire area covered by high abundance values (e.g., 2.5 Ma, Figure 12G).

The observation of abundance maps in multiple time slices allows for a three-dimensional evaluation of abundance values. The areas of concentrated abundance shift through time. We interpret these shifts to be associated with the changing geometry of sediment transport through the minibasin in a series of channel-levee systems and related environments. A three-dimensional view of abundance reveals how correlating abundance peaks alone on well logs (one and two dimensions) might cause interpretation problems. Correlations based on abundance values, therefore, may be problematic. Miscorrelation of abundance peaks could cause problems at the field scale, which could lead to a misinterpreted distribution of reservoirs and connectivity.

### Pitfalls

Even though meaningful results were derived from this technique in this case study, there are five possible errors that one must be careful to avoid. (1) Different paleontologists may prepare and identify the calcareous nannoplankton differently, which will affect the relative abundance. However, for this data set, two paleontologists worked together and applied the same interpretation techniques. (2) The sampling interval in the wells for biostratigraphy was 30 m (90 ft). This may not offer enough resolution to make detailed interpretation of the high-resolution results. The application of the method may be pushed beyond the limits of resolution. Closer sampling intervals (less than 10 m or 30 ft) are recommended in utilizing this technique. (3) Variable dilution of samples from drilling muds and variability in smear slide density during preparation can affect the magnitude of observed fossil abundance values. Recognizing how samples are collected and the mud system used in the well is important in utilizing this technique. (4) There is a potential error for taking abundance values at time horizons based on the highest observed occurrence in a well. If a biostratigraphic top is picked one or two samples too low, it could make a big difference in the abundance value used. Careful interpretation of the sample density must be kept in mind in utilizing this technique. (5) The curves generated in wells reflect relative abundance (e.g., Figure 4); therefore, error may come into treating the data in an absolute sense.

## BENTHIC BIOFACIES TIME-SLICE MAPS

### Technique

Traditional biofacies maps show the average biofacies of a sedimentary unit. In sedimentary basins that are strongly affected by syndepositional tectonics with abrupt bathymetric variations, traditional biofacies maps commonly smooth out the effects of the bathymetric variations.

Early in the study of our area, the potential effects of bathymetric highs associated with salt tectonics were identified in averaged biofacies maps of depositional sequences (Acosta, 1994; Martinez, 1994). Based on structural restorations and analogies to the modern slope (Rowan, 1995; McBride et al., 1998, this issue), we considered the paleoslope to have also consisted of prominent bathymetric highs (salt features) and lows (minibasins), and the topography of the slope to be constantly changing associated with sedimentation and deformation. The proposed benthic biofacies time-slice maps reflect paleobathymetric effects of the slope and can be integrated with independent analyses (e.g., Weimer et al., 1998, this issue; Rowan and Weimer, 1998, this issue).

The technique presented here requires several iterations among biostratigraphy, lithostratigraphy, and structural geology to arrive at the best interpretation for benthic biofacies time-slice maps. The technique allows us to evaluate the effects of bathymetric changes within basins and develop more realistic maps. The maps were generated using the following approach. (1) The biofacies zones were interpreted and plotted on a map of the wells at 29 calcareous nannoplankton and planktonic foraminifera extinctions or LAD in the Pliocene and Pleistocene. Biofacies boundaries were then contoured as a first approximation. (2) The interpreted depositional fairways showing how sediment was transported into the basin (Weimer et al., 1998, this issue) were then integrated with the biofacies maps to evaluate the effects of displaced shallow-water fauna. (3) The results of qualitative palinspastic restorations of the sea floor and the changing position of bathymetric highs and lows through time were integrated (Rowan and Weimer, 1998, this issue). (4) Biofacies boundaries were then recontoured based on the input from the other data sets. This iterative approach allows for a much more robust and accurate interpretation.

The biofacies zonations (ecozones) used here are listed in Figure 5. Most maps shown here reflect sediments deposited in bathyal settings, which are subdivided into ecozone 4 (upper bathyal), ecozones 5 and 5.5 (lower bathyal), and ecozone 6, (lower bathyal).

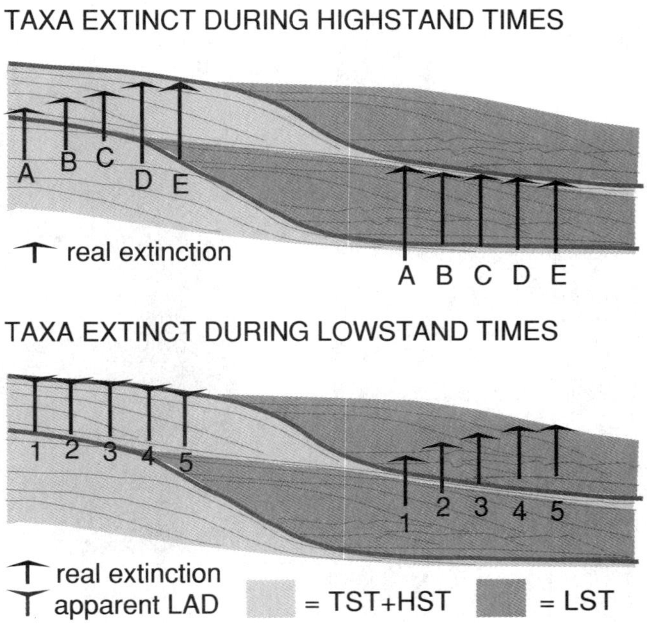


**Figure 11—Conceptual biostratigraphic model for taxa that become extinct during times of lowstand vs. highstand. The extinction of taxa during highstand (true biostratigraphic tops) is regional in character, whereas the true extinctions during lowstands are restricted to sites of deposition of the lowstand systems tract.**

## Results

Shifts in biofacies distribution within the 3.8–3.0-Ma sequence are used to illustrate the resolution of time-slice biofacies maps (Figure 13A–G). Interval averaged biofacies map generally shows ecozones 5 and 5.5 in the area (Figure 13A). Boundaries between the ecozones are generally smooth. In contrast, the biofacies time-slice maps show considerably more variation in the ecozone boundaries, reflecting the effects of paleobathymetry (salt highs and lows) and sediment input through depositional axes.

The 3.8–3.0-Ma interval (Figures 13B–F) shows a good example of changing biofacies boundaries. Between *Globerigerina nepenthes* (3.8 Ma) and *Sphenolithus abies* B (3.75 Ma) LAD (Figures 13B, C), a clear transition in biofacies zones is present where they are seen to shift south or downslope. After that, biofacies boundaries continue to shift south through *Globotoralia margaritae* LAD (3.52 Ma) and *Reticulofenestra pseudoumbilica* LAD (3.5 Ma) (Figures 13D, E). The shift in biofacies can be determined by tracking the position of the boundary between ecozones 5 and 5.5. The biofacies time slice for 3.4 Ma (*Sphenolithus abies*) suggests an upslope migration of the 5–5.5 biofacies boundary (Figure 13F). After 3.5 Ma, ecozone 6 is no longer in the area, having shifted farther downslope.

Based on these patterns and their integration with the regional sequence stratigraphy (Weimer et al., 1998, this issue), we interpret the changes in the biofacies boundaries from 3.8 to 3.4 Ma to reflect both the regression and transgression within this sequence, and the regional shift in the shallow marine depocenter. The basinward shift of the ecozones between 3.75 and 3.5 Ma reflects the lower positions of relative sea level (see Figures 3, 5). The upslope shift in the biofacies at 3.4 Ma corresponds both to the eustatic rise in sea level (Figures 3, 5), and to the major westward shift in the shallow-marine depocenter that began at this time.

The maps illustrate the general basinward movement of biofacies through time, which reflects the overall progradation of the margin. Changes in the sediment patterns and paleobathymetry are also present, reflecting the evolving nature of the intraslope area in terms of tectonics and sedimentation.

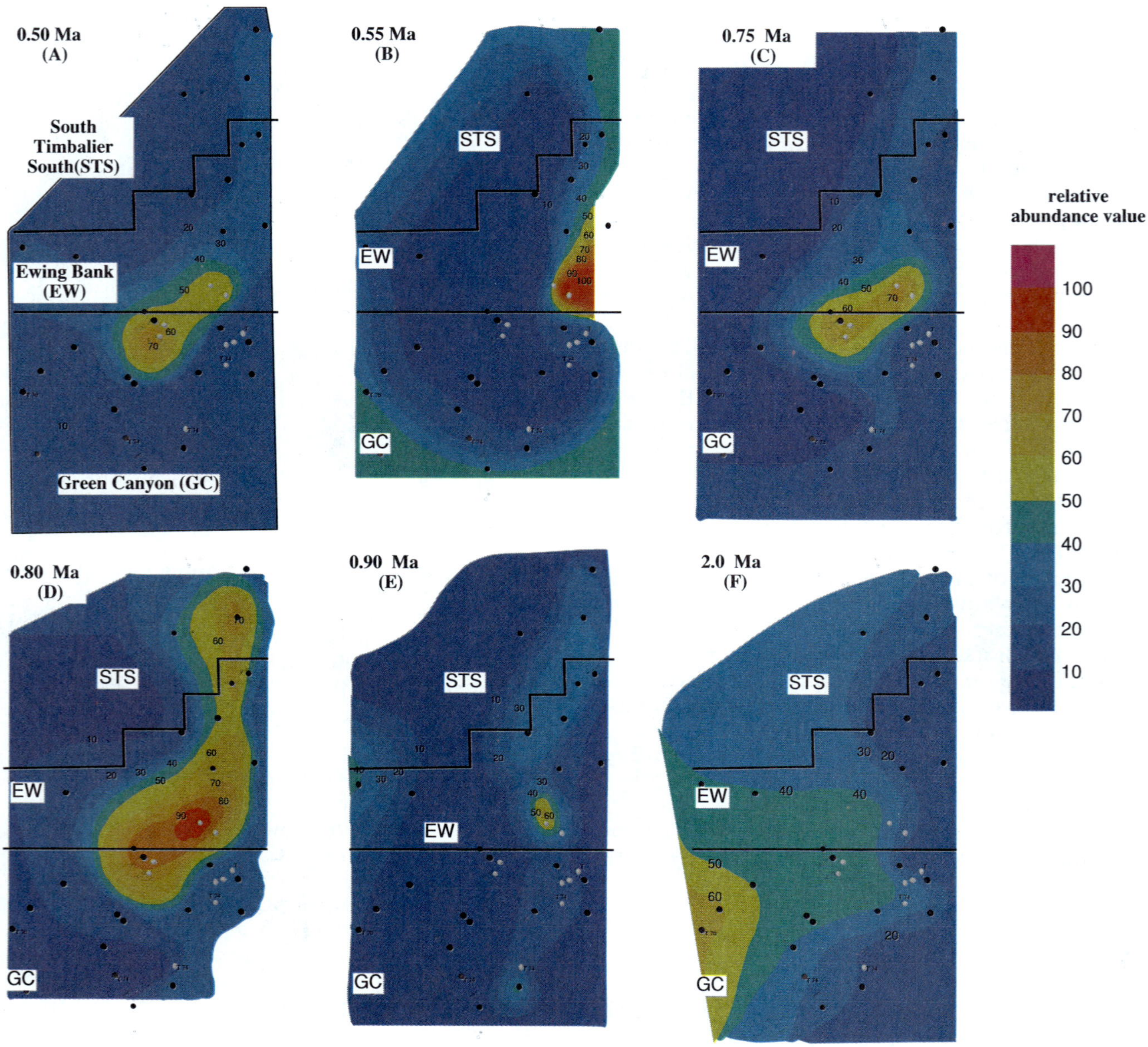


**Figure 12—Selected abundance maps for one minibasin in east-central Ewing Bank and northeastern Green Canyon. See Figure 1 for location. The data are contoured; light colors represent high abundance values; dark colors reflect low values. Abundance is noted by numbers on contour lines. (A) 0.50 Ma, (B) 0.55 Ma, (C) 0.75, (D) 0.80 Ma, (E) 0.90 Ma, (F) 2.0 Ma, (G) 2.5 Ma, (H) 3.0 Ma, (I) 3.4 Ma.**

These maps are an important part of regional basin analysis. They help in evaluating the stratigraphic and structural evolution of the basin, which is essential for exploration success.

## Pitfalls

The shortcomings of this technique rest primarily in the accuracy of the determination of extinction data. Extinctions or LAD for calcareous nannoplankton and planktonic foraminifers ideally represent time horizons, but because of interval sampling, local last occurrences, and other factors that affect the distribution of planktonic species, LAD probably more accurately reflects a relatively short interval of time (±0.05 m.y.). The palinspastic restorations of the sea floor also should be viewed as generalized, and not as a rigorous three-dimensional restoration. In addition, the boundaries

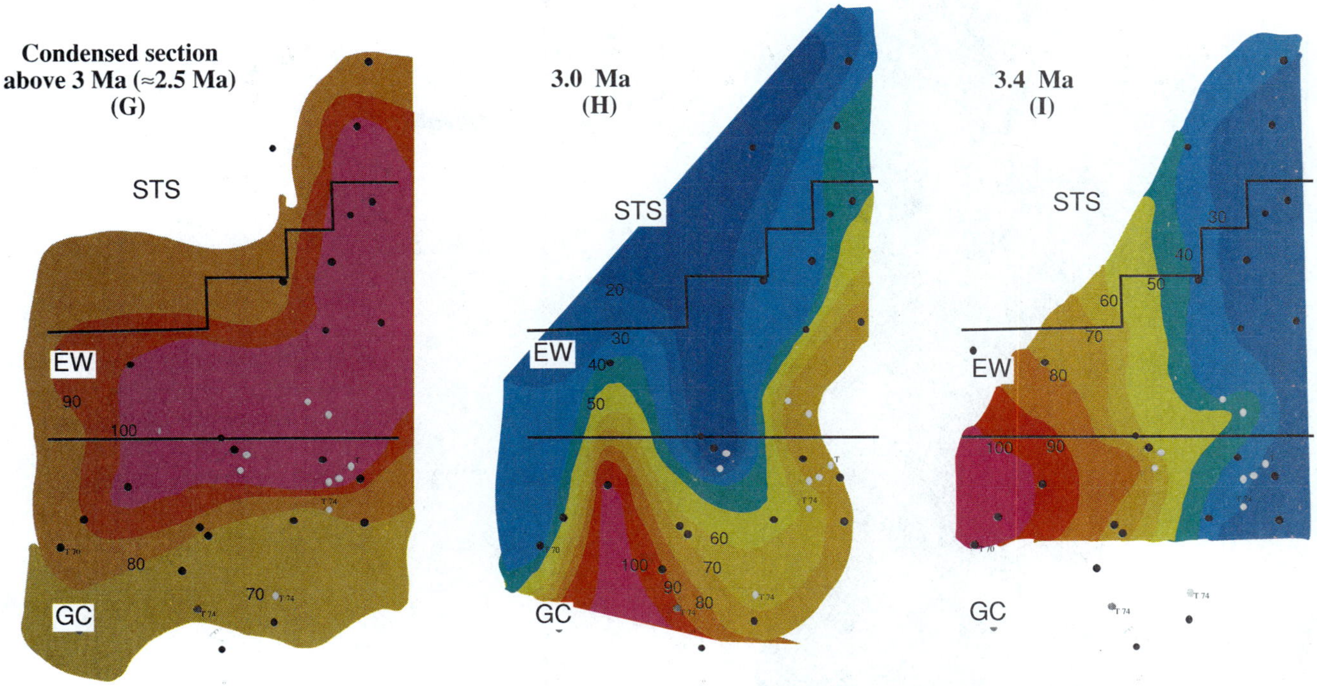


**Figure 12—Continued.**

between biofacies, which are drawn as lines, are in reality diffuse. We used consistent criteria to interpret biofacies based on the deepest benthic foraminifer present; however, a component of transported shallower water fauna in the assemblage may yield a shallower water interpretation (e.g., Armentrout, 1991). This potential error can be minimized by constructing time-slice maps for times of highstand.

## MODIFIED BENTHIC BIOFACIES MODEL FOR SLOPE SEDIMENTS, NORTHERN GULF OF MEXICO

An alternative benthic biofacies model is needed for bathyal settings with significant paleobathymetric relief because traditional models represent a smooth and gently dipping topography from shelf to abyssal plain. Traditional benthic biofacies models are divided into 5–7 ecozones based on benthic foraminifer associations (Figures 5, 14A). These models are an oversimplification of benthic foraminifer associations occupying a relatively fixed environment on a sloping sea bottom. The distribution of benthic foraminifers is related to the depth of the sea bottom in an indirect manner. Benthic foraminifers occupy niches that are defined by a combination of physical, chemical, thermal, and biological characteristics (Corliss and Fois, 1990) and surface water productivity. In addition, benthic biofacies are controlled by other factors including sedimentation rate and type as well as additional physical and chemical bottom conditions (Loubere et al., 1993). A combination of these controlling factors generally corresponds to changes in bathymetry and sediment supply. The geometry of simplified benthic biofacies models does not reflect the highly variable present-day bottom topography, particularly in the northern Gulf of Mexico region where many controls on the topography take place at the same time. These controls on sea bottom topography comprise the presence of salt structures, salt withdrawal features, active growth faults, contractional folds, and differential distribution of small-scale depocenters superimposed onto the shelf-slope-basin gradient. In addition to a gross correlation to sea bottom topography, the distribution of benthic biofacies is markedly affected by the distribution of active sites of sediment delivery.

Mapping benthic biofacies at extinction or LAD provides a regional view of the distribution of biofacies at a particular time surface (Figure 13). A view of biofacies with the method used in this study is the closest approach to high-resolution paleobiogeography, particularly if structural deformation has been taken into account. In general,

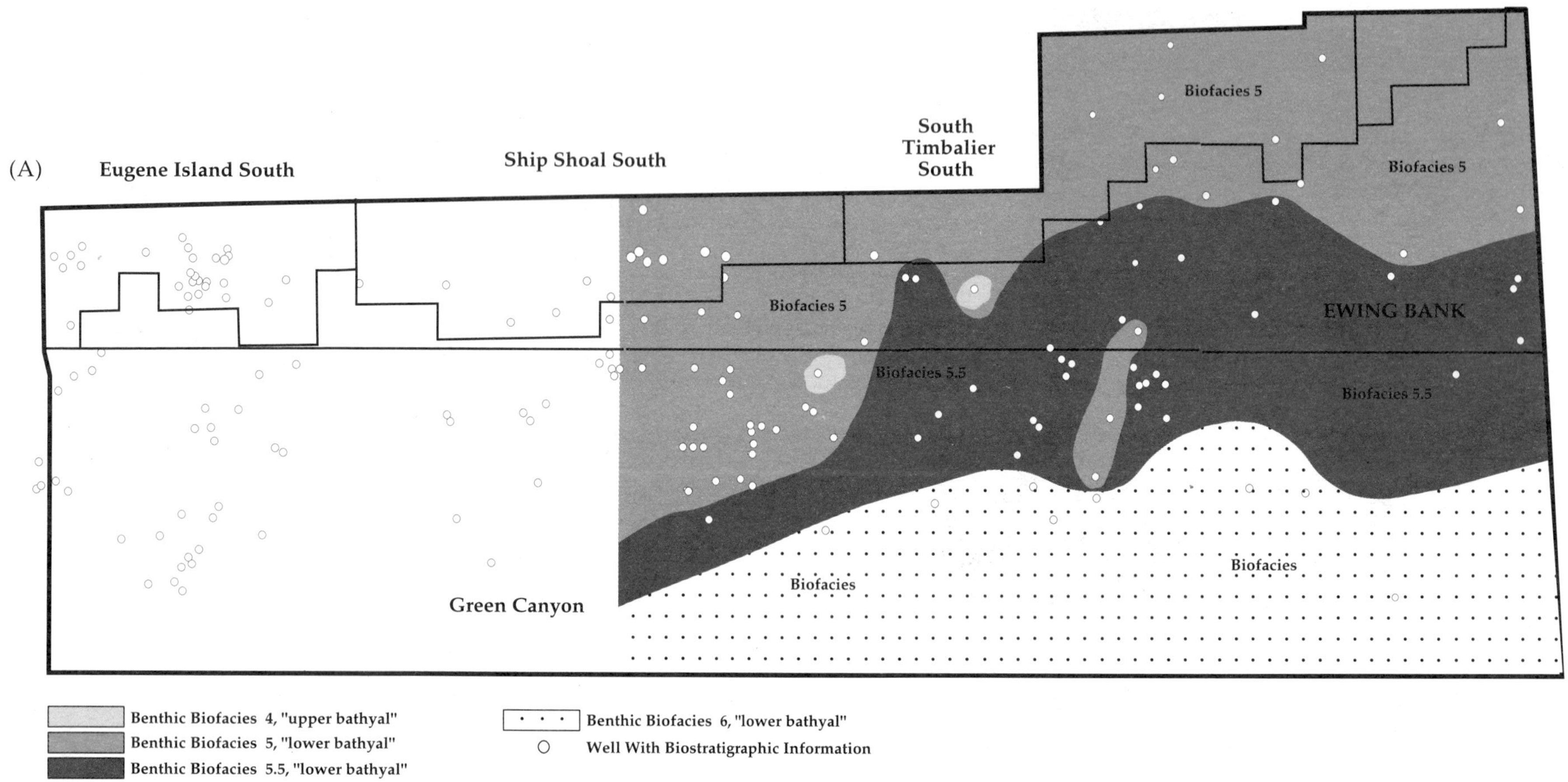


**Figure 13—Benthic biofacies time-slice maps for the 3.8–3.0-Ma depositional sequence. (A) Interval average map (3.8–3.0-Ma sequence), (B) *Globerigerina nepenthes* extinction datum (3.8 Ma), (C) *Sphenolithus abies* B extinction datum (3.75 Ma), (D) *Globorotalia margaritae* extinction datum (3.52 Ma) (E) *Reticulofenestra pseudoumbilica* extinction datum (3.50 Ma), (F) *Sphenolithus abies* extinction datum (3.40 Ma). See Figure 5 for biostratigraphic zonation and text for discussion. Palinspastic reconstructions of salt and fault geometry are from Rowan and Weimer (1998, this issue).**

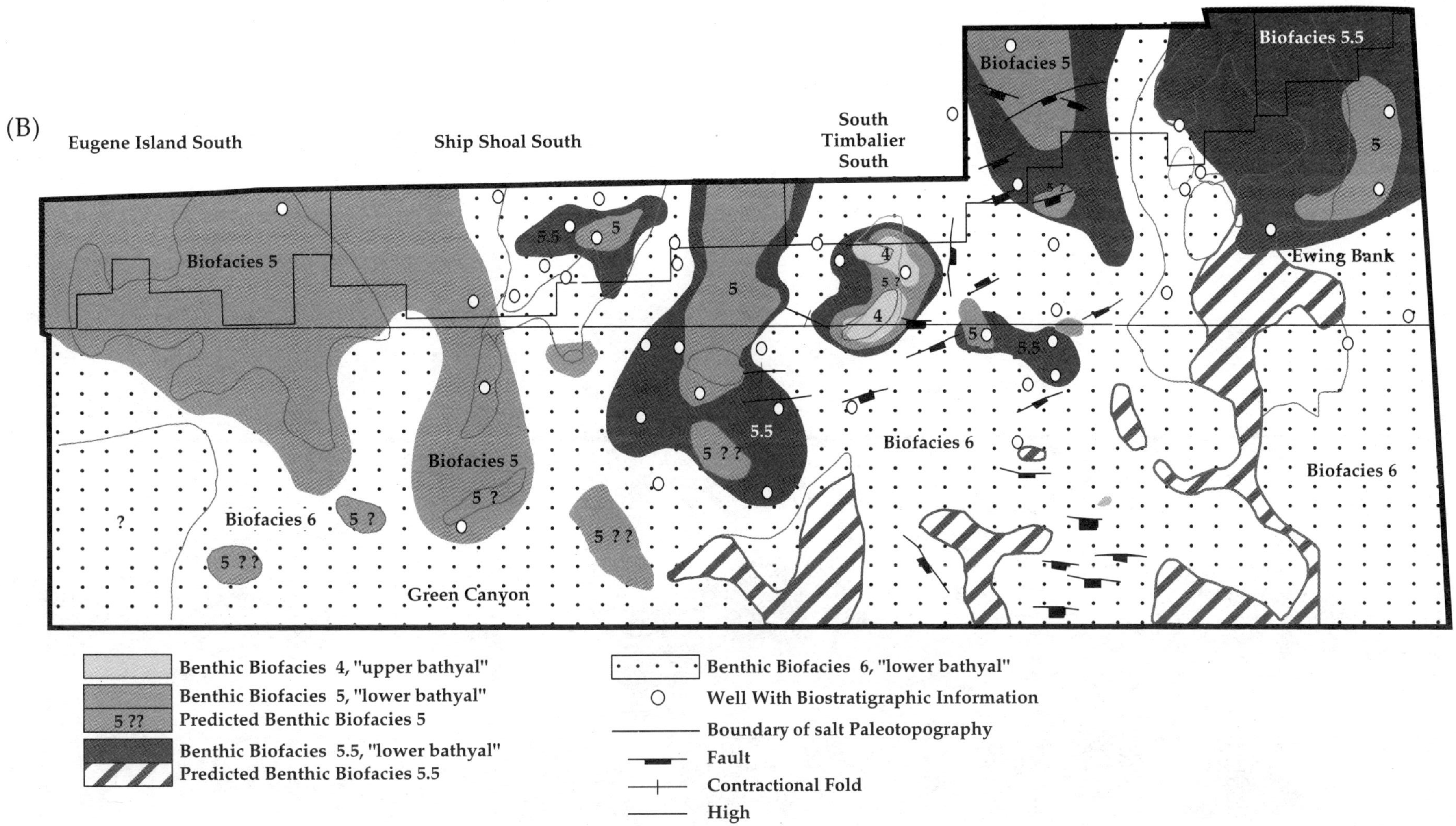


**Figure 13—Continued.**

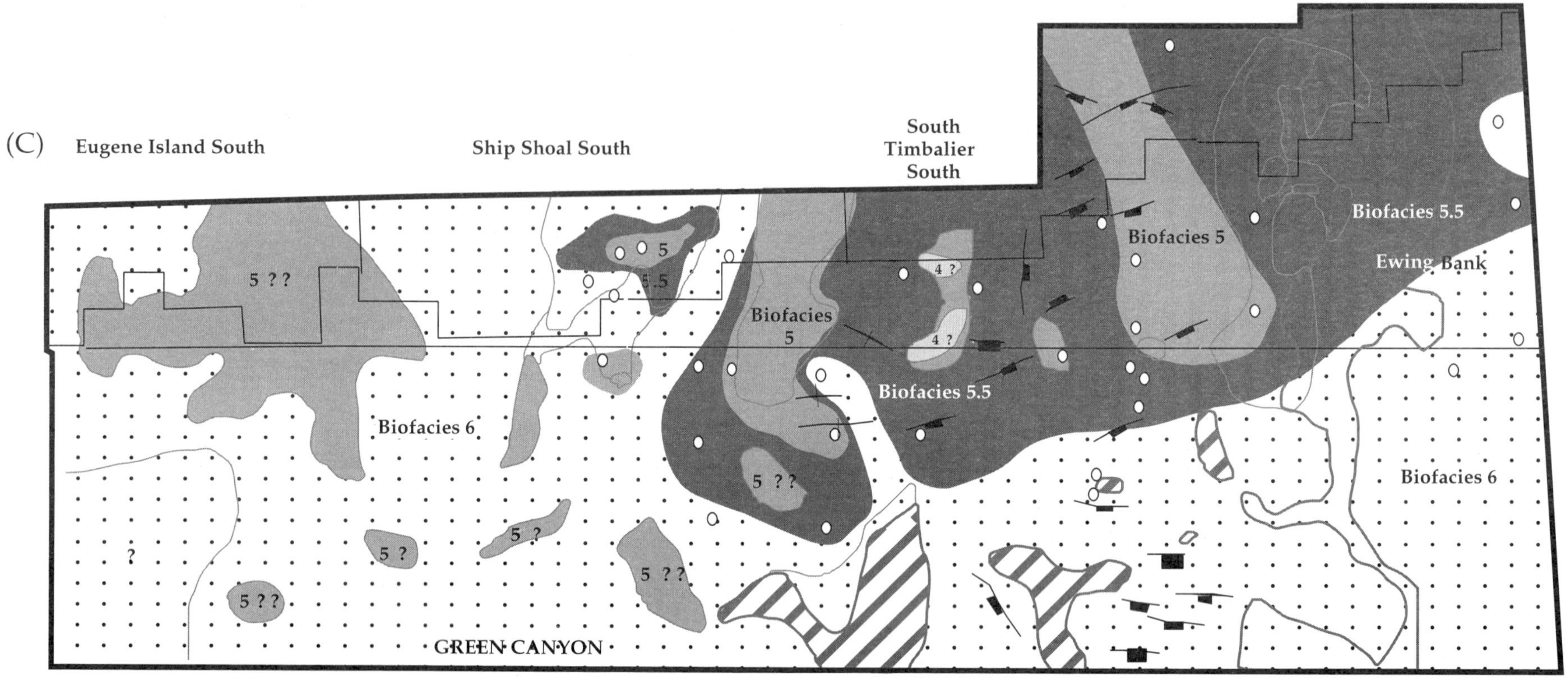


Figure 13—Refer to Figure 13B for legend.

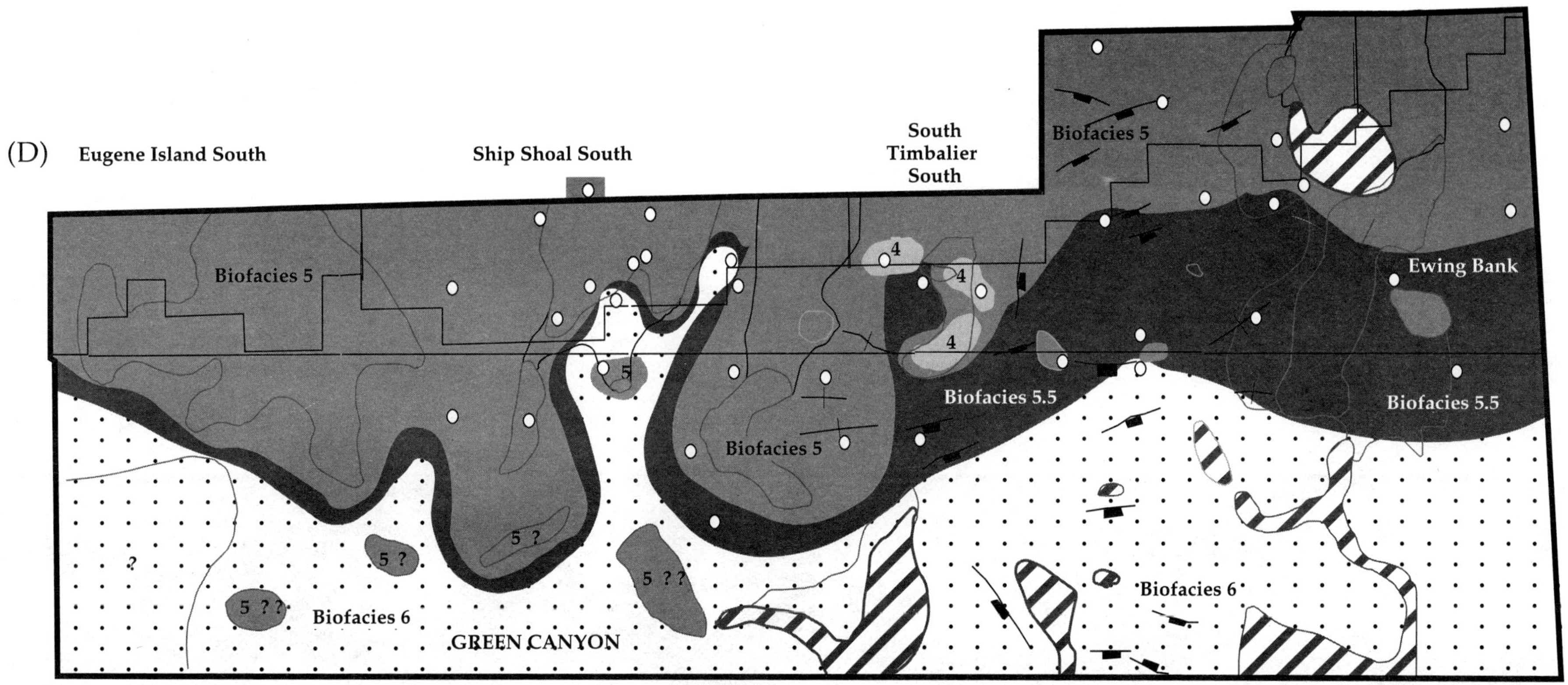


**Figure 13—Refer to Figure 13B for legend.**

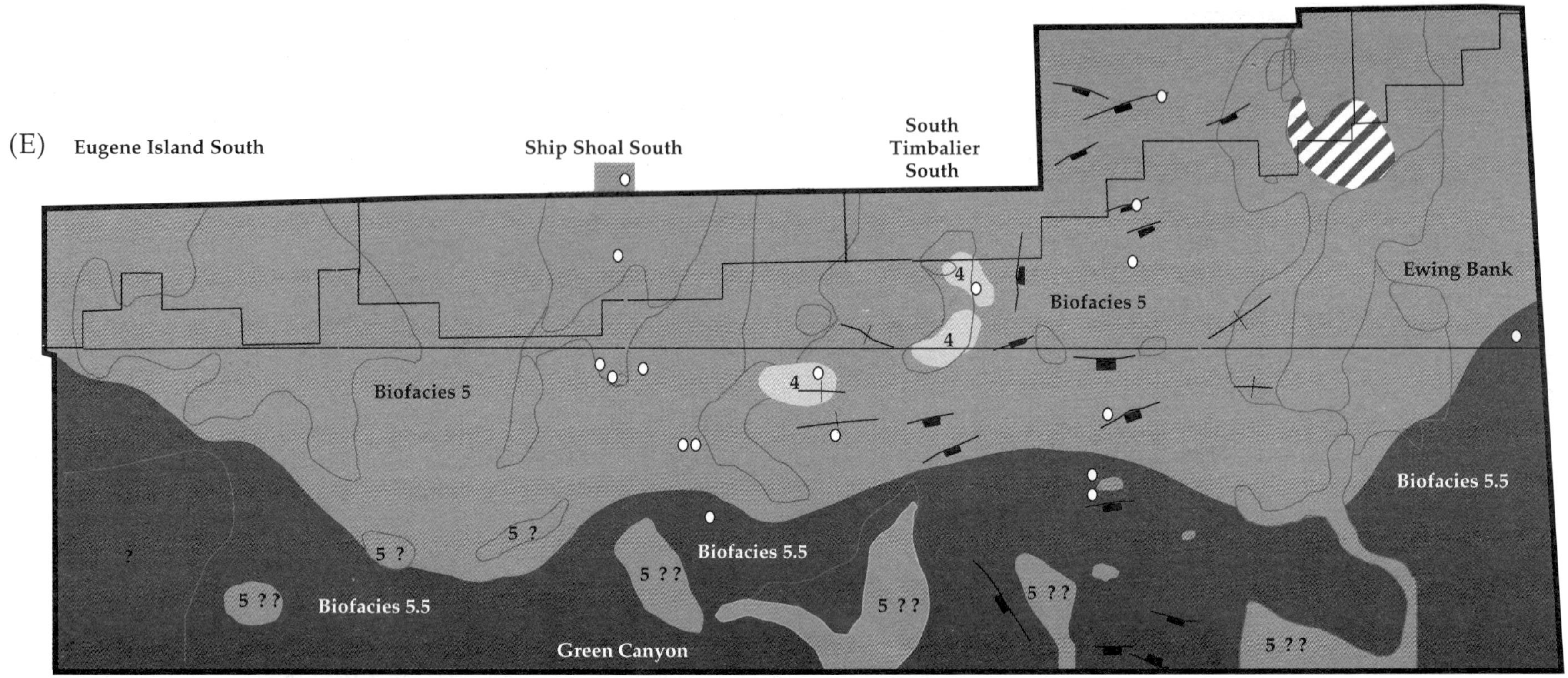


**Figure 13—Refer to Figure 13B for legend.**

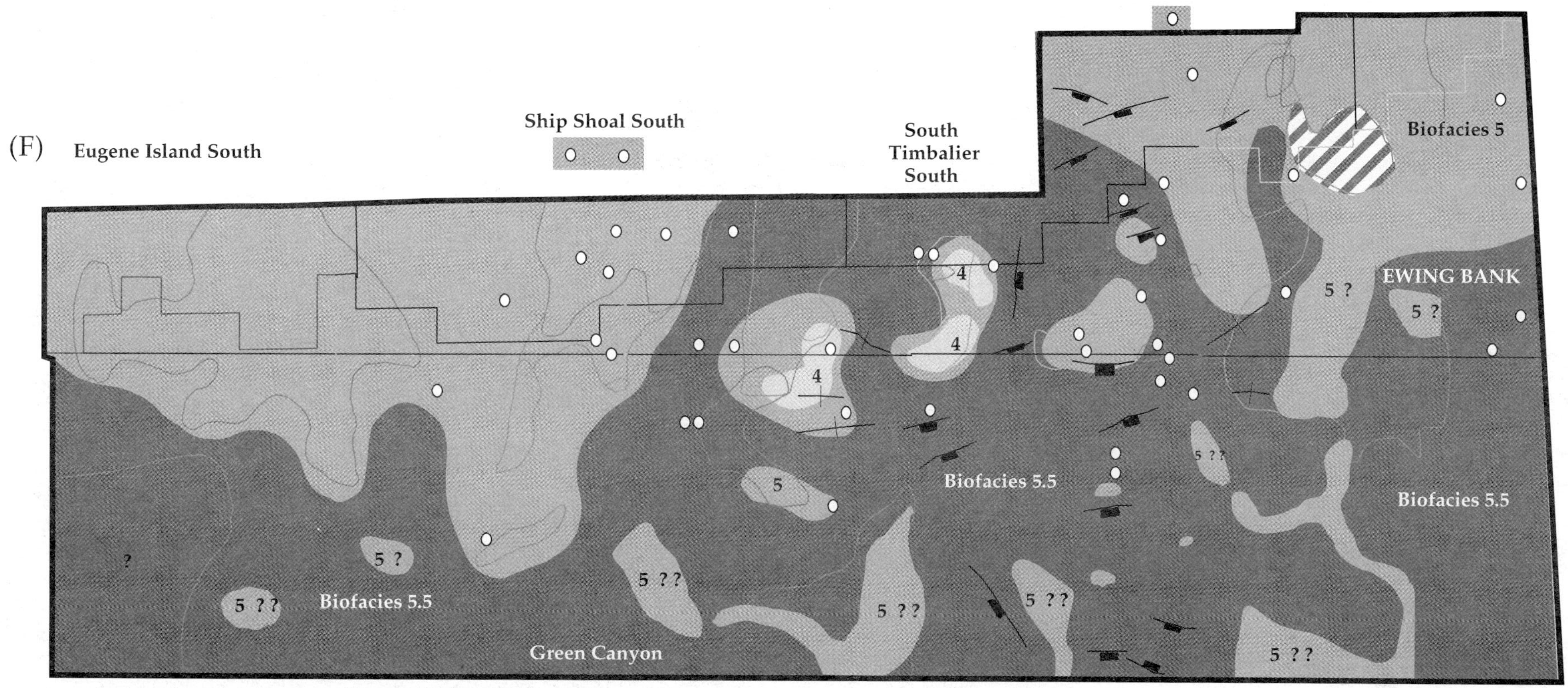


Figure 13—Refer to Figure 13B for legend.

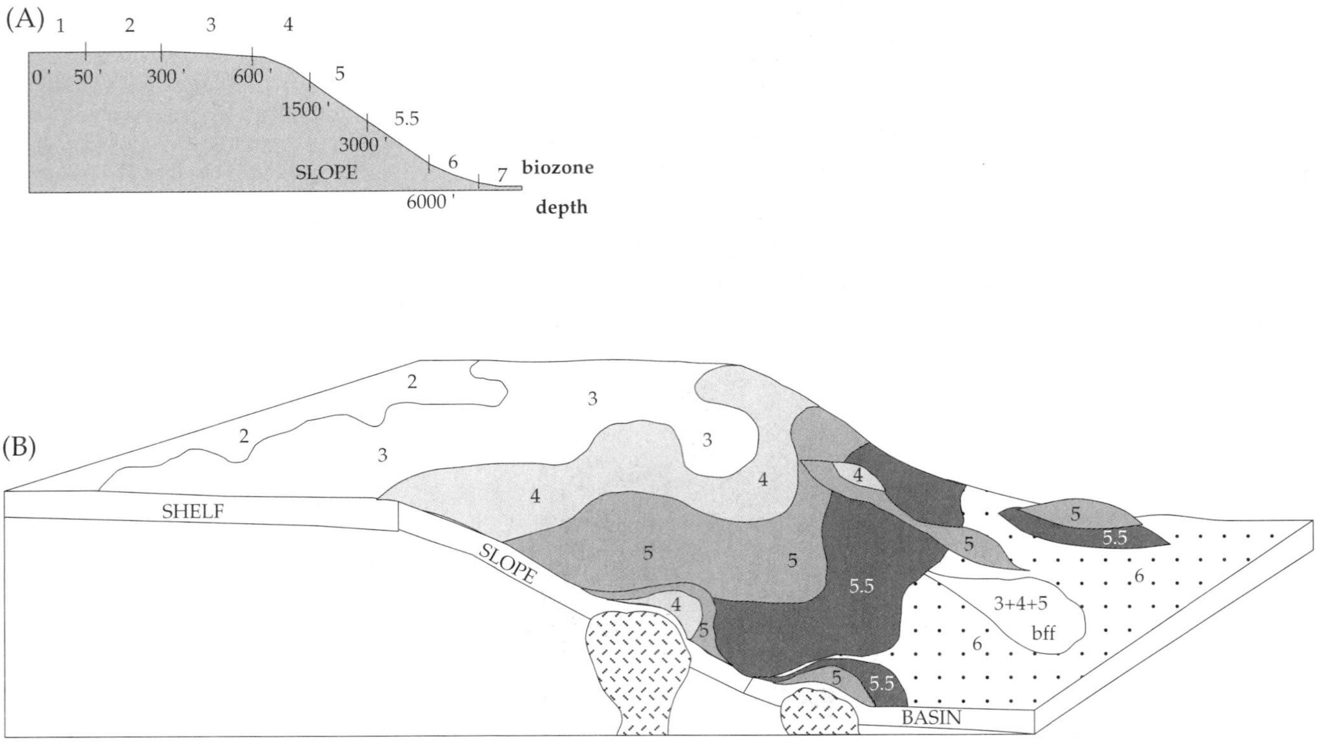


**Figure 14—(A) Traditional benthic biofacies model and (B) modified benthic biofacies model for the northern Gulf of Mexico. See Figure 5 for biofacies zonation.**

benthic biofacies are displaced basinward by active sediment supply (Figure 13B–F).

The benthic biofacies model presented in Figure 14B is compared to traditional models in Figure 14A. This biofacies model is more comprehensive because it (1) is three dimensional, (2) considers lateral variations in sea bottom topography, which may have been the consequence of sea bottom structure, differential sediment supply, and stratigraphy, and (3) emphasizes the importance of salt features because their presence can control sedimentation pathways and accumulation sites. The model presented in Figure 14B takes into account several variables that affect the distribution of benthic foraminifers. The model's application favors the integration of independent data and the interaction of seismic and well-log interpreters with paleontologists. Quality, reliability, and robustness of interpretations are a necessary consequence of integrating independent data and interaction between different disciplines. The final result, hopefully, is a reduction of risk in stratigraphic prediction in petroleum exploration.

## CONCLUSIONS

(1) Biostratigraphy is an essential portion of integrated basin analysis, but it remains underused in petroleum geology. The geology of the northern Green Canyon and Ewing Bank areas reflects complex interaction of syndepositional tectonics (salt deformation, faulting) and abrupt lithofacies changes in bathyal turbidite systems. Biostratigraphy is essential for unraveling such a complex basin history by establishing a chronostratigraphic framework and for its role in the construction of the regional stratigraphic framework.

(2) Sediment accumulation plots are derived from plotting the depths from a well onto an absolute time scale. The resulting curve has a changing slope, reflecting the changes in rates of deposition both between and within depositional sequences. Condensed sedimentation is reflected by steep curves (i.e., decreased sedimentation rates). When several plots are analyzed together in a regional sense, the relative significance of each condensed section can be evaluated. Areally extensive condensed sections are interpreted to be caused by allocyclic controls (i.e., eustatically driven, switches in the regional depocenter). In contrast, condensed sections with a more local distribution are interpreted to be caused by autocyclic factors, primarily switches in the sediment pathway within one minibasin or between adjacent minibasins.

Two regional periods of condensation can be interpreted in the area studied: one from approximately 3.5 to 1.45 Ma and the other from approximately 1.2 to 0.87 Ma. The condensed section from 3.5 to 1.45 Ma is better developed in eastern regions of the area and loses its character to the west where active sedimentation was taking place. The condensed section from 1.2 to 0.87 Ma is developed in western regions of the basin and loses character to the east where active sedimentation was taking place. This change in condensation reflects a large avulsion of the shallow-marine sediment source from west to east.

Sediment accumulation plots are useful in the identification of sequence boundaries and transgressive surfaces. Sequence boundaries can be identified by a change from steep to gentle slope in the sediment accumulation plot; this change is especially marked where basin floor fans overlie condensed deposits. Transgressive surfaces are identified by a change from gentle to steep slopes in the plot.

(3) Detailed analyses of nannofossil abundance in one minibasin from 3.0 to 0.45 Ma show local variations in condensation through time. These changes are probably associated with the changing position of the sites of active sediment supply. The results are important because they point out potential problems in using abundance peaks for regional correlation.

(4) A new technique was developed to construct more accurate biofacies maps at the time of deposition. The approach consists of plotting biofacies of the wells at extinctions or LAD of selected planktonic foraminifers and calcareous nannoplankton. These maps were then integrated with the lithostratigraphic and structural results from the area. The resulting time-slice maps reflect the irregular paleobathymetry at the time of deposition. Time-slice benthic biofacies maps show transgressive and regressive stratigraphic patterns that may be beyond the resolution of seismic and well logs.

(5) As a result of the benthic biofacies time-slice maps, a modified biofacies model for the area was developed because previous models do not take into consideration the many factors that affect the distribution of benthic foraminifers. The new model is a three-dimensional approach and comprises variations in the distribution of biofacies caused by sediment supply and the irregular bathymetry of the sea bottom created by salt deformation.

## REFERENCES CITED


Acosta, Z. M., 1994, Sequence stratigraphy of Pliocene-Pleistocene sediments in north central Green Canyon and western Ewing Bank, northern Gulf of Mexico: M.S. thesis, University of Colorado, Boulder, Colorado, 187 p.

Armentrout, J. M., 1991, Paleontologic constraints on depositional modeling: examples of integration of biostratigraphy and seismic stratigraphy, Pliocene-Pleistocene, Gulf of Mexico, *in* P. Weimer and M. H. Link, eds., Seismic facies and sedimentary processes of submarine fans and turbidite systems: New York, Springer-Verlag, p. 137-170.

Corliss, B. H., and E. Fois, 1990, Morphotype analysis of deep-sea benthic Forminfera from the northwestern Gulf of Mexico: Palaios, v. 5, p. 589-605.

Loubere, P., A. Gary, and M. Lagoe, 1993, Sea-bed biogeochemistry and benthic foraminiferal bathymetric zonation of the slope of the northwestern Gulf of Mexico: Palaios, v. 8, p. 439-449.

Loutit, T. S., J. Hardenbol, P. R. Vail, and G. R. Baum, 1988, Condensed sections: the key to age dating and correlation of continental margin sequences, *in* C. K. Wilgus, B .S. Hastings, C. G. St. G. Kendall, H. W. Posamentier, C. A. Ross, J. C. Van Wagoner, eds., Sea level changes: an integrated approach: SEPM Special Publication 42, p. 183-215.

Mann, K. O., and H. R. Lane, eds., 1995, Graphic correlation: SEPM Special Publication 53, 263 p.

Martinez, R. E., 1994, Sequence stratigraphy of late Pliocene and Pleistocene sediments of northwestern Green Canyon area, Northern Gulf of Mexico: M.S. thesis, University of Colorado, Boulder, Colorado, 138 p.

McBride, B. C., M. G. Rowan, and P. Weimer, 1998, The evolution of allochthonous salt systems, northern Green Canyon and Ewing Bank (offshore Louisiana), northern Gulf of Mexico: AAPG Bulletin, v. 82, no. 5B, p. 1013-1036.

Mitchum, R. M., Jr., J. B. Sangree, P. R., Vail, and W. W. Wornardt, 1993, Recognizing sequences and systems tracts from well logs, seismic data and biostratigraphy: examples from the late Cenozoic of the Gulf of Mexico, *in* P. Weimer and H. W. Posamentier, eds., Siliciclastic sequence stratigraphy: AAPG Memoir 58, p. 163-199.

Paleo-Data Inc./Schlumberger, 1993, Gulf of Mexico chronostratigraphic correlation chart: Schlumberger/Geco-Prakla.

Pulham, A. J., 1993, Variations in slope deposition, Pliocene-Pleistocene, offshore Louisiana, northeast Gulf of Mexico, *in* P. Weimer and H. W. Posamentier, eds., Siliciclastic sequence stratigraphy: AAPG Memoir 58, p. 199-233.

Pulham, A. J., J. A. D'Ablang, D. S. Epps, and R. K. Kasten, 1993, Condensed zones on the deep water Gulf of Mexico: their recognition and stratigraphic significance (abs.): AAPG Annual Convention Program with Abstracts, p. 169.

Rowan, M. G., 1995, Structural styles and evolution of allochthonous salt, central Louisiana outer shelf and upper slope, *in* M. P. A. Jackson, D. G. Robert, and S. Snelson, eds., Salt tectonics: a global perspective: AAPG Memoir 65, p. 199-228.

Rowan, M. G., and P. Weimer, 1998, Salt-sediment interaction, northern Green Canyon and Ewing Bank (offshore Louisiana), northern Gulf of Mexico: AAPG Bulletin, v. 82, no. 5B, p. 1055-1082.

Vail, P. R., 1987, Seismic stratigraphy interpretation using sequence stratigraphy, part 1, *in* A. W. Bally, ed., Atlas of seismic stratigraphy: AAPG Studies in Geology 27, p. 1-10.

Vail, P. R., and W. W. Wornardt, 1991, An integrated approach to exploration and development in the 90's: well log-seismic sequence stratigraphy analysis: Transactions Gulf Coast Association Geological Societies, v. 41, p. 630-650.

Weimer, P., P. Varnai, F. M. Budhijanto, Z. M. Acosta, R. E. Martinez, A. F. Navarro, M. G. Rowan, B. C. McBride, T. Villamil, C. Arango, J. R. Crews, and A. J. Pulham, 1998, Sequence stratigraphy of Pliocene and Pleistocene turbidite systems, northern Green Canyon and Ewing Bank (offshore Louisiana), northern Gulf of Mexico: AAPG Bulletin, v. 82, no. 5B, p. 918-960.

## ABOUT THE AUTHORS

### Tomas Villamil

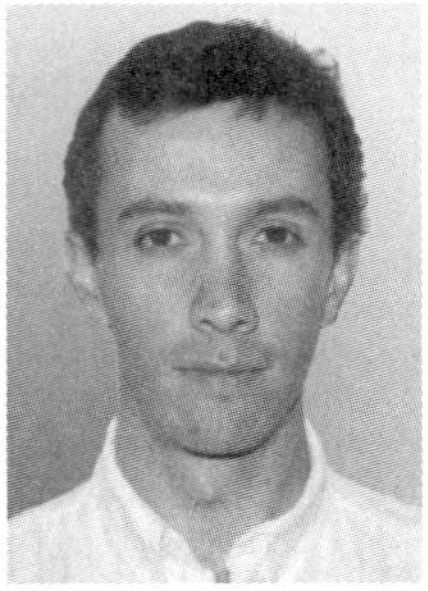

Tomas Villamil is an upstream exploration geologist in Conoco's Advance Exploration Organization. His expertise is in stratigraphy and tectonostratigraphy of northern South America and the Gulf of Mexico. He graduated from the National University of Colombia (Bogotá, 1988), then received a Ph.D. from the University of Colorado at Boulder (1994) and a postdoctoral degree from the same school. He is a member of AAPG and SEPM.

### Claudia Arango

Claudia Arango is a geologist in Marathon Oil Company's Gulf of Mexico exploration group. Her expertise is in stratigraphy and biostratigraphy of Colombia and the Gulf of Mexico. After graduating from the National University of Colombia (Bogotá, 1992), she received an M.S. degree from the University of Colorado at Boulder (1995). She is a member of AAPG.

### Paul Weimer

Paul Weimer is an associate professor and director of the Energy and Minerals Applied Research Center at the University of Colorado at Boulder. His research interests are in sequence stratigraphy and basin analysis, three-dimensional seismic interpretation, and reservoir geology.

### Arthur Waterman

Arthur Waterman is executive vice president of Paleo-Data, Inc., a biostratigraphic consulting firm in New Orleans. Previously, Art was a development geologist and paleontologist with Texaco and a consulting paleontologist with Paleo-Data, Inc. specializing in Cenozoic foraminifera. Since 1985, Art has specialized in calcareous nannoplankton biostratigraphy. He received a B.S. degree from St. Lawrence University in 1973 and an M.A. from Indiana University in 1975.

### Mark G. Rowan

Mark Rowan is a graduate of Caltech (1976), University of California, Berkeley (1982), and the University of Colorado (1991), where he is currently an assistant research professor in structural geology. He has also worked in the petroleum industry, both as an exploration geologist for Sohio and as a consultant with Geo-Logic Systems and Alastair Beach Associates. His research interests focus on the three-dimensional geometry and evolution of both extensional and contractional structures, with a recent emphasis on salt tectonics.

### Peter Varnai

Peter Varnai received his B.S. degree (1985) in geology from Eötvös Loránd University, Budapest, Hungary; an M.S. degree (1991) in geophysics from Miskolc University, Miskolc, Hungary; and a Ph.D. (1996) in geology from the University of Colorado, Boulder, Colorado. He worked for the Hungarian National Oil Company in Budapest as an exploration geologist for six years. He was responsible for seismic interpretation, stratigraphic and tectonic models, and basin analysis in the Pannonian basin. During his graduate studies in Colorado, he participated in the Gulf of Mexico industry consortium. In 1996, he joined Exxon Exploration Company as a senior geologist.

## Andrew J. Pulham

Andy Pulham has a B.Sc. in physical geography and geology from Liverpool University, United Kingdom and a Ph.D. in geology from the University of Wales, United Kingdom. After graduating, he spent 12 years as an international petroleum sedimentolgist for British Petroleum. He has worked in the North Sea and South America and spent over three years in Houston working the deep-water Gulf of Mexico. Currently, Pulham is leading a Reservoir Geology Research Consortium within the Energy and Minerals Applied Research Center in Boulder, Colorado. This is an industry supported project and is a collaborative venture with S. George Pemberton and his Ichnology Research Group at the University of Alberta.

## Jennifer R. Crews

Jennifer Crews received her bachelor's degree in geology from the University of Colorado at Boulder in 1995; she will receive her M.S. degree during the summer of 1998. Her thesis topic is High Resolution Biostratigraphic Analysis of Green Canyon and Ewing Bank Areas, Northern Gulf of Mexico. Upon graduation, she will be employed by Texaco Exploration and Production, Inc., New Orleans. Research interests include applied biostratigraphic methods and sequence analysis.

# Three-Dimensional Seismic Stratigraphic Expression of Pliocene–Pleistocene Turbidite Systems, Northern Green Canyon (Offshore Louisiana), Northern Gulf Of Mexico[1]

Peter Varnai[2]

## ABSTRACT

Interpreting three-dimensional (3-D) seismic data is of great value in high-grading prospects in frontier exploration areas in the Gulf of Mexico. This technique is now routinely used to reduce the exploration and production risks associated with turbidite reservoirs. A subregional 3-D seismic data set was used in this study to describe the Pliocene-Pleistocene turbidite systems in three intraslope minibasins in the north-central Green Canyon protraction area of the Gulf of Mexico.

The stratal geometries and lithology of the different turbidite elements were interpreted based on the integrated analysis of well-log data, seismic facies observed on the vertical seismic profiles, and amplitude distribution recognized on series of horizon slices and amplitude extraction maps (horizontal seismic facies). The lithologic interpretation, however, was limited by the lack of seismic velocity data; the wells were tied to the seismic data through synthetic seismograms created by integrating the sonic and density logs.

Three turbidite elements were mapped in the selected upper Pliocene–lower Pleistocene interval of the study area: basin-floor fan (amalgamated depositional lobes), erosional and depositional channels, and overbank sediments. Lobe-shaped high-amplitude areas (onlapping high-amplitude continuous reflections on seismic profiles) indicate laterally continuous, sheetlike deposition, interpreted as turbidite lobes. Elongated sinuous high-amplitude areas (high-amplitude discontinuous reflections) correspond to depositional channel systems. Sinuous low-amplitude areas in generally high-amplitude areas (e.g., sequence boundaries) reflect the erosional channels. The depositional channels are interpreted to be filled with sands; the shale-filled erosional channels are interpreted as conduits for sediments bypassing farther downslope. The detailed 3-D seismic interpretation allowed description of the characteristics of the channel systems, analysis of their evolution through 2 m.y., and determination of the controls on the variations in turbidite deposition. Widespread low- to moderate-amplitude areas without dominant amplitude patterns are interpreted as overbank deposits.

The turbidite elements identified in this study, including basin-floor fans (depositional lobes) and channel systems, form the reservoirs in the fields of the study area and in many other fields and discoveries of the northern Gulf of Mexico. The geometries of the sand-rich turbidite elements mapped in this study are considered to be good analogs for these producing fields and some exploration targets. The results of this study also can be used as analogs for the description of reservoir architecture in the subsalt trend of the Gulf of Mexico and in other deep-water exploration areas.

## INTRODUCTION

Neogene turbidite systems are the major reservoirs and exploration targets in the northern deep



[1]Manuscript received September 6, 1996; revised manuscript received June 23, 1997; final acceptance November 20, 1997.

[2]Department of Geological Sciences, Energy and Minerals Applied Research Center, University of Colorado, Boulder, Colorado 80309-0399. Present address: Exxon Exploration Company, P.O. Box 4778, Houston, Texas 77210-4778.

This paper summarizes the results of a portion of my Ph.D. studies at the University of Colorado at Boulder. I express my most sincere gratitude to Paul Weimer, my advisor, for all his encouragement, help, advice, and support throughout my studies. I thank Mark Rowan for his help with the structural interpretation, and David Knapp for helping me negotiate the hurdles of the workstation.

I thank the following companies and their representatives for their support and input as part of the Gulf of Mexico Research Consortium at the University of Colorado: Agip, Amoco, Anadarko, BHP, BP Exploration, Burlington Resources, CXY Energy, Chevron USA, CNG, Conoco, Exxon, Marathon, Mobil, Occidental, Pan Canadian, Pennzoil, Petrobras, Phillips, Shell, Texaco, Total, Union Pacific, and Unocal. I extend special thanks to Halliburton Geophysical (now Western Geophysical), specifically to Bob Graebner, John Anderson, and Gary White, for the seismic data used in this project. I am indebted to PaleoData Inc. and Art Waterman for use of their biostratigraphic data. Software donations from GeoQuest, CogniSeis Development, Zeh Graphics, Landmark Graphics, and Platte River Associates were essential to the success of my study. Acknowledgment is made to the donors of the Petroleum Research Fund, administered by the American Chemical Society for partial support of this work. Special thanks go to Paul Weimer, Laurie Lamar, and to the AAPG reviewers, former Elected Editor Kevin Biddle, Cindy Yeilding, and Jory Pacht, for their careful reviews. Their comments greatly improved the manuscript.



AAPG Bulletin, V. 82, No. 5B (May 1998 Part B), P. 986–1012.

Gulf of Mexico (Weimer et al., 1994a, and the papers therein). The architecture of these deposits varies considerably, including both sheetlike and channelized systems (e.g., Mahaffie, 1994). Few publications have described the geometries or the stratigraphic variabilities of these turbidite systems in detail using three-dimensional (3-D) seismic data (e.g., Rafalowski et al., 1994; Chapin et al., 1996; Risch et al., 1996). Hence, the 3-D seismic characteristics of the sheetlike sands and the channelized turbidite systems presented in this paper create important analogs for other deep-water exploration areas.

This study describes the three-dimensional architecture and distribution of Pliocene–Pleistocene turbidite systems in three intraslope minibasins located 220 km (140 mi) southwest of New Orleans in offshore Louisiana (Figure 1). The study area covers 20 offshore federal lease blocks (460 km²; 180 mi²) in the north-central Green Canyon protraction (lease) area. Present water depth ranges between 610 m (2000 ft) on the northwest and 940 m (3100 ft) on the southeast. The average slope gradient of the present-day sea floor is 9 m/km (0.5°) there.

Based on the results of a regional two-dimensional (2-D) sequence stratigraphic study (Varnai, 1996; Weimer et al., 1998a, this issue), five sequences (3.8-, 3.0-, 2.6-, 2.4- and 1.9-Ma sequences; Figure 2) were selected for detailed analysis using the 3-D seismic data. These sequences were interpreted to contain important turbidite elements, as well as reservoir sands in two fields. The objectives of this paper are to describe the seismic facies based on vertical seismic profiles and on horizon slices and amplitude extraction maps; interpret the stratigraphy of the turbidite elements (basin-floor fan, erosional and depositional channels, overbank sediments); describe the evolution of the channelized turbidite systems; interpret the transportation and depositional processes in turbidite systems; and present these elements as analogs to turbidite systems in other intraslope minibasins.

## DATA SET

The seismic data used in the regional 2-D analysis (Varnai, 1996; Weimer et al., 1998a, this issue) (Figure 1) were collected by GSI (now Western Geophysical) during 1985–1986. The data were collected and processed as a 40-fold 3-D seismic survey for the entire area of the regional research project. During 1989, selected lines from this data set were reprocessed and displayed as 2-D migrated profiles. Forty-eight of these seismic profiles (2560 km; 1600 mi) were interpreted in the initial 2-D phase of this study (Varnai, 1996; Varnai and Weimer, 1997). Well-log information from 39 wells and high-resolution biostratigraphic data from 22 wells also were used in that study.

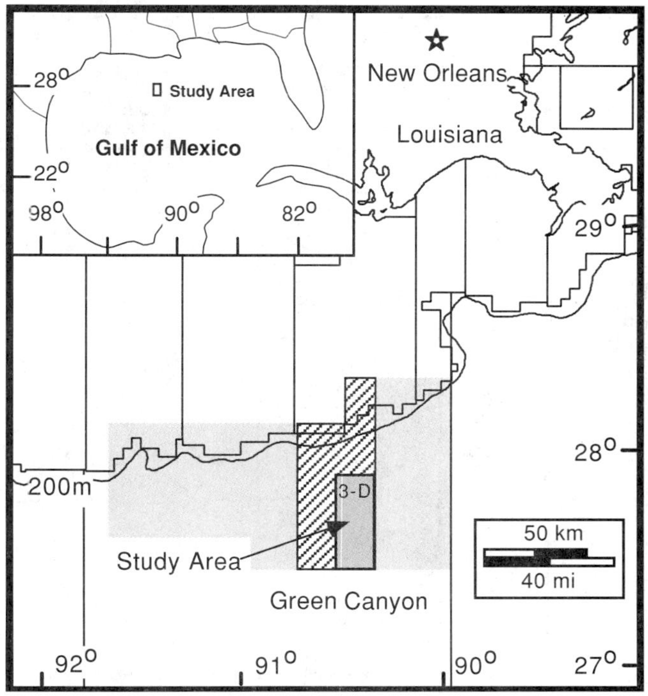


**Figure 1—Map showing location of the study area in the Green Canyon protraction (lease) area, northern Gulf of Mexico. The darker shaded box shows the area of this three-dimensional study, the hachured box shows the two-dimensional work around this three-dimensional study (Varnai, 1996), and the lighter shading shows the area of the entire regional two-dimensional research project (Weimer et al., 1994b, 1998a, this issue).**

The subregional 3-D seismic data set used in this study is a reprocessed subset of the original 3-D survey (Figures 1, 3). The data acquisition parameters were a line spacing of 30 m (100 ft) in inline and 15 m (50 ft) in crossline direction; bin size was 7.5 × 15 m (25 × 50 ft); a single 3.6-km- (2.2-mi-) long streamer was used; and data were recorded on 240 channels. Header information containing seismic processing data was not available; hence the seismic data parameters were estimated from the seismic profiles and the synthetic ties. The stratigraphic interval that was studied rests at depths of 3500–4300 m (11,500–14,000 ft) below sea level (4–5 s two-way traveltime). The seismic data have 20–30-Hz frequency and minimum phase wavelet (Figures 4, 5). Based on the integration of sonic and density logs, the average interval velocity is 2400 m/s (8000 ft/s) in this interval, indicating an average seismic wavelength of 50 m (160 ft). Hence, the vertical resolution and the tuning thickness are calculated as 12–15 m (40–50 ft).

A few data quality problems made interpretation more difficult in some parts of the area. The seismic data were collected in the north-south direction, and many amplitude extraction maps show north-south

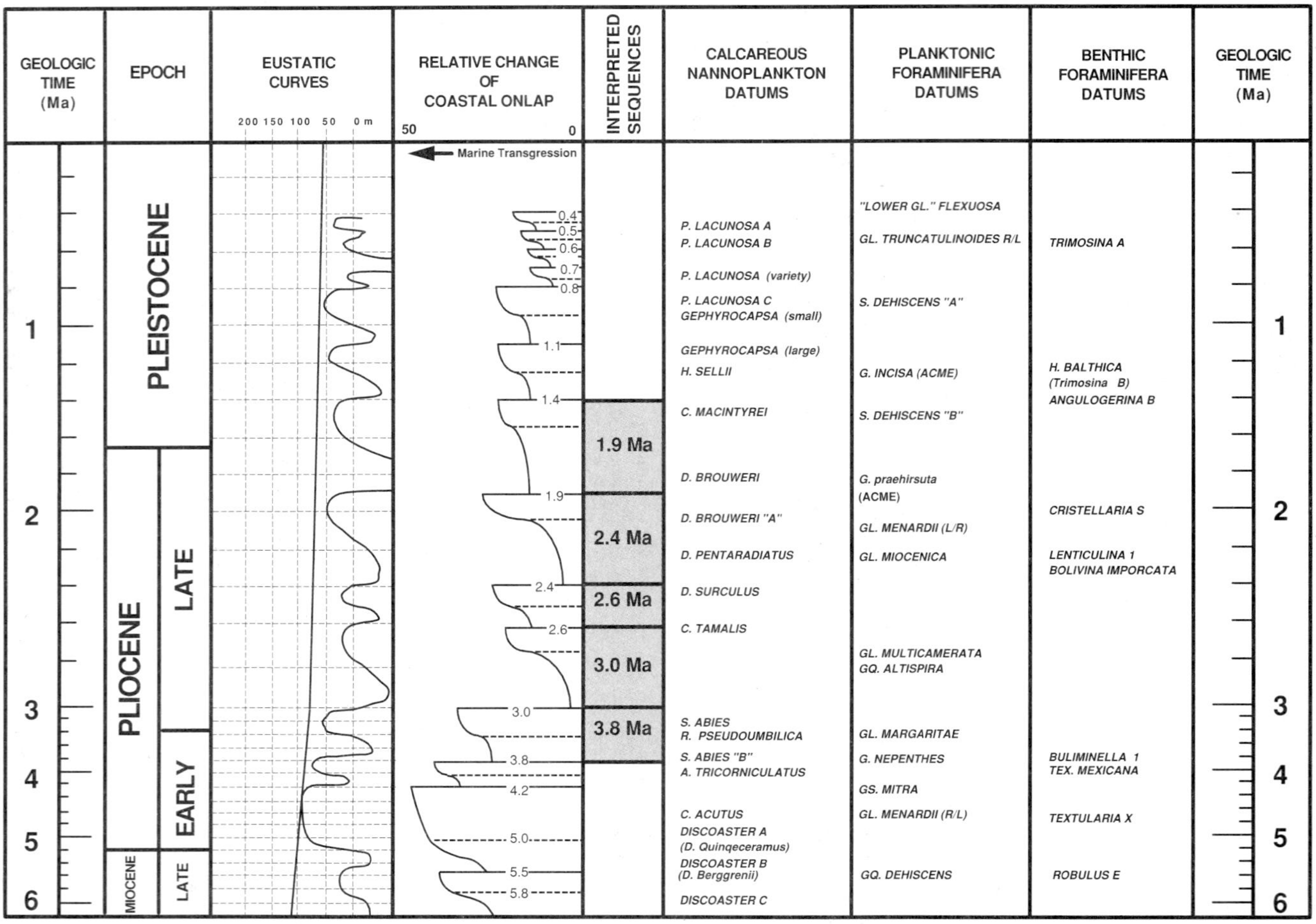


**Figure 2—Pliocene–Pleistocene coastal onlap curve from the Gulf of Mexico [modified after Paleodata–Schlumberger/Geco Prakla (1993) and Pacht et al. (1990)], including the biostratigraphic data. The 3.8-, 3.0-, 2.6-, 2.4-, and 1.9-Ma sequences were interpreted in this subregional three-dimensional study. For the regional two-dimensional interpretation of all the sequences, see Varnai (1996).**

amplitude anomalies, parallel to the inlines. These anomalies are artifacts of the data and do not show the geometry of the turbidites (Figure 6). Because of east-west seismic stacking problems in steeply dipping areas, the "shingled" reflections on the crosslines are interpreted as data acquisition and processing artifacts and not as geologic information (Figures 4, 5). In the northern part of the area, there are large gaps in the 3-D seismic data (Figure 3). To correct map locations between the loaded seismic data batches, four stripes of blank seismic lines had to be added (e.g., Figures 3–7).

Thirteen wells have been drilled in the area, and wireline-log data were available from nine of these wells (Figure 3). Synthetic seismograms and time-based logs were created for every well by integrating the sonic and density logs. No velocity data, including check-shot information, were available. No cores, drill cuttings, or imaging logs (e.g., FMI) were available from the wells, to corroborate the lithologic and depositional environment interpretation. Rafalowski et al. (1994) analyzed numerous cores from the Genesis field (GC-160/161/205, lease blocks; Figure 3); their conclusions were used in this study.

Five wells have high-resolution biostratigraphic data (both faunal abundance curves and calcareous nannoplankton data) (Figure 3). These wells are located nearby the shallow salt in the north-central part of the area or to the north, where seismic data quality is generally poorer. In addition to the well data, the chronostratigraphic framework was established from regional studies by integrating the lithologic, biostratigraphic, and seismic information (Figure 2) (Varnai, 1996; Weimer et al., 1998a, this issue).

The area has two producing fields, Genesis (GC-160/161/205 lease blocks) (Rafalowski et al., 1994) and Popeye (GC-72/73/116 lease blocks), and the King Kong discovery (GC-472/473 lease blocks) (Weimer et al., 1998b, this issue) (Figure 3). Upper Pliocene-lower Pleistocene turbidite sands are the

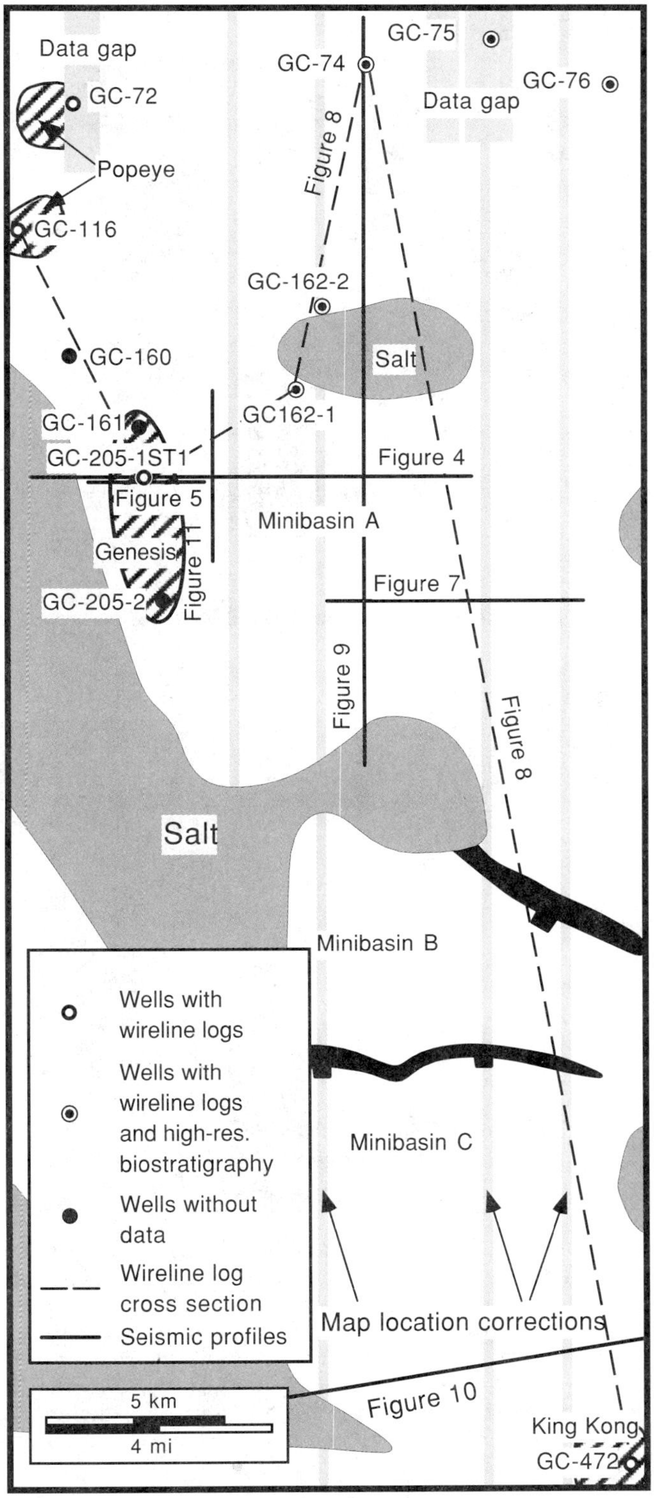


**Figure 3—Detailed location map of the subregional three-dimensional study. Dark shading shows shallow salt bodies; light shading shows gaps in the three-dimensional seismic data on the north and stripes of blank seismic lines that correct map locations (see text for explanation). Hachured areas show the Genesis (GC-205/161 lease blocks), the Popeye (GC-116/72/73 lease blocks) fields, and the King Kong discovery (GC-472/473 lease blocks). The Popeye field has two separate areas, which will be produced independently (Weimer et al., 1998b, this issue). The area was informally subdivided into three minibasins: A, B, and C. Locations of Figures 4, 5, and 7–11 are shown. Location of study area is shown on Figure 1.**

main reservoirs in these fields (2.4- and 1.9-Ma sequences in Genesis field, GC-205-1ST1 well; 1.9- and 1.4-Ma sequences in Popeye field, GC-116-1 well; and 2.4-Ma sequence in King Kong discovery, GC-472-1 well; Figure 8).

## GEOLOGIC SETTING

Deep-water turbidite sedimentation in intraslope minibasins characterized the study area during the Pliocene–Pleistocene (Weimer et al., 1994b, 1998a, this issue; Varnai, 1996). Intraslope minibasins are small localized basins on the slope and represent only a small part of the regional Gulf of Mexico slope system (McBride, 1998, this issue). These basins are bounded by bathymetric highs created by shallow salt bodies and faults. The geometry of these minibasins resulted in the deposition of localized turbidite sediments within the overall slope setting. Several models describe the sequence stratigraphy of these turbidite systems (e.g., Pacht et al., 1992; Mitchum et al., 1993; Pulham, 1993; Weimer et al., 1994b). The modified combination of these models was used to interpret the large-scale stratigraphic framework (Varnai, 1996; Weimer et al., 1998a, this issue).

The entire northern Green Canyon region, including this study area, has a highly complex structural framework consisting of shallow salt bodies, faults, and minibasins (Rowan, 1995; McBride, 1998, this issue; McBride et al., 1998, ths issue). Allochthonous salt forms two shallow bodies (Figures 3, 9). Structural restorations indicate that these salt bodies were deformed extensively during the last 5 m.y.; this deformation changed the shapes of the minibasins (Weimer et al., 1994b, 1998a, this issue; Rowan, 1995; Rowan et al., 1995; McBride et al., 1998, this issue). Changes in the rates and location of salt withdrawal resulted in the development of a large turtle structure south of the area (Figure 10). The shallow salt bodies formed bathymetric highs during the deposition of the studied sequences, focusing sedimentation into the central parts of the minibasins (Figure 9). Stratigraphic interpretations and structural restorations also show that the study area has been a relatively long-lived sediment fairway for the past 5 m.y. (Rowan, 1995; Rowan et al., 1995; Rowan and

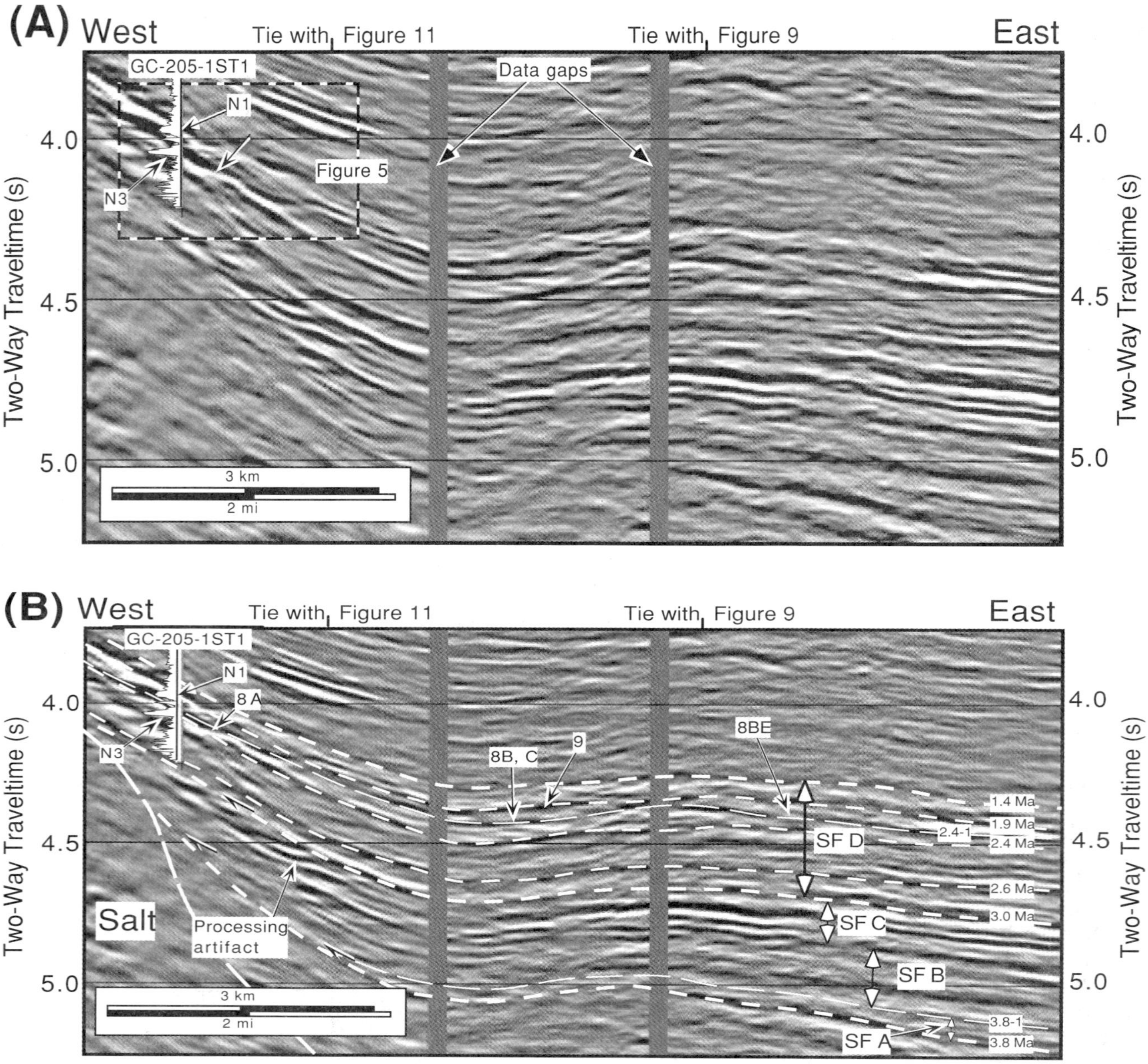


**Figure 4—Depositional strike-oriented three-dimensional seismic profile across the Genesis field (GC-205-1ST1 well) showing the 3.8–1.4-Ma sequences, (A) uninterpreted and (B) interpreted. The discontinuous high-amplitude reflections within the 3.0-1.4-Ma sequences are interpreted as depositional channels. Note the seismic data processing related "shingled" reflections, which made the interpretation difficult around shallow salt bodies. Anomalously high-amplitude reflections around the field are caused by the presence of hydrocarbons. See text for description of the SF A–SF D facies (SF = seismic facies). The channels discussed in this paper are numbered according to their relative age, as shown on Table 1 and Figure 13. Dashed box shows the area of Figure 5. Location of profile is shown on Figures 3, 6, 12, 14–16, and 19.**

Weimer, 1998, this issue; Weimer et al., 1998a, this issue). For this study, the area was divided informally into three minibasins, from north to south, A, B, and C (Figure 3).

Based on the results of the regional 2-D study (Varnai, 1996; Weimer et al., 1998a, this issue), five upper Pliocene and lower Pleistocene sequences (3.8–1.4 Ma) were chosen for the detailed 3-D study (Figures 1, 2). Well control and seismic data quality are better in this area than most of the regional 2-D data set (Figure 1). Two major units were deposited in this interval with markedly different turbidite architecture. The 3.8-Ma sequence consists of sheetlike sediments (lobes) at the base,

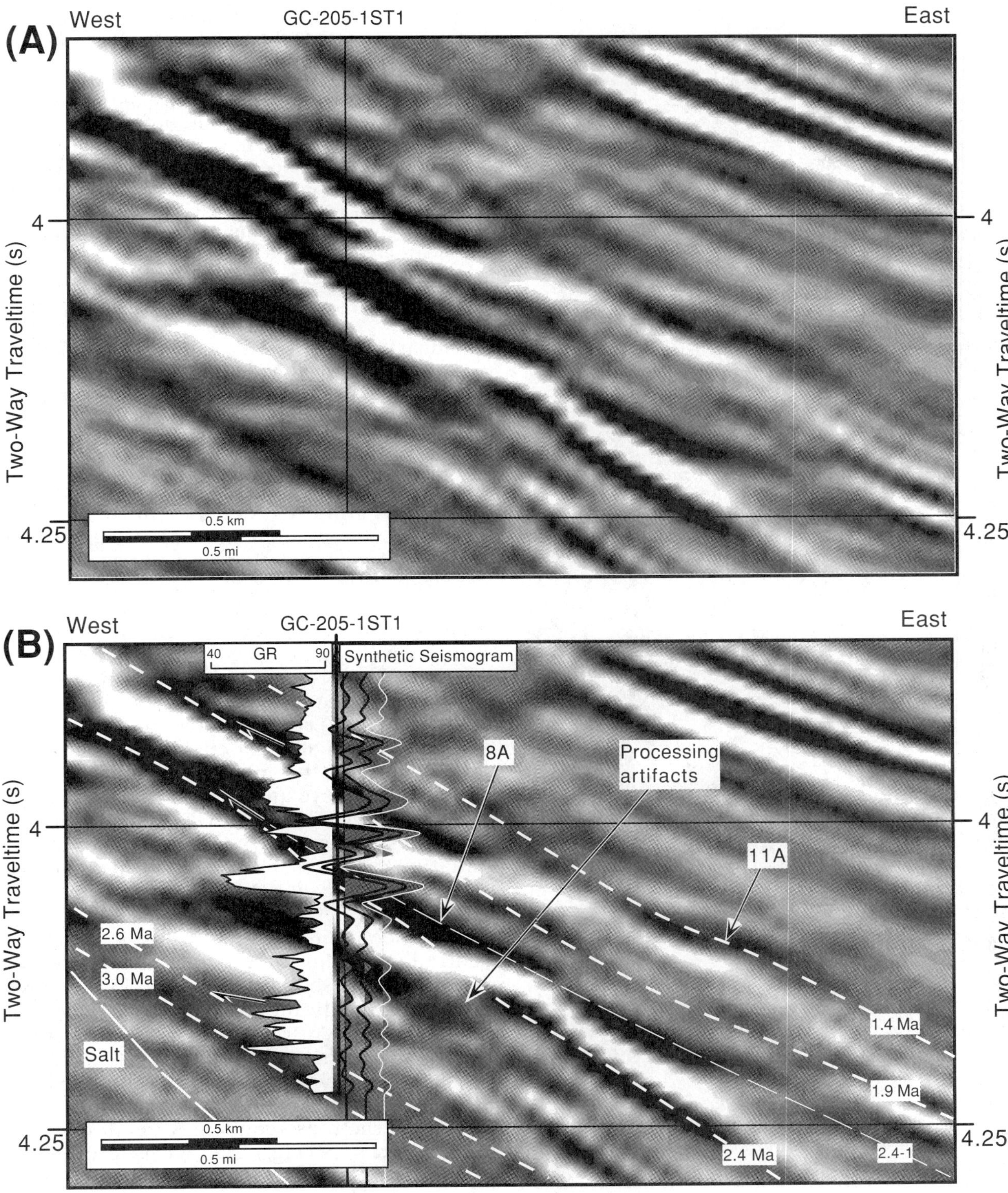


**Figure 5**—Depositional strike-oriented three-dimensional seismic profile across the Genesis field (GC-205-1ST1 well); enlarged detail of Figure 4, (A) uninterpreted and (B) interpreted. The synthetic seismogram and the time-based gamma-ray log show the well-to-seismic tie. The synthetic traces were created using 35-Hz normal polarity Ricker wavelet. Note that the positive amplitudes of the seismogram are at the base of the hydrocarbon bearing reservoir sands that are interpreted as channel-fill sands. These amplitudes correspond to the high-amplitude reflections of the seismic profile. This well-to-seismic tie suggests that the discontinuous high-amplitude reflections observed on the seismic profiles, where they have sinuous shape on the horizon slices (Figure 14), correspond to sand-rich channel-fill deposits. The channels are labeled similar to those in Figure 4. Location of profile is shown on Figures 3, 14–16, and 19.

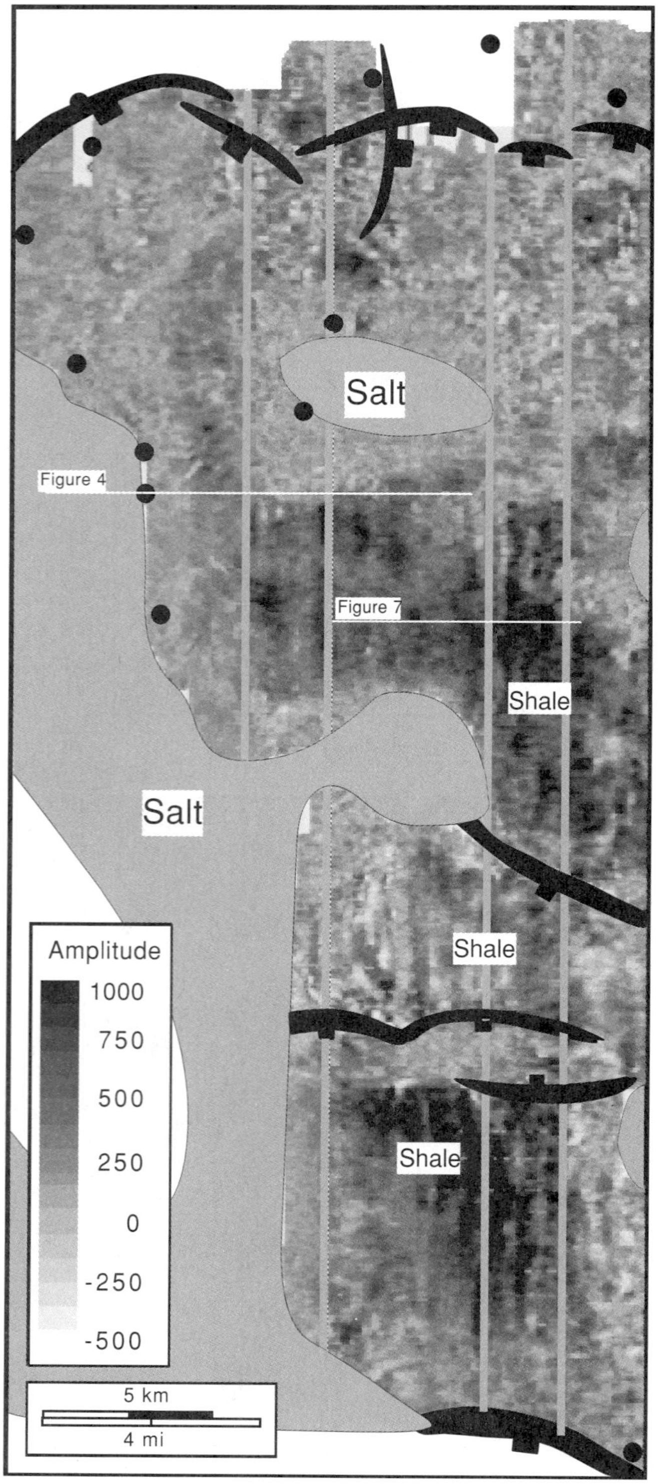


Figure 6—Amplitude extraction map along the 3.8-Ma sequence boundary (Table 1; Figures 4, 7, 9). The map shows widespread, roughly circular high-amplitude areas (black) in the central part of the minibasins. Within these areas is a linear, north-south amplitude pattern, which is an artifact of data acquisition and processing (the survey was collected in a north-south direction). The map was interpreted to represent the condensed section directly below the sequence boundary (correlative conformity). Location of wells is shown by black dots; see Figure 3 for well names. Solid lines show the seismic profiles (Figures 4, 7).

changing upward to more shaly, channelized turbidite systems (Varnai, 1996). The 3.0–1.4-Ma interval contains four depositional sequences (Figure 2) consisting of channel systems and overbank sediment interbedded with hemipelagic shale. These four younger sequences and the individual channel systems were not mappable on the 2-D data set (see Varnai, 1996; Weimer et al., 1998a, this issue) because of the relative thinness of the sequences, the widely spaced seismic lines, and the lower resolution of the seismic data.

The four sequences of the 3.0–1.4-Ma (late Pliocene to earliest Pleistocene) interval are thinner than any other sequence in the area, indicating reduced sedimentation rates. Decrease in sedimentation rates is interpreted to be the result of decreased sediment supply because the ancestral Mississippi Delta avulsed from north of the area to the west (Pulham, 1993). The major sediment source of the area no longer provided large amounts of sediments to this area, resulting in considerably less sediment accumulation in the area. However, significant channel-fill sands were deposited during this interval, forming the reservoirs at the Popeye and Genesis fields and King Kong discovery (GC-116-1, GC-205-151, and GC-472-1 wells; Figures 3, 8). Increased rates of sedimentation resumed about 1.4 m.y. ago and continued to the present (Villamil et al., 1998, this issue).

In this deep-water setting, deposition primarily occurred when the sea level was at a relative low position and the shelf was exposed (lowstand systems tract). Significant condensed sections developed when the entire shelf was flooded (transgressive and highstand systems tracts), and they are the deep-marine equivalent of the maximum flooding surfaces. The sequence boundaries directly overlie these condensed sections as correlative surfaces (Figure 2) with no erosional boundary separating the two sequences.

## METHODOLOGY AND INTERPRETATION TECHNIQUES

The most important objective of the study was to characterize the seismic facies using 3-D seismic data and to analyze the lateral variations within

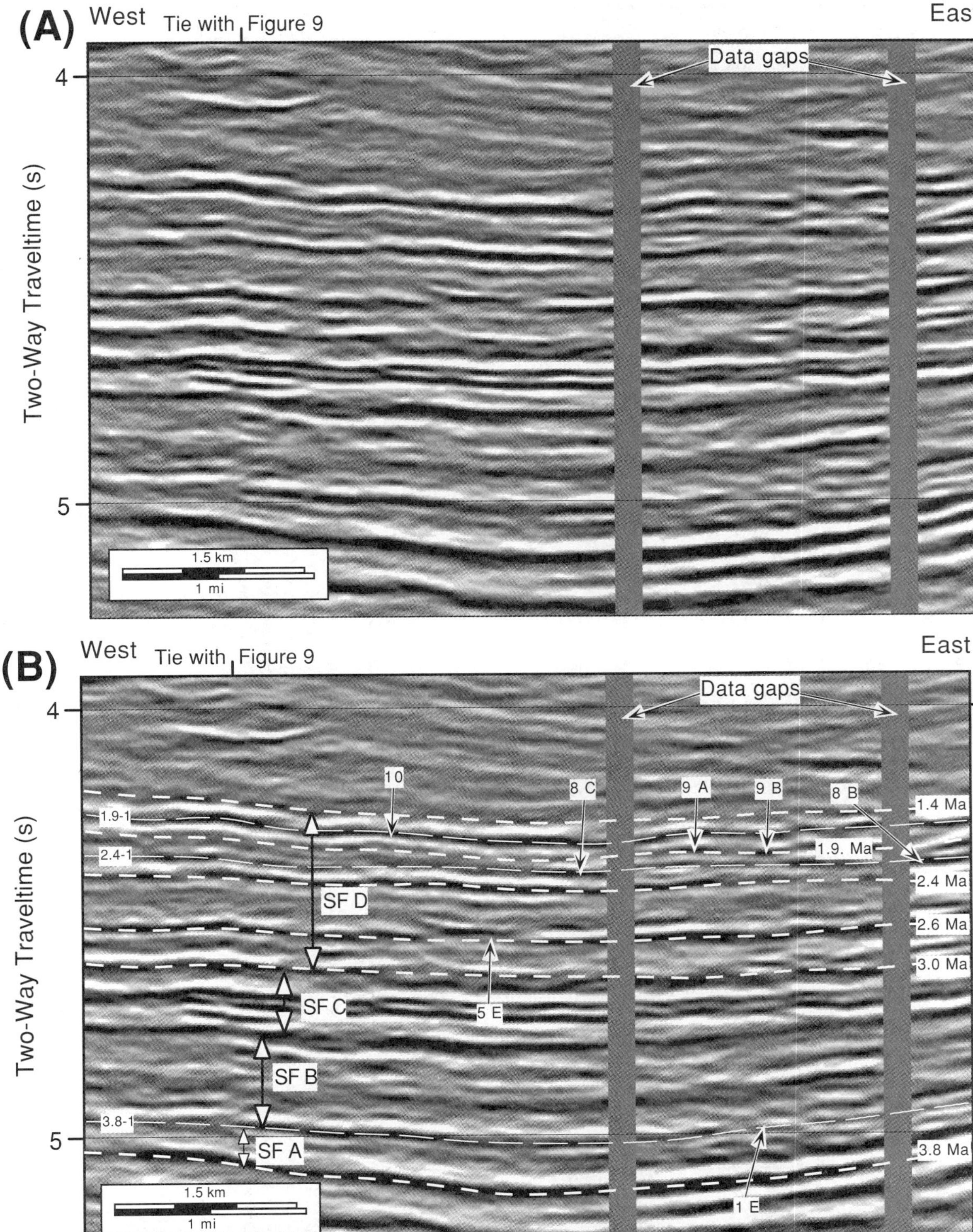


**Figure 7**—Depositional strike-oriented three-dimensional seismic profile showing the 3.8–1.4-Ma sequences, (A) uninterpreted and (B) interpreted. The 2.6-Ma horizon is a high-amplitude continuous reflection across most of the seismic profile interpreted as the condensed section directly below the sequence boundary. The low-amplitude area in the central part of this horizon is interpreted as an erosional channel (5E). The channel fill is represented by a continuous reflection because the seismic profile is almost parallel to the axial direction of channel 10 (Figure 13). See text for description of the SF A–SF D facies (SF = seismic facies). The channels are labeled similar to those in Figure 4. Location of profile is shown on Figures 3, 6, 12, 14–16, and 19.

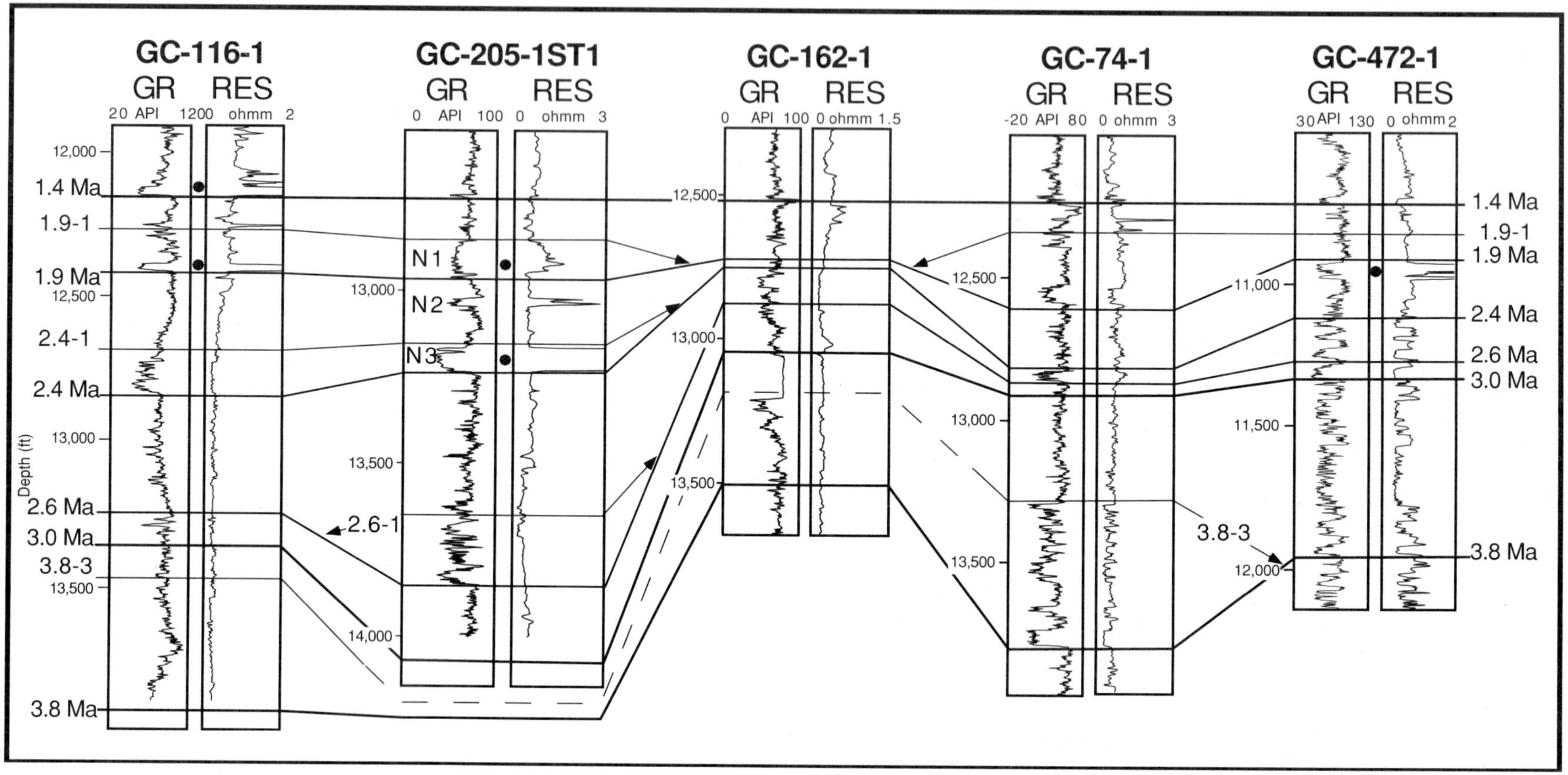


**Figure 8**—Wireline log cross section of five wells of the area shows the different lithology of the interpreted sequences. For example, in the GC-74-1 well, the 3.8-Ma sequence has thick sands with blocky gamma-ray log pattern at the base (sheet sands), overlain by shale with only a few thinner sands (channel sands in overbank shales). The other wells penetrate this sequence on the flanks of the minibasin in shale-dominated areas. The thick sands with blocky gamma-ray log pattern overlying the 2.4- and 1.9-Ma sequence boundaries in the GC-205-1ST1 well (N1 and N3 sands, Genesis field) were interpreted as amalgamated channel-fill sands (Rafalowski et al., 1994). Similar sands overlie the 1.9- and 1.4-Ma sequence boundaries in the GC-116-1 well (Popeye field) and are found in the upper part of the 2.4-Ma sequence in the GC-472-1 well (King Kong discovery). These sands, marked by black dots, comprise the main reservoirs in these fields. Note the low-resistivity, salt-water saturated sands (e.g., GC-472-1 well). The section is not drawn to scale and does not show the shallow salt bodies and the growth faults. See Table 1 for the explanation of correlated horizons. Location of wells is shown on Figure 3.

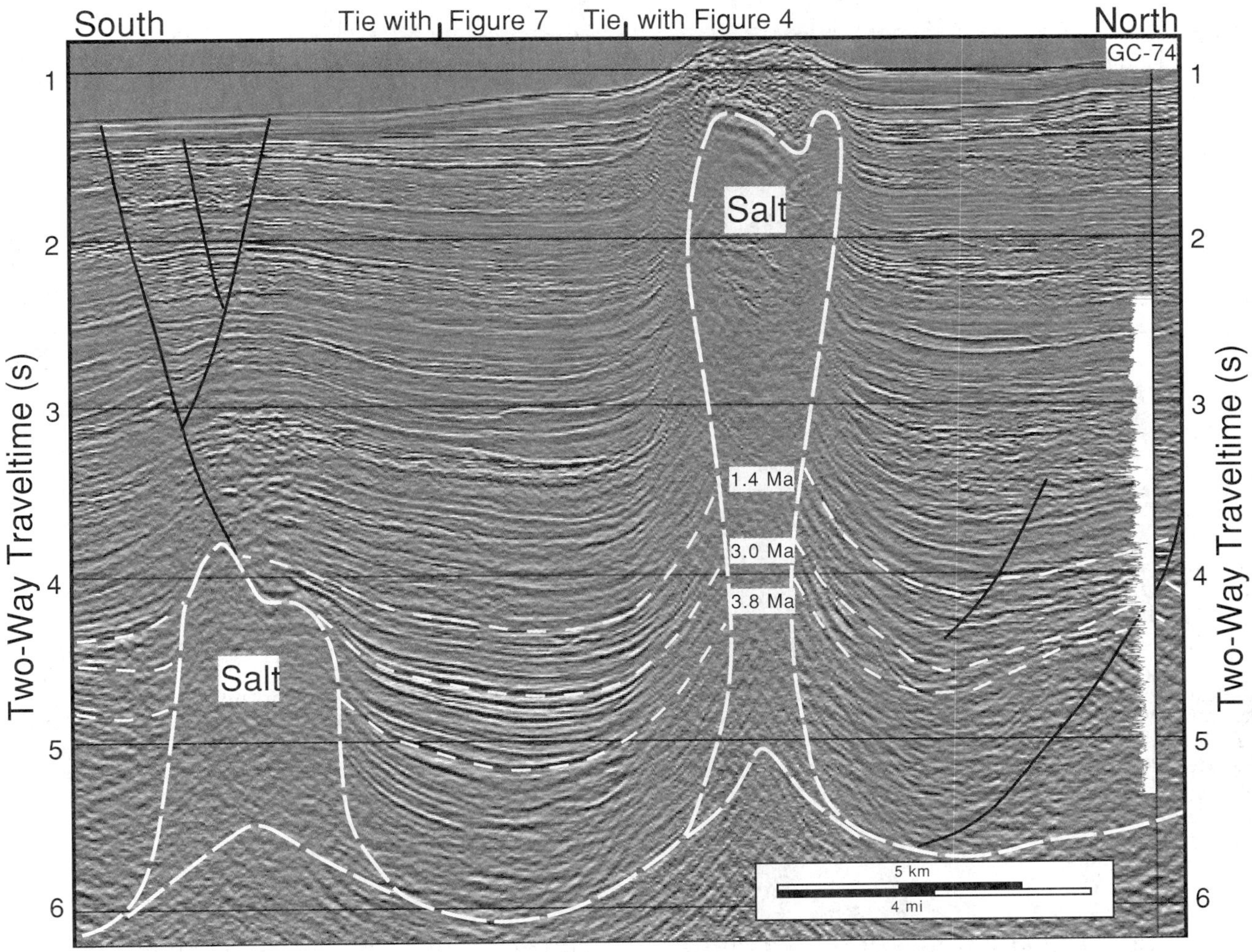


**Figure 9—Depositional dip-oriented subregional three-dimensional seismic profile shows the complex structural setting of the area: shallow salt bodies, growth faults, and the intraslope minibasins filled with thick Pliocene–Pleistocene strata. The studied interval (3.8–1.4 Ma) consists of subparallel reflections with moderate amplitude and moderate to fair continuity. Time-based gamma-ray log of the GC-74 well shows the thick sand bodies at the lower part of the 3.8-Ma sequence and the characteristic shale in the younger sequences. Notice that the salt body in the center is a bathymetric high on the present sea floor. See Figure 3 for location of profile.**

the different turbidite systems. Turbidite sands form the reservoirs in many fields and discoveries in the Gulf of Mexico and in other deep-water exploration areas. Therefore, understanding the geometries of these sand bodies is important for better reservoir prediction and modeling. To achieve this goal, some standard and some new interpretation techniques were used.

## Sequence Boundary Definition

Initially, the sequence boundaries were identified on the 2-D regional data set based on seismic reflection terminations and configuration and on biostratigraphic data (Varnai, 1996; Varnai and Weimer, 1997). The sequence boundaries were then transferred to the 3-D seismic data for detailed correlation. The 2-D seismic profiles are a subset of the original 3-D data, hence there was a good correlation between the regionally identified sequence boundaries and the 3-D data.

In interpreting 3-D seismic data, it is important to distinguish two kinds of reflection discontinuities. The first kind of reflection discontinuities are reflections that onlap onto older horizons. These reflections commonly onlap onto sequence boundaries or condensed sections of the underlying sequence (e.g., sequence boundaries on Figures 4 and 11). The second type of reflection discontinuity is caused by lateral facies and lithology changes within the sequences, as in the

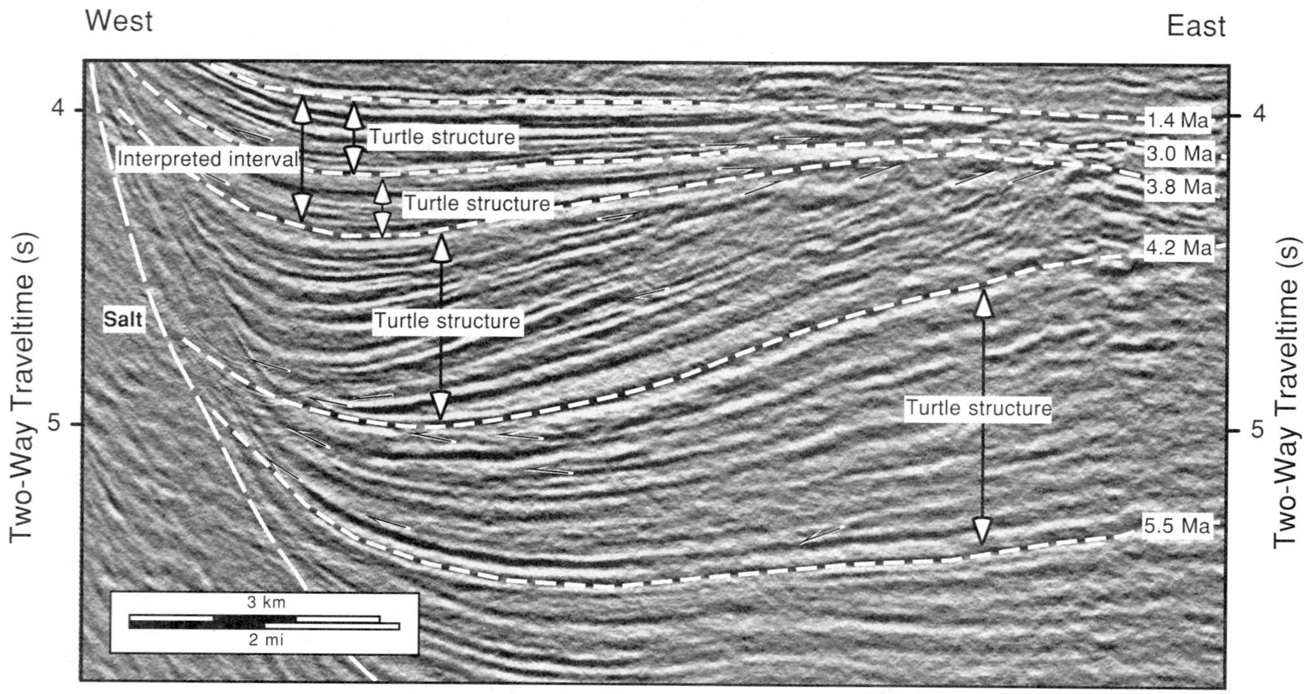


**Figure 10—Depositional strike-oriented subregional two-dimensional seismic profile from the southern part of the area showing the 5.5–1.4-Ma sequences. Turtle structures show the migration of the depocenter in this minibasin from the east to the west. The divergent reflections and external wedge shape in all sequence boundaries indicate syndepositional salt movement. Onlapping reflections onto the 5.5-, 4.2-, and 3.8-Ma sequences boundaries, and reflections truncated below the 4.2- and 3.8-Ma sequence boundaries are associated with salt tectonics. Notice that the sequences within the 3.0–1.4-Ma depositional interval were not differentiated on the two-dimensional seismic profiles. See Figures 3, 17, and 18 for the location of the profile.**

lateral thinning and pinch-out of channel-fill sands in overbank sediment (e.g., horizon 2.4-1 on Figures 4B, 7B, and 11B), or low-amplitude areas in the reflections caused by channel erosion (e.g., horizons 3.8-1, 2.6, and 1.9-1 on Figure 7B). These discontinuities generally are not associated with reflection terminations and are not associated with the sequence boundaries.

## Amplitude Extraction Maps

The methodology used in this study to describe the 3-D geometries of the different seismic facies is based on the detailed correlation of individual reflections. The high density of the 3-D data allowed this type of mapping. All sequence boundaries within the 3.8–1.4-Ma interval and many internal reflections within the sequences were correlated (Table 1; Figures 4, 7, 9). The peak (positive maximum) amplitude values were extracted along these horizons, and a series of 13 amplitude maps was created. The next step was analyzing amplitude distribution in these maps. In this study high- or low-amplitude areas are defined based on their relative amplitude compared to their surroundings on the amplitude extraction maps (Figure 12).

## Horizon Slices

A series of horizon slices was created to describe the stratal geometries between the correlated horizons and to analyze the distribution of high- and low-amplitude areas through the studied section. Similar to the amplitude extraction maps, the amplitude distributions of horizon slices (high- and low-amplitude areas) are interpreted to reflect the geometries of the turbidite elements. Horizon slices are horizontal cuts of flattened 3-D seismic volumes. The seismic data are typically flattened on major condensed sections or sequence boundaries (reference horizon), removing the postdepositional structural overprint. This technique allows one to directly image the depositional surfaces and strata. Risch et al. (1996) studied the overlying 1.4–1.1-Ma sequence on the northwestern part of this study area, and showed that horizon slices are an effec-

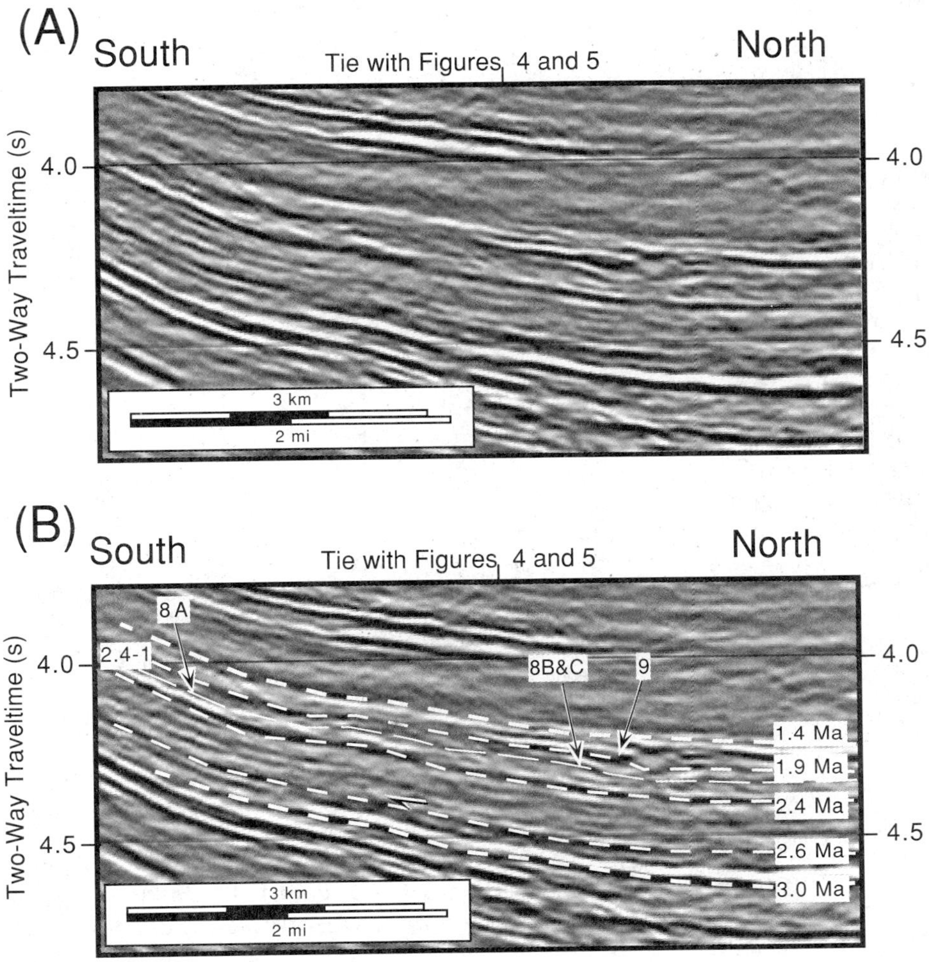


**Figure 11—Depositional dip-oriented three-dimensional seismic profile showing the 3.0–1.4-Ma sequences, (A) uninterpreted and (B) interpreted. The channels are labeled similar to Figure 4. Location of profile is shown on Figures 3, 14-16, and 19.**

tive method to filter out most of the structural effects of steep dips.

In the study area, syndepositional salt withdrawal caused significant thickness changes within all sequences, limiting the effective thickness of the flattening to about 100–150 ms below the reference horizons. In this data set, horizon slices deeper than 150 ms below the reference horizons show oblique cuts of reflections, and not the true stratigraphic geometries; hence, separate sets of horizon slices were created for each sequence. The major condensed sections and the overlying sequence boundaries form a single seismic reflection in the area; therefore, all the correlated sequence boundaries were used as reference horizons (Table 1).

Five sets of seismic volumes were created; they were flattened on the 1.4,- 1.9-, 2.4-, 2.6-, and 3.0-Ma sequence boundaries. Horizon slices were then created every 4 ms below the flattened horizon, creating sets of horizon slices. The results were a separate series of horizon slices for each sequence; in each case, the seismic data were flattened on the overlying sequence boundary.

## Lithologic Ties

The available well data were integrated with the seismic profiles through synthetic seismograms and time-based logs (e.g., GC-205-1ST1 well; Figure 5). No data from velocity surveys were available to check these ties. The synthetic seismograms, however, showed good correlation with the seismic data, which suggests correct time-depth conversions (Figure 5). Lithologic interpretation also was limited by the few well penetrations (Figure 3). Therefore, lithologic estimates in most cases were based on the interpretation of the 3-D reflection geometries and on geophysical considerations. Reflections are derived from acoustic impedance contrasts, such as lithological boundaries (shale-sand), and reflection-free areas may represent no or subtle lithologic changes (shale-shale).

## Depositional Environment and Lithologic Interpretations

The depositional environments of different turbidite strata are routinely interpreted from wireline-log patterns (e.g., Mitchum et al., 1993; Chapin et al., 1994; Mahaffie, 1994). Based on these studies, three primary kinds of turbidite sands were interpreted: (1) sheet sands (basin-floor fan), (2) channel-fill sands, and (3) thin overbank sands.

Elongate, slightly sinuous low- and high-amplitude areas, lobe-shaped high-amplitude areas, and low- to moderate-amplitude areas without any recognizable patterns were observed on the amplitude extraction maps and horizon slices (Table 2). For the depositional environment interpretation, for the identification of the different turbidite elements, these geometries were integrated with the isopach maps, lithologic and biostratigraphic information from well logs, and seismic facies information from the vertical profiles (Table 2; Figure 13).

**Table 1. Correlated Sequence Boundaries and Horizons Within Sequences Used to Describe Turbidite System Three-Dimensional Geometries***

| Sequence Boundaries | Internal Horizons | Mapped Channels |
|---|---|---|
| 1.4 Ma | | 11a, b |
| | 1.9-1 | 10, 10E |
| 1.9 Ma | | 9A, B |
| | 2.4-1 | 8A, B, C, 8BE** |
| 2.4 Ma | | 7a, b |
| | 2.6-1 | 6A, B |
| 2.6 Ma | | 5E |
| | 3.0-1 | 4 |
| 3.0 Ma | | 3E |
| | 3.8-3 | None |
| | 3.8-2 | 2 |
| | 3.8-1 | 1E |
| 3.8 Ma | | None |

*Channels in the 2.6-Ma, 2.4-Ma, and 1.9-Ma sequences (6, 8, and 9, respectively) bifurcate downslope. In the mapped channels column, "A" indicates the oldest channel and "C" the youngest channel; lowercase "a" and "b" do not indicate relative age (coeval channels; e.g., 7a, b). "E" indicates erosional channels; all other channels are depositional channels.

**Channel 8BE is an erosional channel, its relative age is 8B.

## Pitfalls in Interpretation of Amplitude Patterns

The patterns observed on the amplitude maps and horizon slices, combined with well logs and vertical seismic profiles, were interpreted to indicate the stratal geometries of the turbidite systems. However, several other factors also affected the distribution of reflection amplitude. The presence of hydrocarbons anomalously increased the seismic amplitude around the Genesis field (Figures 5, 14–16). No anomalously high amplitudes were found in the central parts of the minibasins, so this fluid effect is not interpreted to be present there.

Based on the frequency and interval velocity of the seismic data, the tuning thickness for the studied interval is estimated to be 12–15 m (40–50 ft). The sand bodies penetrated in the wells (e.g., GC-205-1ST1; Figure 8) are up to 30 m (100 ft) thick, which is considerably thicker than the tuning thickness; therefore, the amplitude distribution is interpreted to represent lithologic changes rather than thickness variations.

# DEFINITION OF TURBIDITE ELEMENTS

Mutti and Normark (1987, 1991) defined five primary turbidite elements common to modern and ancient turbidite systems: major erosional features (including slides), channels, overbank deposits, depositional lobes, and channel-lobe transition deposits. Most of the sediments in the studied sequences are interpreted to have been deposited in basin-floor fans (depositional lobes), channels, and overbanks, hence, their definitions are briefly reviewed here. Slides affected only relatively small portions of the sequences, and channel-lobe transition deposits are rarely resolvable on conventional multifold seismic data used in this study.

Depositional lobes are defined as roughly tabular, nonchannelized sand bodies that have an individual thickness of generally 3–15 m (10–50 ft) (Mutti and Normark, 1991). In ancient systems individual sandstone lobes commonly form vertical successions up to several hundred meters thick. In these sand packages the sand layers alternate with variable thicknesses of mudstone and thinner bedded and finer grained sandstone.

Channels are defined as relatively long-lived erosional features associated with sediment transport pathways (Mutti and Normark, 1991). As the channels ceased to transport sediments, they were filled. Three main types of channel-fill deposits are commonly observed in ancient turbidite systems: (1) erosional channel-fill deposits, where basal coarse-grained strata are overlain by fine-grained channel abandonment deposits, (2) depositional channel-fill deposits, which form sand beds with characteristics similar to the depositional lobes on the wireline logs, and (3) mixed channel-fill deposits, a transition between the two end members. Both erosional and depositional channel-fill deposits were recognized in the studied sequences.

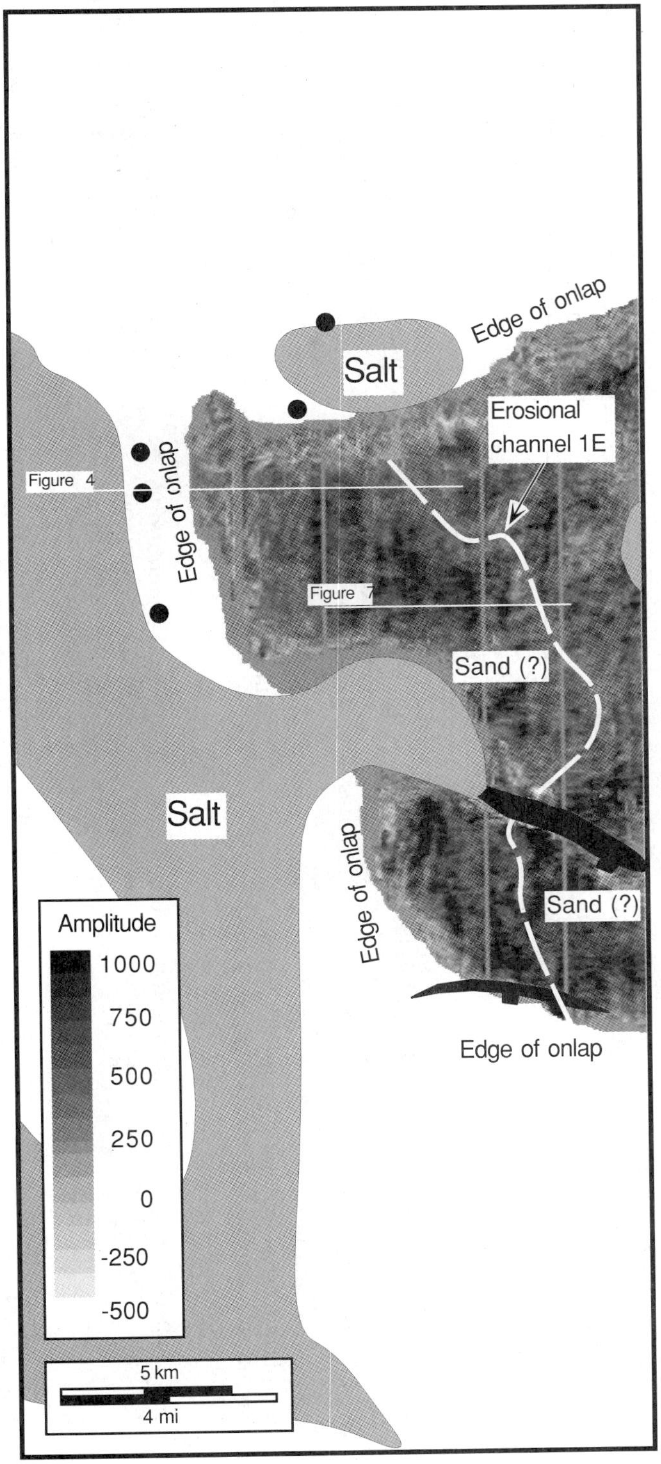


**Figure 12—Amplitude extraction map along the 3.8-1 horizon (Table 1; Figures 4, 7). The map shows lobe-shaped high-amplitude anomalies (black) in the lower part of the 3.8-Ma sequence. The horizon was interpreted as a basin-floor fan (stacked turbidite lobes), that onlaps onto the underlying sequence boundary on the north, west, and south. There is a slightly sinuous low-amplitude (light gray) area in the central part of mini-basin A. The feature was interpreted as an erosional channel, which eroded parts of the basin-floor fan and then was filled with shale. Location of wells is shown by black dots; see Figure 3 for well names. Solid lines show the seismic profiles (Figures 4, 7).**

Overbank deposits are laterally extensive, fine-grained, thin-bedded turbidite sediments adjacent to the channels in a turbidite system (Mutti and Normark, 1991). The overbank deposits are divided into two main groups: (1) channel-related overbank deposits (levee) and (2) overbank wedges. Minor depositional channels and associated thin-bedded sandstones may form a volumetrically small component of the overbank sediments.

## THREE-DIMENSIONAL FACIES DESCRIPTION OF SEQUENCES

The vertical 2-D and 3-D seismic profiles show similar seismic facies for all sequences between 3.8 and 1.4 Ma; these are primarily subparallel reflections with moderate to low amplitude and continuity (Table 2; Figures 4, 7, 9, 11). The 3.8–3.0-Ma sequence has three discrete, stacked vertical seismic facies (Figures 4, 7). At the base, subparallel reflections have moderate to high amplitude and continuity (SF A, Figures 4, 7). In the middle they have subparallel to hummocky reflections with fair amplitude and continuity (SF B, Figures 4, 7), and in the upper part they have subparallel reflections with high amplitude and good continuity (SF C, Figures 4, 7). The 3.0–1.9-Ma sequences also have subparallel reflections with moderate to fair amplitude and continuity (SF D, Figures 4, 7), which is similar to the middle part (SF B, Figures 4, 7) of the underlying sequence. Seismic facies B and D are widespread, consisting of most of the sequences, whereas facies A and C are restricted to the central part of the area at the base and top of the 3.8 Ma sequence (Figures 4, 7, 11).

The previously described methods of amplitude mapping show the lateral variations of the seismic facies. These 3-D geometries showed distinct differences interpreted to reflect the different turbidite elements.

Due to space limitations, the 3.8- and 2.4-Ma sequences are described to illustrate the variations in seismic facies, turbidite elements, and interpreted depositional processes in the minibasins. An example of a basin-floor fan (amalgamated depositional lobes) is shown from the 3.8-Ma sequence (Figure 12), and examples of different channels and overbank sediments are shown from the 2.4-Ma sequence

**Table 2. Seismic Facies and Characteristics of Turbidite Elements**

| Turbidite Element | Horizontal Seismic Facies | Vertical Seismic Facies* | Lithology** | Thickness (m) | Size |
|---|---|---|---|---|---|
| Erosional channel | Slightly sinuous low-amplitude anomaly | Low-amplitude area cut into a continuous reflection | No well penetration; channel fill is shale; interpreted from seismic | 25–35 | 2 to >30 km in length |
| Depositional channel | Slightly sinuous high-amplitude anomaly | High-amplitude discontinuous reflections (SF B, D) | 15–37-m-thick sands with blocky and crescent gamma-ray log pattern | 25–50 | 20 to >40 km in length |
| Overbank deposits | Widespread moderate-amplitude areas without significant amplitude variations | Subparallel, moderate-amplitude, and continuous reflections (SF B, D) | Shale with interbedded thin sands | Hundreds | Widespread |
| Depositional lobes (basin-floor fan) | Lobe-shaped high–amplitude areas | Onlapping high-amplitude continuous reflections (SF A) | No well penetration; sand-rich; based on well analogies of adjacent areas | 150 | Larger than 142 $km^2$ in areal extent |
| Condensed section overlain by sequence boundary | Widespread high-amplitude areas | High-amplitude continuous reflections (SF C) | Shale; well penetrations show increased faunal abundance | 13–15 | Widespread |

*SF = seismic facies.
**Well ties or seismic interpretation.

(Figures 13–16). The detailed description of the three other sequences (Figure 2) are included in Varnai (1996).

### 3.8–3.0-Ma Sequence

The 3.8–3.0-Ma sequence has the greatest variations in thickness within the 3.8–1.4-Ma interval (Figure 17). The average thickness of sediments in this sequence is 360 m (1200 ft). The sequence is the thickest on the southern part of minibasin A, where a large symmetric subbasin formed, with sediments up to 610 m (2000 ft). Minibasin B has asymmetric basin fill that thins significantly toward the west; minibasin C has asymmetric fill with thinning to the east (Figure 17). The later structural overprint on the basin caused some interpretation problems by greatly decreasing the effective flattening thickness of the horizon slices; therefore, they were not used in interpreting this sequence.

A high-amplitude, laterally continuous reflection was mapped at the 3.8-Ma sequence boundary on the seismic profiles (Figures 4, 7, 9). This horizon has relatively even high-amplitude distribution on the amplitude extraction map (Figure 6). Based on this laterally continuous seismic character and the increased faunal abundance values and shale observed in wells (Varnai, 1996), this horizon was interpreted as a condensed section (Figure 8). The sequence boundary (correlative conformity) directly overlies the condensed section.

High-amplitude, continuous reflections were mapped at the base of the 3.8-Ma sequence to the southeast (SF A; Figures 4, 7). The amplitude extraction map (horizon 3.8-1, Table 1; Figures 4, 7) shows lobe-shaped, high-amplitude areas in the central part of the minibasins, and lower amplitude areas on the flanks (Figure 12). The horizon onlaps onto the 3.8-Ma sequence boundary to the north, west, and south. The only well penetrates a low-amplitude area close to a shallow salt body, indicating shale on the flanks of the minibasin (GC-162-1 well; Figure 8). No well penetrates the high-amplitude area; therefore, lithologic interpretation is based on analogs from the areas to the east, such as the reservoir in the Allegheny discovery (GC-254 well) and in the GC-301 well (see Weimer et al., 1998a, b, this issue). These wells show thick sand bodies with blocky to crescentic gamma-ray log patterns (sensu Mitchum et al., 1993) in that interval. Hence, the high-amplitude reflection areas are interpreted as laterally continuous stacked sheet sands (amalgamated depositional lobes) in the central parts of the minibasins, changing laterally to shaly sediments on the flanks of the basin.

Using the sequence stratigraphic terminology, this seismic stratigraphic character and position of this unit would be interpreted as a basin-floor fan (Mitchum et al., 1993). However, due to the lack of well penetration, the basin-floor fan terminology is used only as a description of the observed seismic character and does not imply depositional processes or internal sediment architecture. The maximum thickness of the sediment that corresponds to these reflections is 150 m (500 ft), decreasing toward the flanks of the minibasins. The mapped areal extent of the fan is 142 $km^2$ (55 $mi^2$) (Figure 13A), but the total area is larger, because the eastern portion of the fan extends out of the study area (Figures 4, 12).

The amplitude extraction map also shows a slightly sinuous low-amplitude area in the central part of minibasin A (Figure 12), which corresponds to a low-amplitude area on the seismic profiles (Figure 7). This feature was interpreted as an erosional channel, that eroded parts of the basin-floor fan. No wells penetrate this channel. Based on the low reflection amplitude associated with the channel fill, the channel is interpreted to be filled with shale. This shale has lower acoustic impedance contrast to the overlying overbank sediment than the sands of the lobe around the channel. Such erosional channels are recognized both in modern (e.g., Navy fan) (Mutti and Normark, 1991) and ancient turbidite systems (e.g., Clark and Pickering, 1996) (Mississippi Canyon Mars minibasin, M. Mahaffie, 1996, personal communication). Similar erosional channels also were found in the overlying sequences (Table 3; Figure 13A).

The isopach map indicates that the thickness distribution in this sequence is controlled by active syndepositional salt withdrawal (Figure 17) (see also Varnai, 1996). The sediments in minibasin C form the upper part of a turtle structure (Figure 10). Sediments there are considerably thinner than in the northern part of the area, which may indicate the decreasing rate of salt withdrawal in the upper part of the turtle structure (Figure 10). Higher rates of salt withdrawal continued on the north.

The basin-floor fan is located around the thickest part of the area (deepest part of the minibasin at the time of deposition), and the edge of the lobe correlates with the 300 m (1000 ft) isopach contour (Figure 17). Both the depositional and erosional channels are located in the thickest parts of minibasins A and B. The southern part of the area was a relative high during this sequence, and both the lobe and the channels were deflected to the east from this minibasin (Figure 13A).

### 2.4–1.9-Ma Sequence

This sequence is relatively thin, with an average thickness of 90 m (300 ft) and a maximum of 150 m (500 ft) in minibasin A (Figure 18). The sediments in

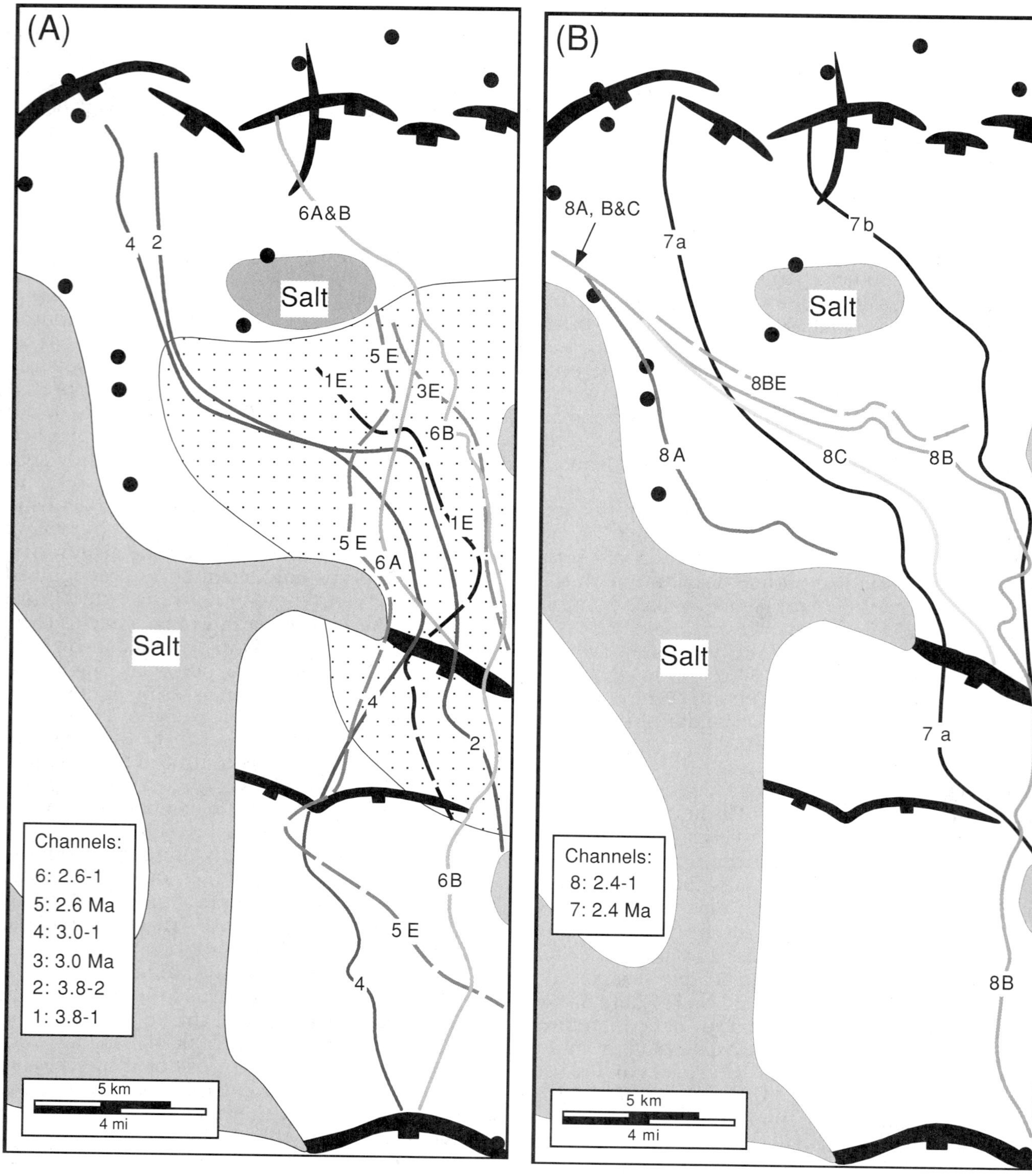


Figure 13—Summary of interpreted channel axes of the (A) 3.8–2.6-Ma, (B) 2.4-Ma, and (C) 1.9-Ma sequences. Dotted area on (A) shows the basin-floor fan (amalgamated turbidite lobes) at the base of the 3.8-Ma sequence (Figure 12). Solid lines show depositional, and dashed lines show erosional (E) channels. Increasingly lighter shading shows the younger channels on each figure. Channels in most sequences are focused in the central parts of the minibasins between the shallow salt bodies. These salt bodies were bathymetric highs during the deposition of those sequences (Figure 9). In several sequences, one updip channel bifurcates downslope (channels 6, 8, 9). The relative ages of the bifurcating channels were determined based on the series of horizon slices (A = oldest, C = youngest). See Table 1 for the correlation of horizons and channels, and the ages of the channel systems; see Table 3 for the summary of channel geometries.

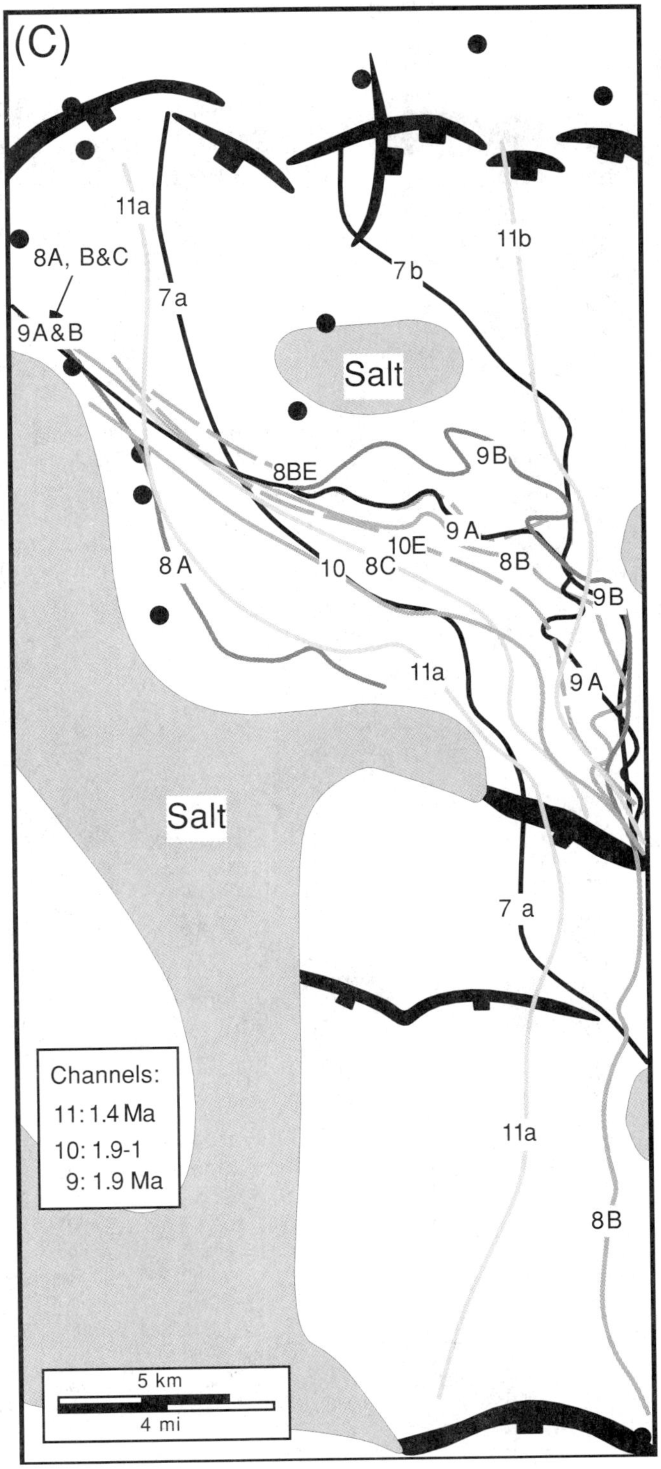


**Figure 13—Continued.**

minibasins B and C are less than half as thick as the sediments in A. Similar to the 3.8-Ma sequence, minibasin C has asymmetric fill with thinning to the east, but the thickness change is considerably less.

On the seismic profiles, high-amplitude, discontinuous reflections characterize this sequence (SF D; Figures 4, 7, 11). The amplitude extraction map from the upper part of the 2.4-Ma sequence shows three slightly sinuous high-amplitude patterns in minibasin A (channels 8A, 8B, 8C, horizon 2.4-1; Figure 14). Anomalously high amplitudes are concentrated around the GC-205-1ST1 well, Genesis field area (channel 8A), and are interpreted as the result of the hydrocarbon content of the field (Rafalowski et al., 1994). This high-amplitude area, however, extends farther south and forms a slightly sinuous feature.

The series of horizon slices from 0 to 80 ms below the reference horizon (1.9-Ma sequence boundary) illustrates the development of the four sinuous amplitude areas (Figures 13B, 15, 16). On the deepest slice (76 ms below the 1.9-Ma sequence boundary), the 8A high-amplitude area is visible on the west side; the horizon slice at 44 ms shows 8B and 8BE (Figure 15); and the uppermost slice (32 ms; Figure 16) shows the 8C high-amplitude area. The relative age of the reflections, therefore, is 8A (oldest) to 8C (youngest). On the 3-D vertical seismic profiles, this sequential migration of the high-amplitude reflections is not visible because they form a single horizon (Figures 4, 7, 11).

The sinuous, high-amplitude patterns of this sequence were penetrated by the GC-205-1ST1 well on the west (N3 sand; Figures 5, 8) and by the GC-472 well on the south (Figure 8); the other wells (e.g., GC-116) penetrated the low-amplitude areas. The GC-205-1ST1 well penetrated a 30-m- (100-ft-) thick sand body with blocky gamma-ray log response, which was interpreted as amalgamated channel-fill sands by Rafalowski et al. (1994). These authors interpreted similar sands in all the other wells of the Genesis field (see figure 3 in Rafalowski et al., 1994). The GC-472 well penetrated a 20-m- (70-ft-) thick sand bed with blocky gamma-ray log shape, which is also interpreted as channel-fill sand, whereas the GC-116 well shows dominantly shale above the 2.4-1 horizon, interpreted as overbank sediment (Figure 8). Sands with a blocky gamma-ray log pattern were penetrated below the 2.4-1 horizon in the GC-116 well. These channel-fill sands probably correspond to the slightly older channel 8A. These sands, however, have low resistivity and are not filled with hydrocarbons.

Based on the characteristic geometries and the lithologic ties, the sinuous high-amplitude patterns were interpreted as bifurcating depositional channel systems with sand-rich channel-fill sediments (channels 8A–8C). The low-amplitude area is similar to the example from the underlying sequence, and is interpreted as an erosional channel filled with shale (channel 8BE; Figures 13–15). No coeval overbank sediments are interpreted to be present for the erosional channel.

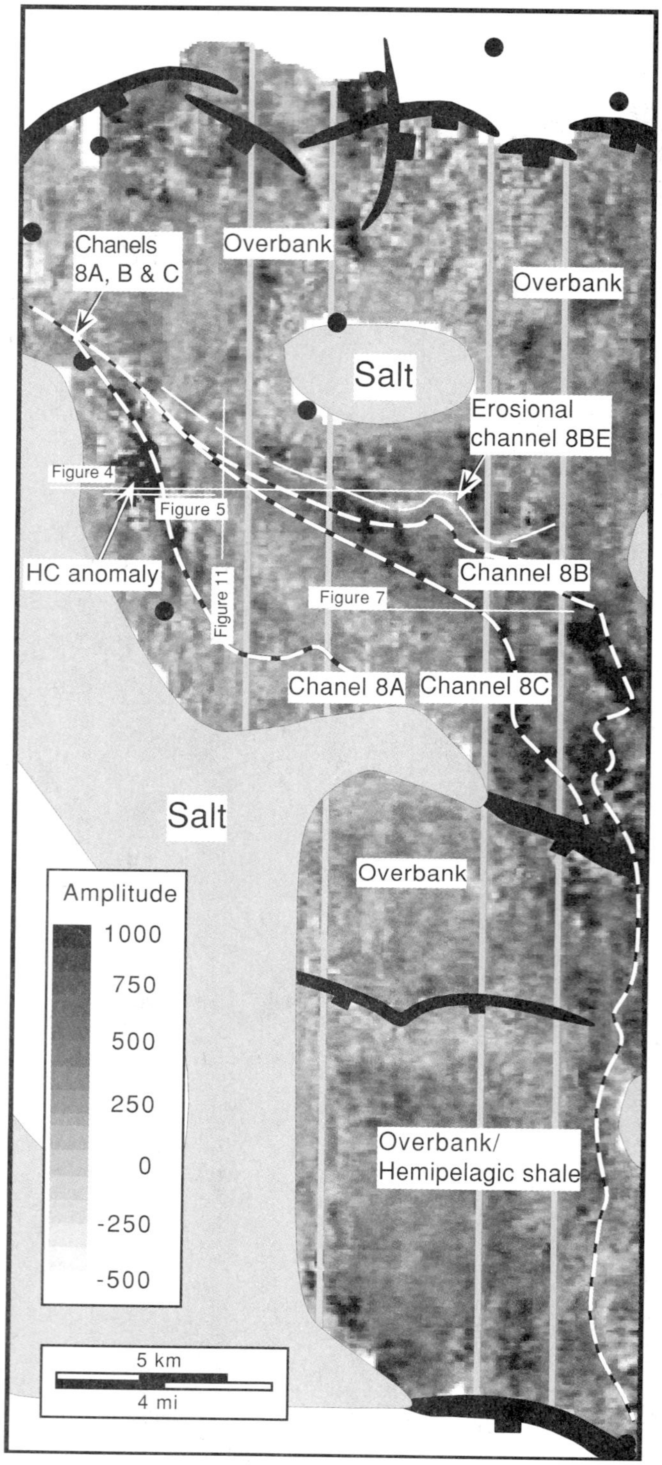


**Figure 14—Amplitude extraction map along the 2.4-1 horizon (Table 1; Figures 4, 5, 7, 11). The map shows sinuous high-amplitude areas (black) within the 2.4 Ma sequence. Using well-log data (GC-205-1ST1 well; Figure 8) and the characteristic geometry, these features were interpreted as depositional channel systems. The series of horizon slices (Figures 15, 16) indicate that channel 8A is the oldest and channel 8C is the youngest member of a bifurcating channel system. The sinuous low-amplitude area (light shading) to the north of channel 8B was interpreted as a possible erosional channel (8BE). Note that the hydrocarbon content of the Genesis field (GC-205-ST1 well; Figure 8) causes increased amplitude around the field (Figure 5). Location of wells is shown by black dots; see Figure 3 for well names. Solid lines show the seismic profiles (Figures 4, 6, 7, 11).**

Most wells penetrate this sequence in areas with widespread low to moderate amplitude (Figures 14–16) and found shale with 3–10 m (10–30 ft) of interbedded sand and silt (Figure 8). Therefore, the low-amplitude areas without recognizable patterns around the channels were interpreted as overbank sediment.

The thickness variations on the isopach map indicate that salt withdrawal influenced the thickness distribution of the sediment fill (Figure 18). The sediments in minibasins B and C are considerably thinner than the northern part of the area, possibly indicating the differential salt withdrawal between minibasins A and B and C. Salt withdrawal was apparently less in the upper part of the turtle structure on the south (Figure 10) than in minibasin A.

The channels within the sequence generally correspond to the greatest sediment thickness on the southwest of minibasin A (Figures 13, 14, 18), which suggests that the decreasing salt withdrawal only generally determined the position of the channels. Autocyclic depositional processes (avulsion) became more dominant in controlling the position of the channels. The channels within the sequence, especially channel 8B, show greater sinuosity compared to the channels of the underlying sequences (Table 3; Figures 13B, 14). Channels 7b and 8C are deflected to the east at the southeast part of minibasin A, possibly by a local bathymetric high on the southern part of the area (Figures 13B, 14, 18).

## DISCUSSION

### Basin-Floor Fan (Amalgamated Depositional Lobes)

Onlapping high-amplitude continuous reflections at the base of the 3.8-Ma sequence were interpreted as basin-floor fans (amalgamated depositional lobes). It is difficult to identify how the sediments of the fan were transported into the area because the eastern side of the fan extends beyond the boundaries of this study area, and the lobe onlaps onto the sequence boundary in all other directions. The most likely

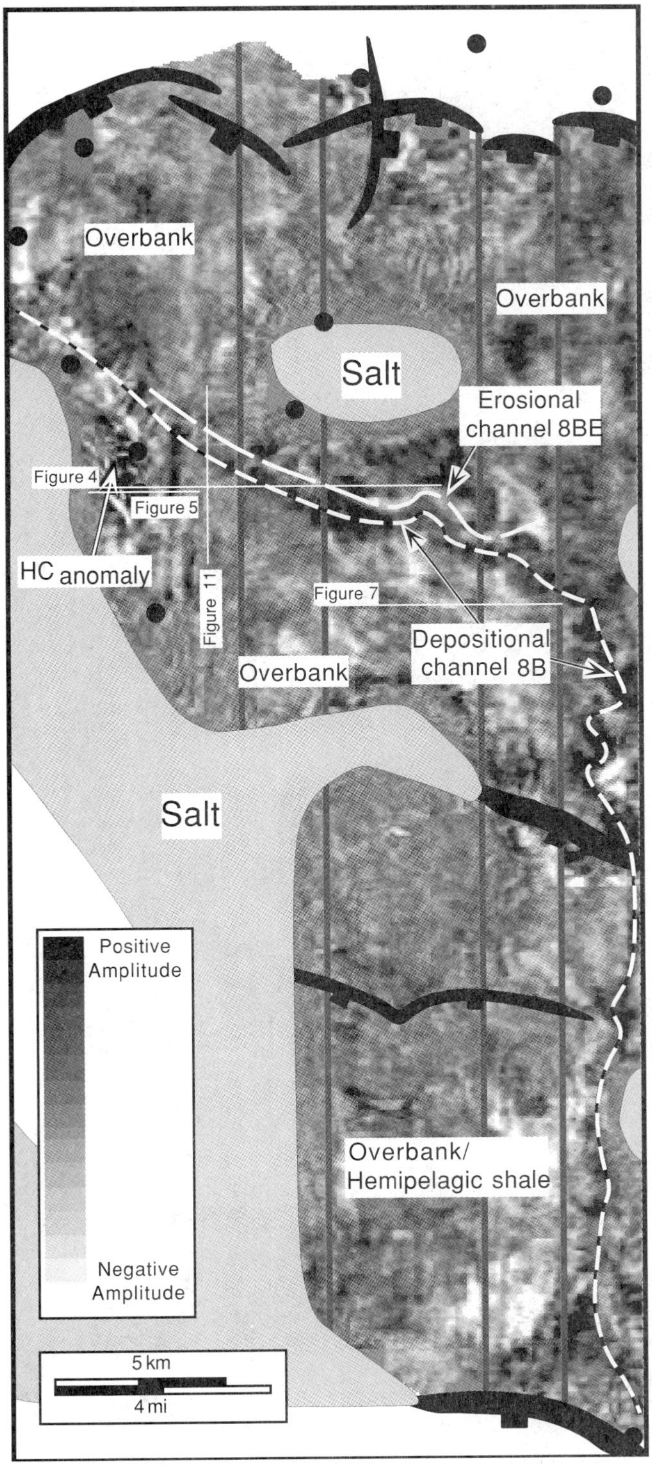


**Figure 15—Horizon slice 44 ms below the 1.9-Ma sequence boundary is roughly parallel to the 2.4-1 horizon (Table 1; Figures 4, 5, 7, 11, 14). The horizon slice shows a similar sinuous amplitude pattern to 8B of Figure 14. This high-amplitude area was interpreted as a depositional channel (8B). The feature extends to the southern part of the area, and the channel is mappable across the entire area. The rest of the area has relatively even low-amplitude distribution, interpreted as overbank sediments. Location of wells is shown by black dots; see Figure 3 for well names. Solid lines show the seismic profiles (Figures 4, 5, 7, 11).**

transport process was turbidity currents carrying sands through erosional channels. An erosional channel is mapped on the top of this fan and is interpreted as a possible sediment conduit for additional basin-floor fans farther downdip to the southeast (discussed in following paragraphs). The sands deposited in the fan of minibasins A and B probably were transported through similar erosional channels/sediment conduits updip. The channel systems of the overlying units show sediment transport directions from the north-northwest, but no erosional channel was mapped on the 3.8-Ma sequence boundary north of the fan. However, an erosional channel might be located to the northeast where the fan extends beyond the study area. The sands may have been carried through that channel to this fan.

The asymmetric salt withdrawal resulted in a relative bathymetric low in the central part of the area and a high in the south. The turbidity currents arriving from the northeast were ponded in the depocenter in the central part of the area. This sediment ponding caused the deposition of the turbidity flow's coarser material here, forming the interpreted sheet sands.

## Depositional Channels

Numerous well penetrations indicate that most channels have sand-rich channel fills and are interpreted as depositional channels in this study. Well data show that these channels are filled with 10–30-m- (30–100-ft-) thick sands with blocky to crescentic gamma-ray log patterns (Figure 8). The channel-fill sands are interbedded with thin-bedded shales; shale content may reach 30%. The thickness of the channel-fill sands was also estimated from the seismic data; the thickness of the sediments corresponding to the reflections are interpreted to represent the thickness of the sand bodies. The thickness of the channel fill ranges between 25 and 50 m (80 and 160 ft) (Table 3) and increases toward the central parts of the minibasins. The wells penetrated the channel-fill sands on the minibasin flanks; hence, the wells probably have thinner sands than those indicated on the seismic data. In general, the channel fills in the 2.4 Ma sequence are the thickest, possibly indicating the increased volume and height of the turbidite flows (Flood and Damuth, 1987).

The channel systems are easier to identify in the younger sequences (Figure 19). Better seismic data quality and thicker channel-fill sands (Figure 8) are the most likely cause of this apparent better channel continuity in the shallower, younger sequences.

## Erosional Channels

Elongated, sinuous low-amplitude areas on the amplitude extraction maps and seismic profiles are interpreted as erosional channels. The thickness of the erosional channels (that is, the amount of erosion at their base) ranges between 25 and 35 m (80 and 110 ft) (Table 3). No wells penetrate these channels, so the lithology of their fill is interpreted based on the seismic character. The channel fill is a low-amplitude zone on the seismic profiles (e.g., horizons 3.8-1, 2.6 Ma; Figure 7). There are no reflections from the base or the top of the fill, indicating that there is no significant acoustic impedance contrast between the fill and the sediment below and above. The channel fill is shale, similar to the overbank shale above the channels.

The three oldest erosional channels are mappable across the entire study area (Table 3; Figure 13A). These channels are interpreted as possible sediment transport axes carrying sands for depositional lobes farther downslope. These channels may be sand conduits similar to those described as type I turbidite systems (Mutti, 1985). However, no possible basin-floor fans were found in the study area; the fans possibly were deposited farther south.

## Channel Geometry and Evolution

The mapped channels are 0.5–1.6 km (0.3–1.0 mi) wide. Some channels, especially in the younger sequences, seem to increase in width downdip (Table 3). Most of the channels are mappable across the entire area; their mapped length ranges between 16 and 45 km (10 and 28 mi), depending on their orientation and location (Table 3; Figure 13). In the older sequences, usually only one channel was active at a given time, similar to channels studied in modern submarine fans (Amazon Fan, Flood and Damuth, 1987; Indus Fan, Kenyon et al., 1995; Mississippi Fan, Weimer, 1990). In the younger sequences, two or three channels were mapped on each amplitude extraction map. Most of these channels are the downfan bifurcations of one updip channel and differ slightly in age. Only one channel was active at a given time. Some of these channels, however, trend parallel to each other, and even the series of horizon slices could not identify a relative age difference between them. They possibly are coeval channels.

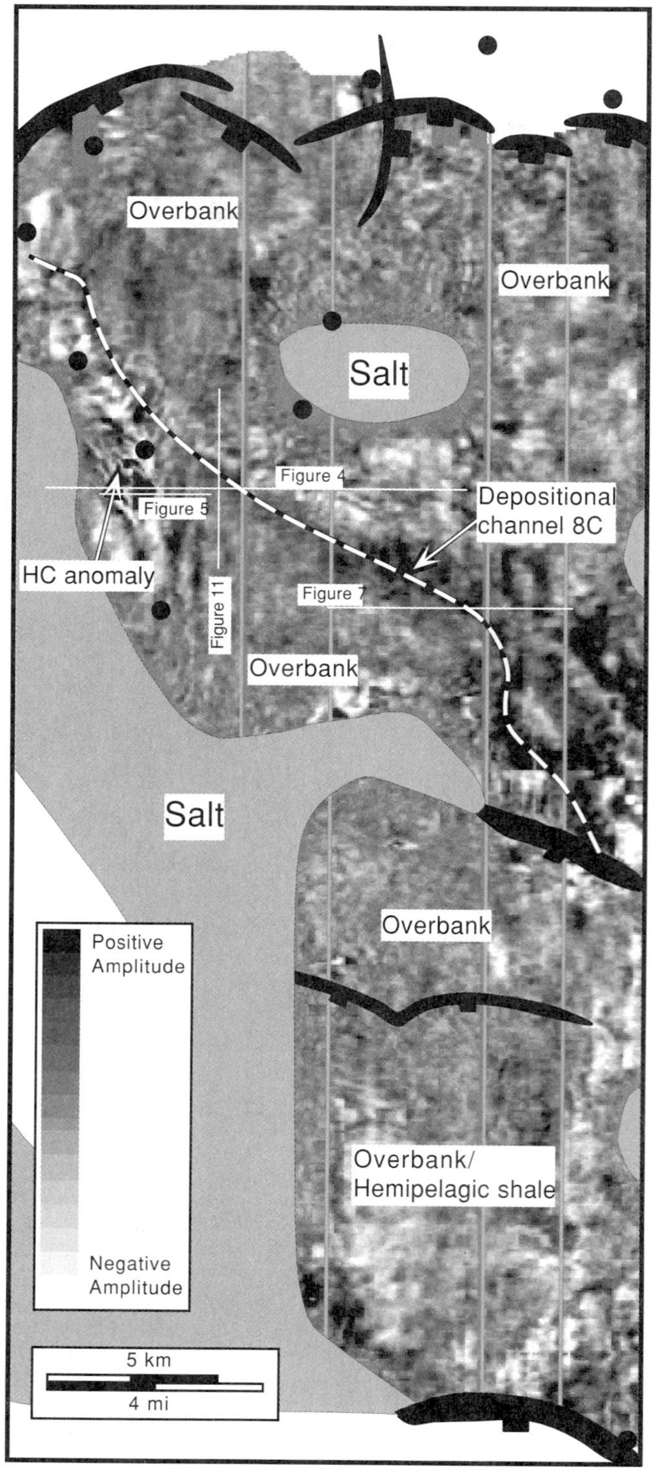


**Figure 16—Horizon slice 32 ms below the 1.9-Ma sequence boundary is roughly parallel to the 2.4-1 horizon (Table 1; Figures 4, 5, 7, 11, 14). The horizon slice shows a slightly sinuous high-amplitude pattern similar to feature 8C of Figure 14, which was interpreted as a depositional channel (8C). Location of wells is shown by black dots; see Figure 3 for well names. Solid lines show the seismic profiles (Figures 4, 5, 7, 11).**

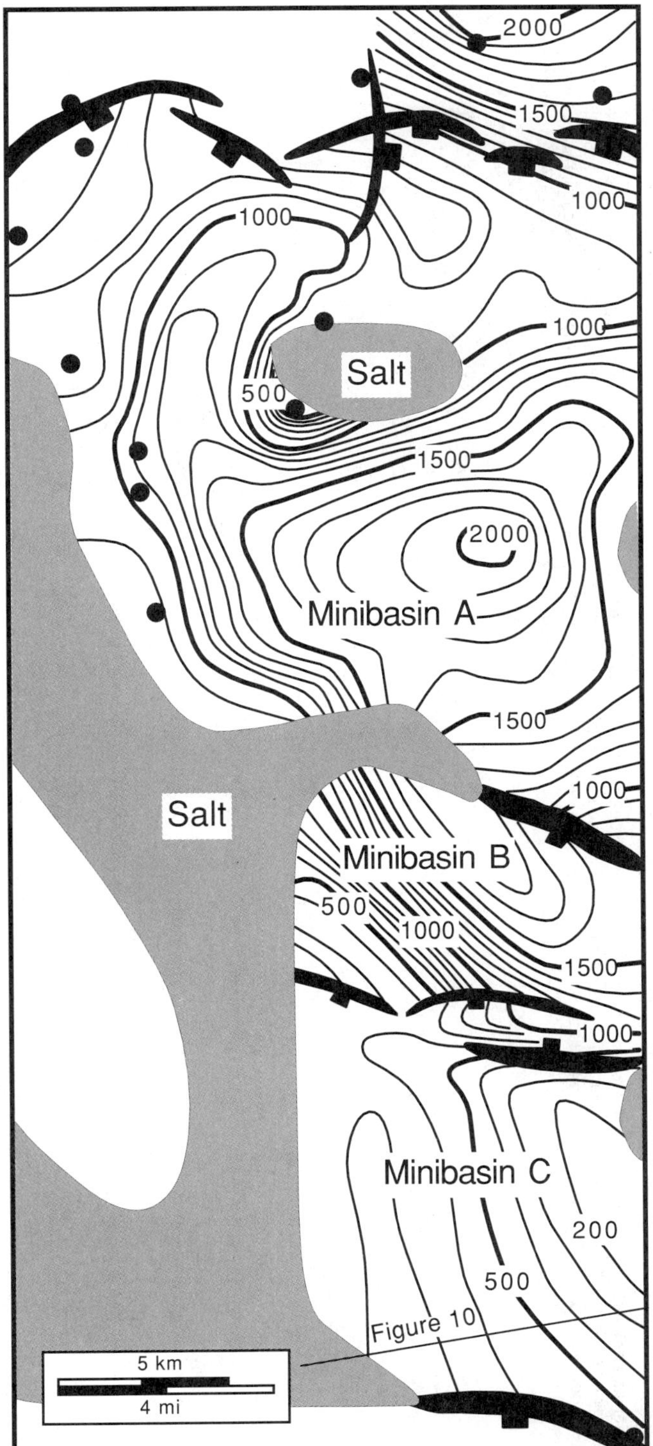


**Figure 17—Isopach map of the 3.8-Ma sequence; contour interval is 100 ft (30 m). The thickness distribution in this sequence was greatly influenced by salt withdrawal. Notice that sediments in minibasin C are considerably thinner than the sediments in minibasins A and B. The asymmetric fill of minibasin C is caused by salt withdrawal, forming a turtle structure (Figure 10). See text for discussion of the correlation of thickness distribution and channel position. Well locations are shown by black dots; see Figure 3 for well names.**

The channels mapped in this study have varying amounts of sinuosity; in general the younger channels are more sinuous (Table 3; Figure 13). Several factors apparently affect the sinuosity of channels. Flood and Damuth (1987) showed a strong correlation between sinuosity and slope gradient in the Amazon Fan. They noted that an increased slope angle results in increased sinuosity to maintain a relatively constant channel gradient. However, the volume of turbidite flow and sediment load also affect channel sinuosity in the Amazon Fan (Flood and Damuth, 1987). Changes in grain size or in the amount of the supplied sediment also may have caused increased sinuosity, similar to the youngest Mississippi Fan channels (Stelting et al., 1985).

An important characteristic in the evolution of the channel systems is the continuous switching of the channels through deposition (Figure 13). The channels abruptly shifted their positions by 1.6–5 km (1–3 mi). No gradual switching (channel meandering) was observed in this study, probably due to the limited resolution of the seismic data. This lack of meandering indicates that the channels switched to a new position by avulsion. A similar relationship has been noted on several modern fans (e.g., Flood and Damuth, 1987). The individual channels may have meandered (below seismic resolution), but the major shifts in channel position (bifurcation) were accomplished by avulsion.

The initial axis of sediment transport was to the south-southwest through the eastern and occasionally the western part of the area (channels 1–7; Figure 13A, B). The transport direction changed within the 2.4-Ma sequence to the east-southeast through the central part of the area (channels 8–10; Figure 13B, C). About 1.4 Ma, the transportation direction shifted back to the south (channel 11; Figure 13C). Risch et al. (1996) described similar shifts in the sediment transport axis in the 1.9-Ma and 1.4-Ma sequences on the northwestern part of the area.

The change in channel entry points might have been the result of the westward shift in the position of the sediment source (Mississippi Delta) at about 3.0 Ma (Pulham, 1993), and then back to north of the study area (1.4 Ma).

Channel axes generally occupied the central parts of the minibasins. The shallow salt bodies on the flanks of the minibasin formed syndepositional bathymetric highs, focusing the channels toward the center of the minibasin. This concentration of channel systems indicates that

**Table 3. Summary of Geometry, Position, and Evolution of Channel Systems***

| Sequence Boundary | Internal Horizon | Depositional Channel | Erosional Channel | Length (km) | Width (km) | Thickness (m) | Channel Direction | Sinuosity | Position on Isopach |
|---|---|---|---|---|---|---|---|---|---|
| 1.4 | | 11b | | 21 | 0.8 | 26 | N-S | 1.05 | Side |
| 1.4 | | 11a | | 42 | 0.8–1.6 | 29 | N-S | 1.07 | Side |
| | 1.9-1 | | 10E | 21 | 0.8–2 | 34 | NW-SE | 1.02 | Side |
| | 1.9-1 | 10 | | 23 | 0.5–1.3 | 33 | NW-SE | 1.06 | Thick |
| 1.9 | | 9B | | 32 | 0.5 | 24 | NW-SE | 1.31 | Side |
| 1.9 | | 9A | | 31 | 0.5 | 31 | NW-SE | 1.21 | Side |
| | 2.4-1 | 8C | | 23 | 0.5–1.3 | 29 | NW-SE | 1.06 | Thick |
| | 2.4-1 | | 8BE | 13 | 0.3 | 27 | NW-SE | 1.08 | Side |
| | 2.4-1 | 8B | | 45 | 0.5–1.3 | 41 | NW-SE | 1.15 | Thick |
| | 2.4-1 | 8A | | 18 | 0.8 | 48 | NW-SE | 1.08 | Side |
| 2.4 | | 7b | | 24 | 0.8 | 34 | N-S | 1.1 | Nose |
| 2.4 | | 7a | | 32 | 0.8–1.3 | 29 | N-S | 1.08 | Nose |
| | 2.6-1 | 6B | | 40 | 0.8–1.3 | 41 | N-S | 1.16 | Side |
| | 2.6-1 | 6A | | 23 | 0.8–1.3 | 27 | N-S | 1.09 | Nose |
| 2.6 | | | 5E | 34 | 1.1 | 31 | N-S | 1.08 | Nose |
| | 3.0-1 | 4 | | 42 | 0.8 | 24 | N-S | 1.07 | Thick |
| 3.0 | | | 3E | 13 | 1.6 | 30 | N-S | 1.05 | Thick |
| | 3.8-2 | 2 | | 32 | 1.6 | 34 | N-S | 1.11 | Thick |
| | 3.8-1 | | 1E | 21 | 1.3 | 24 | N-S | 1.12 | Thick |
| 3.8 | | | | | | | | | |

*The length and width of channels are estimates based on the dimensions of the corresponding high- and low-amplitude areas observed on the amplitude extraction maps and horizon slices. The values in the width column that are ranges (e.g., 0.5–1.3) indicate channel widening related to meandering in minibasin A. The maximum thicknesses of the depositional channels are estimates based on the thickness of the sediment corresponding to the seismic reflections. The thicknesses of the erosional channels are estimated based on the thickness of the strata into which the channel was cut. See text for discussion of channel geometry and evolution.

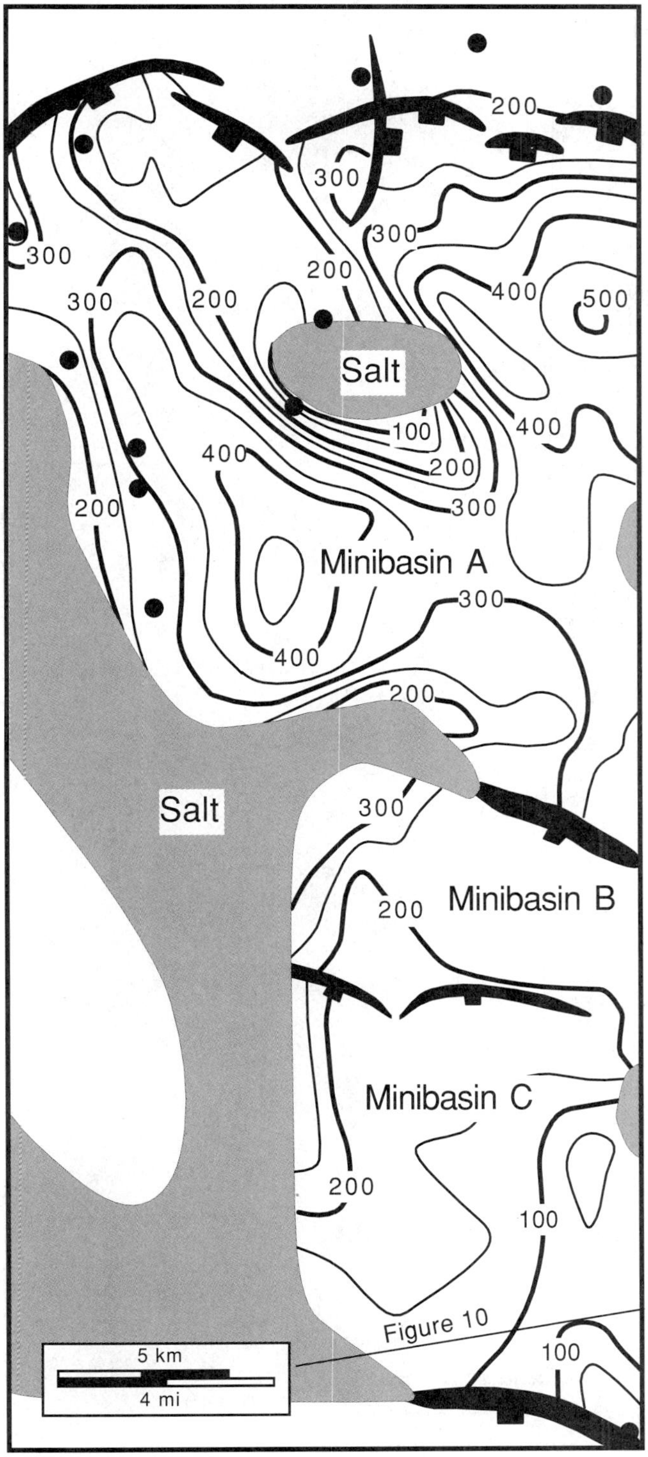


**Figure 18—Isopach map of the 2.4-Ma sequence; contour interval is 50 ft (15 m). The influence of salt withdrawal is still recognized in sediment thickness variations in minibasin A, but is less important than in the underlying sequences. Westward thickening of sediments associated with the turtle structure in minibasin C is less in this sequence than in the 3.8-Ma sequence (Figures 10, 17). See text for discussion of the correlation of thickness distribution and channel position. Well locations are shown by black dots; see Figure 3 for well names.**

this area was a sediment transport axis for at least 2 m.y.

The series of isopach maps shows the changes in the position of the thickest sediment fills of the sequences, which are interpreted to reflect the changes in salt withdrawal through the interpreted sequences (Figures 17, 18). These maps also reveal that the rate of salt withdrawal changed in different parts of the area within each sequence. Salt tectonics was an important factor in changing the slope gradient, which then affected the changing position of the channels, and the variations in channel bifurcation and sinuosity. The changes in channel positions compared to the isopach thicks (deepest parts of minibasins) show the decreasing influence of salt withdrawal on the positioning of channels in the younger sequences.

In summary, four factors controlled the position and the evolution of the channel systems. The channel entry points determined whether the channels crossed the area in a general north-south or northwest-southeast direction. Syndepositional salt withdrawal determined the general positioning of the channels within the minibasins. The constant changes in channel locations probably were caused by autocyclic depositional processes (avulsion). The bathymetric effects of the underlying channels (local highs) may have influenced the positions of the channels.

The sinuosity, avulsion, and possible meandering of these interpreted Pliocene-Pleistocene channels show remarkable similarities to the channel systems of modern submarine fans.

## Overbank

Overbank deposits form most of the sediments in the interpreted sequences. The well penetrations indicate that this turbidite element is dominantly shale with a few thin interbedded sands (Figure 8). The overbank shales probably were deposited as the fine-grained component of the turbidite flows. The sediments were transported through channels, and then they overflowed the levees and were spread out through the area. The thin sand layers are interpreted as minor channels; however, due to the limited resolution of the seismic data, these channels could not be identified, even on the 3-D data set.

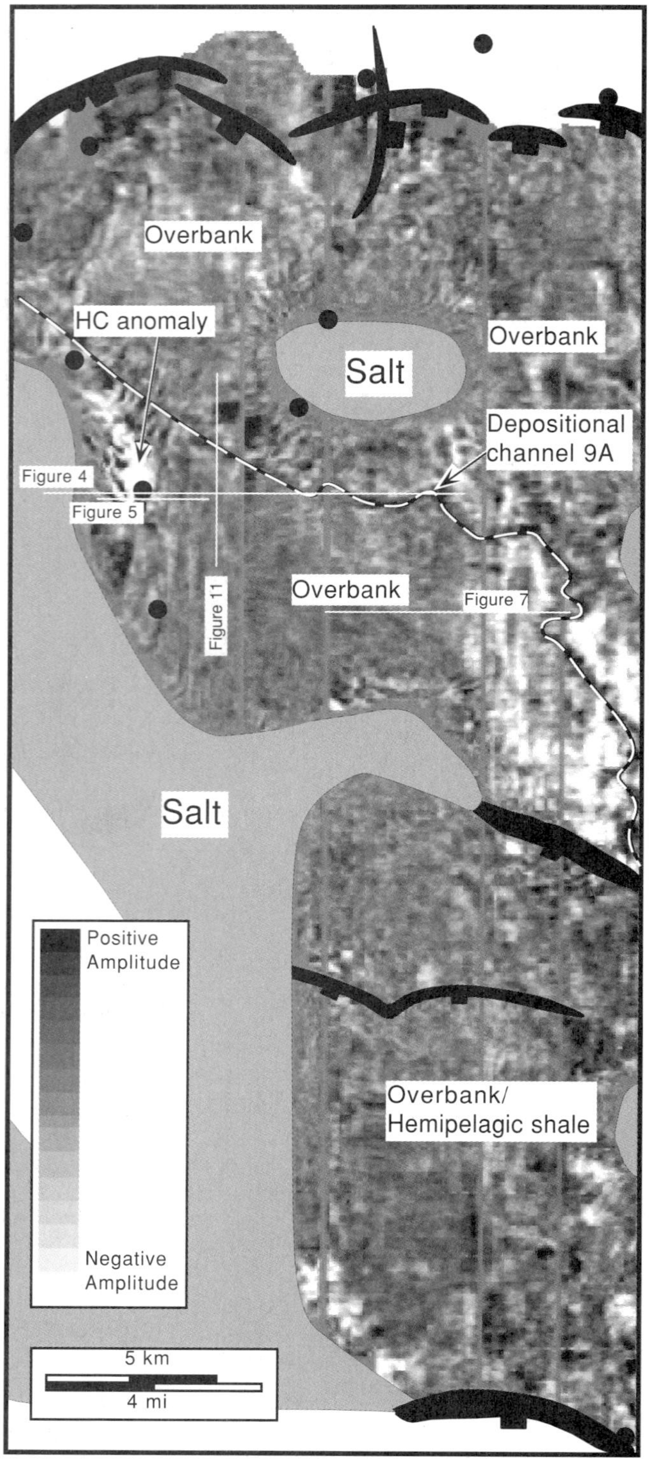


**Figure 19—Horizon slice 12 ms below the 1.9-Ma sequence boundary is roughly parallel to the 1.9-Ma sequence boundary (Table 1; Figures 4, 7, 11). The slice shows a highly sinuous high-amplitude area in mini-basin A. Based on its characteristic shape, the feature is interpreted as a depositional channel system. The rest of the area has relatively even low-amplitude distribution. Based on the shales penetrated in the wells, these areas are interpreted as overbank sediments (Figure 8). Location of wells is shown by black dots; see Figure 3 for well names. Solid lines show the seismic profiles (Figures 4, 5, 7, 11).**

## Significance for Exploration and Development

Two fields and a discovery in this study area (Popeye GC-72/73/116 lease blocks, Genesis GC-161/205 lease blocks, and King Kong GC-472/473 lease blocks) and the many other petroleum fields and discoveries of the Gulf of Mexico produce from turbidite systems, primarily from amalgamated sheet and sheet sands (basin-floor fan) and channel-fill and levee sands (Weimer et al., 1994a; Weimer et al., 1998b, this issue). Several discoveries north of the area are found below salt sheets (e.g., Mahogany, Teak) and have primarily Miocene–Pliocene turbidite-related reservoirs (Harrison et al., 1995; Moore and Brooks, 1995; Snyder and Nugent, 1995; Montgomery and Moore, 1997; Weimer et al., 1998b, this issue).

The geometries of the sandy turbidite elements mapped in this study (Tables 2, 3) are considered to be good analogs for producing fields and some exploration targets. The results of this study also can be used as analogs for reservoir architecture description in the subsalt trend of the Gulf of Mexico and in other deep-water exploration areas.

## CONCLUSIONS

Comparing the results of the regional 2-D (two-dimensional) (Varnai, 1996; Weimer et al., 1998a, this issue) and subregional 3-D (three-dimensional) studies shows the advantages of using 3-D seismic data in stratigraphic interpretation. Series of horizon slices and amplitude extraction maps incorporated with vertical seismic facies and well-log data allowed the analysis of the turbidite elements in three dimensions. Facies maps based on 3-D reflection continuity and amplitude patterns were created that illustrated the geometries of the different elements. The series of facies maps showed the changes in these geometries, the evolution of the turbidite systems within each sequence, and the variations between the sequences. Finally, this detailed analysis allowed the interpretation of the transportation and deposition processes of these turbidites.

Three different turbidite elements were identified in the late Pliocene in the area. Lobe-shaped stratal geometry characterizes the onlapping basin-floor fan (amalgamated turbidite lobes) at the base of the 3.8-Ma sequence. This basin-floor fan is inferred to consist of sand interpreted to have been transported through erosional channels northeast of the area. Sinuous stratal geometries were interpreted as sand-filled depositional and shale-filled erosional channel systems. The mapped channel systems are interpreted to be the major conduits of turbidite sediments in the area. The erosional channels are interpreted as possible sediment transport axes carrying sand farther downslope. Significant sands were deposited in these channel systems, forming the main reservoirs of the area.

The mapped and discussed geometry of the turbidite lobe and the channelized turbidite systems may present an analog for other deep-water exploration areas, including the subsalt trend of the Gulf of Mexico.

## REFERENCES CITED


Chapin, M. A., P. Davies, J. L. Gibson, and H. S. Pettingill, 1994, Reservoir architecture of turbidite sheet sandstones in laterally extensive outcrops, Ross formation, western Ireland, *in* P. Weimer, A. H. Bouma, and B. F. Perkins, eds., Submarine fans and turbidite systems: Gulf Coast Section SEPM 15th Annual Research Conference, p. 53–68.

Chapin, M. A., G. M. Tiller, and M. J. Mahaffie, 1996, 3-D architecture modeling using high-resolution seismic data and sparse well control; example from the Mars field, Mississippi Canyon area, Gulf of Mexico, *in* P. Weimer and T. L. Davis, eds., Application of 3-D seismic data to exploration and production: AAPG Studies in Geology 42, p. 123–132.

Clark, J. D., and K. T. Pickering, 1996, Architectural elements and growth patterns of submarine channels: application to hydrocarbon exploration: AAPG Bulletin, v. 80, p. 194–221.

Flood, R. D., and J. E. Damuth, 1987, Quantitative characteristics of sinuous distributary channels on the Amazon deep-sea fan: Geological Society of America Bulletin, v. 98, p. 728–738.

Harrison, H. L., D. Moore, and P. Hodkins, 1995, The Mahogany subsalt discovery: a unique hydrocarbon play, offshore Louisiana, *in* C. J. Travis, H. Harrison, M. R. Hudec, B. C. Vendeville, F. J. Peel, and B. E. Perkins, eds., Salt, sediment and hydrocarbons: Gulf Coast Section SEPM 16th Annual Research Conference, p. 95–97.

Kenyon, N. H., A. A. Amir, and A. Cramp, 1995, Geometry of the young sediments of the Indus Fan, *in* K. T. Pickering, R. N. Hiscott, N. H. Kenyon, F. Ricci Lucchi, and R. D. A. Smith, eds., Atlas of deep water environments, architectural style in turbidite systems: London, Chapman and Hall, p. 89–93.

Mahaffie, M. J., 1994, Reservoir classification for turbidite intervals at the Mars discovery, Mississippi Canyon 807, Gulf of Mexico, *in* P. Weimer, A. H. Bouma, and B. F. Perkins, eds., Submarine fans and turbidite systems: Gulf Coast Section SEPM 15th Annual Research Conference, p. 233–244.

McBride, B. C., 1998, The evolution of allochthonous salt along a megaregional profile across the northern Gulf of Mexico Basin: AAPG Bulletin, v. 82, this issue, p. 1037–1054.

McBride, B. C., M. G. Rowan, and P. Weimer, 1998, The evolution of allochthonous salt systems, northern Green Canyon and Ewing Bank (offshore Louisiana), northern Gulf of Mexico: AAPG Bulletin, v. 82, this issue, p. 1013–1036.

Mitchum, R. M., Jr., J. B. Sangree, P. R Vail, and W. W. Wornardt, 1993, Recognizing sequences and systems tracts from well logs, seismic data, and biostratigraphy: examples from the late Cenozoic, *in* P. Weimer and H. W. Posamentier, eds., Siliciclastic sequence stratigraphy: AAPG Memoir 58, p. 163–199.

Montgomery, S. C., and D. C. Moore, 1997, Subsalt play, Gulf of Mexico: a review: AAPG Bulletin, v. 81, p. 871–896.

Moore, D. C., and R. O. Brooks, 1995, The evolving exploration of the subsalt play in the offshore Gulf of Mexico: Gulf Coast Association of Geological Societies Transactions, v. 45, p. 8–12.

Mutti, E., 1985, Turbidite systems and their relation to depositional sequences, *in* G. G. Zuffa, ed., Provenance of arenites: NATO-ASI series, Dordrecht, Reidel, p. 65–93.

Mutti, E., and W. R. Normark, 1987, Comparing examples of modern and ancient turbidite systems: problems and concepts, *in* J. K. Leggett and G. G. Zuffa, eds., Marine clastic sedimentology: concept and case studies: London, Graham and Trotman, p. 1–38.

Mutti, E., and W. R. Normark, 1991, An integrated approach to the study of turbidite systems, *in* P. Weimer and M. H. Link, eds., Seismic facies and sedimentary processes of submarine fans and turbidite systems: New York, Springer-Verlag, p. 75–106.

Pacht, J. A., B. E. Bowen, J. H. Beard, B. L. Shaffer, and W. R. Pottorf, 1990, Sequence stratigraphy of Plio-Pleistocene strata in the offshore Louisiana Gulf Coast: applications to hydrocarbon exploration, *in* Sequence stratigraphy as an exploration tool—concepts and practices in the Gulf Coast: Gulf Coast Section SEPM 11th Annual Research Conference, p. 269–285.

Pacht, J. A., B. E. Bowen, J. H. Beard, B. L. Shaffer, and W. R. Pottorf, 1992, Systems tracts, seismic facies, and attribute analysis within a sequence stratigraphic framework—example from the offshore Louisiana Gulf Coast, *in* E. G. Rhodes and T. F. Moslow, eds., Marine clastic reservoirs: New York, Springer-Verlag, p. 27–38.

Paleodata Inc.-Schlumberger/Geco-Prakla, 1993, Gulf of Mexico chronostratigraphic correlation chart:Schlumberger/Geco-Prakla.

Pulham, A. J., 1993, Variations in slope deposition, Pliocene-Pleistocene, offshore Louisiana, northeast Gulf of Mexico, *in* P. Weimer and H. W. Posamentier, eds., Siliciclastic sequence stratigraphy: AAPG Memoir 58, p. 199–236.

Rafalowski, J. W., B. W. Regel, D. J. Jordan, and D. O. Lucidi, 1994, Green Canyon Block 205 lithofacies, seismic facies, and reservoir architecture, *in* P. Weimer, A. H. Bouma, and B. F. Perkins, eds., Submarine fans and turbidite systems: Gulf Coast Section SEPM 15th Annual Research Conference, p. 293–306.

Risch, D. L., B. E. Donaldson, and C. K. Taylor, 1996, Deep water facies analysis using 3-D seismic sequence stratigraphy and workstation techniques: an example from Plio-Pleistocene strata, northern Gulf of Mexico, *in* P. Weimer and T. L. Davis, eds., Application of 3-D seismic data to exploration and production: AAPG Studies in Geology 42, p. 143–148.

Rowan, M. G., 1995, Structural styles and evolution of allochthonous salt, central Louisiana outer shelf and upper slope, *in* M. P. A. Jackson, D. G. Roberts, and S. Snelson, eds., Salt tectonics: a global perspective: AAPG Memoir 65, p. 199–228.

Rowan, M. G., and P. Weimer, 1998, Salt-sediment interaction, northern Green Canyon and Ewing Bank (offshore Louisiana) northern Gulf of Mexico: AAPG Bulletin, v. 82, this issue, p. 1055–1082.

Rowan, M. G., B. C. McBride, and P. Weimer, 1995, The role of extension in the evolution of allochthonous salt, central Louisiana outer shelf and upper slope, *in* C. J. Travis, H. Harrison, M. R. Hudec, B. C. Vendeville, F. J. Peel, and B. E. Perkins, eds., Salt, sediment and hydrocarbons: Gulf Coast Section SEPM 16th Annual Research Conference, p. 223–233.

Snyder, F. C., and J. Nugent, 1995, Teak—testing a subsalt hydrocarbon trap geometry, South Timbalier Block 260, Gulf of

Mexico, *in* C. J. Travis, H. Harrison, M. R. Hudec, B. C. Vendeville, F. J. Peel, and B. E. Perkins, eds., Salt, sediment and hydrocarbons: Gulf Coast Section SEPM 16th Annual Research Conference, p. 257-267.

Stelting, C. E., and the DSDP Leg 96 Shipboard Scientists, 1985, Migratory characteristics of a mid-fan meander belt, Mississippi Fan, *in* A. H. Bouma, W. R. Normark, and N. E. Barnes, eds., Submarine fans and related turbidite systems: New York, Springer-Verlag, p. 283-290.

Varnai, P., 1996, Integrated 2-D and 3-D sequence stratigraphic analysis of Plio-Pleistocene turbidite systems, northern Green Canyon and central Ewing Bank protraction areas, northern Gulf of Mexico: Ph.D. dissertation, University of Colorado, Boulder, Colorado, 487 p.

Varnai, P., and P. Weimer, 1997, Sequence stratigraphic analysis of Plio-Pleistocene turbidite systems, northern Green Canyon and central Ewing Bank areas, northern Gulf of Mexico: Gulf Coast Association of Geological Societies Transactions, v. 47, p. 599-610.

Villamil, T., C. Arango, P. Weimer, A. C. Waterman, M. G. Rowan, P. Varnai, J.R. Crews and A. J. Pulham, 1998, Biostratigraphic techniques for analyzing benthic biofacies, stratigraphic condensation, and key surface identification, Pliocene and Pleistocene sediments, northern Green Canyon and Ewing Bank (offshore Louisiana), northern Gulf of Mexico: AAPG Bulletin, v. 82, this issue, p. 961-985.

Weimer, P., 1990, Sequence stratigraphy, facies geometries, and depositional history of the Mississippi Fan, Gulf of Mexico: AAPG Bulletin, v. 74, p. 425-453.

Weimer, P., A. H. Bouma, and B. F. Perkins, eds., 1994a, Submarine fans and turbidite systems: Gulf Coast Section SEPM 15th Annual Research Conference, 440 p.

Weimer, P., P. Varnai, Z. Acosta, F. M. Budhijanto, R. Martinez, A. Navarro, M. G. Rowan, B. McBride, and T. Villamil, 1994b, Sequence stratigraphy of a Neogene turbidite system, northern Green Canyon and Ewing Bank, northern Gulf of Mexico, *in* P. Weimer, A. H. Bouma, and B. F. Perkins, eds., Submarine fans and turbidite systems: Gulf Coast Section SEPM 15th Annual Research Conference, p. 383-396.

Weimer, P., P. Varnai, F. M. Budhijanto, Z. Acosta, R. Martinez, A. Navarro, T. Villamil, C. Arango, M. G. Rowan, B. McBride, and A. Pulham, J.R. Crews, 1998a, Sequence stratigraphy of Pliocene and Pleistocene turbidite systems, northern Green Canyon and Ewing Bank (offshore Louisiana), northern Gulf of Mexico: AAPG Bulletin, this issue, p. 918-960.

Weimer, P., J. Crews, R. Crow, and P. Varnai, 1998b, Atlas of petroleum fields and discoveries, northern Green Canyon, Ewing Bank, and southern Ship Shoal and South Timbalier areas (offshore Louisiana), northern Gulf of Mexico: AAPG Bulletin, v. 82, this issue, p. 878-917.

## ABOUT THE AUTHOR

### Peter Varnai

Peter Varnai received his B.S. degree (1985) in geology from Eötvös Loránd University, Budapest, Hungary; an M.S. degree (1991) in geophysics from Miskolc University, Miskolc, Hungary; and a Ph.D. (1996) in geology from the University of Colorado, Boulder, Colorado. He worked for the Hungarian National Oil Company in Budapest as an exploration geologist for six years. He was responsible for seismic interpretation, stratigraphic interpretation of well logs, construction of stratigraphic and tectonic models, and basin analysis in the Pannonian basin. During his graduate studies in Colorado, he participated in the Gulf of Mexico industry consortium. In 1996 he joined Exxon Exploration Company as a senior geologist.

# The Evolution of Allochthonous Salt Systems, Northern Green Canyon and Ewing Bank (Offshore Louisiana), Northern Gulf Of Mexico[1]

**Barry C. McBride,[2] Mark G. Rowan,[3] and Paul Weimer[3]**


## ABSTRACT

The discovery that major episodes of subhorizontal, allochthonous salt flow have occurred in the Gulf of Mexico Basin requires a means of quantifying the evolution of allochthonous salt and associated structures to conduct both basin and petroleum systems analyses. Sequential structural restorations of allochthonous salt systems provide an evolving structural framework for integrating stratigraphic, geophysical, and geochemical data sets. In this study, interpretation of more than 10,000 km (6200 mi) of multifold seismic data, and sequential restoration of eleven profiles, were used to determine the geometry and evolution of allochthonous salt structures within Ewing Bank and northern Green Canyon protraction areas.

The results illustrate the complex geometry of the multilevel salt system and the types of interactions between counterregional and salt-stock canopy models of allochthonous salt system evolution. Sedimentary loading is accommodated by salt sheet extrusion, gravity spreading, gravity gliding, extension, salt evacuation, and contraction. Salt geometry commonly changes dramatically through time because it provides much of the accommodation for sediments and absorbs much of the extension and contraction within its overburden. The positioning and kinematics of extensional and contractional structures are controlled by salt body geometries, salt system interactions, and, most importantly, the topography of the base salt or equivalent salt weld.

The structural restorations also constrain the timing of salt sheet and salt weld formation and document the positive correlation among sedimentation rates, salt flow, and structural deformation. Cross-sectional salt area generally decreases through time in areas of salt evacuation and minibasin formation, but increases in sections crossing growing salt bodies. Three-dimensional restoration is required to determine the three-dimensional kinematics and balance of allochthonous salt tectonics.


## INTRODUCTION

A proper understanding of the evolution of allochthonous salt in the northern Gulf of Mexico is essential for determining relationships among salt tectonics, sedimentation, and petroleum accumulations. Sequential structural restorations of interpreted seismic profiles provide the framework for evaluating the evolving depositional environments and facies development associated with salt deformation. When integrated with stratigraphic analyses, restorations provide the foundation for evaluating petroleum systems through detailed thermal maturation modeling and petroleum migration analysis (McBride, 1995, 1998, this issue; McBride et al., 1998, this issue).

Recent advances in our understanding of salt tectonics, especially within the Gulf of Mexico Basin, has been one of the most significant recent developments in structural geology. These advances have been accomplished primarily through



[1]Manuscript received July 22, 1996; revised manuscript received April 15, 1997; final acceptance November 19, 1997.

[2]Energy and Minerals Applied Research Center, Department of Geological Sciences, Campus Box 250, University of Colorado, Boulder, Colorado 80309. Present address: HS Resources Inc., 1999 Broadway, Suite 3600, Denver, Colorado 80202.

[3]Energy and Minerals Applied Research Center, Department of Geological Sciences, Campus Box 250, University of Colorado, Boulder, Colorado 80309.

We thank the following companies and their representatives for their tremendous support and input as part of the Gulf of Mexico research consortium at the University of Colorado: AGIP, Amoco, Anadarko, BHP, BP Exploration, Burlington Resources, CXY Resources, Chevron USA, CNG, Conoco, Exxon, Marathon, Mobil, Occidental, PanCanadian, Pennzoil, Petrobras, Phillips Enterprise, Shell, Texaco, Total, Union Pacific, and Unocal. We extend special thanks to Halliburton Geophysical (now Western Geophysical), specifically to Bob Graebner, John Anderson, and Gary White, for the seismic data used in the project. We are indebted to PaleoData Inc. and Art Waterman for use of biostratigraphic data. Software donations from GeoQuest, CogniSeis Development Inc., Landmark, Platte River Associates, and Zeh Graphics are essential to the success of our program and very much appreciated. We appreciate reviews by Albert Bally, Harry Roberts, and, especially, Martin Jackson for his particularly thorough review. We thank Jonathan Irick, Jennifer Crews, and Ryan Crow for drafting. Acknowledgment is made to the Donors of the Petroleum Research Fund, administered by the American Chemical Society, for partial support for this research.

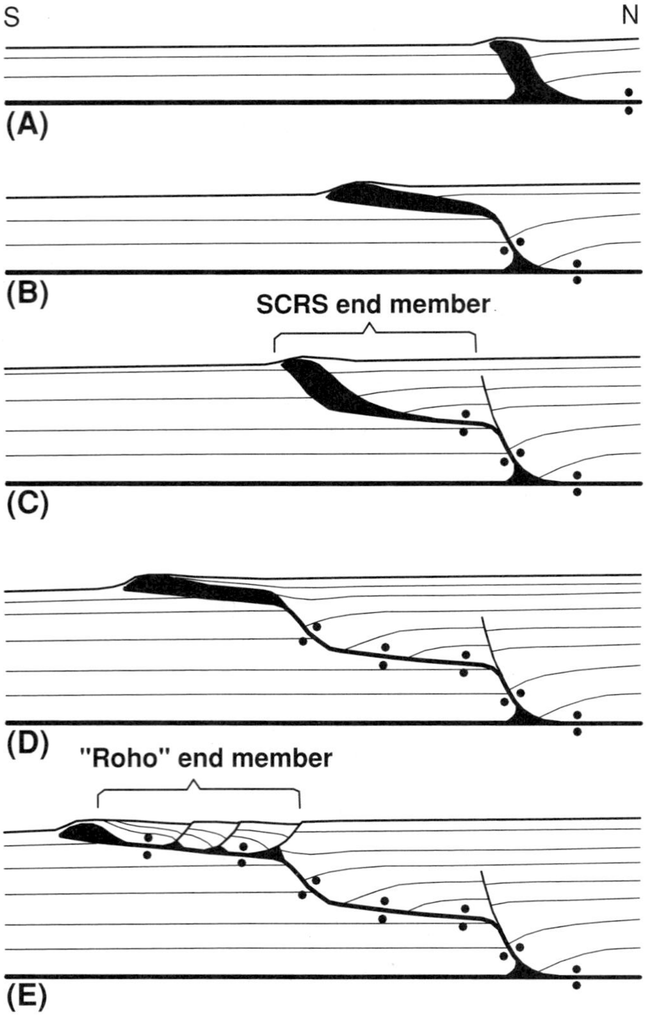


**Figure 1—Schematic evolution of counterregional feeder model for allochthonous salt combining both the (C) stepped counterregional system (SCRS) (middle salt level) and (E) roho system (upper salt level) end-member models of Schuster (1995). Counterregional systems are characterized by north-dipping feeders and approximately horizontal salt sheets. Salt is in black, salt welds are indicated by pairs of dots, and bedding geometries are shown by thin lines. No ages are inferred; section not to scale (from Rowan, 1995).**

improved seismic imaging (e.g., Ratcliff and Weber, 1997), better physical modeling (e.g., Vendeville and Jackson, 1992a, b), and the application of sequential structural restorations (e.g., Worrall and Snelson, 1989; Moretti et al., 1990; Diegel et al., 1995; Peel et al., 1995; Rowan, 1995; Schuster, 1995). The presence of tabular allochthonous salt sheets has been documented by numerous workers and has led to various models explaining their formation and evolution (e.g., Humphris, 1978; West, 1989; Worrall and Snelson, 1989; Wu et al., 1990; Nelson, 1991; Fletcher et al., 1995). These observations and models led to the recognition that ancient allochthonous salt sheets have also existed throughout the northern Gulf of Mexico basin and have undergone various stages of evacuation (e.g., Worrall and Snelson, 1989; Diegel and Cook, 1990; Diegel et al., 1995; Peel et al., 1995; Schuster, 1995).

Two end-member models have emerged that explain the emplacement and evacuation of allochthonous salt as a result of counterregional feeder systems (Figure 1) (Worrall and Snelson, 1989; Moretti et al., 1990; Wu et al., 1990; Diegel et al., 1995; Schuster, 1995) or salt-stock canopies (Figure 2) (Jackson and Talbot, 1989; Hodgkins and O'Brien, 1994; House and Pritchett, 1994; Rowan et al., 1994; Rowan, 1995). These models may have similar geometries of the top salt surface, but differ dramatically in respect to the topography of the base salt surface. Counterregional feeder systems generally extruded with a base salt surface that was subhorizontal with minor relief except for salt feeders and local ramps that cut upsection basinward and laterally (Schuster, 1995). Two end-member models of salt evacuation for the counterregional feeder systems have been proposed by Schuster (1995): stepped counterregional systems and roho systems. Stepped counterregional systems are identified by their ramp-flat-ramp geometry along the base salt or equivalent salt weld with relatively unfaulted overlying sediments dipping basinward (Figure 1C). Roho systems are distinguished by numerous basinward-dipping listric normal growth faults that merge into a detachment along the base salt or weld surface with overlying sediments dipping landward (Figure 1E). In contrast, salt-stock canopies have significant structural relief (up to 4 s two-way traveltime) along the salt weld surface, with ramps commonly cutting upsection in all directions (Rowan et al., 1994; Rowan, 1995). Elliptical lows of the salt weld surface are interpreted to identify the location of the original feeder stocks (Rowan et al., 1994; Rowan, 1995); however, the quality of the seismic used here is insufficient to delineate much of the deeper portions of the tertiary weld, let alone the actual feeder stock.

The emplacement and eventual evacuation of allochthonous salt appears to vary spatially and temporally throughout the northern Gulf of Mexico Basin. Neither of the two end-member models can explain all of the geometries observed, and a combination of the two is commonly required (Rowan, 1995). Complex three-dimensional (3-D) geometries result from the interaction of salt derived from these two types of systems. The original salt geometries and the manner in which they interact

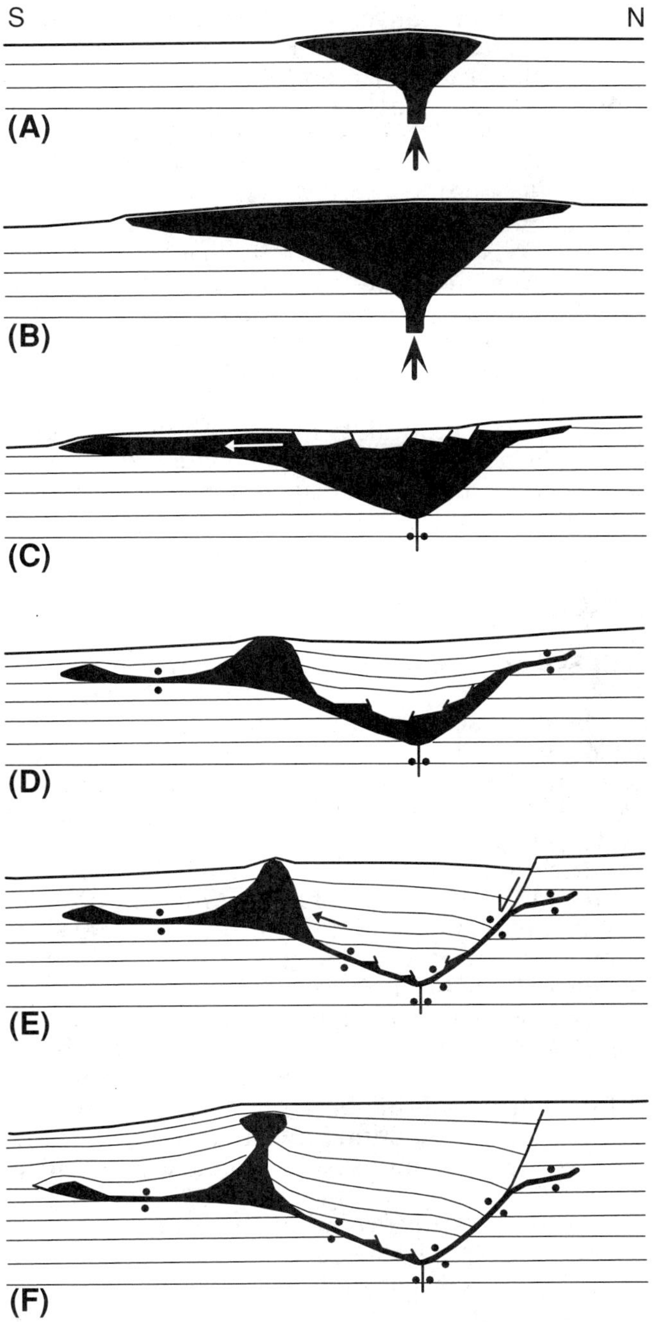


**Figure 2—Schematic evolution of a bulb-shaped salt stock that undergoes evacuation to form a bowl-shaped minibasin. The merger of two or more stocks creates a salt-stock canopy (not shown), and the eventual withdrawal of salt forms a deep weld system with significant relief and elliptical lows. Salt is in black, salt welds are indicated by pairs of dots, and bedding is shown by thin lines. No ages are inferred; section not to scale (from Rowan, 1995).**

dictate the positions of later minibasins, remnant salt diapirs, extensional growth faults, contractional structures, and strike-slip deformation.

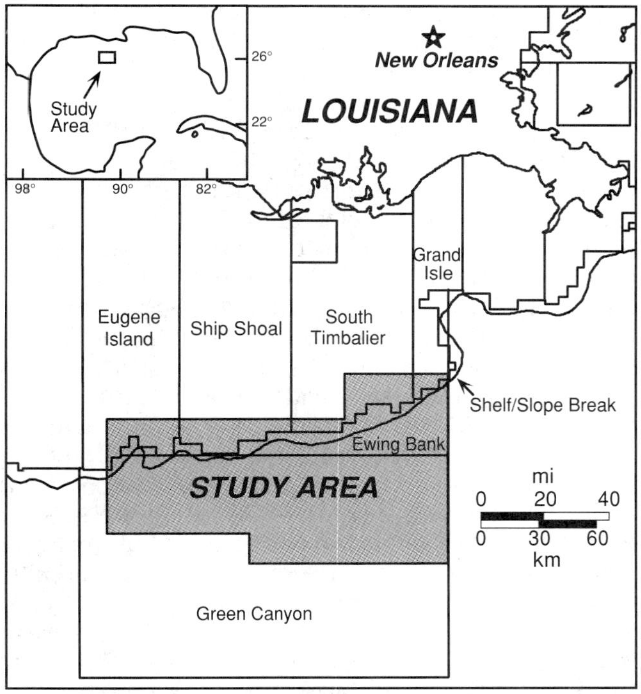


**Figure 3—Location map of study area in the northern Gulf of Mexico.**

In this paper, we summarize the structural geometry and evolution of allochthonous salt systems in northern Green Canyon, Ewing Bank, and the southern parts of Grand Isle, South Timbalier, Ship Shoal, and Eugene Island lease areas (Figures 3–5). The summary is based on integrating seismic interpretation, models for allochthonous salt extrusion and evacuation, and sequential restorations of eleven regional profiles. We first describe the regional salt system geometry, the distribution of minibasins, and both extensional and contractional structures that are illustrated by the eleven depth profiles. Next, we use the sequential structural restorations to document the evolution of salt-stock canopies, tabular salt sheets, and associated structures. This paper presents new structural restorations that illustrate the evolution of salt-stock canopies and how their evolution changes because of interaction with tabular salt sheets and the formation of composite salt canopies. Finally, we discuss new data that summarize the timing and rates of sedimentation, salt flow, and deformation relative to various structural styles.

The study area has undergone a complex 3-D structural evolution that would ideally be described by conducting 3-D restorations; however, the large size of the study area (~12,000 km$^2$; 4600 mi$^2$) makes this impossible, and we rely on two-dimensional (2-D) restorations evenly spaced across the study area. Three-dimensional restorations have been conducted on smaller regions within the central part of the study area (McBride et al., 1995; McBride, 1997) and are consistent with the results of 2-D restorations illustrated there.

The study area is located in water depths between 60 and 1400 m (200 and 4500 ft). The entire region is a mature petroleum province that includes part of the area of active subsalt exploration and three subsalt discoveries: Mahogany (Ship Shoal 349), Agate (Ship Shoal 361), and Teak (South Timbalier 260) (Figure 4). The study area includes 42 other fields and discoveries, and is considered to have excellent potential for additional discoveries in the suprasalt, subsalt, and deeper water regions. Turbidite-related facies are the primary reservoirs throughout the study area.

The database consists of a 10,000-km (6300-mi) grid of exploration 2-D multifold reflection seismic data (Figure 4). The data were collected originally in 1985–1986 as a 3-D grid to 7.5 s with offsets of 3000 ft (900 m), and partially reprocessed in 1989 as a 2-D survey with a spacing of 1.6–4.8 km (1–3 mi). In addition, digital well logs from 175 exploration and production wells and biostratigraphy reports from 180 wells were used to constrain the seismic interpretations (see Weimer et al., 1994, 1998, this issue).

## STRATIGRAPHY

The regional stratigraphic setting of the study area is described in detail by Weimer et al. (1994, 1998, this issue) and is briefly summarized here. Salt withdrawal created a series of minibasins that have up to 6100 m (20,000 ft) of Pliocene–Pleistocene strata (Figures 6–9). These strata consist primarily of turbidite systems deposited in bathyal water depths with abrupt lateral and vertical facies changes. Significant reservoir sands are locally developed.

Four main stratigraphic intervals have been identified in the region and reflect different types of turbidite systems based on lithofacies and seismic facies: lower Pliocene (5.5–3.0 Ma), upper Pliocene (3.0–1.4 Ma), lower Pleistocene (1.4–0.7 Ma), and upper Pleistocene (0.7 Ma to Holocene) (Figures 6–9). The lower Pliocene interval contains the 5.5-, 4.2-, and 3.8-Ma sequence boundaries and consists primarily of sand-rich, ponded turbidite systems. The upper Pliocene interval includes the 3.0-, 2.6-, 2.4-, and 1.9-Ma sequence boundaries and consists of a well-developed stratigraphic section to the west and a thinner section to the east. The lower Pleistocene interval includes the 1.4-, 1.1-, and 0.8-Ma sequence boundaries and is primarily shale rich with localized sands developed in channel-levee systems and basin-floor fans. The upper Pleistocene interval contains shale-dominated turbidite systems with submarine canyons developed in the east. By 0.8 Ma, deltas prograded into the northern portion of the study area, and neritic water depths existed there until the present. Extensive Pliocene–Pleistocene sea level fluctuations controlled the timing of deposition (Weimer et al., 1998, this issue).

The interplay between salt deformation and sedimentation is complex. There is a general stacking pattern in which early sand-rich turbidite systems within ponded settings evolved to bypass settings dominated by shale-rich facies (Weimer et al., 1994, 1998, this issue); however, detailed analysis of the relationships suggests that facies distribution, both spatial and temporal, is more complex and controlled by a number of factors ranging from regional sediment supply patterns to local salt deformation (Rowan and Weimer, 1998, this issue).

## STRUCTURE

### Salt System Geometry

The Green Canyon/Ewing Bank region comprises a structurally complex, multilevel salt system consisting of a series of allochthonous, amalgamated salt sheets, rootless salt bodies, and salt welds (Figures 5–9). Welds are surfaces that mark the location of evacuated salt such that original suprasalt and subsalt strata are now in contact (Jackson and Cramez, 1989). Weld surfaces are identifiable as subhorizontal to steeply dipping structural discontinuities that link isolated, rootless salt bodies throughout the study area (Figures 5–9). Welds are commonly (but not everywhere) associated with discontinuous, high-amplitude reflections on seismic profiles and can be differentiated from shale detachment systems by structural restoration and isostatic calculations (Diegel et al., 1995).

The time structure map is a composite map (Figure 5). Where there is a discrete thickness of salt, the top salt has been mapped; where salt is interpreted to have been evacuated, the resultant salt weld has been mapped. The map (Figure 5) and depth sections (Figure 6) illustrate the 3-D structure of the salt system. Although there is significant variability and structural relief of the top salt, the key to understanding the salt system is the geometry of the base salt or equivalent salt weld. Two distinct types of geometry are observed: (1) areas of subhorizontal base salt or equivalent salt weld and (2) elliptical depressions separated by saddles. These correspond, respectively, to counterregional sheet systems and salt-stock canopies that are in various stages of emplacement and evacuation (Rowan, 1995).

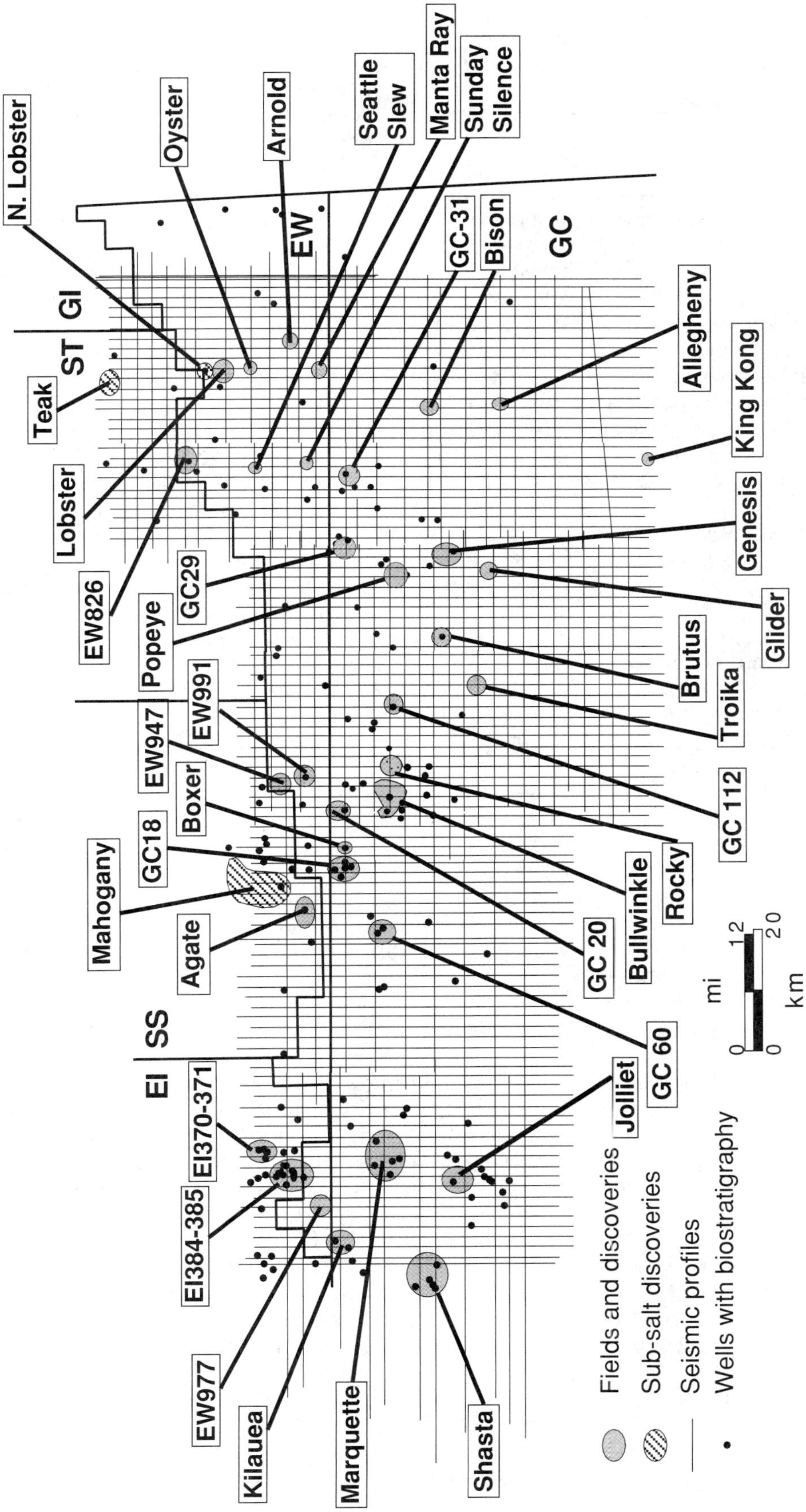


Figure 4—Map showing database, discoveries, and fields.

Three levels of salt sheets are present. To the north is a series of shallow sheets (at 3–5 km, 2.5–4.1 s two-way traveltime) that extend beyond the study area to the north (Figures 5; 6A–E, G, H, J, K; 7–9). The presumed counter-regional or salt-stock feeders would be located landward along the outer shelf of the southern additions. These sheets have remnant tabular salt in places and welds in others. The second level is deeper (at 6–9 km, 4.9–6.9 s two-way traveltime) and is observed in parts of the northern half of the area (Figures 5; 6C, D, F, H, I; 7). This level may be more widespread, but cannot be imaged beneath shallower salt. This deep level is marked by salt welds with only localized salt rollers (e.g., Figure 6I). The third level consists of a series of shallow sheets (base salt at 3–4 km, 2.5–3.4 s two-way traveltime) that occur along the southern margin of the study area (Figures 5; 6C, E, G, K; 7; 8). These are young features that have experienced only minor salt evacuation; the top salt is often just below the sea floor.

In contrast to these various levels of sheets are numerous elliptical depressions in the deep salt weld scattered throughout the study area (Figures 5; 6A, C, E–G, J, K; 7–9). The bottoms of the depressions (low points in the deep welds) are at 8–10 km (6.5–7.5 s two-way traveltime). Adjacent subweld reflections, when imaged, commonly are subhorizontal and truncated by the welds. Rowan (1995) documented such truncation beneath the south-dipping portions of several of these salt weld depressions and referred to the dipping welds as ramps (regardless of orientation). Similar reflection truncations can be seen on seismic profiles with less certainty beneath the depressions traversed by Figures 7–9. Overlying the elliptical depressions in the salt weld are elliptical minibasins. Within the minibasins, sequences may mimic the bowl-shaped geometry of the underlying salt weld (Figure 6A, J, K), or they may be rotated landward (Figure 6C, F) or basinward (Figure 6E, G, K) (see also Rowan and Weimer, 1998, this issue).

Salt-weld depressions and subhorizontal sheets have varying relationships to each other. Adjacent elliptical lows commonly are connected by saddles in the base salt or equivalent weld (Figure 5). Shallow salt sheets along the northern margin of the study area, in places, are isolated and overhang deeper salt weld levels (Figures 5; 6C, D, H; 7). Elsewhere, these sheets merge with the northern margins of elliptical lows at the tops of south-dipping ramps (Figures 5; 6A, E, F, J, K; 8; 9). The young, shallow sheets along the southern edge of the area are fed by counterregional feeders from either deeper sheets (Figures 5, 6E) or elliptical depressions (Figures 5; 6C, G; 7; 8). The result of these varying interactions is a highly complex, 3-D geometry in which different types and levels of salt are either separated or connected by ramps of all orientations.

Salt diapirs in this complex system are not located randomly; diapirs are generally found either along the frontal or lateral edges of shallow sheets or over the ridges and saddles between elliptical lows in the salt weld surface; therefore, these diapirs commonly do not have deep roots. Exceptions exist in the eastern part of the study area, where several stocks extend upward from a deep salt weld underlying a major north-northwest–trending depositional trough (Figures 5, 6I). Most salt bodies are irregular in map view and show variable geometries in profile view (Figure 6); they may have gently sloping, vertical, or overhanging flanks, with smooth to uneven tops.

A transition zone runs approximately east-west through the center of the study area, differentiating allochthonous salt styles. To the north, the system is dominated by the southern portions of shallow, partially evacuated salt sheets that extend north onto the shelf. These sheets generally have thin remnant salt linking small, shallow salt bodies. To the south, salt bodies are broader and thicker, with roots to deeper elliptical salt welds or sheets. The salt body in the southeastern corner of the study area represents the northern edge of the Sigsbee nappe system of shallow tabular salt (Wu et al., 1990; Seni, 1992). The north-south transition is partly an effect of different salt systems (counter-regional to the north, salt-stock canopy to the south) and partly a result of differing maturity in evolution (more mature to the north). Although the boundary roughly parallels the shelf margin, the transition does not reflect simply the differences between shelf and slope salt tectonics because an evacuated salt-stock canopy system also has been identified beneath the shelf in the Eugene Island lease area (Rowan, 1995).

## Extensional Structures

Normal faulting is the primary fault style within the study area and is intimately related to the salt system. Small planar faults commonly are found associated with the crests of folds and salt bodies, along the tops of tabular salt sheets, within the hanging walls of larger growth faults as antithetic faults, and along the flanks of large monoclines (Figure 6). These small faults usually are accommodating local strain associated with the folding and bending of strata. In contrast, the larger listric normal growth faults are dominant and occur in three basic structural styles.

South-dipping listric growth faults merge into a deep, subhorizontal detachment surface in the eastern portion of the study area (Figure 6I). The east-west-striking faults contain salt rollers in their footwalls. This "roho" style of listric normal faulting

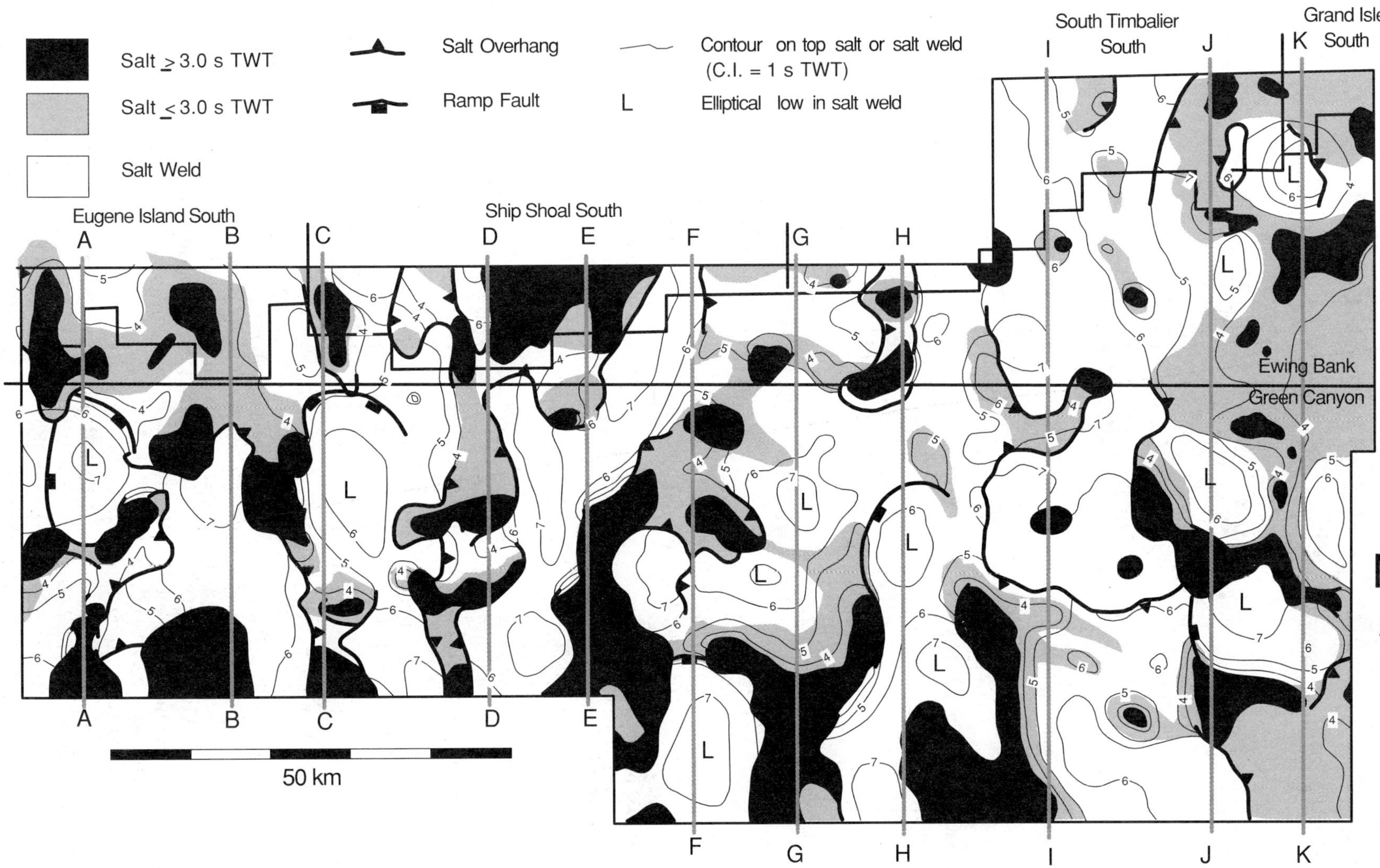


Figure 5—Time structure map on top salt or equivalent salt weld (contour interval = 1 s two-way traveltime). The salt system is characterized by large structural relief with different levels of salt either separated by overhangs or connected by ramps in the salt weld. Major salt diapirs, overhangs, and elliptical lows on the composite salt weld are shown, as are locations of restored seismic profiles (see Figure 6) (modified from Rowan, 1995). s TWT = seconds two-way traveltime.

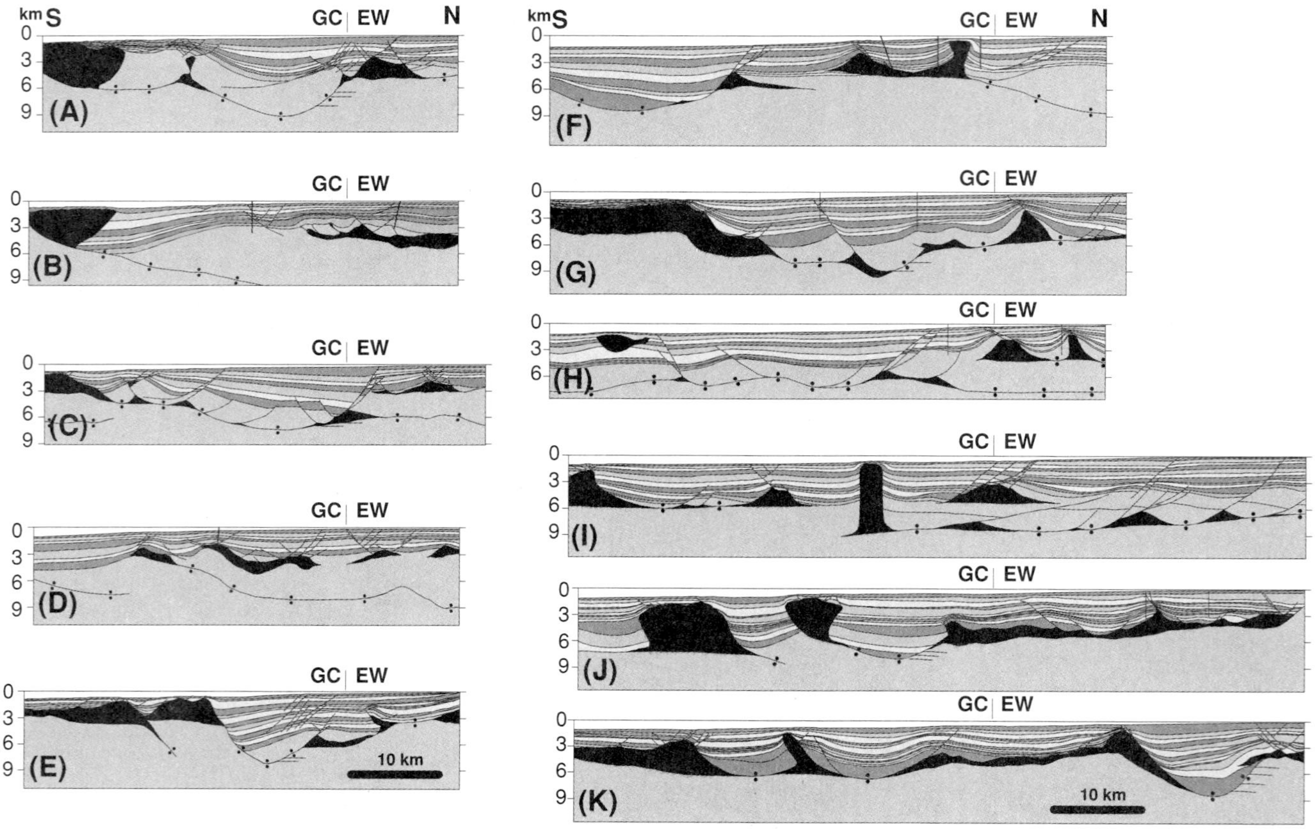


**Figure 6—Eleven regional true-scale (1:1) depth sections of interpreted seismic profiles used for structural restoration (locations are shown in Figure 5). The sections emphasize the dramatic lateral variability of the salt system. Stratigraphic patterns are arbitrary, salt is in black, salt welds are indicated by pairs of dots, wells are vertical lines, lease boundaries are indicated by GC (Green Canyon)/EW (Ewing Bank). Stratigraphic interpretations from Martinez (1994), Navarro (1994), Acosta (1994), Varnai (1996), Budhijanto (1995), and Weimer et al. (1994, 1998, this issue). Seismic data courtesy of Halliburton (now Western).**

(Sumner et al., 1991; Diegel et al., 1995; Schuster, 1995) appears to be uncommon within the study area and more representative of the onshore and continental shelf (Rowan et al., 1995). It is unclear whether this deep system is only a local occurrence or extends farther west and is poorly imaged beneath shallow salt bodies.

The second type of large normal fault is north-dipping or counterregional; these faults also are rare in the study area. Some apparent counterregional faults probably have only minor fault displacement and, in fact, are salt body edges or salt welds; collapse and growth of the hanging wall is due to subsidence into salt rather than extension (Figure 6E) (see Schuster, 1995). Other counterregional faults, however, are large-displacement faults where hanging-wall strata have been offset from corresponding condensed footwall intervals.

The most common large normal faults in the study area are ramp faults (Rowan, 1995). They have semicircular fault traces, mimicking the ellipticity of the salt weld depressions within the study area (Figure 5). In profile view, the faults extend upward from the landward margins of the elliptical lows (Figure 6C, F). These relationships suggest that topography on the salt weld is controlling the orientation and location of faults, as well as the location and size of many suprasalt minibasins. The absence of ramp faults around some of the salt weld depressions appears related to interaction with shallow tabular salt sheets in the northern part of the study area (Figure 6E, G, J, K). These zones of interaction are characterized either by an overhanging shallow sheet or by contractional structures that helped inhibit growth-fault formation (see also Rowan and Weimer, 1998, this issue).

## Contractional Structures

Contractional structures (both folds and thrusts) commonly are recognized on allochthonous salt

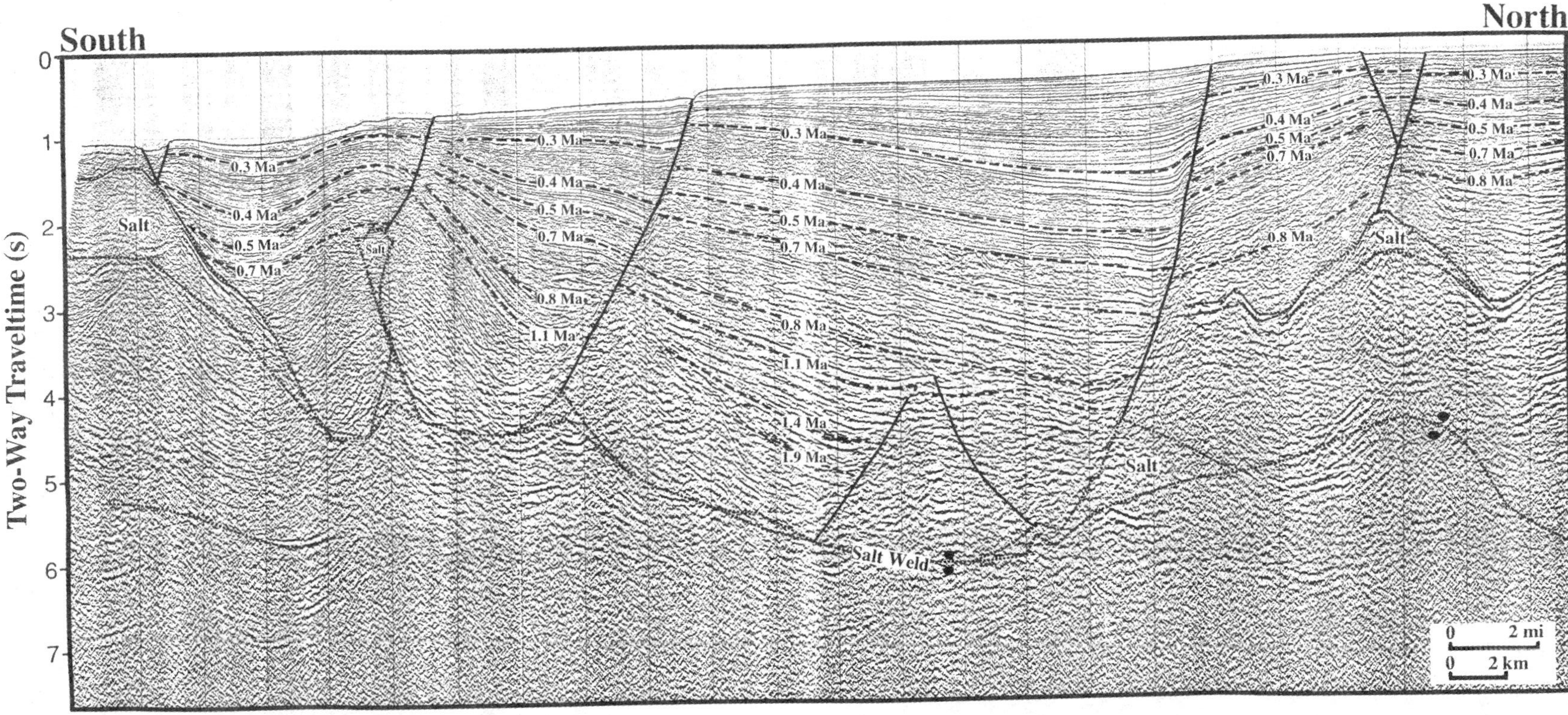


**Figure 7—Regional dip-oriented seismic profile C. Profile location shown on Figure 5; data courtesy of Halliburton (now Western Geophysical). Stratigraphic interpretation from Navarro (1994).**

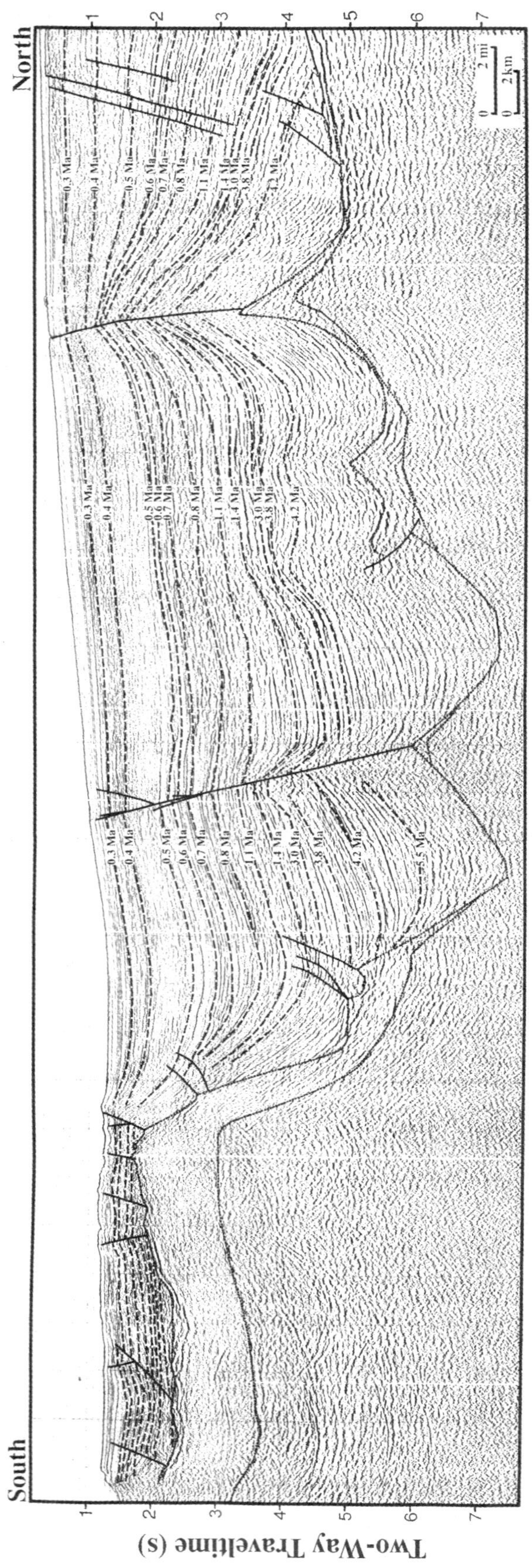


Figure 8—Regional dip-oriented seismic profile G. Profile location shown on Figure 5; data courtesy of Halliburton (now Western Geophysical). Stratigraphic interpretation from Acosta (1994).

sheets within the Gulf of Mexico Basin (e.g., Huber, 1989; Hossack and McGuinness, 1990; Cook and D'Onfro, 1991; Harrison and Patton, 1995; Hossack, 1995; Peel et al., 1995; Rowan, 1995). In the study area, contractional structures are found in three different structural settings (described in following paragraphs). The contractional structures use detachment surfaces that also are used by listric normal faults farther north. This linkage between contractional and extensional deformation along common detachment surfaces suggests that the contractional structures are partially compensating for the extension above a salt sheet or salt weld (see Hossack, 1995; Peel et al., 1995).

The first contractional setting is along the basinward termination of salt sheets (Figure 6B, C, H). Strata commonly appear to be modified by contraction and uplift that are characterized by asymmetric folds verging basinward, stratigraphic horizons raised above regional elevation, structural inversion of growth strata, angular unconformities, and thrust faults. Contractional structures such as these also have been recognized along the basinward margin of other salt sheets by Wu et al. (1990), Harrison and Patton (1995), Hossack (1995), and Peel et al. (1995). Thrust faults routinely are found adjoining rejuvenated salt bodies (Rowan, 1995; Vendeville and Nilsen, 1995). These contracted salt bodies are bounded by more strongly folded and steeply dipping strata than are typical salt domes (Figure 6F–H).

Similar contractional structures also are found along the northern flanks of elliptical depressions in the salt system, where they have linked with shallow salt sheets extending from the north. Thrust faults and contractional structures verge basinward and tend to be concentrated along the inflection point, where subhorizontal salt welds ramp downward and basinward into the elliptical depressions (Figures 6E, G, J; 8; 9). The "Ewing Bank thrust" observed by Huber (1989) is an analogous structure.

The third structural setting for contraction is along the southern margins of some of the elliptical depressions within the salt system (Figures 6A, C; 7). These structures also verge basinward and occur only where the elliptical low is bounded by a ramp fault along its northern margin. These structures, similar to the rejuvenated salt bodies previously described, tend to be contracted salt walls instead of individual domes. The contraction appears to have pinched off the salt stem (creating a vertical salt weld) within the original salt wall. Similar contracted salt structures within the Kwanza basin, Angola, have been discussed by Duval et al. (1992) and Vendeville and Nilsen (1995).

## EVOLUTION

### Restoration Method

Eleven interpreted seismic profiles (Figures 7–9) were converted to depth (Figure 6) using vertical ray paths and depth-dependent velocity functions obtained from numerous nearby wells. Locally, base salt was not always interpreted on the seismic profiles because of velocity pull-ups and poor imaging. In these areas, geometries were more easily estimated in the depth section by projecting regional dips both north and south of the bad data area when possible. In addition, subsalt geometries were rarely interpreted. The depth conversions and subsequent sequential restorations were performed using the software package GeoSec (Kligfield et al., 1986).

The restored sections (Figures 10–12) were constructed following a method of restoration for salt structures outlined by Rowan (1993), which is briefly described here. For each stage of restoration, the top layer of section was stripped off and the underlying section decompacted according to standard porosity-depth functions (e.g., Baldwin and Butler, 1985); rock types were determined from analyses of well-log data. The thickness and average density of the removed layer were then used to calculate and restore the magnitude of the sedimentary loading subsidence using a smoothed Airy isostatic model. Airy calculations were made in locations that did not experience salt withdrawal or inflation during the time interval being restored. Thermal subsidence due to postrift cooling was not incorporated into the restorations because rifting, which formed the Gulf of Mexico Basin, occurred during the Late Triassic–Jurassic (ending by 160 Ma) (Sawyer et al., 1991), and its effects were negligible in the Pliocene–Pleistocene.

The results of decompaction and isostatic correction were used to adjust the base salt and to construct a sea floor template for structural restoration of the suprasalt section. Although these specific paleobathymetric templates were not checked using biostratigraphy data, the technique has been confirmed elsewhere (Rowan et al., 1996). Fault displacements were restored by removing offsets of the top layer. Folding was removed by restoring the suprasalt sequences to the sea floor template, and any material removed by erosion during the interval being restored was estimated and replaced. The resulting gap between the restored top salt and adjusted base salt represents the preexisting salt geometry. Thus, the reconstructions show our best estimate of the restored fault and fold geometries, changes in shape of any salt bodies, changes in section length caused by extension or contraction, and

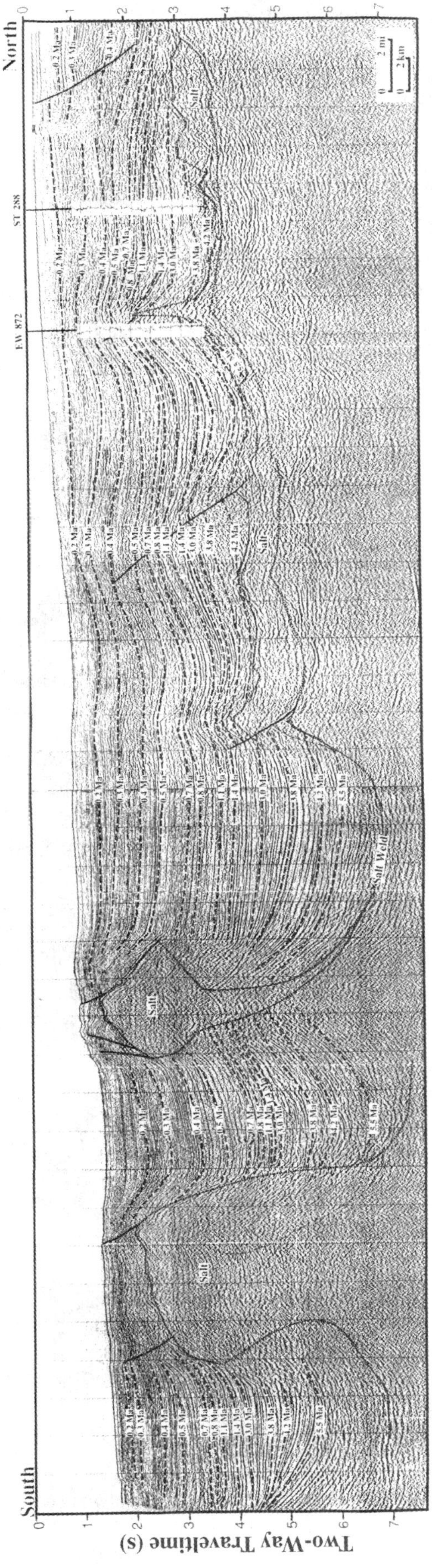


**Figure 9**—Regional dip-oriented seismic profile J. Profile location shown on Figure 5; data courtesy of Halliburton (now Western Geophysical). Stratigraphic interpretation from Budhijanto (1995).

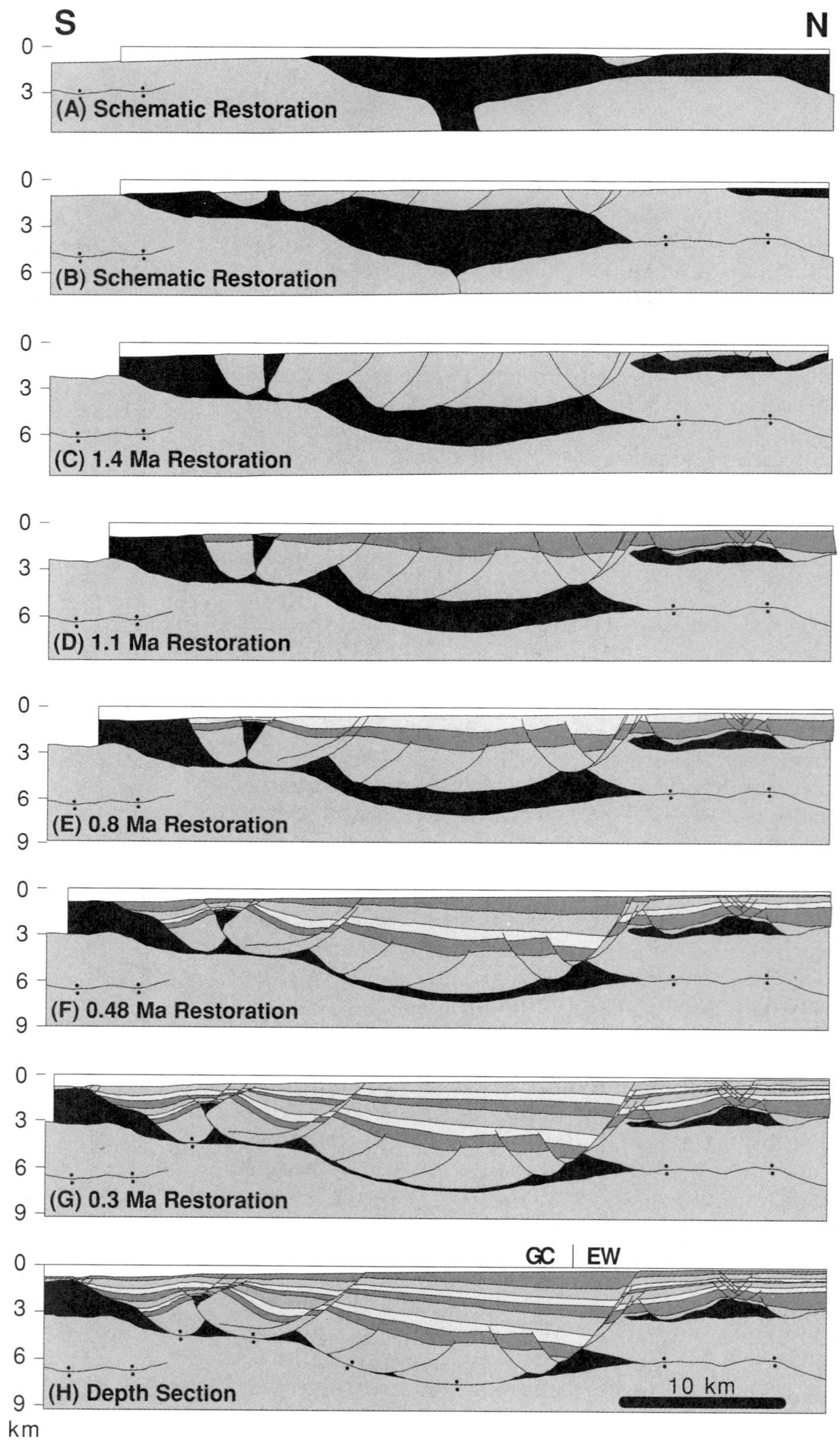


**Figure 10—Sequential restoration (using GeoSec computer program) of depth-converted interpretation of seismic profile C (Figures 6C, 7; see Figure 5 for profile location): (A, B) schematic restorations; (C–G) restorations to indicated times; (H) depth section. The changing length of the suprasalt section boundaries indicates the amount of extension or contraction restored. Vertical scale in kilometers; no vertical exaggeration. See text for discussion.**

variations in basin subsidence and paleobathymetry. Present-day sea level was maintained, and eustatic fluctuations were not incorporated into the restorations. Eustatic fluctuations can be included if desired; however, they would not affect the structural evolution, just the relative water

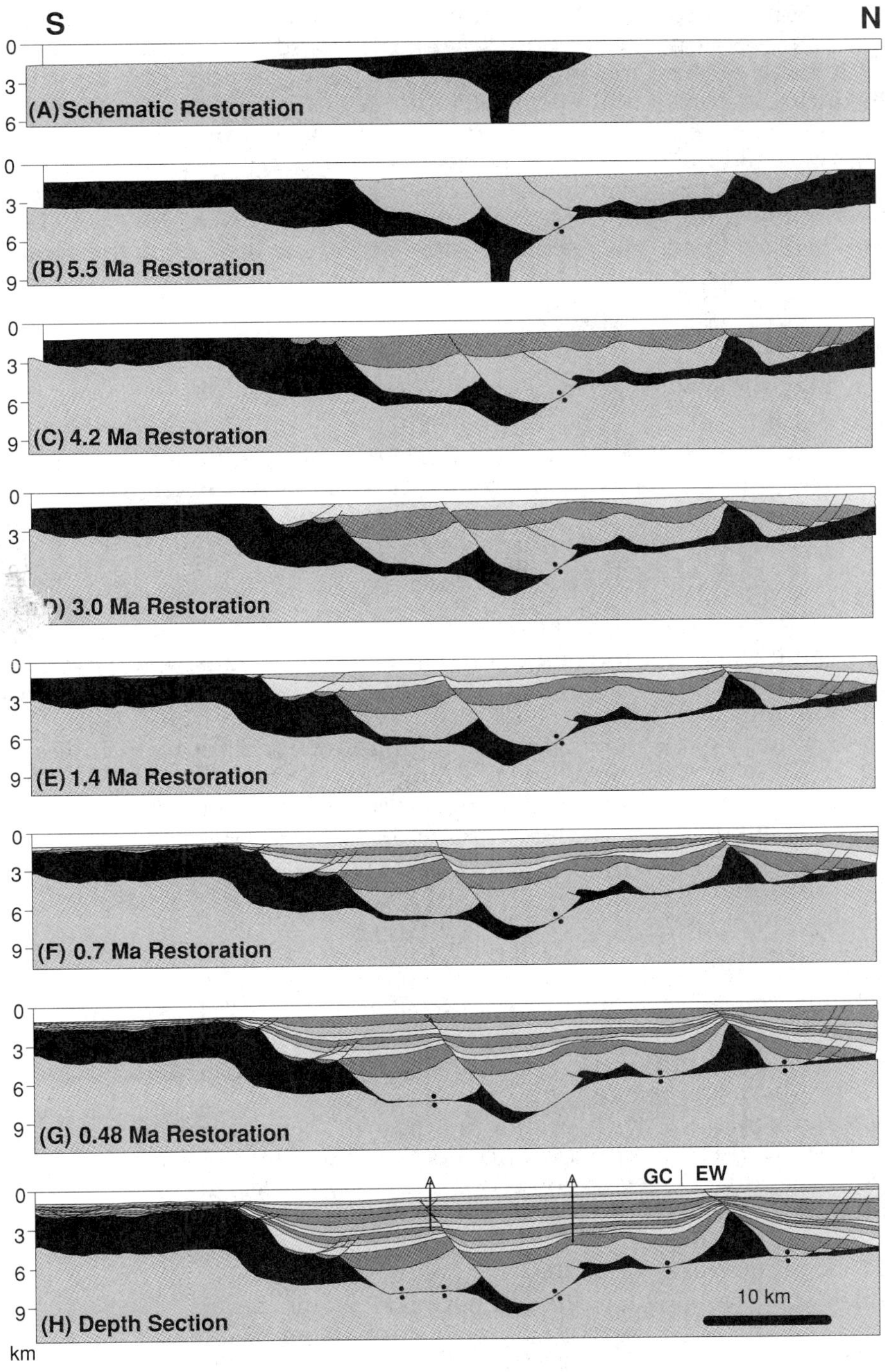


**Figure 11—Sequential restoration (using GeoSec computer program) of depth-converted interpretation of seismic profile G (Figures 6G, 8; see Figure 5 for profile location): (A) schematic restoration; (B–G) restorations to indicated times; (H) depth section. The changing length of the suprasalt section boundaries indicates the amount of extension or contraction restored. Vertical scale in kilometers; no vertical exaggeration. See text for discussion.**

depth through time. The change in water depth (a few tens of meters) would be negligible compared to the scale of the sections.

The restoration technique makes no assumptions about the kinematics of salt deformation. Because the top salt and the base salt are restored separately, salt area is usually not maintained (see following paragraphs). Salt is assumed to flow into or out of the section because of the complex 3-D kinematics of salt deformation (e.g., Worrall and Snelson, 1989; Diegel and Cook, 1990; Moretti et al., 1990; Hossack, 1995; Rowan, 1996). In addition, salt can leave the system as a result of dissolution, especially when it is close to the sea floor. Salt flow may occur parallel or oblique to the plane of section because of the various orientations of salt weld ramps and local variations in the transport direction of extension and contraction. Likewise, the kinematics of salt overburden are believed to be 3-D and not always along the plane of section locally or subregionally; therefore, the area of the overburden may need to vary locally during restoration to compensate for out-of-plane movement.

Where the footwall cutoff of a horizon did not exist because of a footwall salt body, no displacement was assumed along the salt-sediment interface. Although this assumption of no displacement is probably incorrect, there is no practical way of estimating such displacement along a single 2-D section. Thus, no widening of footwall salt bodies is shown during extension (cryptic extension) (Vendeville and Jackson, 1992a). Late subsalt faulting could have occurred and offset base salt locally, but no evidence was seen in the data to suggest regional extension, and there was no need conceptually to apply it to restore the sections.

Different restoration algorithms were used for different locations and time intervals (see Rowan and Kligfield, 1989; Schultz-Ela, 1992). Block rotation and bed-length restoration were used on contractional structures to remove shortening caused by folding. Vertical simple shear was applied in areas of extensional or vertical movement. Each section contains local pins at the northern and southern boundaries from the base of the section upward to the base of the shallowest allochthonous salt body or salt weld active during the sequential restorations; therefore, the length of the base of each section does not vary during the restoration. The location of a true pin (location of no lateral displacement) that extends from the sea floor downward to the base of the section varies depending on the geometries of individual salt bodies and salt welds within each section. Such "master" pins control the displacement restored above allochthonous salt bodies or salt welds shown at the northern and southern boundaries (loose lines) of some sections.

Restorations were constructed back to the oldest dated horizon, which ranges between 1.4 and 5.5 Ma from section to section. In addition, schematic restorations have been included to illustrate our best estimate of the precursor salt geometries responsible for the evolution depicted in the sequential restorations. No absolute ages are inferred for the schematic restorations.

## Restoration Results

Because of space limitations, we show seismic interpretations and structural restorations for only three of the cross sections through the study area (Figures 7–12). These three sections illustrate most of the various structural relationships discussed. Interpretations and restorations of the remaining seismic profiles are in McBride (1997).

Many of the salt geometries and deformation styles described correspond well to the two end-member models proposed for the evolution of allochthonous salt: counterregional systems and salt-stock canopies. Because complete stepped counterregional and roho systems are not observed within the study area, their evolution can be best understood by consulting Schuster (1995). In the next sections, we concentrate instead on the evolution of salt-stock canopies and combinations of the two models.

### *Salt-Stock Canopy Evolution*

The salt-stock canopy model (Rowan et al., 1994; Rowan, 1995) explains the evolution of minibasins overlying elliptical lows in the deep salt weld. The structural relief of the salt weld is considered a primary structural feature showing the original geometry of precursor, bulb-shaped salt stocks. Modern analogs, still filled with salt, have been presented by McGuinness and Hossack (1993), Hodgkins and O'Brien (1994), House and Pritchett (1994), and Fletcher et al., (1995). The ramps surrounding the lows are interpreted to represent the original overhanging interface between roughly horizontal bedding and the base of a preexisting salt-stock canopy that has since been evacuated to form the deep minibasins observed today (Figures 2; 6A, C, E–G, J, K; 7–12).

Evacuation of individual isolated salt stocks appears to begin with vertical subsidence and symmetrical sedimentary loading that commonly switches to asymmetrical growth faulting along its landward margin (Figures 2, 10). The position of these growth faults corresponds to the northern portion of elliptical ramps along the base salt surface. The evacuation process forms a minibasin with sediments that thicken into the ramp fault with a pronounced landward dip. Early minibasin evolution is dominated by vertical accommodation; extension and lateral basin growth is a relatively late process (Rowan, 1995).

Loading of the salt stock causes salt to extrude along its southern and lateral margins. The salt and overburden undergo a combination of gravity spreading and gravity gliding in response to loading. Salt movement is predominantly basinward as a result of the sea-floor topography and differential load. The southern boundary of the restorations in Figure 10 is shown offset because of the displacement restored within the suprasalt strata. The relative positions of the upper part of the southern boundary during the restoration document the magnitude of downslope translation through time above the salt system. This loose line is not intended to represent the geometry of the toe of the allochthonous salt sheet at any time. Allochthonous

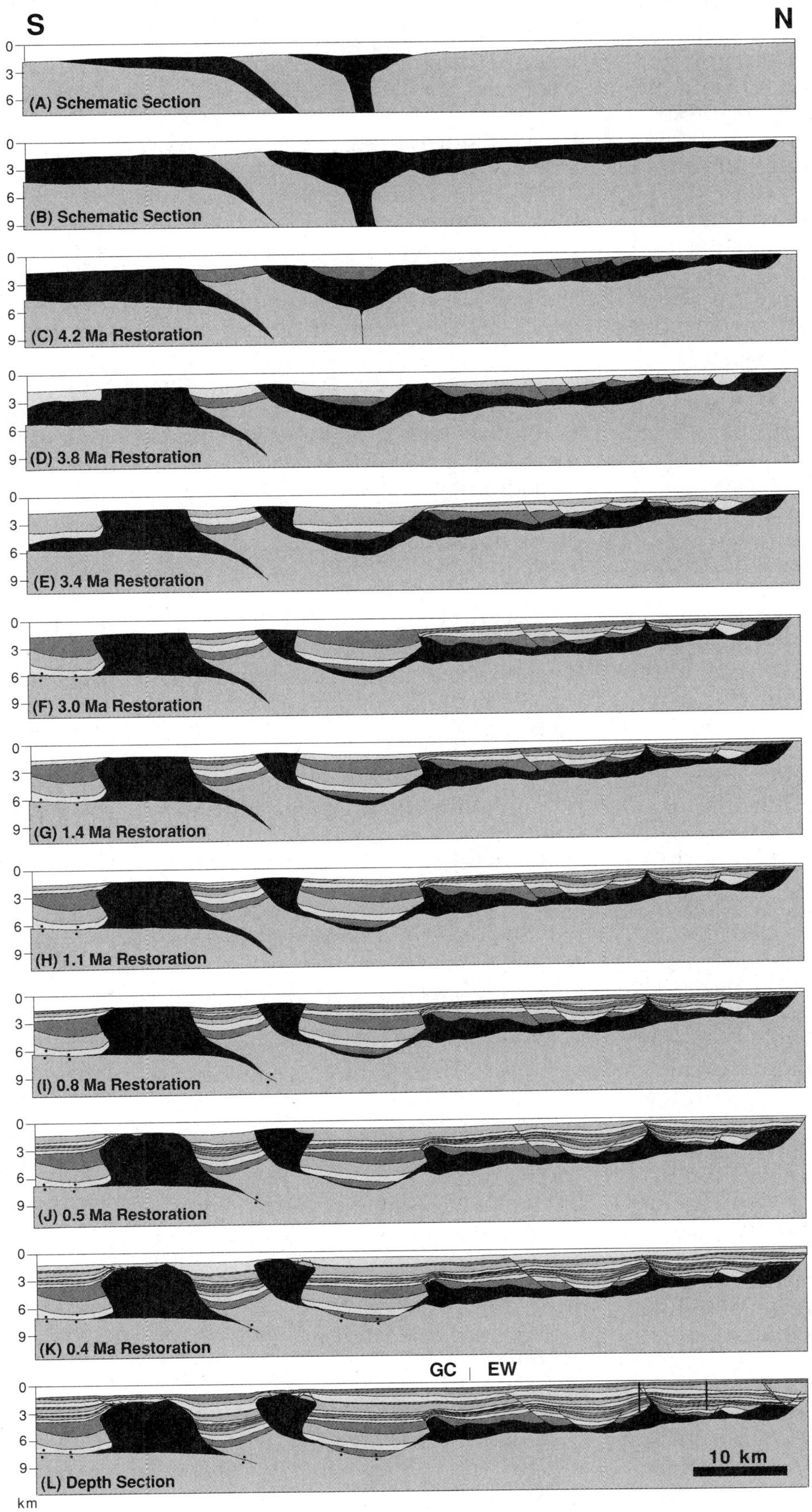


Figure 12—Sequential restoration (using GeoSec computer program) of depth-converted interpretation of seismic profile J (Figures 6J, 9; see Figure 5 for profile location): (A, B) schematic restorations; (C–K) restorations to indicated times; (L) depth section. The changing length of the suprasalt section indicates the amount of extension or contraction restored. Vertical scale in kilometers; no vertical exaggeration. See text for discussion.

salt undoubtedly continued basinward of this loose line, but the salt and possible overburden thicknesses are unconstrained. A component of the extension from the ramp fault and the other listric fault also was accommodated in part by contracting and rejuvenating the small diapir along the southern margin of the minibasin during the last 0.8 m.y.

The shallow salt sheet in the northern portion of Figure 10 did not interact with this particular bulb-shaped salt stock and does not appear to have affected its evolution. The shallow sheet is significantly younger than the salt stock and did not advance far enough basinward to inhibit growth faulting or salt evacuation. Other elliptical depressions within the study area that are isolated (e.g., southern part of Figure 6F) have undergone evolutionary histories similar to that shown in Figure 10.

### *Composite Salt-Stock/Salt Sheet Canopy Evolution*

The evolution of other elliptical salt-weld depressions in the study area was different than that shown in Figure 10 because of interactions with salt sheets advancing from the north. Figures 11 and 12 illustrate the evacuation of salt stocks where they have amalgamated with shallow salt sheets advancing from the north. Both examples show how an originally isolated salt stock expands in all directions and joins with a shallow salt sheet to form a composite salt canopy prior to salt evacuation. The composite salt canopy probably amalgamated during the late Miocene.

The amalgamation of salt-stock canopies and salt sheets has inhibited the formation of ramp faults along the landward margins of the original salt stocks. Although initial subsidence was again vertical and symmetrical, subsequent evacuation was asymmetrical, with growth sediments expanding basinward to form a pronounced southern dip of the suprasalt strata (Figure 11) (asymmetric growth is not seen in Figure 12 because the growth occurred out of the plane). Similar evolution occurred in other composite salt canopies (e.g., Figure 6E).

Salt extruded from these salt stocks (those without ramp faults) appears to move into counterregional feeders located along their southern and lateral margins (Figures 6E, G, K; 11). These feeders may form new, shallow, tabular salt sheets in a manner similar to the stepped counterregional model (Schuster, 1995); however, salt flow out of salt stocks can vary depending on the spacing of adjacent stocks and the topography of the base salt surface. As described, some salt bodies are located above the saddles separating some of the salt weld depressions. Both basinward and lateral salt flow occurs during the evacuation of salt-stock canopies.

The evolution of the composite salt canopy at the northern end of section K (Figure 6K) was slightly different than the others (see Rowan, 1995). This canopy experienced phases of both evacuation styles for salt stocks. In this case, an early ramp fault formed along the landward margin of the salt stock prior to the advancement of the shallow salt sheet from the north. The tabular salt sheet then advanced past the salt-stock margin, inhibiting further faulting; subsequent evacuation of the minibasin was asymmetrical, with basinward expansion and growth. Such varying patterns of evacuation are a result of different types of salt bodies evolving at different times and rates, and amalgamating at different stages in their evolution. Such complexities appear commonplace in this portion of the northern Gulf of Mexico.

### *Linked Deformation*

As mentioned previously, the existence of listric normal faults and contractional folds and thrusts using common detachment surfaces suggests that the deformation is linked, and that contractional features are partially compensating for extension farther north. This style of deformation is common with gravity gliding and spreading and has been recognized and described at various scales within the Gulf of Mexico (Buffler et al., 1979; Wu et al., 1990; Diegel et al., 1995; Fletcher et al., 1995; Harrison and Patton, 1995; Hossack, 1995; Peel et al., 1995). The restorations (Figures 10–12) clearly show this relationship, but document that contractional structures are only partially compensating for the extension. The remainder of the displacement is absorbed by salt (see also Hossack and McGuinness, 1990; Diegel et al., 1995; Hossack, 1995; Peel et al., 1995). Salt absorbs the displacement by some combination of downslope flow as a salt glacier, stratigraphic climb, lateral flow into and out of the plane of section, and dissolution. The evolution of the salt stock within Figure 10 nicely illustrates how extension farther upslope is compensated for downdip by a combination of contractional structures and salt flow.

In map view, the linkage of extensional and contractional structures above a common detachment requires strike-slip faulting (tear faults) along the system's lateral boundaries. Such linkage has been documented by Harrison and Patton (1995) and Peel et al. (1995). Although strike-slip faults are not shown on the map (Figure 5) or along any of the cross sections (Figure 6), they

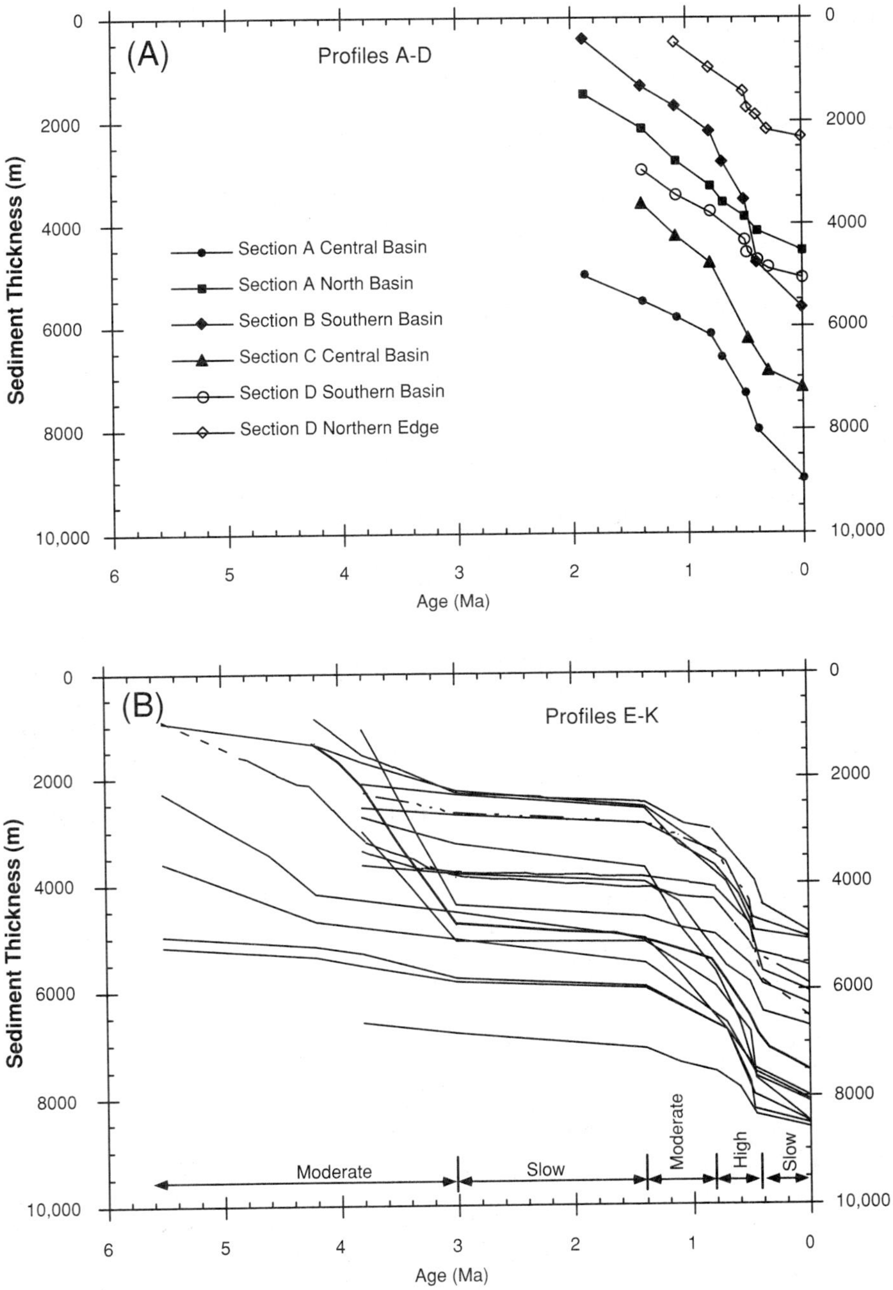


**Figure 13—Net sediment accumulation from minibasins along (A) profiles A–D beginning at 1.9 Ma and (B) profiles E–K beginning at 5.5 Ma (see Figure 5 for profile locations). Accumulation data measured from sequential restorations in minibasin centers.**

have been mapped bounding several of the minibasins within the study area (McBride et al., 1995; McBride, 1997). Snyder and Nugent (1995) discussed how strike-slip faulting has occurred west of the Teak discovery and controls the field's western closure. The positioning of the extensional, contractional, and strike-slip faults is controlled by the geometry of the salt system. Strike-slip faulting is a requirement for both counterregional and salt-stock canopy models in which downslope translation of the overburden occurs during salt evacuation.

## DISCUSSION

### Error

The restorations presented here were constructed using the systematic approach of Rowan (1993, 1996) that tries to account for all processes that influenced the structural evolution; however, numerous assumptions and limitations are inherent in any cross section restoration, and these must be understood to fully evaluate structural restorations.

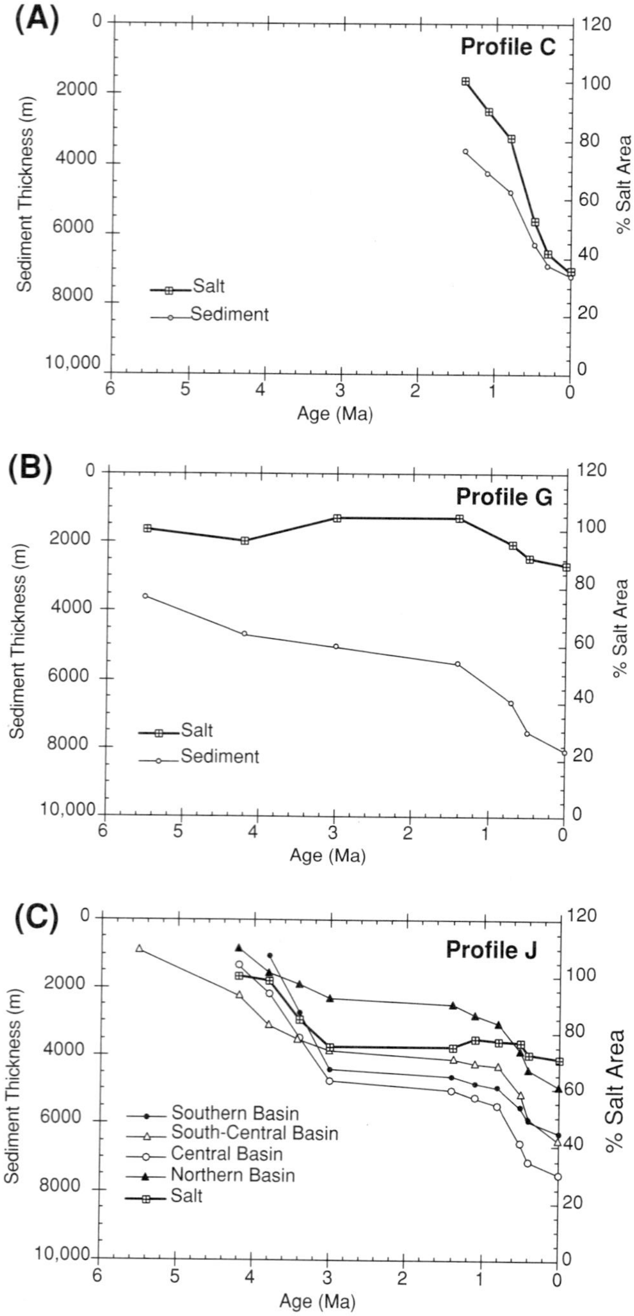


**Figure 14—Plots of net sediment accumulation and percent salt area for (A) profile C, (B) profile G, and (C) profile J (see Figure 5 for profile locations). Accumulation data and salt area measured from sequential restorations; salt area is measured for the entire section as a percentage of the value for the oldest restoration (100%).**

Error may result from many different sources and can become compounded during the restoration process. The initial interpretation of the geometries to be restored has perhaps the largest effect on the outcome of a restoration. Individual geologists may interpret the same data differently because of varying concepts of how structures are related. We commonly tested multiple interpretations iteratively by restoration. The restorations presented are from what we considered to be our best interpretations based on the critical evaluation of 3-D geometries and their evolution. Validation of an interpretation may be accomplished for large-scale features in this type of terrane, but cannot generally be used to test and refine the details of a specific interpretation (Rowan, 1996). Restorations of multiple interpretations, however, readily display the differences in their implied evolutionary histories, which, in turn, may be assessed relative to lithofacies trends, thermal maturity data, or petroleum systems.

Potential error of the initial interpretation can be further compounded by errors associated with depth conversion. For example, the assumption of vertical ray paths in this type of terrane is clearly a simplification. During the restoration process, errors also may occur because of assumptions made in (1) the smoothed Airy calculation of loading subsidence, (2) decompaction, (3) reconstruction of eroded material, (4) generation of the seafloor restoration template, (5) the plane-strain constraints, (6) the choice of restoration algorithms, (7) the absence of extension across certain salt bodies, (8) the lack of subsalt faulting, and (9) the linkage of some contractional structures to extension north of the profiles. Experience has shown, however, these errors to be relatively minor on the scale of the processes of bedding and salt deformation being evaluated here.

## Timing and Rates of Processes

Allochthonous salt sheet extrusion throughout the study area occurred mostly before the oldest dated stratigraphic horizons and therefore is not well documented with the current data set; however, the timing of formation of the small, shallow allochthonous salt body located in the southern portion of section H (Figure 6H) is well constrained and occurred during the Pleistocene, between 1.1 and 0.4 Ma. The restorations suggest that the majority of the allochthonous, tabular salt sheets present along the southern margin of the study area extruded during the late Miocene to early Pliocene. The timing was estimated by the relative thickness of the undated suprasalt sediments remaining within the minibasins at the time of the last age-constrained restoration. Sedimentation rate variations could affect this timing estimate significantly. The timing of salt-stock canopy growth also is not constrained by this data set; although it appears to have ended during the late Miocene, its inception may have been much older.

The timing of sedimentation pulses back to 5.5 Ma is well constrained. These pulses appear to control the timing of salt flow and fault deformation within the study area. The four main stratigraphic intervals described were based upon lithofacies and seismic facies, but also help delineate five distinct periods of sediment accumulation east of cross section D (Figure 13B): (1) moderate accumulation prior to 3.0 Ma, (2) slow, condensed sedimentation from 3.0 to 1.4 Ma, caused by a westward avulsion of the ancestral Mississippi River (Piggot and Pulham, 1993; Pulham, 1993), (3) renewed moderate accumulation between 1.4 and 0.8 Ma, (4) rapid sedimentation associated with continental glaciation between 0.8 and 0.4 Ma, and (5) slow accumulation during the past 0.4 m.y. Sections west of profile E show the last three stages less clearly (Figure 13A); earlier stages are not constrained due to lack of well data, and the interval from 3.0–1.4 Ma is not as condensed farther east due to closer proximity to the source deltas.

The pulses of sedimentation correspond to the times of the largest percent change in salt area along most of the cross sections (Figures 14, 15). The thicker the amount of sediment backstripped and removed from the section, the more salt that was required to replace it during the restoration process. Likewise, the removal of a thin layer during restoration required less salt to replace it, resulting in lower rates of salt area change. This relationship is a function of the restoration method, but also is believed to be real because most salt bodies are interpreted as passive diapirs (see Vendeville and Jackson, 1992a). A complete set of graphs illustrating the variations in sedimentation and salt flow vs. time for each cross section is available in McBride (1997).

The change in salt area varied significantly between sections. Although most sections showed a decrease in salt area forward in time, this is not universally the case. Decreases in salt area ranged from 26 to 65% (Figure 14A) where profiles cross minibasin centers vs. decreases of 3–13% (Figure 14B) where profiles cross minibasin margins. In contrast, section E had a 44% increase in salt area due to its position over a ridge of salt that grew in the Pliocene–Pleistocene (Figures 5, 6E) (McBride, 1997). Interestingly, during periods of low sediment accumulation, salt area was maintained or increased (Figure 15C), implying active flow and inflation probably because the existing differential load would continue to drive salt flow in the absence of active sedimentation. These variations in salt area are a function of how 3-D salt flow varies spatially and temporally relative to phases of active extrusion or evacuation. Statistical analysis comparing these values of percent change in salt area is not appropriate because each section represents random

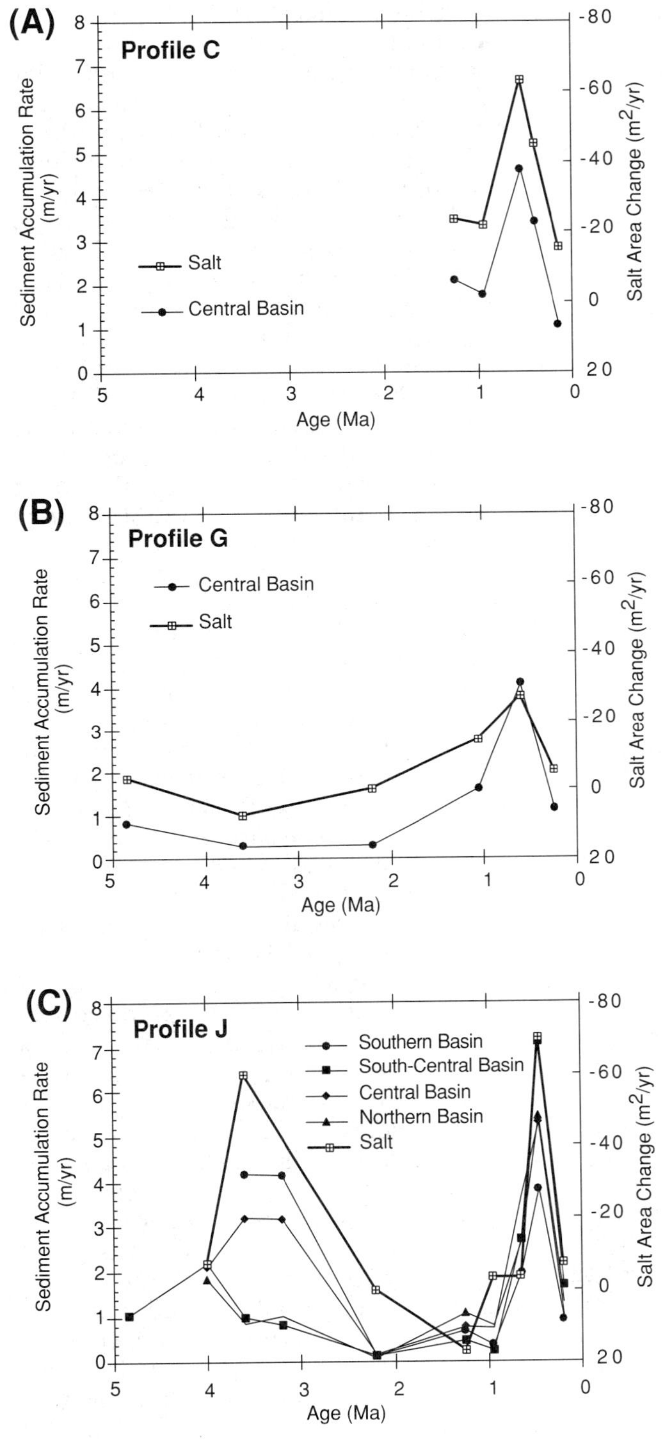


**Figure 15. Plots of finite sediment accumulation rates and rates of finite salt area change for (A) profile C, (B) profile G, and (C) profile J (see Figure 5 for profile locations). Rate values are plotted at the midpoint between dated restorations. Negative values of salt area change reflect a decrease in area.**

profiles through salt bodies at different phases of extrusion or evacuation relative to each other. The values of salt area change are presented to emphasize the large areal changes of salt that have occurred within a short period of time, and the variations that occurred laterally. Because these sections are restored only back to 5.5 Ma, they do not depict the initial extrusion phase of allochthonous salt evolution. The area of salt within an individual allochthonous salt body tends to progressively increase forward through time during the extrusion phase, reach a maximum, then progressively decrease during the evacuation phase (McBride, 1997; 1998, this issue). Depending on the age and the phase of extrusion or evacuation of individual salt sheets and salt-stock canopies within any particular cross section, the timing and manner of salt area change should be expected to change dramatically laterally from one section to another.

The structural restorations indicate that salt bodies within the study area were either exposed along the sea floor or covered by only a thin sheath of sediments for long periods of time (Figures 10–12). Such geometries suggest that large-scale salt dissolution probably occurred. The amounts and effects of dissolution have been discussed by Diegel et al. (1995), Hossack (1995), and Fletcher et al. (1995). They suggested that as much as 50% of the salt may have been lost by dissolution in some areas. Because it is impossible to calculate how much salt leaves the system by dissolution or salt flow through the section in three dimensions, calculated rates of salt flow (based on 2-D structural restorations) are minimum values.

The timing of salt weld formation within the salt-stock canopies varies throughout the study area (Figures 10–12), but commonly was synchronous with burial of adjacent salt stocks (Figure 12). Salt weld formation prevented salt flow into the adjacent stocks, which stopped the stocks from any further upward growth. Because of the elliptical, bowl-shaped, base salt surface of the precursor salt stocks, local welds commonly occurred along the landward ramps prior to evacuation of the centermost positions. Rowan (1995) illustrated that rates of extension along these margins increased after weld formation occurred there, possibly because salt could no longer provide vertical accommodation for sediments. Subsequent space was provided by the overburden moving downslope along the basinward dipping salt weld ramp, evacuating salt from the center of the minibasin. This relationship between salt weld formation and extension occurs in some of the cross sections, but not others. The lack of a clear relationship probably is a function of the placement of the cross sections relative to the minibasins and the positions where initial salt weld formation occurred. Thus, initial salt weld formation within minibasins may be out of the plane of the section, and its timing would not be accurately measured by the restoration.

Diegel et al. (1995) calculated that for shale-based detachment systems, such as the Vicksburg, more than 600% extension can accumulate across a fault zone. In contrast, they illustrated that fault systems associated with withdrawal from tabular salt sheets accumulate a maximum of about 20% extension across a fault zone. In this study, salt-stock canopy systems averaged 12% extension (e.g., Figure 10). In the shale-based detachment system, vertical motion (subsidence) was only about 7% of the horizontal motion (translation) compared with 70% recorded in the withdrawal of salt from a tabular sheet (Diegel et al., 1995). In the salt-stock canopy systems of northern Green Canyon, vertical motion was greater, averaging 114% of the horizontal motion (136% for section C; Figure 10). These values suggest that restorations of fault systems may quantitatively distinguish not only between shale-based detachment and salt-withdrawal fault systems, but also between salt-stock canopies and tabular salt sheets; however, more displacement data are necessary to make these measurements statistically meaningful. Easier qualitative ways of distinguishing the structural styles exist.

### Restoration Applications

Sequential restorations not only provide important structural information, but also serve as templates for other types of studies. For example, the restorations quantify the evolution of allochthonous salt and constrain salt thickness variations through time. Because of the high thermal conductivity of salt, constraining salt thickness variations enables more accurate thermal maturation modeling of subsalt source rocks. In addition, changes in salt geometries through time have impacted petroleum migration pathways, creating zones of concentration for subsalt petroleum migration (McBride, 1995; McBride et al., 1998, this issue). The profound impact of the structural evolution of allochthonous salt on the petroleum systems in this area is discussed in a companion manuscript by McBride et al. (1998, this issue), which illustrates the critical need for structural restorations to accurately conduct petroleum systems analyses in complex salt provinces.

## SUMMARY

Although salt tectonics is a complex three-dimensional process, regional two-dimensional structural restorations can provide significant insight into understanding the evolution of allochthonous salt. Structural restorations document a dynamic evolution of allochthonous salt in the northern Green Canyon and Ewing Bank region that resulted from a complex interplay among sedimentation, salt flow, and associated deformation. Basin and petroleum systems analyses of the area cannot be modeled properly without first establishing the structural geometry and evolution of allochthonous salt.

The northern Green Canyon and Ewing Bank region consists of a three-dimensionally complex, multilevel salt system. The multiple levels are not always distinct and overlapping; they commonly are connected in a complex manner by ramps of all orientations. The area is characterized by the interaction of two types of allochthonous salt systems: the salt-stock canopy and counterregional feeder models (Rowan, 1995; Schuster, 1995). Elliptical lows in the mapped salt weld surface record the geometry of precursor salt stocks from which salt was later evacuated. Above isolated salt stocks, minibasin formation was initiated by symmetric, vertical aggradation, then switched to asymmetrical growth during extensional growth faulting and northward rotation along the landward margins. Other salt stocks interacted and amalgamated with tabular salt sheets flowing from the north before salt withdrawal was completed to form composite salt canopies. Such interactions inhibited ramp faulting along the landward margins of the original salt stocks; minibasins developed with basinward collapse and rotation.

Contractional folds and thrusts concentrated where salt-stock canopies and tabular salt sheets merged at the northern margins of the lows. The positioning of extensional and contractional structures associated with precursor salt stocks is controlled by the original base salt geometry. Contractional folds, thrusts, and diapirs also developed along the southern margins of isolated salt sheets that did not merge with salt-stock canopies. These structures formed in partial response to linked extension farther upslope as a result of downslope translation of the overburden.

A positive correlation among sediment accumulation, extensional faulting, and salt flow exists throughout the study area. The main stratigraphic intervals represent not only differences in lithofacies and seismic facies, but also periods with different rates of sediment accumulation. Periods of relatively higher sediment accumulation rates generally correspond with periods of greater allochthonous salt flow and extensional faulting (5.5-3.0 and 1.4-0.4 Ma). A lack of correlation between sediment accumulation and salt flow occurred when a partial salt weld formed and accommodation space was provided primarily by listric normal faults detaching along the salt weld rather than by simple salt withdrawal. Quantitative analysis of extension and subsidence may help distinguish between salt-stock canopies and tabular salt sheets. The timing of salt weld formation within minibasins varied and was commonly synchronous with burial of adjacent salt stocks, suggesting the stocks were sourced locally.

A net decrease in salt area occurred during the evolution of allochthonous salt structures within the study area, although it varied from section to section. The greatest decreases in salt area occurred within profiles that cross minibasin centers. Smaller decreases in salt area occurred in profiles that cross minibasin margins. A significant increase in salt area through time was observed along one profile that avoided minibasin depocenters and recorded the progressive growth of a shallow salt body. Increases in salt also occurred temporally during periods of low sediment accumulation. Such variations in salt area reflect the various phases of salt extrusion or evacuation, as well as the randomness of the section locations. Because these sections are restored only back to 5.5 Ma, they do not depict the initial extrusion phase of allochthonous salt evolution. A complete cycle of allochthonous salt flow should consist of a progressive increase in salt area forward in time during the extrusion phase, then a progressive decrease in salt area in the evacuation phase.

## REFERENCES CITED


Acosta, Z. M., 1994, Sequence stratigraphy of Plio-Pleistocene sediments in north-central Green Canyon and western Ewing Bank, northern Gulf of Mexico: M.S. thesis, University of Colorado, Boulder, Colorado, 187 p.

Baldwin, B., and C. O. Butler, 1985, Compaction curves: AAPG Bulletin, v. 69, p. 622-625.

Budhijanto, F. M., 1995, Sequence stratigraphy of Plio-Pleistocene sediments, northeastern Green Canyon and eastern Ewing Bank, Gulf of Mexico: M.S. thesis, University of Colorado, Boulder, Colorado, 140 p.

Buffler, R. T., F. J. Shaub, J. S. Watkins, and J. L. Worzel, 1979, Anatomy of the Mexican ridges, southwestern Gulf of Mexico; *in* J. S. Watkins, L. Montadert, and P. W. Dickerson, eds., Geological and geophysical investigations of continental margins: AAPG Memoir 29, p. 319-327.

Cook, D., and P. D'Onfro, 1991, Jolliet field thrust fault structure and stratigraphy, Green Canyon Block 184, offshore Louisiana: Gulf Coast Association of Geological Societies, Transactions, v. 41, p. 100-121.

Diegel, F. A., and R. W. Cook, 1990, Palinspastic reconstruction of salt-withdrawal growth-fault systems, northern Gulf of Mexico (abs.): Geological Society of America Abstracts with Programs, v. 22, p. A48.

Diegel, F. A., J. F. Karlo, D. C. Schuster, R. C. Shoup, and P. R. Tauvers, 1995, Cenozoic structural evolution and tectono-stratigraphic framework of the northern Gulf Coast continental margin, *in* M. P. A. Jackson, D. G. Roberts, and S. Snelson, eds.,

Salt tectonics: a global perspective: AAPG Memoir 65, p. 109-151.

Duval, B. C., Cramez, and M. P. A. Jackson, 1992, Raft tectonics in the Kwanza basin, Angola: Marine and Petroleum Geology, v. 9, p. 389-404.

Fletcher, R. C., M. R. Hudec, and I. A. Watson, 1995, Salt glacier and composite sediment/glacier models for the emplacement and early burial of allochthonous salt sheets, *in* M. P. A. Jackson, D. G. Roberts, and S. Snelson, eds., Salt tectonics: a global perspective: AAPG Memoir 65, p. 77-108.

Harrison, H., and B. Patton, 1995, Translation of salt sheets by basal shear: Gulf Coast Section SEPM Foundation Sixteenth Annual Research Conference, p. 99-107.

Hodgkins, M. A., and M. J. O'Brien, 1994, Salt sill deformation and its implications for subsalt exploration: The Leading Edge, v. 13, p. 849-851.

Hossack, J. R., 1995, Geometrical rules of section balancing of salt structures, *in* M. P. A. Jackson, D. G. Roberts, and S. Snelson, eds., Salt tectonics: a global perspective: AAPG Memoir 65, p. 29-40.

Hossack, J. R., and D. B. McGuinness, 1990, Balanced sections and the development of fault and salt structures in the Gulf of Mexico (GOM) (abs): Geological Society of America Abstracts with Programs, v. 22, p. A48.

House, W. M., and J. A. Pritchett, 1994, Salt deformation modeling through the use of enhanced seismic imaging techniques: The Leading Edge, v. 13, p. 844-848.

Huber, H. F., 1989, Ewing Bank thrust fault zone Gulf of Mexico and its relationship to salt sill emplacement: Gulf Coast Section SEPM Foundation Tenth Annual Research Conference, p. 60-65.

Humphris, C. C., Jr., 1978, Salt movement on continental slope, northern Gulf of Mexico, *in* A. H. Bouma, G. T. Moore, and J. M. Coleman, eds., Framework, facies, and oil-trapping characteristics on the upper continental margin: AAPG Studies in Geology 7, p. 69-85.

Jackson, M. P. A., and C. Cramez, 1989, Seismic recognition of salt welds in salt tectonics regimes: Gulf Coast Section SEPM Foundation Tenth Annual Research Conference, p. 66-71.

Jackson, M. P. A., and C. J. Talbot, 1989, Salt canopies: Gulf Coast Section SEPM Foundation Tenth Annual Research Conference, p. 72-78.

Kligfield, R., P. Geiser, and J. Geiser, 1986, Construction of geologic cross sections using microcomputer systems: Geobyte, v. 1, p. 60-69.

Martinez, R. E., 1994, Sequence stratigraphy of Plio-Pleistocene sediments, northwestern Green Canyon, Gulf of Mexico: M.S. thesis, University of Colorado, Boulder, Colorado, 138 p.

McBride, B. C., 1995, Evaluation of subsalt petroleum potential using structural restorations (abs.): AAPG Annual Convention Program, v. 4, p. 62A.

McBride, B. C., 1997, Geometry and evolution of allochthonous salt and its impact on petroleum systems, northern Gulf of Mexico Basin: studies in three- and four-dimensional analysis: Ph.D. dissertation, University of Colorado, Boulder, Colorado, 276 p.

McBride, B. C., 1998, The evolution of allochthonous salt along a megaregional profile across the northern Gulf of Mexico Basin: AAPG Bulletin, v. 82, no. 5B, p. 1037-1054.

McBride, B. C., M. G. Rowan, P. Weimer, and R. Ratliff, 1995, Three-dimensional restoration of allochthonous salt structures and its impact on understanding reservoir trends and petroleum migration: preliminary results from Green Canyon and Ewing Bank, northern Gulf of Mexico: Gulf Coast Section SEPM Foundation Sixteenth Annual Research Conference, p. 171-175.

McBride, B. C., P. Weimer, and M. G. Rowan, 1998, The effect of allochthonous salt on petroleum systems of northern Green Canyon and Ewing Bank (offshore Louisiana), northern Gulf of Mexico: AAPG Bulletin, v. 82, no. 5B, p. 1083-1112.

McGuinness, D. B., and J. R. Hossack, 1993, The development of allochthonous salt sheets as controlled by rates of extension, sedimentation, and salt supply, *in* J. M. Armentrout, R. Bloch, M. C. Olson, and B. F. Perkins, eds., Rates of geological processes: Gulf Coast Section SEPM Foundation Fourteenth Annual Research Conference, p. 127-189.

Moretti, I., S. Wu, and A. Bally, 1990, Computerized balanced cross section LOCACE to reconstruct an allochthonous salt sheet, offshore Louisiana: Marine and Petroleum Geology, v. 7, p. 371-377.

Navarro, A., 1994, Sequence stratigraphy of Plio-Pleistocene sediments, northwestern Green Canyon and western Ewing Bank, Gulf of Mexico: M.S. thesis, University of Colorado, Boulder, Colorado, 150 p.

Nelson, T. H., 1991, Salt tectonics and listric normal faulting, *in* A. Salvador, ed., The geology of North America; the Gulf of Mexico basin: Geological Society of America Decade of North American Geology, v. J, p. 73-89.

Peel, F. J., C. J. Travis, and J. R. Hossack, 1995, Genetic structural provinces and salt tectonics of the Cenozoic offshore U.S. Gulf of Mexico: a preliminary analysis, *in* M. P. A. Jackson, D. G. Roberts, and S. Snelson, eds., Salt tectonics: a global perspective: AAPG Memoir 65, p. 153-175.

Piggot, N., and A. J. Pulham, 1993, Sedimentation rate as the control on hydrocarbon sourcing, generation and migration in the deepwater Gulf of Mexico: Gulf Coast Section SEPM Foundation Fourteenth Annual Research Conference, p. 179-191.

Pulham, A. J., 1993, Variations in slope deposition, Plio-Pleistocene, offshore Louisiana, north-east Gulf of Mexico, *in* P. Weimer and H. Posamentier, eds., Siliciclastic sequence stratigraphy: recent developments and applications: AAPG Memoir 58, p. 199-234.

Ratcliff, D. W., and D. J. Weber, 1997, Geophysical imaging of subsalt geology: The Leading Edge, v. 16, p. 115-142.

Rowan, M. G., 1993, A systematic technique for the sequential restoration of salt structures: Tectonophysics, v. 228, p. 331-348.

Rowan, M. G., 1995, Structural styles and evolution of allochthonous salt, central Louisiana outer shelf and upper slope, *in* M. P. A. Jackson, D. G. Roberts, and S. Snelson, eds., Salt tectonics: a global perspective: AAPG Memoir 65, p. 199-228.

Rowan, M. G., 1996, Benefits and limitations of section restoration in areas of extensional salt tectonics: an example from offshore Louisiana, *in* P. G. Buchanan and D. A. Nieuwland, eds., Modern developments in structural interpretation, validation and modelling: Geological Society of London Special Publication 99, p. 147-161.

Rowan, M. G., and R. Kligfield, 1989, Cross section restoration and balancing as an aid to seismic interpretation in extensional terranes: AAPG Bulletin, v. 73, p. 955-966.

Rowan, M. G., and P. Weimer, 1998, Salt-sediment interaction, northern Green Canyon and Ewing Bank (offshore Louisiana), northern Gulf of Mexico: AAPG Bulletin, v. 82, no. 6, 1055-1082.

Rowan, M. G., P. Weimer, and P. B. Flemings, 1994, Three-dimensional geometry and evolution of a composite, multi-level salt system, western Eugene Island, offshore Louisiana: Gulf Coast Association of Geological Societies, Transactions, v. 44, p. 641-648.

Rowan, M. G., B. C. McBride, and P. Weimer, 1995, The role of extension in the evolution of allochthonous salt, central Louisiana outer shelf and upper slope: Gulf Coast Section SEPM Sixteenth Annual Research Conference, p. 223-233.

Rowan, M. G., T. Villamil, P. B. Flemings, and P. Weimer, 1996, Use of cross-section restoration to determine paleobathymetry and sea-floor paleotopography in the Gulf of Mexico basin: Geology, v. 24, p. 299-302.

Sawyer, D. S., R. T. Buffler, and R. H. Pilger, Jr., 1991, The crust under the Gulf of Mexico basin, *in* A. Salvador, ed., The geology of North America; the Gulf of Mexico basin: Geological Society of America Decade of North American Geology, v. J, p. 53-72.

Schultz-Ela, D. D., 1992, Restoration of cross-sections to constrain deformation processes of extensional terranes: Marine and Petroleum Geology, v. 9, p. 372–388.

Schuster, D. C., 1995, Deformation of allochthonous salt and evolution of related salt/structural systems, eastern Louisiana Gulf Coast, *in* M. P. A. Jackson, D. G. Roberts, and S. Snelson, eds., Salt tectonics: a global perspective: AAPG Memoir 65, p. 177–198.

Seni, S. J., 1992, Evolution of salt structures during burial of salt sheets on the slope, northern Gulf of Mexico: Marine and Petroleum Geology, v. 9, p. 452–468.

Snyder, F. C., and J. A. Nugent, 1995, Teak—testing a subsalt hydrocarbon trap geometry, South Timbalier Block 260, Gulf of Mexico: Gulf Coast Section SEPM Foundation Sixteenth Annual Research Conference, p. 257–267.

Sumner, H. S., B. A. Robison, W. K. Dirks, and J. C. Holliday, 1991, Structural style of salt/minibasin systems: lower shelf and upper slope, central offshore Louisiana (abs.): Gulf Coast Association of Geological Societies, Transactions, v. 41, p. 582.

Varnai, P., 1996, Integrated 2-D and 3-D sequence stratigraphic analysis of Plio-Pleistocene turbidite systems, northern Green Canyon and central Ewing Bank protraction areas, northern Gulf of Mexico: Ph.D. dissertation, University of Colorado, Boulder, Colorado, 502 p.

Vendeville, B. C., and M. P. A. Jackson, 1992a, The rise of diapirs during thin-skinned extension: Marine and Petroleum Geology, v. 9, p. 331–352.

Vendeville, B. C., and M. P. A. Jackson, 1992b, The fall of diapirs during thin-skinned extension: Marine and Petroleum Geology, v. 9, p. 354–371.

Vendeville, B. C., and K. Nilsen, 1995, Episodic growth of salt diapirs driven by horizontal shortening: Gulf Coast Section SEPM Foundation Sixteenth Annual Research Conference, p. 285–295.

Weimer, P., P. Varnai, Z. M. Acosta, F. M. Budhijanto, R. E. Martinez, A. F. Navarro, M. G. Rowan, B. C. McBride, and T. Villamil, 1994, Sequence stratigraphy of Pliocene-Pleistocene turbidite systems, northern Green Canyon and Ewing Bank, northern Gulf of Mexico: Gulf Coast Section SEPM Foundation Fifteenth Annual Research Conference, p. 383–396.

Weimer, P., P. Varnai, F. M. Budhijanto, Z. M. Acosta, R. E. Martinez, A. F. Navarro, M. G. Rowan, and B. C. McBride, T. Villamil, C. Arango, J. R. Crews, and A. J. Pulham, 1998, Sequence stratigraphy of Pliocene and Pleistocene turbidite systems, northern Green Canyon and Ewing Bank (offshore Louisiana), northern Gulf of Mexico: AAPG Bulletin, v. 82, no. 5B, p. 918–960.

West, D. B., 1989, Model for salt deformation on deep margin of Gulf of Mexico basin: AAPG Bulletin, v. 73, p. 1472–1482.

Worrall, D. M., and S. Snelson, 1989, Evolution of the northern Gulf of Mexico, with emphasis on Cenozoic growth faulting and the role of salt, *in* A. W. Bally and A. R. Palmer, eds., The geology of North America; an overview: Geological Society of America Decade of North American Geology, v. A, p. 97–138.

Wu, S., A. W. Bally, and C. Cramez, 1990, Allochthonous salt, structure and stratigraphy of the northeastern Gulf of Mexico, part II: structure: Marine and Petroleum Geology, v. 7, p. 334–370.

## ABOUT THE AUTHORS

### Barry C. McBride

Barry McBride is a geologist at HS Resources Inc. He received his Ph.D. in geology at the University of Colorado, his M.S. degree in geology from Western Michigan University, and his B.S. degree in geology from Furman University. From 1988 to 1993 he worked for Mobil Exploration and Producing U.S. Inc. and Mobil Global Interpretation Support as an exploration geologist. His research interests focus on complex structures, petroleum exploration, and petroleum systems.

### Mark G. Rowan

Mark Rowan is a graduate of Caltech (1976), University of California, Berkeley (1982), and the University of Colorado (1991), where he is currently an assistant research professor in structural geology. He has also worked in the petroleum industry, both as an exploration geologist for Sohio and as a consultant with Geo-Logic Systems and Alastair Beach Associates. His research interests focus on the three-dimensional geometry and evolution of both extensional and contractional structures, with a recent emphasis on salt tectonics.

### Paul Weimer

Paul Weimer is an associate professor and director of the Energy and Minerals Applied Research Center at the University of Colorado at Boulder. His research interests are in sequence stratigraphy and basin analysis, three-dimensional seismic interpretation, and reservoir geology.

# The Evolution of Allochthonous Salt Along a Megaregional Profile Across the Northern Gulf of Mexico Basin[1]

Barry C. McBride[2]

## ABSTRACT

Sequential restorations of a north-south megaregional cross section across the north-central Gulf of Mexico Basin from east-central Louisiana to the abyssal plain define a dynamic, complex history of sedimentation, salt flowage, and salt evacuation. Proprietary composite seismic profiles (590 km), 33 wells, and published depth-to-basement maps were used to constrain the section in depth. Thirteen sequential structural restorations, incorporating both decompaction and isostatic subsidence (thermal and tectonic), were then constructed from the Late Cretaceous to the Holocene. The restorations highlight and constrain a protracted history of deformation that is primarily controlled by gravity and the progradation of Cenozoic sediments over salt.

Early stages of the tectonic history of the northern Gulf of Mexico Basin were related to differential thermal subsidence resulting from Early-Middle Jurassic rifting. During the Cenozoic, the evolution of the basin was dominated by the influx of large clastic depocenters, which caused the basinward evacuation of autochthonous Jurassic salt. Salt extrusion from the autochthonous layer was accomplished by inclined salt bodies, which flowed into salt glaciers or sheets near the sea floor. Evacuation of allochthonous salt layers provided significant sediment accommodation, and unusually thick sedimentary sections were deposited, such as the Terrebonne Trough of southern Louisiana (3–7 km of Miocene strata). Salt sheet formation and evacuation occurred progressively basinward through time in response to basinward shifts of major Cenozoic sedimentary depocenters. As a salt sheet neared complete evacuation, the underlying autochthonous salt layer would begin to evacuate, providing additional sediment accommodation that caused autochthonous salt flowage basinward, and the formation of the next allochthonous salt sheet basinward of the depocenter.

The area of autochthonous salt progressively decreased through time and currently represents at most 45% of its maximum along this transect. The total area of salt through time was more stable, with variations of only 30% from its maximum. This relationship is a function of lateral salt flow into and out of the plane of section, and possible salt dissolution. The restorations indicate that very little translation or extension (1.46%) occurred at the autochthonous salt level during the evolution of the basin. The majority of translation or extension occurred above allochthonous salt sheets (25%) and was compensated laterally by salt flow. Displacements above allochthonous salt sheets were driven by gravitational instabilities caused by the slope gradient.



[1]Manuscript received February 11, 1997; revised manuscript received October 20, 1997; final acceptance November 19, 1997.

[2]Energy and Minerals Applied Research Center, Department of Geological Sciences, University of Colorado, Boulder, Colorado 80309-0250. Present address: HS Resources Inc., 1999 Broadway, Suite 3600, Denver, Colorado 80202.

Many thanks are in order to the following employees of Mobil Exploration and Producing U.S. Inc. for their loyal support during the initiation, compilation, and completion of this project: Arlene Anderson, Al Koch, Bob Hiseler, John Krueger, Sybil Calloway, Lori Denning, Allison Pease, Jim Carpenter, Catherine Langley, and Dave Walz. I also thank the following companies and their representatives for their tremendous support and input as part of the Gulf of Mexico research consortium at the University of Colorado: AGIP, Amoco, Anadarko, BHP, BP Exploration, Burlington Resources, Canadian Occidental, Chevron USA, CNG, Conoco, Enterprise, Exxon, Marathon, Mobil, Occidental, Pan Canadian, Pennzoil, Petrobras, Phillips Petroleum, Shell Oil Co., Texaco, Total, Union Pacific, Unocal, Halliburton, and PaleoData Inc. Software donations from CogniSeis Development Inc., GeoQuest, Landmark, Platte River Associates, and Zeh Graphics were essential for the success of this project and are very much appreciated. Special thanks to Bob Ratliff, Mark Rowan, Paul Weimer, Jim Geiser, Roy Kligfield, and Andy Pulham for their support, advice, discussions, training, and input during the development of this project. I appreciate reviews by Mike Hudec, Dick Buffler, and an anonymous reviewer. Acknowledgment is made to the Donors of the Petroleum Research Fund, administered by the American Chemical Society, for partial support for this research.

## INTRODUCTION

Our understanding of the evolution of allochthonous salt bodies in the northern Gulf of Mexico Basin has undergone tremendous advancements during the past decade. Although Amery

(1969), Lehner (1969), and Humphris (1978) first proposed or suggested the presence of allochthonous salt in the northern Gulf of Mexico, it was not until the middle to late 1980s that this concept began to be applied (West, 1989; Worrall and Snelson, 1989; Diegel and Cook, 1990; Diegel and Schuster, 1990; Wu et al., 1990a, b; Sumner et al., 1991). The first significant conceptual breakthrough was the recognition that shallow salt bodies were not necessarily rooted directly into the autochthonous Louann Salt layer as vertical diapirs or massifs with thicknesses of up to 15 km. The second breakthrough was the observation that allochthonous salt bodies have undergone various phases and styles of formation and evacuation throughout the late Mesozoic–Cenozoic history of the Gulf of Mexico Basin. These revisions of long-standing beliefs were accomplished primarily through improved seismic imaging (e.g., Ratcliff, 1993), better physical modeling (e.g., Vendeville and Jackson, 1992a, b), and the application of sequential structural restorations that illustrated the incremental structural evolution of allochthonous salt (e.g., Worrall and Snelson, 1989; Diegel et al., 1995; Fletcher et al., 1995; Peel et al., 1995; Rowan, 1995; Schuster, 1995).

The use of structural restorations to study the evolution of allochthonous salt has been applied primarily on a local or subregional scale in the Gulf of Mexico Basin (Worrall and Snelson, 1989; Diegel and Cook, 1990; Diegel and Schuster, 1990; Moretti et al., 1990; McGuinness and Hossack, 1993; Diegel et al., 1995; Fletcher et al., 1995; Rowan, 1995; Schuster, 1995; McBride et al., 1998a, this issue). These studies are beneficial in explaining the evolution of particular allochthonous salt bodies or general allochthonous salt structural styles; however, documenting the evolution of allochthonous salt at the scale of the entire basin was beyond the scope of most of these investigations. Only two previous studies have included restorations across the entire northern Gulf of Mexico Basin as a method to address the evolution of allochthonous salt on a basin scale (megaregional) (Diegel et al., 1995; Peel et al., 1995) (Figure 1).

The goal of this contribution is to use sequential structural restorations to illustrate the progressive loading and evacuation of both autochthonous and allochthonous salt relative to the major late Mesozoic and Cenozoic depocenters along a megaregional transect in the northern Gulf of Mexico Basin. A 1:1 megaregional cross section and associated sequential structural restorations were constructed across the entire northern Gulf of Mexico Jurassic salt basin from east-central Louisiana to the abyssal plain (Figure 1). The interpretation is based on integrating industry reflection seismic surveys (~590 km; 365 mi), 33 wells, published depth-to-basement maps (Sawyer et al., 1991), various models of allochthonous salt extrusion and evacuation (Rowan, 1995; Schuster, 1995), and 13 sequential structural restorations back to the Jurassic (Figure 2).

The restorations are important for several reasons: (1) they place the subregional structural analyses of allochthonous salt in the Green Canyon lease area by McBride et al. (1995, 1998a, this issue) and Rowan (1995) in perspective to the evolution of the entire northern Gulf of Mexico Basin; (2) they provide an additional megaregional reconstruction across the basin in an area where no published examples exist; (3) they help to demarcate a major transition zone in the structural styles and timing of allochthonous salt formation between western and eastern offshore Louisiana when compared to adjacent transects by Diegel (1995) and Peel et al. (1995); (4) they illustrate genetic relationships between sedimentary depocenters, autochthonous salt flow, and allochthonous salt extrusion and evacuation; (5) they cross a region of intense petroleum exploration containing numerous supra- and subsalt discoveries; and (6) they provide the foundation for improved burial history and thermal maturation modeling of potential petroleum source rocks for petroleum systems analysis associated with the evolution of allochthonous salt (McBride et al., 1998b, this issue).

The salt system geometry and the distribution of extensional growth faults, minibasins, and contractional structures observed along the megaregional depth profile are described first. Next, sequential structural restorations are used to document the evolution of allochthonous salt and associated structures. Similarities and critical differences with interpretations made by other workers in nearby or comparable study areas are discussed where appropriate. Finally, the relationships among the timing and rates of sedimentation, salt flow, and deformation are discussed.

## MEGAREGIONAL CROSS SECTION

### Construction Technique

The megaregional cross section is based on the interpretation of a north-south–trending, composite seismic profile approximately 590 km (365 mi) in length. The seismic data extend from the Louisiana–Mississippi state border, across offshore east-central Louisiana, to the abyssal plain (Figure 1). The seismic reflection data are standard industry two-dimensional time migrated events collected from 1984 to 1988. Record lengths ranged from 13.0 s in the abyssal plain to 6.0 s in Louisiana. Processing of the seismic data also varied along the length of the profile, with folds ranging from 20 to 60. The paper

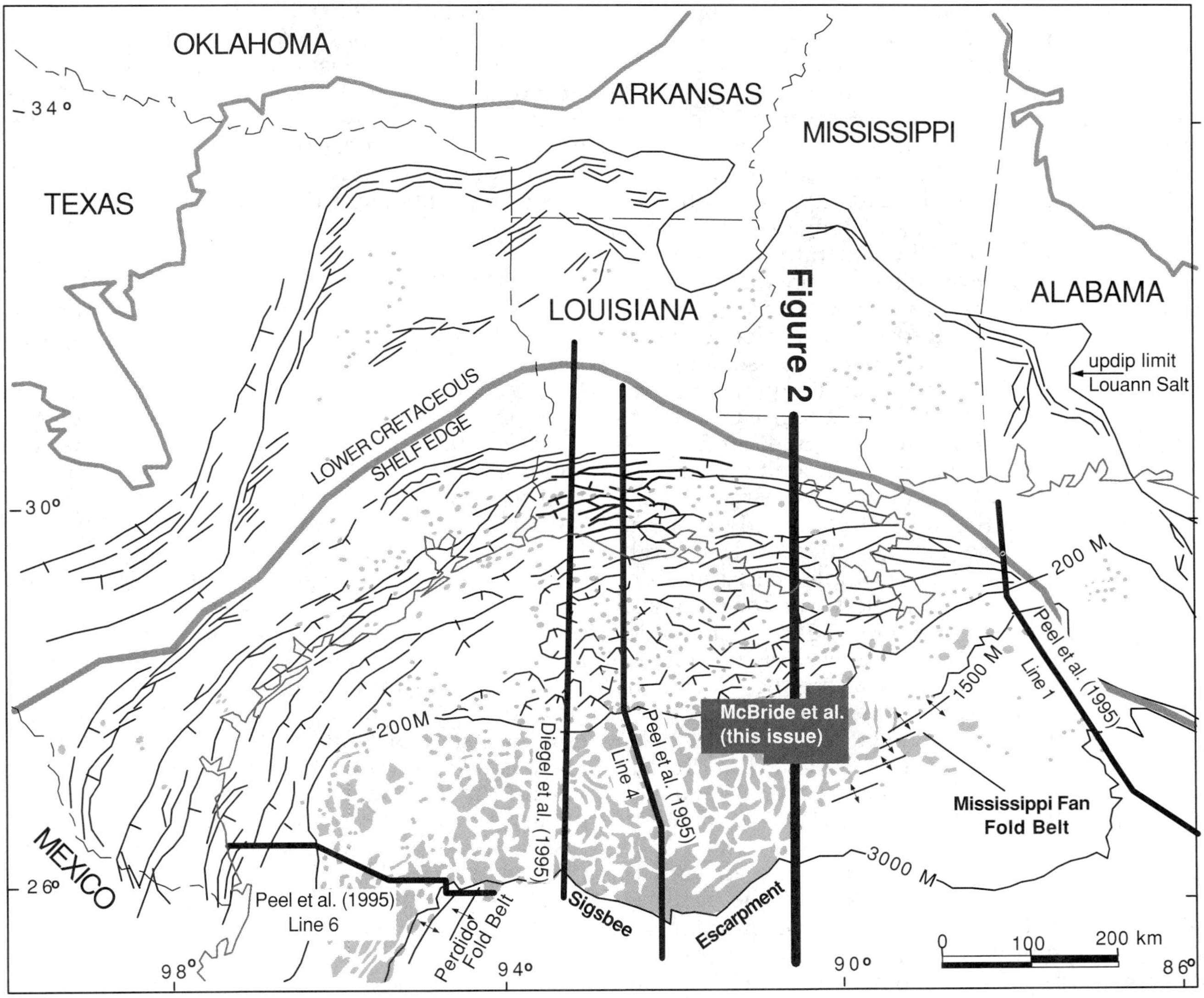


**Figure 1—Regional map of the northern Gulf of Mexico Basin showing major structural and stratigraphic features (modified from Worrall and Snelson, 1989). Locations of the megaregional reconstruction (Figure 2) of this study, the restorations of Diegel et al. (1995), Peel et al. (1995), and the subregional study area of McBride et al. (1998a, this issue) are shown. Shaded areas are shallow salt bodies; barbed lines are faults.**

seismic profiles are more than 15 m (50 ft) long at a scale of 2.5 in./s and are impractical to include here.

Biostratigraphic markers from 33 wells were tied to the seismic using proprietary time-depth curves. Major sequence boundaries and biostratigraphic markers were correlated across the profile by integrating the well tops with previously published seismic interpretations in the upper slope (Acosta, 1994) and abyssal plain (Feng and Buffler, 1991; Feng, 1995). The Mesozoic and early Cenozoic horizons are well imaged onshore (extreme north, Figure 2a) and in the abyssal plain (extreme south, Figure 2A), and were interpreted as far as possible to the south and north, respectively, as a function of seismic quality. In addition, any good deep reflections of unknown age were highlighted in the central portion of the basin. The seismic interpretations were then digitized into the geologic cross section restoration software GEOSEC® (Kligfield et al., 1986), and depth converted using a laterally varying time-depth function for strata and an interval velocity of 4.6 km/s (15,000 ft/s) for salt.

Once the seismic interpretation was converted to depth, the interpretation was merged with top basement and Moho depth profiles extracted from the published maps of Sawyer et al. (1991). In addition, the location of the oceanic/transitional continental crust boundary (Buffler, 1989, 1991) and beta (crustal thinning ratio) values determined by Sawyer et al. (1991) were transferred to the cross section. The basement geometry shown in Figure 2A is extremely smooth and does not show any fault blocks associated with initial rifting of the basin. Basement faults undoubtedly exist along this profile, but they cannot be resolved with present seismic or gravity data sets because of the great depth to basement and the interference of shallow salt for seismic and gravity data interpretation.

The deep Mesozoic and Paleogene strata were then projected across the central portion of the cross section (where seismic data were of poor quality or did not penetrate deep enough). Extrapolation of the Mesozoic and Paleogene strata was accomplished by using the basement profile and the depth-converted deep reflections (representing dip domains) as a template for projection. In addition, a thickness gradient for each interpreted horizon was incorporated into the lateral projection, based on thickness variations calculated between the abyssal plain and southernmost Louisiana. This extrapolation of deep strata is speculative and certainly simplistic as a result of the limited data constraint; however, the extrapolation is thought to be representative of the first-order structures across the area. There is no support in the data for thickness variations within the Cretaceous or early Cenozoic strata associated with early salt deformation along this transect; however, Wu (1993) documented local thickness variations in the Cretaceous and early Cenozoic strata associated with early salt deformation in the northeastern Gulf of Mexico Basin.

The unknown rift geometries and amount of continental crustal extension render the probable thickness variations in the deep postrift sediments interpreted unlikely. These unknown geometries and thickness variations may control the locations and thicknesses of early autochthonous Louann salt basins (isopach thicks). Variations in the rift geometries may therefore control the locations of both counterregional salt feeders and salt stock canopies.

## Present-Day Geometry

The present-day depth section (Figure 2A) is a north-south dip line across the basin (Figure 1), starting near the Early Cretaceous shelf margin (Figure 2A, location 21) of east-central Louisiana at the Louisiana–Mississippi state border. The section extends southward offshore across the South Pelto, South Timbalier, Green Canyon, and Walker Ridge lease areas to the abyssal plain of the Gulf of Mexico. The section consists of a complex mosaic of various allochthonous and autochthonous salt bodies, counterregional normal faults, salt welds, detachment faults, listric normal fault imbricates, and local thrust faults and contractional folds (Figure 2A). The section is described from south to north (left to right) following the nomenclature of tectonic-stratigraphic provinces described by Diegel et al. (1995). The boundaries and names of some of the tectonic-stratigraphic provinces of Diegel et al. (1995) have been modified slightly to better reflect the geometries and evolution of Figure 2A.

### *Abyssal Plain*

The abyssal plain region of the section is represented by relatively uniform, gently dipping Late Jurassic–Holocene strata that show slight thickening landward (north). The abyssal region is interpreted to be underlain primarily by oceanic crust (Buffler, 1989, 1991; Sawyer et al., 1991; Marton and Buffler, 1994). Horizon tops in this area were tied to published seismic interpretations of the deep basin strata by Feng and Buffler (1991) and Feng (1995). The only recognizable fault within the crust imaged by the seismic profile is immediately basinward (south) of the oceanic/transitional continental crust boundary as shown (Figure 2A, location 1). The basinward depositional limit of autochthonous Jurassic Louann Salt was interpreted to be coincident with the oceanic/transitional continental crust boundary (Buffler, 1989, 1991). Marton and Buffler (1994) suggested that the oceanic/transitional continental crust boundary occurs much farther north (Figure 2A, location 9); resolving the placement of the oceanic/transitional continental crust boundary along the megaregional profile is beyond the scope of this paper.

The autochthonous Louann Salt forms a large pillow structure (Figure 2A, location 2) near the salt's basinward depositional limit. This large salt pillow is along strike with the Mississippi Fan fold belt (Figure 1) (Weimer and Buffler, 1992) and has experienced episodic movement since the Late Jurassic, as documented by well-imaged onlapping and thinning of all the overlying strata on seismic data (Fiduk et al., 1995). This cross section represents the westernmost position of a recognizable, basinward depositional limit of the autochthonous Louann Salt on seismic reflection data along offshore Louisiana. To the west, the basinward salt pinch-out is hidden beneath the allochthonous salt of the Sigsbee Escarpment. The basinward limit of autochthonous salt is not imaged again on seismic reflection data until east of the Perdido fold belt in Alaminos Canyon, approximately 300 km (190 mi) to the west (Fiduk et al., 1995).

### *Tabular Salt–Minibasin Province*

Farther north, the section becomes dominated by large, shallow, thick, allochthonous salt sheets and suprasalt minibasins (Figure 2A). Diegel et al. (1995) referred to this portion of the basin, from the Sigsbee Escarpment to the present shelf/slope break, as the tabular salt-minibasin province. These allochthonous salt sheets and sediments in the suprasalt minibasins reach thicknesses up to approximately 3 km (10,000 ft). Suprasalt sediments above these salt bodies were correlated from the south (Feng and Buffler, 1991) and north (Acosta, 1994) and appear to be Pliocene-Pleistocene in age (exact correlations were not possible). In addition, the age of subsalt strata with truncations against the base of these allochthonous salt bodies ranges from late Miocene to Pleistocene. The southernmost salt body (Figure 2A, location 3) displays a more steeply dipping base salt surface than those of the next three allochthonous salt sheets to the north (Figure 2A, locations 4, 6, 8). A vertical salt stock feeder is interpreted to be out of the plane of section for the southernmost allochthonous salt body (Figure 2A, location 3). The other salt bodies were interpreted as sheets with counterregional normal faults along their landward margins.

This particular transect through the tabular salt-minibasin province is important because the allochthonous salt bodies are separate and discrete. Intervening minibasins have continuous strata down to the autochthonous Louann Salt without any recognizable subhorizontal detachments or salt welds (Figure 2A, locations 5, 7. In contrast, Diegel et al. (1995) interpreted a regionally extensive salt weld through this province at the level of Paleogene strata along their regional transect, located approximately 200 km (124 mi) west of this study (Figure 1). They argued that a salt weld within the Paleogene age strata is a better model in their study area based on assessing various structural interpretations and restorations. The interpretation in this study (Figure 2A) is more similar to interpretations of Wu et al. (1990a, b), Wu (1993), and Fletcher et al. (1995) for the northeastern Gulf of Mexico, which show that shallow allochthonous salt sheets are sourced directly from the autochthonous Louann Salt. A major transition zone in the structural styles and timing of allocthonous salt formation must therefore exist between this transect (Figure 2A) and that of Diegel et al., (1995) and Peel et al. (1995) (see Figure 1).

The strata separating these allochthonous salt sheets dip southward and appear to define rollover and drag associated with large counterregional normal faults (Figure 2A, locations 5, 7). Jurassic-late Miocene strata beneath the sheets are subhorizontal, with some large-wavelength and low-amplitude folding. Suprasalt strata record evacuation of the allochthonous salt. The positions of these suprasalt minibasins are probably the result of differential loading caused by focused sediment pathways within the slope. Local contractional structures are present near the basinward margins of several of these salt sheets, as indicated by thrust faults, strata elevated above the regional structural datum, angular unconformities at the sea floor, and inverted growth structures. Extensional faults above the northernmost salt sheet of this province (Figure 2A, location 8) are interpreted to represent late stage inflation of the sheet because of the presence of inversion structures and elevation of the overlying sea floor above its regional level.

### *Pliocene–Pleistocene Detachment Province*

The upper slope region of northern Green Canyon and Ewing Bank represents the northernmost portion of the tabular salt-minibasin province and its transition into the Pliocene-Pleistocene detachment province of Diegel et al. (1995). This region is also the area covered in more detail by Rowan (1995) and McBride et al. (1998a, this issue). Upper Miocene through Pliocene-Pleistocene sediments locally reach thicknesses greater than 9 km (30,000 ft) in this area (Figure 2A, location 10), which is the major Pliocene-Pleistocene depocenter in the northern Gulf of Mexico (Winker, 1982).

A deep minibasin is centered along this profile and displays strong angular discordance across a concave-upward, bowl-shaped salt weld (Figure 2A, location 10). Sediments immediately above this weld have dips to the south approaching 30°, whereas underlying sediments are subhorizontal, a geometry that is common in northern Green Canyon (Rowan, 1995). Contractional folds and thrust faults are present directly north of this minibasin (Figure 2A, location 11). These folds and thrusts appear to detach along an allochthonous salt sheet that contains local salt welds and merges basinward with the deep bowl-shaped salt weld. The contractional structures raise strata above their regional structural datum, exhibit stratigraphic onlapping and thinning, and contain erosional unconformities along fold crests (Figure 2E, location 11).

### *Late Miocene–Pliocene Detachment Province*

The South Timbalier lease area (Figure 2A) is dominated by imbricate listric growth faults detaching along a subhorizontal decollement (a "roho" system; see Diegel et al., 1995; Schuster, 1995) at a depth varying from 6 to 8 km (18,000 to 25,000 ft). The oldest strata above this detachment are upper Miocene. Most of the growth appears to be late Miocene-early Pliocene and corresponds to when the major shallow marine depocenter of the Gulf of Mexico was located here (Winker, 1982). Upper Pliocene-Pleistocene sediments are subhorizontal and display little syndepositional growth across the faults. An allochthonous salt sheet is present basinward of the southernmost listric fault block in this imbricate system, and the salt is interpreted to have been fed from the detachment surface (Figure 2A, location 12). Two salt rollers are interpreted as part of this roho detachment system (Figure 2A, locations 14, 16).

Beneath the late Miocene-Pliocene roho system, several large counterregional normal faults are visible and appear to merge with the overlying detachment and the underlying autochthonous Louann Salt (Figure 2A, location 13). The Mesozoic and lower Cenozoic strata within the hanging walls of these counterregional faults exhibit strong southerly dips and broad anticlines. A large regional normal fault is also interpreted to merge with the autochthonous Louann Salt and offset the overlying detachment (Figure 2A, location 15). This fault is coincident with the southern flank of the South Pelto dome, a large allochthonous salt roller (Figure 2A, location 16). Immediately south of 29° north latitude, another counterregional normal fault is visible that also appears to merge with both the autochthonous Louann Salt and the northern termination of the Miocene-Pliocene roho system (Figure 2A, location 17). Hanging-wall strata above this counterregional fault contain southerly dips in the Mesozoic and lower Cenozoic strata, and northerly dips in the lower to middle Miocene strata. Upper Miocene-Pleistocene strata are relatively subhorizontal above this counterregional normal fault.

### *Oligocene–Miocene Detachment Province*

East-central Louisiana, south of the Mississippi River, is dominated by the Terrebonne Trough and associated structures, and is representative of the Oligocene-Miocene detachment province described by Diegel et al. (1995) (Figure 2A). This province is dominated by a roho detachment fault with an overlying, southerly dipping, imbricated listric growth fault system. Growth strata are Oligocene-late Miocene in age. The southern margin of the Terrebonne Trough is defined by a large counterregional listric normal fault (Figure 2A, location 18) that merges at depth into the same detachment (weld of the Oligocene Terrebonne salt sheet) as the southward dipping listric imbricate faults described. The Terrebonne Trough consists of approximately 7 km (22,000 ft) of Oligocene-Miocene strata and is located within the major depocenter of this age within the northern Gulf of Mexico (Winker, 1982). Beneath the detachment, numerous large counterregional normal faults appear to merge upward into the subhorizontal roho detachment system and downward into the autochthonous Louann Salt (Figure 2A, location 19).

### *Salt Dome/Minibasin Province*

Along this transect, the region of east-central Louisiana north of the Mississippi River is part of the salt dome/minibasin province described by Diegel et al. (1995); however, along this transect, neither salt domes nor salt-related minibasins are present. The province is bounded to the north by the Late Jurassic-Early Cretaceous shelf margin (Figure 2A, location 21) and is dominated by the lowstand systems tracts of the Upper Cretaceous Tuscaloosa Formation and the general aggradation and progradation of subhorizontal Cenozoic strata. This transect shows some minor detachment faulting (Figure 2A, location 20), no salt domes, and no recognizable salt evacuation structures. Several wells in this area have penetrated the Lower Cretaceous strata and have provided the critical tie of the Mesozoic section to seismic data and their correlation basinward. All pre-Louann Salt stratigraphy is interpreted as undifferentiated crust. Although older Mesozoic and Paleozoic strata may exist at depth, they could not be resolved on the seismic data, and they have not been drilled in the vicinity of this transect.

## STRUCTURAL EVOLUTION

### Restoration Method

The megaregional cross section (Figure 2A) was restored following a systematic method for salt structures outlined by Rowan (1993), which is briefly described here. For each stage of restoration, the top layer was removed and the underlying section decompacted according to standard porosity-depth functions (e.g., Baldwin and Butler, 1985); lithologies were estimated based on nearby well penetrations, regional net sand trends, and regional stratigraphic trends (Salvador, 1991a; Piggott and Pulham, 1993; Pulham, 1993; A. Pulham, 1996, personal communication). The thickness and average density of the removed layer were then used to calculate and restore the magnitude of the sedimentary loading subsidence using a smoothed Airy isostatic model. Airy calculations were made in locations that did not experience salt withdrawal or inflation during the time interval being restored. Thermal subsidence was incorporated into the restorations because rifting occurred during the Late Triassic-Middle Jurassic (ending by 160 Ma) (Sawyer et al., 1991). Although the thermal effects were negligible in the Pliocene-Pleistocene, they were progressively greater during pre-Pliocene time. The amount of thermal subsidence incorporated into these restorations varied laterally as a function of beta (crustal thinning ratio) and varied temporally as a function of the amount of time since rifting

(Parsons and Sclater, 1977; McKenzie, 1978; Jarvis and McKenzie, 1980). The time of initial rifting used in calculating thermal subsidence curves was 190 Ma, following Sawyer et al. (1991).

The results of decompaction and subsidence corrections were used to adjust the base salt and to construct a sea-floor template for structural restoration of the suprasalt section. The accuracy of this technique has been demonstrated by Rowan et al. (1996). Ignoring compaction, subsidence, and paleobathymetry would have had a major impact on the restored thickness of salt through time and, thus, the timing of salt weld formation (McBride et al., 1998b, this issue).

Fault displacements were removed by restoring offsets of the top horizon. Where the footwall cutoff of a horizon did not exist because of a footwall salt body, no extension was assumed along the salt–sediment interface; in other words, salt bodies were not assumed to widen by extension during downbuilding. Folding was removed by restoring the suprasalt intervals to the sea-floor template, and any sediments removed by erosion during the interval being restored were estimated and replaced. The resulting space between the restored top salt and adjusted base salt represents the preexisting salt geometry. Thus, the reconstructions show the restored fault and fold geometries, changes in shape of any salt bodies, and variations in basin subsidence and paleobathymetry. Present-day sea level was maintained, and eustatic fluctuations were not incorporated into these restorations. Eustatic fluctuations can be included if desired; however, they would not affect the structural evolution, only the relative water depth through time.

The restoration technique makes no assumptions about the kinematics of salt deformation. Because the top salt and the base salt are restored separately, salt area is usually not maintained (see following paragraphs). Salt is assumed to flow into or out of the section because of the complex three-dimensional kinematics of salt deformation (e.g., Worrall and Snelson, 1989; Diegel and Cook, 1990; McGuinness and Hossack, 1993; Hossack, 1995; Rowan, 1996). In addition, salt may leave the system entirely as a result of dissolution. Likewise, the kinematics of salt overburden is believed to be three dimensional and not always along the plane of section locally or subregionally; therefore, the area of the overburden may need to vary locally during restoration to compensate for out-of-plane movement of both salt and sediments (Rowan, 1996).

Different restoration algorithms were used for different locations and time intervals (see Rowan and Kligfield, 1989; Schultz-Ela, 1992). Block rotation and bed-length (flexural slip) restoration were used on contractional structures to remove shortening caused by folding. Vertical simple shear was applied in areas of extensional or vertical movement. The northern and southern boundaries of the restorations are regional pin lines for the restorations and define a closed system in the dip direction. In addition, local pin lines were also used as necessary to separate regions restored using vertical simple shear from regions restored using flexural slip.

## Restoration Results

The structural restorations, generated using software package GEOSEC®, illustrate that salt within the northern Gulf of Mexico Basin has undergone a dynamic structural evolution through time. Restorations of the megaregional cross section document the timing and locations of allochthonous salt extrusion and amalgamation across the basin. Allochthonous salt extrusion formed salt-stock canopies, salt sheets, and a composite salt-stock/salt-sheet canopy that were subsequently loaded and evacuated by overlying sediments (Figure 2B–J). Although these restorations were constructed sequentially backward through time from the present to the Late Cretaceous, the description of the megaregional restorations is more easily understood when presented in a forward manner through time. Because of space limitations, not all of the sequential restorations have been included within this report. The complete set of restorations consists of restorations to 0.5, 1.1, 3.0, 5.5, 8.8, 10.5, 12.5, 15.5, 21.5, 40.0, 66.4, 97.5, and 163.0 Ma. For this report, the sections that represented the most dramatic or kinematically significant changes were included (Figure 2B–J). Key structural and stratigraphic features are labeled and described here in detail. Note that key features in each restoration are numbered independently.

### *Late Jurassic–Cretaceous*

The late Mesozoic phase of the evolution of the northern Gulf of Mexico Basin was dominated by differential thermal subsidence as a result of a rapid rate of cooling of oceanic crust and attenuated continental crust following rifting. These relationships caused the section to subside more rapidly in the south, causing a regional tilt southward of all Mesozoic strata (Figure 2J). During the Late Cretaceous, the regional dip of the top of salt reached its maximum of 0.5° to the south. The autochthonous Louann Salt appears to have flowed laterally to the south in response to this regional southward tilt. Salt pillows formed along the basinward depositional limit of the autochthonous salt in response to this flow of salt southward (Figure 2J, location 1). There also may have been a component of gravitational gliding toward the south of the suprasalt strata, as suggested by the numerous extensional faults that were active during this time in the region of present-day Louisiana (e.g., Figure 2J, location 5). Some of this extension may have been partially absorbed by the autochthonous salt and by lateral contraction of salt stocks or salt pillows (Figure 2J, locations 1, 2, 3, 4).

The thickness of salt in the section may be overestimated because of the restoration method. If pre-Jurassic strata are present anywhere, which is probable, they were not decompacted throughout the restorations because the strata were interpreted as undifferentiated crust. Because undifferentiated crust was assigned a solidity function that does not decompact, the restored base of autochthonous salt is therefore at a maximum depth. Pre-Jurassic strata, if present and properly decompacted, would raise the restored depth of the base of autochthonous salt.

The placement, orientation, and width of the salt stocks (Figure 2J, locations 2, 3, 4) are primarily a function of the interpretation, as well as the restoration method, and are considered to be accurate. The salt stocks grew by downbuilding (Barton, 1933) and lean southward as a result of the salt's propensity to flow basinward because of the gravitational potential head. In addition, this basinward flow is accentuated by lower sedimentation rates on the basinward side than on the landward side, which sets up and maintains a differential load on the autochthonous Louann Salt. The placement of these asymmetric, southward-leaning salt stocks may be linked to some sort of basement faulting (probably down to the north). Potential fault scarps along the base of salt may have initiated the growth of these salt bodies because the faults would have perturbed the basinward flow of autochthonous salt (e.g., Koyi et al., 1993; Ge et al., 1997). The salt stocks may alternatively be the result of reactive diapirism associated with regional extension (e.g., Vendeville and Jackson, 1992a). In addition, it is unclear why some salt stocks lean southward when other coeval salt stocks appear to have been vertical. Variations in salt stock styles may be the result of the basement geometry partitioning the autochthonous Louann Salt into salt subbasins that were not in communication laterally (Rowan et al., 1995a), as well as a result of variations in sedimentation. Resolving this issue is beyond the scope of this paper and requires several restorations of parallel sections, or three-dimensional restorations to constrain possible lateral displacement variations.

### *Late Cretaceous–Early Miocene*

In the region of present-day Louisiana, a large allochthonous salt sheet was interpreted to have flowed into the section from out of plane sometime during the middle to late Eocene (Figure 2I, location 6). Because its position corresponds with the present-day Terrebonne Trough, this salt body will be referred to here as the Terrebonne salt sheet. The Terrebonne salt sheet may have originated from the autochthonous Louann Salt horizon out-of-plane of the cross section from the northwest. Flow from the northwest is proposed to have been in response to the major Paleocene and Eocene depocenters in southern Texas (Winker, 1982); however, Diegel et al. (1995) and Schuster (1995) also have restored sections in this region across the Terrebonne Trough and show the allochthonous Terrebonne salt sheet to be sourced by feeders along counterregional faults within the plane of their sections. Such faults are comparable to the subsalt faulting shown beneath the Terrebonne salt sheet within Figure 2I. The restoration technique used in this study indicated these particular faults are not evacuated feeders. If these faults were originally feeders that evacuated after Terrebonne salt sheet formation, the overlying salt weld should have obvious inflections associated with each fault scarp. Because the salt weld is unaffected by these faults, they are interpreted to be extensional in origin and to predate the formation of the Terrebonne salt sheet; however, they may merge laterally with counterregional feeders. The extension associated with these subsalt normal faults (Figure 2I, location 6) was interpreted to be linked to contraction of the adjacent basinward salt stock (Figure 2I, location 5).

The Terrebonne salt sheet appears to have flowed slowly basinward during the late Eocene–Oligocene, as indicated by the subsalt truncations of strata along its basinward margin. By the end of the Eocene, the Terrebonne salt sheet started to become buried by prograding sediments onto its landward margin. Thickness variations of the overlying strata indicate that the salt sheet was providing accommodation for the sediments by evacuating as a result of differential sedimentary loading and basinward salt flow.

Throughout the Paleogene, the asymmetric, southward-leaning salt stocks continued to grow by downbuilding (Figure 2I, locations 2, 3, 5). In addition, a bulb-shaped salt stock flowed into the plane of section (Figure 2I, location 4). The base-salt surface of this stock cuts upsection basinward and landward. The feeder for this stock is interpreted to be out of the plane of the section, below the deepest part of the present-day equivalent salt weld. The timing of initial lateral allochthonous salt flow forming this bulb-shaped salt stock was approximately late Paleocene or early Eocene. The salt feeder is interpreted to be

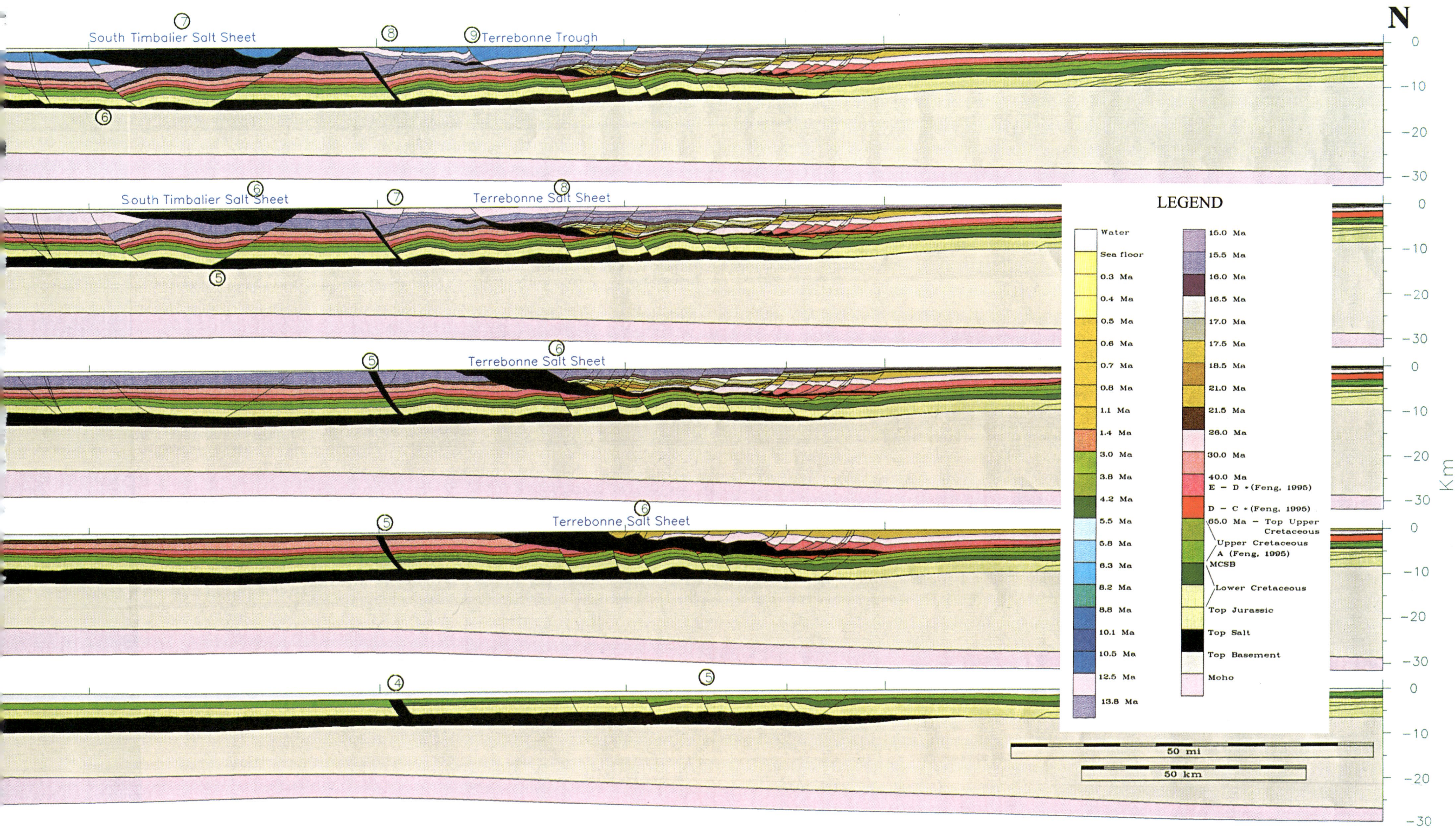
N
South Timbalier Salt Sheet
Terrebonne Trough
Terrebonne Salt Sheet
Km
LEGEND
Water
Sea floor
0.3 Ma
0.4 Ma
0.5 Ma
0.6 Ma
0.7 Ma
0.8 Ma
1.1 Ma
1.4 Ma
3.0 Ma
3.8 Ma
4.2 Ma
5.5 Ma
5.8 Ma
6.3 Ma
8.2 Ma
8.8 Ma
10.1 Ma
10.5 Ma
12.5 Ma
13.8 Ma
15.0 Ma
15.5 Ma
16.0 Ma
16.5 Ma
17.0 Ma
17.5 Ma
18.5 Ma
21.0 Ma
21.5 Ma
26.0 Ma
30.0 Ma
40.0 Ma
E – D *(Feng, 1995)
D – C *(Feng, 1995)
65.0 Ma – Top Upper Cretaceous
Upper Cretaceous
A (Feng, 1995)
MCSB
Lower Cretaceous
Top Jurassic
Top Salt
Top Basement
Moho
50 mi
50 km

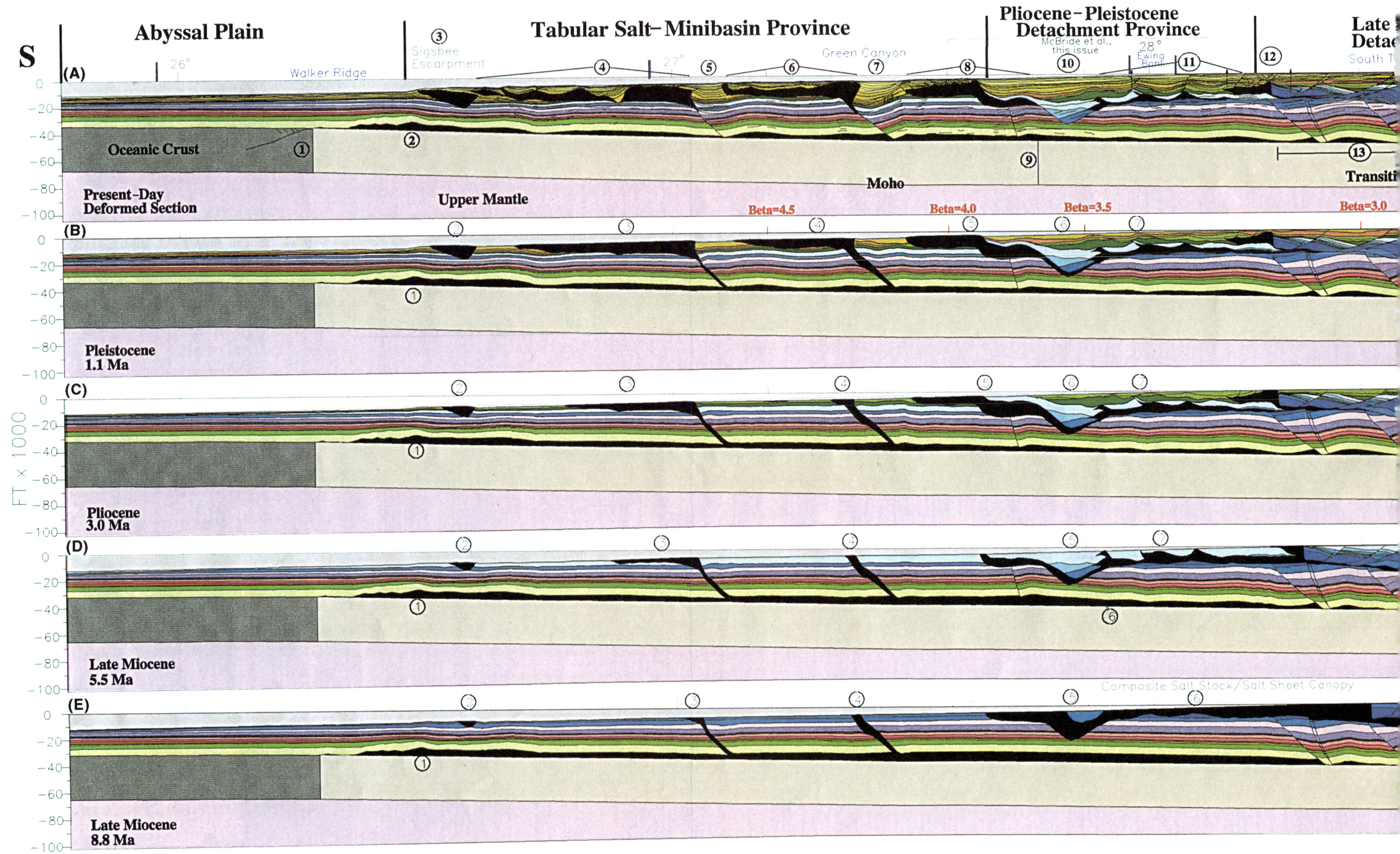


Figure 2—Present-day megaregional north-south depth cross section and sequential structural restorations across the northern Gulf of Mexico Basin from east-central Louisiana to the abyssal plain. Sections are shown without vertical exaggeration and key structures are numbered and described within the text. Note that key features in each restoration are numbered independently. Depth to basement, Moho, and beta (crustal thinning ratios) are from Sawyer et al. (1991). (A) Present-day depth section with tectonic and stratigraphic provinces (modified from Diegel et al., 1995), offshore lease areas, and significant structures labeled, (B) 1.1-Ma (Pleistocene) restoration, (C) 3.0-Ma (Pliocene) restoration, (D) 5.5-Ma (late Miocene) restoration, (E) 8.8-Ma (late Miocene) restoration, (F) 10.5-Ma (late Miocene) restoration, (G) 12.5-Ma (late Miocene) restoration, (H) 15.5-Ma (middle Miocene) restoration, (I) 21.5-(early Miocene) Ma restoration, and (J) 66.4-Ma (Late Cretaceous) restoration.

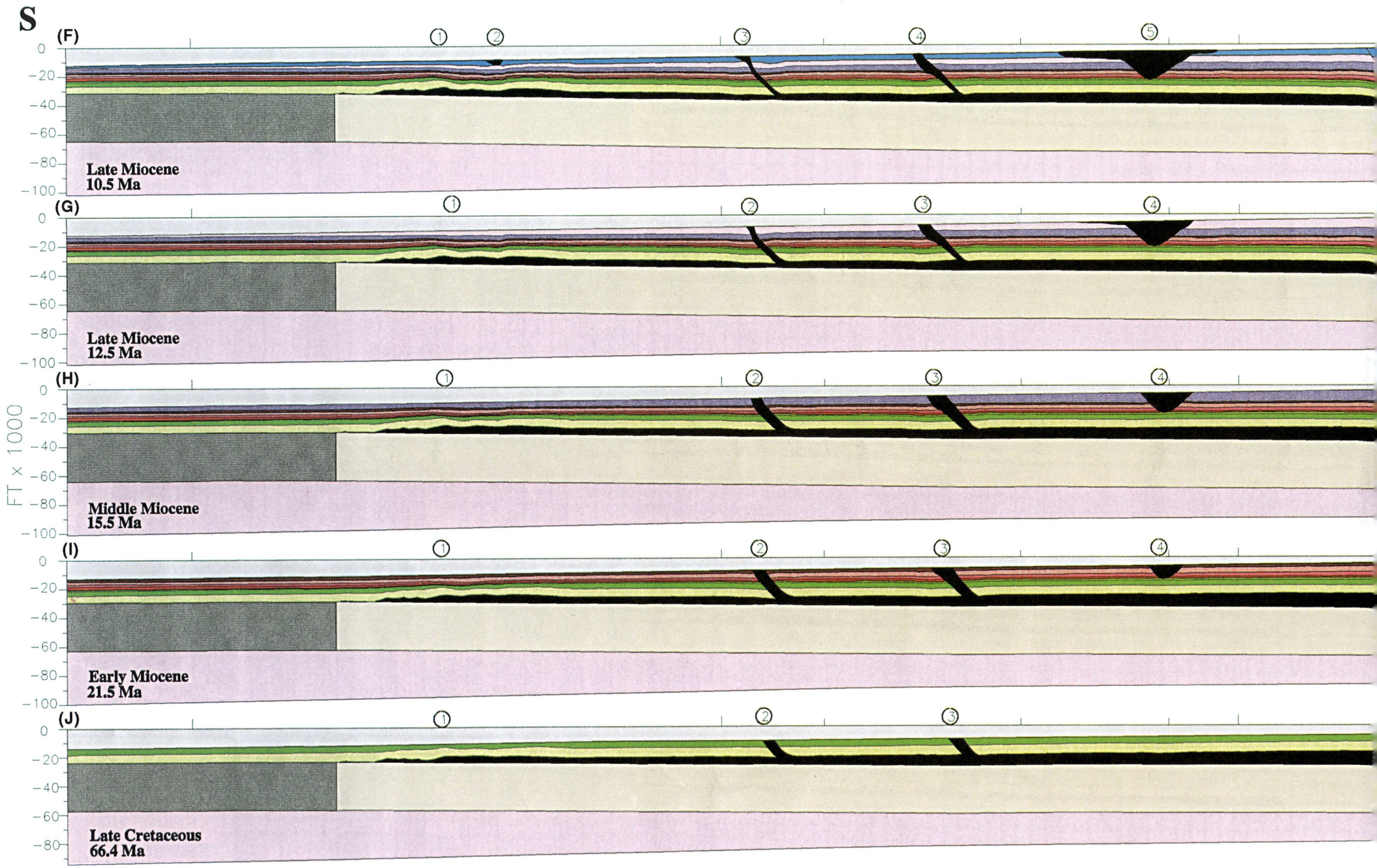


Figure 2—Continued.

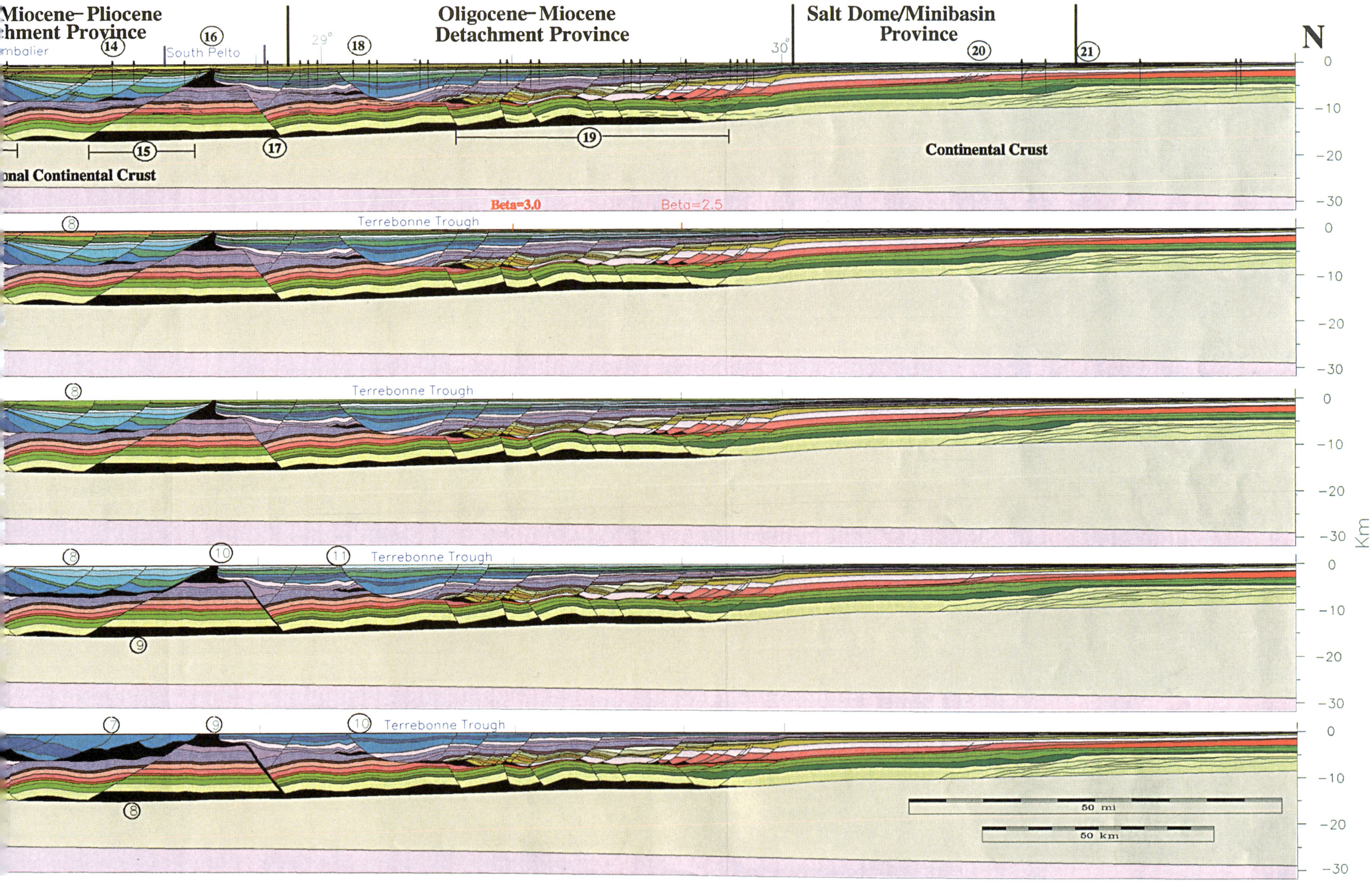

Miocene–Pliocene
hment Province
mbalier
South Pelto
Oligocene–Miocene
Detachment Province
Salt Dome/Minibasin
Province
N
29°
30°
14
15
16
17
18
19
20
21
onal Continental Crust
Continental Crust
Beta=3.0
Beta=2.5
Terrebonne Trough
8
9
10
11
7
0
-10
-20
-30
Km
50 mi
50 km

sourced directly from the autochthonous Louann Salt, similar to the salt glacier model proposed by Fletcher et al. (1995) in Mississippi Canyon and interpretations of allochthonous salt by Wu et al. (1990a, b), Wu (1993), and Rowan (1995). A direct feeder to the autochthonous salt horizon is also consistent with the interpretations of the southward-leaning salt stocks (Figure 2I, location 2, 3, 5).

### *Early–Late Miocene*

The major depocenter for the Gulf of Mexico was located in the region of present-day south-central Louisiana during the early–middle Miocene (Winker, 1982). Loading from this depocenter evacuated the Terrebonne salt sheet (Figure 2G, location 8; H, location 6). The Terrebonne salt sheet continued to flow basinward during most of this time, as indicated by the subsalt truncations of strata. Part of the basinward translation of suprasalt strata of the Terrebonne salt sheet is balanced by movement into salt. As the Terrebonne salt sheet neared the final stages of evacuation, the next major allochthonous salt sheet (South Timbalier salt sheet) was extruded basinward in the vicinity of the present-day South Timbalier lease area (Figure 2G, location 6). It appears that once the Terrebonne salt sheet could no longer provide all of the accommodation for the overlying sediments, the autochthonous Louann Salt began to provide the additional accommodation by flowing basinward into the most proximal south-leaning salt stock and extruding as a salt glacier near the sea floor (Figure 2G, locations 6, 7).

Diegel et al. (1995) used the evacuation history of the Terrebonne salt sheet to compare and contrast two models of salt evacuation (autochthonous vs. allochthonous) to explain the anomalous thickness of Miocene sediments within the Terrebonne Trough. They concluded that evacuation of allochthonous salt avoids having to impose an unrealistic autochthonous salt history or an unusual basement subsidence event. Although Diegel et al. (1995) treated the two models as end members, the restoration in this paper suggests that there is a transition from initially allochthonous salt evacuation to later autochthonous salt evacuation.

The bulb-shaped salt stock (Figure 2H, location 4) continued to grow symmetrically during the early Miocene. A relatively rapid pulse of lateral spreading during the middle Miocene produced a prominent salt overhang along the basinward margin of the bulb-shaped salt stock (Figure 2G, location 4). The south-leaning salt stocks (Figure 2G, locations 2, 3; H, locations 2, 3) continued to grow basinward during this time, and the large salt pillow (Figure 2G, location 1; H, location 1) continued to grow also. Growth of the salt pillow during this time is indicated by the thinning of the overlying strata and the increase in salt thickness through time. Some of this growth may be related to a component of contraction associated with the landward extensional faulting active beneath the South Timbalier salt sheet during this time (Figure 2G, location 5). The remaining contraction needed to compensate this extension was absorbed equally between the two remaining south-leaning salt stocks by making them thinner (Figure 2G, locations 2, 3; H, locations 2, 3). Stresses were interpreted to have been transmitted basinward around these stocks, because they were interpreted to be isolated feeders, not laterally continuous salt walls.

### *Late Miocene*

The late Miocene is represented by three restorations: 5.5, 8.8, and 10.5 Ma (Figure 2D–F, respectively). During this time, the Terrebonne salt sheet experienced its final stages of evacuation, with sedimentary growth primarily concentrating along the basinward bounding counterregional fault (Figure 2D, location 11; E, location 10; F, location 9) and the three southernmost regional listric normal faults. Along strike, several large allochthonous salt domes (e.g., Bay Marchand dome) were preserved adjacent to this counterregional fault (Schuster, 1995). Although the basinward margin of the Terrebonne salt sheet is bounded by a counterregional fault, this surface is primarily a result of salt withdrawal, rather than a fault with significant lateral displacement of sediments (Schuster, 1995).

During this same period, the South Timbalier salt sheet continued to spread basinward until it amalgamated with the bulb-shaped salt stock to the south to form a composite salt-stock/salt-sheet canopy sometime between 10.5 and 8.8 Ma (Figure 2E, location 6). The similarity in timing between the spreading of the salt stock and the basinward advance of the salt sheet suggests a similar origin, probably a regional acceleration of salt flow due to updip differential loading of the autochthonous salt and or slower sedimentation rates. This accelerated allochthonous salt flow is coeval with the continuation of extensional faulting and subsidence of subsalt strata beneath the South Timbalier salt sheet into the autochthonous Louann Salt horizon (Figure 2F, location 6).

Evacuation of the allochthonous South Timbalier salt sheet began during the early late Miocene (Figure 2f, location 7) and continued through the late Miocene (~8.8–5.5 Ma), with large imbricated listric growth faults forming a roho style of allochthonous salt evacuation (Figure 2D, location 8; E, location 7). Minibasin formation above the

South Timbalier salt sheet appears to have occurred after most of the accommodation of the Terrebonne salt sheet and its underlying autochthonous salt had been used. The northernmost growth fault of the South Timbalier salt sheet forms the southern flank of the South Pelto dome (Figure 2D, location 10), and its location appears influenced by the deeper subsalt faulting beneath the South Timbalier salt sheet (Figure 2D, location 9; E, location 8; F, location 6). Modeling by Schultz-Ela and Jackson (1996) indicated that such subsalt faulting may control the positioning of suprasalt minibasins.

The South Timbalier salt sheet absorbed most of the extension displayed within its suprasalt strata, but some contractional structures also formed along the landward margin of the bulb-shaped salt stock about 5.5 Ma (Figure 2D, location 7). The position of the contractional structures is interpreted to have been controlled by the geometry of the base salt, where the subhorizontal South Timbalier salt sheet merged with the south-dipping northern margin of the bulb-shaped salt stock (Figure 2D, location 6). As suprasalt strata glided basinward, thrusting was initiated at this inflection point along the base salt.

The bulb-shaped salt stock began to experience evacuation between 10.5 and 8.8 Ma (Figure 2E, location 5). This interpretation is based on age estimates of the deepest and oldest strata within the present-day minibasin, which have not been drilled and cannot be correlated laterally with confidence. In this interpretation, evacuation of the bulb-shaped salt stock was assumed to not have occurred before, or with a rate greater than, the evacuation of the South Timbalier salt sheet because of their positions relative to the major Gulf of Mexico depocenters during these times. Thus, all of the suprasalt strata above the bulb-shaped salt stock (Figure 2E, location 5) were interpreted to be younger than subsalt strata landward; however, it is possible that the bulb-shaped salt stock experienced partial evacuation earlier and contains some suprasalt strata that are older than subsalt strata landward.

During the late Miocene, the south-leaning salt stocks (Figure 2E, locations 3, 4; F, locations 3, 4) continued to grow by downbuilding, with some lateral flow as a salt glacier around 5.5 Ma (Figure 2D, location 3). The timing of this lateral flow is a function of the accuracy of the subsalt correlations and interpretation and thus speculative. The timing of extrusion and lateral flow from the southernmost south-leaning salt stock (Figure 2D, location 3) could easily be later than or coeval with the next sheet to the north or vice versa. In addition, a new bulb-shaped salt stock began to flow laterally into the section shortly after 12.5 Ma to form the southernmost allochthonous salt body along this transect (Figure 2F, location 2). This bulb-shaped salt stock is analogous to the salt stock in the central portion of the section described above [Figure 2I, location 4; (Figure 2G, location 4; H, location 4; I, location 4)]. Although it appeared in the section only during the late Miocene, it is interpreted to have been sourced from a nearby vertical feeder, out of the plane of this section, directly from the autochthonous Louann Salt. This vertical feeder is interpreted to have been active since the Jurassic, similar to the adjacent southward-leaning salt stocks. The underlying large autochthonous salt pillow at the basinward edge of the salt margin also appears to continue to grow slowly during this period, as shown by its overlying stratigraphic thickness variations (Figure 2D, location 1; E, location 1; F, location 1). This structure is part of the Mississippi Fan fold belt, which was actively growing during this time (Weimer and Buffler, 1992; Rowan, 1997).

### *Pliocene–Pleistocene*

The Pliocene–Pleistocene is represented by two restorations at 3.0 and 1.1 Ma (Figure 2B, C, respectively). During this time period, the major marginal marine depocenter for the Gulf of Mexico prograded basinward and was located near the present-day shelf/slope break area along the line of this section (Winker, 1982). During the Pliocene–Pleistocene, allochthonous salt sourced from the partially evacuated central bulb-shaped salt stock (Figure 2C, location 6) continued to flow basinward as a salt sheet in front of the advancing depocenter; this salt sheet eventually became encased by sediments sometime during the Pleistocene (Figure 2A, location 8). Three-dimensional analysis of this salt sheet determined that the counterregional feeder along its landward margin is relatively narrow and focuses allochthonous salt flow from out of the plane of section (McBride, 1997). The contractional features (Figure 2C, location 7) continued to develop during this time as a result of the last stages of extensional growth faulting associated with the final salt evacuation of the South Timbalier salt sheet (Figure 2B, location 8; C, location 8). These structures were still only partially compensating the extension above the salt sheet farther to the north. Allochthonous salt flow compensated for the rest of the displacement.

The tabular salt and minibasin province was active primarily during the Pliocene–Pleistocene. Southward-flowing salt sheets began to extrude from the southern and central bulb-shaped salt stocks (Figure 2B, locations 2, 5) and the south-leaning salt stocks (Figure 2B, locations 3, 4) to form a series of four salt glaciers near the sea floor. These features are interpreted to be similar to allochthonous salt bodies observed in the northeastern Gulf of Mexico Basin (Wu et al., 1990a, b; Wu, 1993; Fletcher et al., 1995). Subsalt truncations of strata suggest continual and rapid salt flow basinward during this period for many of these salt sheets (Figure 2B, locations 3, 4, 5). The timing of extrusion of the allochthonous salt corresponds with the final phases of Pleistocene sediment loading and evacuation of the bulb-shaped salt stock and the South Timbalier salt sheet (Figure 2B, location 6). Evacuation of this composite allochthonous salt body is interpreted to have caused its underlying autochthonous salt to flow basinward and extrude through the preexisting southern salt feeders (Figure 2B, locations 2, 3, 4).

### *Late Pleistocene–Holocene*

The transition of structural geometries from the Pliocene–Pleistocene to the present day is highlighted by the final evacuation of the central bulb-shaped salt stock (Figure 2A, location 10) to form a deep sedimentary minibasin, and by the progradation of Pleistocene and Holocene sediments basinward above the tabular salt and minibasin province (Figure 2A, B). The late Pleistocene–Holocene sediments in the Green Canyon region form a basinward-thickening wedge adjacent to the northernmost salt sheet of the tabular salt and minibasin province (Figure 2A, location 10). This particular salt sheet is interpreted to be undergoing active inflation as a result of its overlying sediments preventing any further basinward advance (Figure 2A, location 8). Salt evacuating from the deep landward portion of this sheet is flowing southward and inflating the shallow basinward part. This inflation process has created contractional inversion of suprasalt growth strata along the sheet's southern margin (Figure 2A, location 8). In addition, although the suprasalt strata contain extensive normal faults, the salt sheet forms a topographic high along the sea floor. Salt inflation is elevating the suprasalt strata in a counterclockwise manner, and salt will eventually extrude at the sea floor where the overburden is thinnest.

During this interval, the south-leaning feeder stocks that sourced the allochthonous salt of the tabular salt and minibasin province became welded (Figure 2A, locations 5, 7). The welds appear as counterregional normal faults, but are interpreted not to represent extension. These counterregional faults are the result of vertical subsidence and salt evacuation along the south-leaning salt stocks. The southernmost bulb-shaped salt stock continued to grow and formed the present-day Sigsbee Escarpment along this transect (Figure 2A, location 3). The next allochthonous salt sheet to the north is in the process of flowing over the southernmost bulb-shaped salt stock (Figure 2A, locations 3, 4). The relationship of this salt sheet overriding a basinward bulb-shaped salt stock is considered to be a modern analog to the composite salt-stock/salt-sheet canopy that formed in the central portion of the section during the late Miocene (Figure 2D, locations 5, 7). In both of these examples, the salt sheet flowing from the north exhibits contractional structures adjacent to or above the overridden bulb-shaped salt stock (Figure 2D, location 6; A, locations 3, 4), with key seismic horizons elevated above their regional structural datums and angular unconformities of sediments at the sea floor.

## DISCUSSION

### Error Analysis

The restorations presented here were constructed using the systematic approach of Rowan (1993, 1996), which tries to account for all of the various processes that influenced the changing geometry of the evolving basin. McBride et al. (1998b, this issue) illustrated the sensitivity of this approach relative to the predicted salt thicknesses within restorations; however, numerous assumptions and limitations are inherent in any cross section restoration, and these must be understood to fully use the power of structural restorations and evaluate the results.

Error may result from many different sources and can become compounded during the restoration process. The initial interpretation of the geometries to be restored has the largest effect on the outcome of a restoration. Individual geologists may interpret the same data differently because of varying concepts of how structures are related. In the present-day depth section (Figure 2A), the older and deeper strata within the central portion of the basin are the least constrained. The basic premise used to interpret this portion of the basin was to keep the interpretation as simple as possible while honoring all available data. Although this technique minimized the potential of inventing structures arbitrarily, it also smoothed through smaller scale structures representative of the structural evolution of the basin; nevertheless, the first-order structures and their associated evolutions are thought to be sufficiently represented by the interpretation and subsequent restorations.

The basement geometry is certainly more irregular and faulted than shown. Fault blocks and half grabens associated with rifting surely exist and influenced sediment distribution and salt deposition (e.g., Simmons, 1992). Such geometries would affect Louann Salt thickness variations through time, and may have influenced the timing and location of various salt structures. Because Late

Jurassic–middle Miocene strata were interpreted using linear thickness gradients across the central part of Figure 2A, restorations back to these periods show minimal autochthonous salt flow. Differential sedimentary loading, growth faulting, and autochthonous salt flow were probably ubiquitous across the basin during this time interval, thus forming geometries similar to Figure 2J, location 5. Regardless, incorporating these smaller scale structures would not significantly change the overall structural evolution of the basin as depicted by the restorations in Figure 2.

Potential error of the initial interpretation can be further compounded by errors associated with depth conversion. For example, the assumption of vertical ray paths in this type of terrane is clearly a simplification; however, such errors are insignificant at the scale of the entire basin. During the restoration process, errors may also occur because of assumptions made in the (1) Airy calculation of loading subsidence, (2) decompaction, (3) calculation of thermal subsidence, (4) reconstruction of eroded material, (5) paleobathymetry prediction, (6) plane-strain constraints, and (7) choice of restoration algorithms; however, experience has shown these errors to be relatively minor on the scale of the processes of bedding and salt deformation being evaluated here.

Another potential source of error during the restoration process includes the decision of whether individual counterregional faults are the result of welded south-leaning salt stocks or regional extension. Although this decision may appear subjective, it is controlled by the fault's relationship to overlying salt welds and the placement of regional pins (locations of no lateral translation). Faults beneath salt welds were interpreted to have been active before allochthonous salt flow if the faults do not offset the salt weld. Counterregional faults beneath salt welds were interpreted as collapsed feeders if they created inflections in the weld without faulting the overlying strata. Because the restorations are bounded by pins along the northern and southern boundaries, extensional faulting was used only where the amount of extension could be compensated laterally by either folding or salt flow. In addition, the relative motion of the extensional blocks had to be consistent with dip of the detachment.

Errors associated with inaccurate thermal subsidence calculations affect only the relative depth and dip of the basement through time. Such errors would not affect the general evolution of the salt structures illustrated in Figure 2. If the beta values were uniformly overestimated or underestimated, the elevation of the restorations would need to be uniformly lowered or raised, respectively. If the beta values used to calculate thermal subsidence were progressively overestimated basinward, the relative depth and dip of basement would have been greater than the restorations shown in Figure 2, because smaller beta values generate less thermal subsidence. Conversely, if the age of rifting was younger than 190 Ma, the relative depth and dip of basement would be less than the restorations shown in Figure 2, because a younger rifting event would generate greater thermal subsidence for all the restorations.

## Progressive Salt Extrusion or Evacuation

The megaregional depth section and structural restorations presented here combined with an understanding of the location of major depocenters (Winker, 1982) suggest the progressive loading and evacuation of both autochthonous and allochthonous salt relative to the basinward shifts of major depocenters across the northern Gulf of Mexico Basin. The restorations imply that this area of the Gulf of Mexico Basin has undergone three periods of major allochthonous salt extrusion: (1) middle–late Eocene, (2) middle–late Miocene, and (3) late Miocene–Pliocene. In addition, there have been three associated periods of significant salt evacuation: (1) Oligocene–late Miocene, (2) late Miocene, and (3) late Pliocene–Holocene. The positions of these extrusion and evacuation events are localized and are not ubiquitous for the entire basin or the entire transect. The timing and location of these events appear related to the proximal positioning of synchronous shelf-margin/marginal marine sedimentary depocenters (i.e., ancestral Mississippi Delta).

The evolution of the Terrebonne salt sheet (Figure 2I, location 6) illustrates the first of these allochthonous salt extrusion and evacuation events. During formation of the Paleocene and Eocene depocenters in southwestern Texas, autochthonous salt is interpreted to have flowed basinward from beneath these depocenters in response to the differential sedimentary load. The restorations suggest that in southern Louisiana this flow of salt was in the form of an allochthonous salt sheet (Terrebonne salt sheet). The formation of analogous salt sheets has been interpreted by Diegel et al. (1995) and Peel et al. (1995) southwest of this area during this same period. As the sedimentary depocenter of the Gulf of Mexico shifted to southern Louisiana during the early Miocene, the Terrebonne salt sheet provided the majority of accommodation for sediments by salt evacuation.

The evolution of the Terrebonne salt sheet (Figure 2I, location 6) is illustrated in Figure 3A, which is a plot of the cross-sectional area of salt vs. time. The cross-sectional area was measured directly

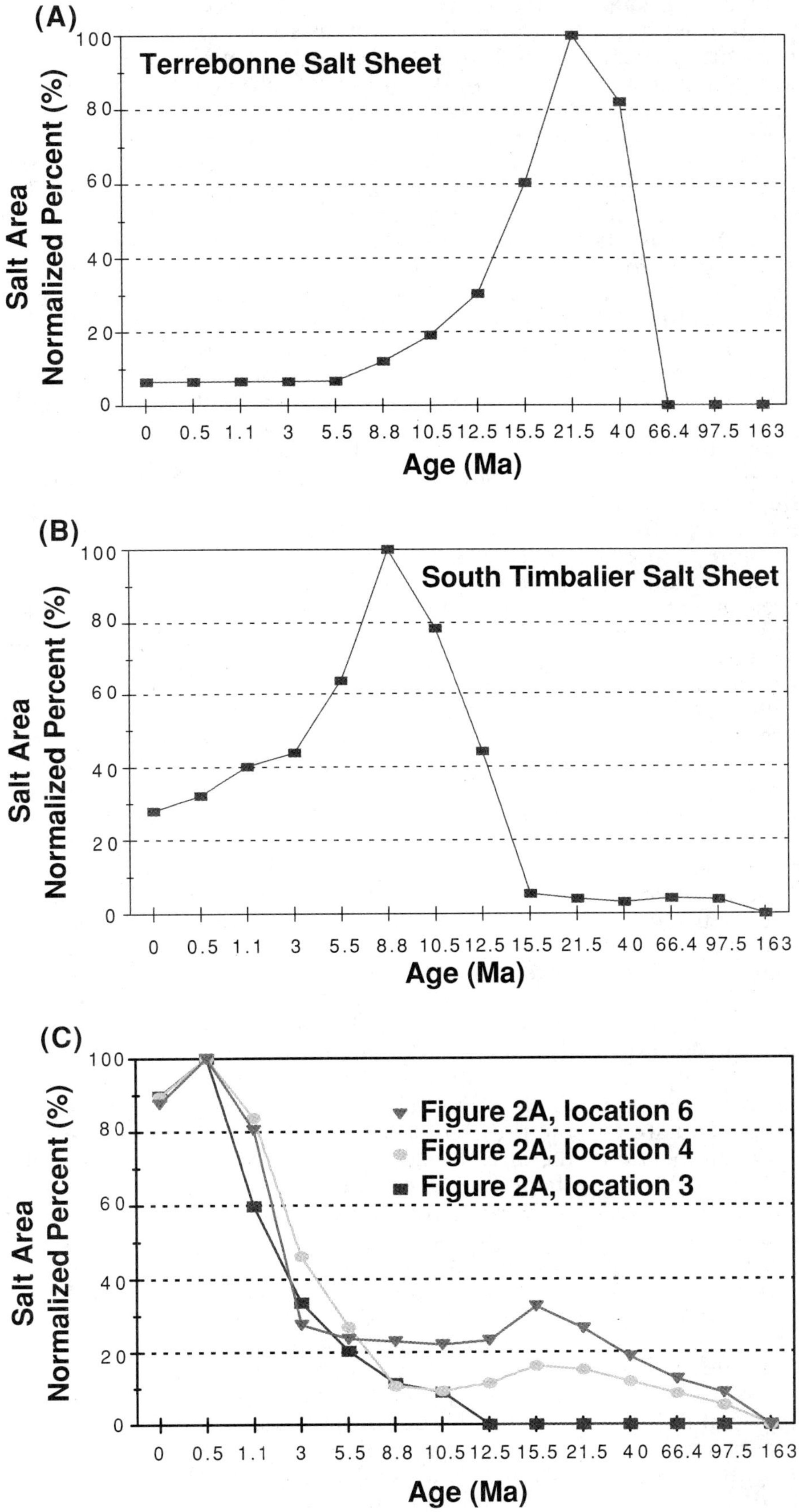


**Figure 3—Cross-sectional area of allochthonous salt sheets normalized and plotted vs. time. (A) Terrebonne salt sheet (Figure 2I, location 6), (B) South Timbalier salt sheet (Figure 2G, location 6), and (C) southern allochthonous salt sheets from tabular salt–minibasin province (Figure 2A). Evolution of sheets is represented by an initial phase of rapid salt extrusion followed by a longer phase of evacuation.**

from the restorations using the software package GEOSEC®, and the values were normalized with the maximum salt area measured representing 100%. Figure 3A illustrates that the cross-sectional area of this salt sheet initially increased during extrusion (64.5–21.5 Ma), reached a maximum (21.5 Ma), and then progressively decreased during evacuation (21.5–0.0 Ma). Only about 7% of the maximum salt sheet area remains along this transect. This value is probably an overestimate, because it assumes that the 21.5 Ma (early Miocene) restoration represents the maximum area of salt for the salt sheet (Figures 2I, 3A). Without additional restorations between 40.0 and 21.5 Ma, it is not possible to confirm that the 21.5 Ma restoration represents the time of maximum salt area.

Although Figure 3A shows salt area, the general trend of the curve is considered to be representative of the changes volumetrically of the salt sheet through time as well, because during the extrusion phase of an allochthonous salt body, only a finite volume of salt is extruded and, therefore, depending on the rate of salt dissolution, a maximum volume of allochthonous salt will be reached at some point in time. During the evacuation phase of an allochthonous salt body, a portion of the salt volume is redistributed in three dimensions into new structures (e.g., salt stocks, salt rollers, etc.), and a portion of the salt volume is lost because of salt dissolution. The amount of salt dissolution during salt sheet evolution is not known and is beyond the scope of this study. Determining the relative percentage of salt dissolution volumetrically requires three-dimensional structural restorations in a closed system (i.e., no salt flow into or out of the study area).

The process of allochthonous salt extrusion and evacuation illustrated by the Terrebonne salt sheet (Figure 3A) was repeated by the South Timbalier salt sheet (Figures 2G, location 6; 3B). The South Timbalier salt sheet extruded as a salt glacier at about the same time the Terrebonne salt sheet had undergone about 60% evacuation along this profile (Figure 4). This transition is interpreted to represent a point in time when the Terrebonne salt sheet could no longer provide all of the accommodation for the overlying sediments. Consequently, the autochthonous Louann Salt began to provide additional accommodation by flowing basinward into the South Timbalier salt sheet. The process of allochthonous salt extrusion and evacuation illustrated by both the Terrebonne and South Timbalier salt sheets (Figure 3A, B) is interpreted to be representative of the complete life cycle of allochthonous salt bodies.

This interpreted relationship between the two salt sheets suggests that, in some cases, an allochthonous salt sheet will undergo evacuation before the underlying autochthonous salt layer. This observation initially seems counterintuitive because allochthonous salt is sourced from underlying levels; however, the source is landward of the allochthonous salt sheet and beneath the major depocenter of the time, and salt flow is more lateral than vertical. This sequence of evacuation between allochthonous and underlying autochthonous salt is interpreted to be controlled by the relative boundary conditions of the two salt horizons. Allochthonous salt glaciers near the sea floor are less confined and can undergo dissolution and flow more rapidly than the underlying autochthonous salt horizon. In addition, the autochthonous salt is overlain with more competent (rigid) strata requiring a greater net differential load to deform. Therefore, it is proposed that allochthonous salt sheets commonly evacuate before the underlying autochthonous salt.

Salt sheets of the tabular salt–minibasin province (Figure 2A) are in the early phase of the allochthonous salt sheet life cycle (Figure 3C). Figure 3C indicates that the area of these sheets reached their maximum by 0.5 Ma, and they are in the initial stages of the allochthonous salt evacuation phase (Figure 3A, B). The southern salt sheets are predicted to experience a comparable evacuation history to both the Terrebonne and South Timbalier salt sheets. Because the majority of the restorations were constructed between 15.5 Ma and the present, the extrusion phase of these younger sheets is better constrained than for the Terrebonne and South Timbalier salt sheets. If the rate of allochthonous salt extrusion recorded for the southernmost salt sheets (Figure 3C) is representative of all allochthonous salt sheets, both the Terrebonne and South Timbalier salt sheets may have experienced similar extrusion phases that also lasted only 5 to 8 m.y.

The factors controlling the positions of the bulb-shaped salt stocks (Figure 2B, location 2; I, location 4) cannot be determined with this data set; however, the bulb-shaped salt stocks appear to be a more distal form of allochthonous salt growth relative to the position of major sedimentary depocenters of the time. Because of this more distal position, bulb-shaped salt stocks appear to be longer-lived structures compared to salt sheets. In addition, a more distal position from a major depocenter would explain the lack of differential sedimentation around the salt stock, which encourages the formation of vertical feeder stocks.

The autochthonous salt area for the entire section decreases through time, with the present-day area along this transect representing approximately 45% of its maximum area. This value is possibly an overestimate, because the calculation assumes that the 97.5 Ma restoration represents the time of maximum salt area. The variations of

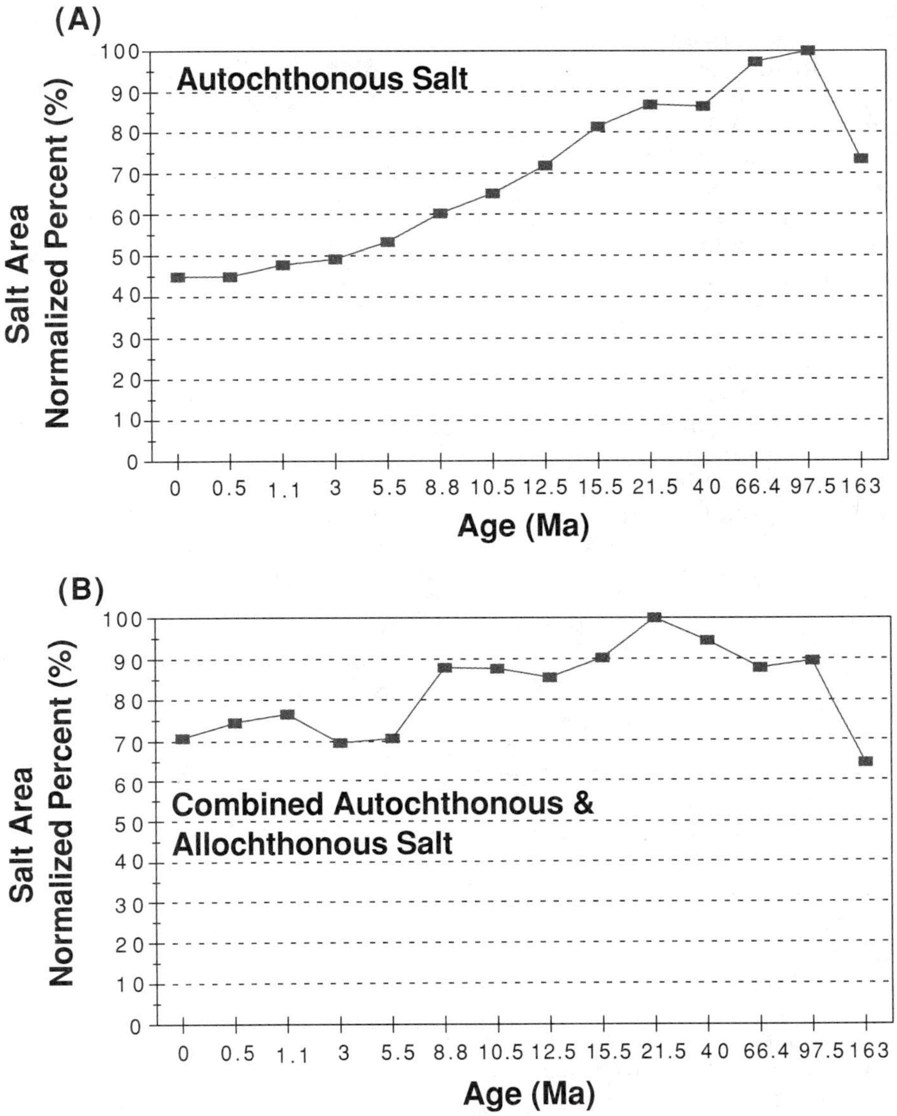


**Figure 4—(A) Cross-sectional area of autochthonous Louann Salt normalized and plotted vs. time. (B) Cross-sectional area of combined autochthonous and allochthonous salt normalized and plotted vs. time.**

the total salt area along the transect (autochthonous and allochthonous) through time are shown in Figure 4B. Even though individual allochthonous salt sheets may undergo dramatic histories of extrusion and evacuation (Figure 3), and the autochthonous salt progressively decreases though time (Figure 4A), Figure 4B suggests that the total salt area is more stable and varies only about 30% from its maximum during the early Miocene (21.5 Ma restoration). The irregularity of the curve in Figure 4B is caused by the three periods of peak allochthonous salt growth at 21.5, 8.8, and 1.1 Ma (compare with Figure 3). Because these analyses are only two dimensional, it is not possible to determine how much of the variations in salt area through time are a result of salt dissolution vs. lateral salt flow into and out of the plane of section.

## Gravitational Spreading and Gliding

Analysis of the complex structural evolution of Figure 2 indicates that gravity and differential sedimentary loading are the primary forces controlling deformation. There was no need to impose crustal-scale extension to explain any of the post-Louann Salt deformation. All of the deformation was a result of gravitational spreading of salt and gravitational gliding of suprasalt strata. Very little translation (5.86 km, or 1.46%) occurred at the autochthonous salt level throughout the history of the basin along this transect. In contrast, significant amounts of translation of suprasalt strata occurred above the various allochthonous salt sheets through time. For example, suprasalt strata above the South Timbalier salt sheet (Figure 2G, location 6) experienced 21.27 km (13 mi) of translation basinward since 8.8 Ma (25%

extension). This lateral translation of suprasalt strata was primarily compensated by salt flow within the South Timbalier salt sheet. Such displacements of suprasalt strata above allochthonous salt sheets were driven by gravitational instabilities caused by the slope gradient, adjusted primarily by deformation above subhorizontal detachments.

The restorations across the northern Gulf of Mexico suggest that reactive diapirism (Vendeville and Jackson, 1992a, b) may play a relatively minor role in the evolution of the basin. This conclusion is in agreement with the results of Rowan et al. (1995b), which were based on analysis of restorations throughout the Ewing Bank, Northern Green Canyon, and Eugene Island lease areas. Significant extension and translation of suprasalt strata at the autochthonous salt level were not interpreted in Figure 2A, and this is open to debate; however, the results of this interpretation indicate that such extension is not necessary to explain the location of suprasalt minibasins above allochthonous salt sheets as modeled by Schultz-Ela and Jackson (1996).

## CONCLUSIONS

(1) The early history of the Gulf of Mexico Basin was dominated by the differential thermal subsidence of the crust as a result of Late Triassic through Early-Middle Jurassic rifting. Subsidence was greatest in the central portion of the basin (south end of profile) and the southward-sloping basement caused autochthonous Louann Salt to flow basinward through time, which formed a large pillow structure near its southern depositional limit.

(2) Allochthonous salt sheets were formed by lateral salt flow basinward from autochthonous salt beneath major sedimentary depocenters of the time. Salt extruded from southward-leaning salt stocks and formed allochthonous salt glaciers near the sea floor. Allochthonous salt sheets experienced a rapid phase of extrusion (5-8 m.y. duration) and a longer phase of evacuation (15-25 m.y. duration). Some allochthonous salt sheets may have experienced evacuation of more than 93% of their maximum salt area through time along two-dimensional transects; volume variations, however, are unknown.

(3) Allochthonous salt sheets may have experienced evacuation before the immediately underlying autochthonous Louann Salt. This sequence of evacuation is controlled by the boundary conditions of the two salt horizons. Salt glaciers are less confined and can flow and undergo dissolution more rapidly; autochthonous salt is bounded by more competent rock. Therefore, I propose that allochthonous salt commonly can evacuate before the underlying autochthonous salt. Allochthonous salt sheets formed from long-distance lateral flow from the autochthonous salt and not by vertical salt flow.

(4) Salt sheet formation and evacuation occurs progressively basinward through time in response to basinward shifts of major Cenozoic sedimentary depocenters.

(5) Bulb-shaped salt stocks commonly grow through time in more distal regions relative to major depocenters and coeval salt sheets. The early growth of bulb-shaped salt stocks relative to adjacent salt sheets creates apparent out-of-sequence allochthonous salt growth basinward. The factors controlling the formation of bulb-shaped salt stocks and canopies vs. allochthonous salt sheets sourced by counterregional feeders are still not clearly understood.

(6) The area of autochthonous salt progressively decreases through time and is at most 45% of its maximum along this transect. The total area of salt through time is more stable, with variations of only 30% from its maximum. This relationship is a function of lateral salt flow into and out of the plane of section, possible salt dissolution, and the estimates of original salt thicknesses and their lateral variations.

(7) The restorations indicate that very little translation or extension (1.46%) is necessary at the autochthonous salt level along this transect to explain the evolution of the basin. The majority of translation or extension occurred above allochthonous salt sheets (25%) and was compensated laterally by salt flow. Displacements above allochthonous salt sheets were driven by gravitational instabilities caused by the slope gradient.

## REFERENCES CITED

Acosta, Z. M., 1994, Sequence stratigraphy of Plio-Pleistocene sediments in north-central Green Canyon and western Ewing Bank, northern Gulf of Mexico: M.S. thesis, University of Colorado, Boulder, Colorado, 187 p.

Amery, G. B., 1969, Structure of Sigsbee scarp, Gulf of Mexico: AAPG Bulletin, v. 53, p. 2480-2482.

Baldwin, B., and C. O. Butler, 1985, Compaction curves: AAPG Bulletin, v. 69, p. 622-625.

Barton, D. C., 1933, Mechanics of formation of salt domes of Texas and Louisiana: AAPG Bulletin, v. 17, p. 1025-1083.

Buffler, R. T., 1989, Distribution of crust, distribution of salt and the early evolution of the Gulf of Mexico Basin: Gulf Coast Section SEPM Tenth Annual Research Conference, p. 25-27.

Buffler, R. T., 1991, Seismic stratigraphy of the deep Gulf of Mexico Basin and adjacent margins, *in* A. Salvador, ed., The Gulf of Mexico Basin: Decade of North American Geology: Geological Society of America, Boulder, Colorado, p. 353-387.

Diegel, F. A., and R. W. Cook, 1990, Palinspastic reconstruction of salt-withdrawal growth-fault systems, northern Gulf of Mexico (abs.): Geological Society of America Abstracts With Programs, v. 22, p. A48.

Diegel, F. A., and D. C. Schuster, 1990, Regional cross sections and palinspastic reconstructions, northern Gulf of Mexico (abs.): Geological Society of America Abstracts With Programs, v. 22, p. A66.

Diegel, F. A., J. F. Karlo, D. C. Schuster, R. C. Shoup, and P. R. Tauvers, 1995, Cenozoic structural evolution and tectono-stratigraphic framework of the northern Gulf Coast continental margin, *in* M. P. A. Jackson, D. G. Roberts, and S. Snelson, eds., Salt tectonics: a global perspective: AAPG Memoir 65, p. 109–151.

Feng, J., 1995, Post mid-Cretaceous seismic stratigraphy and depositional history, deep Gulf of Mexico: Ph.D. dissertation, University of Texas at Austin, Austin, Texas, 253 p.

Feng, J., and R. T. Buffler, 1991, Preliminary age determinations for new deep Gulf of Mexico Basin sequences: Gulf Coast Association of Geological Societies Proceedings, v. 41, p. 283–289 a.

Fiduk, C. J., B. C. McBride, P. Weimer, M. G. Rowan, and B. D. Trudgill, 1995, Defining the basinward limits of salt deposition, northern Gulf of Mexico: Gulf Coast Section SEPM Foundation Sixteenth Annual Research Conference, p. 53–64.

Fletcher, R. C., M. R. Hudec, and I. A. Watson, 1995, Salt glacier and composite sediment/glacier models for the emplacement and early burial of allochthonous salt sheets, *in* M. P. A. Jackson, D. G. Roberts, and S. Snelson, eds., Salt tectonics: a global perspective: AAPG Memoir 65, p. 77–108.

Ge, H., M. P. A. Jackson, and B. C. Vendeville, 1997, Kinematics and dynamics of salt tectonics driven by progradation: AAPG Bulletin, v. 81, p. 398–423.

Hossack, J. R., 1995, Geometrical rules of section balancing of salt structures, *in* M. P. A. Jackson, D. G. Roberts, and S. Snelson, eds., Salt tectonics: a global perspective: AAPG Memoir 65, p. 29–40.

Humphris, C. C., Jr., 1978, Salt movement on continental slope, northern Gulf of Mexico, *in* A. H. Bouma, G. T. Moore, and J. M. Coleman, eds., Framework, facies, and oil-trapping characteristics on the upper continental margin: AAPG Studies in Geology 7, p. 69–85.

Jarvis, G. T., and D. P. McKenzie, 1980, Sedimentary basin formation with finite extension rates: Earth and Planetary Science Letters, v. 48, p. 42–52.

Kligfield, R., P. Geiser, and J. Geiser, 1986, Construction of geologic cross sections using microcomputer systems: Geobyte, v. 1, p. 60–69.

Koyi, H., M. K. Jenyon, and K. Petersen, 1993, The effect of basement faulting on diapirism: Journal of Petroleum Geology, v. 16, p. 285–312.

Lehner, P., 1969, Salt tectonics and Pleistocene stratigraphy on the continental slope of northern Gulf of Mexico: AAPG Bulletin, v. 53, p. 2431–2479.

Marton, G., and R. T. Buffler, 1994, Jurassic reconstruction of the Gulf of Mexico Basin: International Geology Review, v. 36, p. 545–586.

McBride, B. C., 1997, The geometry and evolution of allochthonous salt and its impact on petroleum systems, northern Gulf of Mexico Basin: studies in three- and four-dimensional analysis: Ph.D. dissertation, University of Colorado, Boulder, Colorado, 276 p.

McBride, B. C., M. G. Rowan, P. Weimer, and R. Ratliff, 1995, Three-dimensional restoration of allochthonous salt structures and its impact on understanding reservoir trends and petroleum migration: preliminary results from Green Canyon and Ewing Bank, northern Gulf of Mexico: Gulf Coast Section SEPM Sixteenth Annual Research Conference, p. 171–175.

McBride, B. C., M. G. Rowan, and P. Weimer, 1998a, The evolution of allochthonous salt systems, northern Green Canyon and Ewing Bank, northern Gulf of Mexico: AAPG Bulletin, v. 82, no. 5B, p. 1013–1036.

McBride, B. C., P. Weimer, and M. G. Rowan, 1998b, The effect of allochthonous salt on petroleum systems of the northern Green Canyon and Ewing Bank areas, northern Gulf of Mexico: AAPG Bulletin, v. 82, no. 5B, p. 1083–1112.

McGuinness, D. B., and J. R. Hossack, 1993, The development of allochthonous salt sheets as controlled by the rates of extension, sedimentation, and salt supply: Gulf Coast Section SEPM Fourteenth Annual Research Conference, p. 127–139.

McKenzie, D. P., 1978, Some remarks on the development of sedimentary basins: Earth and Science Planetary Letters, v. 40, p. 25–32.

Moretti, I., S. Wu, and A. Bally, 1990, Computerized balanced cross section LOCACE to reconstruct an allochthonous salt-sheet, offshore Louisiana: Marine and Petroleum Geology, v. 7, p. 371–377.

Parsons, B., and J. G. Sclater, 1977, An analysis of the variation of ocean floor bathymetry and heat flow with age: Journal of Geophysical Research, v. 82, p. 803–827.

Peel, F. J., C. J. Travis, and J. R. Hossack, 1995, Genetic structural provinces and salt tectonics of the Cenozoic offshore U.S. Gulf of Mexico: a preliminary analysis, *in* M. P. A. Jackson, D. G. Roberts, and S. Snelson, eds., Salt tectonics: a global perspective: AAPG Memoir 65, p. 153–175.

Piggott, N., and A. J. Pulham, 1993, Sedimentation rate as the control on hydrocarbon sourcing, generation and migration in the deepwater Gulf of Mexico: Gulf Coast Section SEPM Foundation Fourteenth Annual Research Conference, p. 179–191.

Pulham, A. J., 1993, Variations in slope deposition, Plio-Pleistocene, offshore Louisiana, north-east Gulf of Mexico, *in* P. Weimer and H. Posamentier, eds., Siliciclastic sequence stratigraphy: recent developments and applications: AAPG Memoir 58, p. 199–234.

Ratcliff, D. W., 1993, New technologies improve seismic images of salt bodies: Oil & Gas Journal, September 27, 1993, v. 91, p. 41–49.

Rowan, M. G., 1993, A systematic technique for the sequential restoration of salt structures: Tectonophysics, v. 228, p. 331–348.

Rowan, M. G., 1995, Structural styles and evolution of allochthonous salt, central Louisiana outer shelf and upper slope, *in* M. P. A. Jackson, D. G. Roberts, and S. Snelson, eds., Salt tectonics: a global perspective: AAPG Memoir 65, p. 199–228.

Rowan, M. G., 1996, Benefits and limitations of section restoration in areas of extensional salt tectonics: an example from offshore Louisiana, *in* P. G. Buchanan and D. A. Nieuwland, eds., Modern developments in structural interpretation, validation and modelling: The Geological Society of London Special Publication 99, p. 147–161.

Rowan, M. G., 1997, Three-dimensional geometry and evolution of a segmented detachment fold, Mississippi Fan foldbelt, Gulf of Mexico: Journal of Structural Geology, v. 19, p. 463–480.

Rowan, M. G., and R. Kligfield, 1989, Cross section restoration and balancing as an aid to seismic interpretation in extensional terranes: AAPG Bulletin, v. 73, p. 955–966.

Rowan, M. G., B. C. McBride, and P. Weimer, 1995a, Composite origin of allochthonous salt, offshore Louisiana (abs): AAPG Annual Convention Program, v. 4, p. 83A–84A.

Rowan, M. G., B. C. McBride, and P. Weimer, 1995b, The role of extension in the evolution of allochthonous salt, central Louisiana outer shelf and upper slope: Gulf Coast Section SEPM Sixteenth Annual Research Conference, p. 223–233.

Rowan, M. G., T. Villamil, P. B. Flemings, and P. Weimer, 1996, Use of cross-section restoration to determine paleobathymetry and sea-floor paleotopography in the Gulf of Mexico Basin: Geology, v. 24, p. 299–302.

Sawyer, D. S., R. T. Buffler, and R. H. Pilger, Jr., 1991, The crust under the Gulf of Mexico Basin, *in* A. Salvador, ed., The Gulf of Mexico Basin: Decade of North American Geology, Geological Society of America, Boulder, Colorado, p. 53–72.

Schultz-Ela, D. D., 1992, Restoration of cross-sections to constrain deformation processes of extensional terranes: Marine and Petroleum Geology, v. 9, p. 372–388.

Schultz-Ela, D. D., and M. P. A. Jackson, 1996, Relation of subsalt structures to suprasalt structures during extension: AAPG Bulletin, v. 80, p. 1896–1924.

Schuster, D. C., 1995, Deformation of allochthonous salt and evolution of related salt/structural systems, eastern Louisiana Gulf Coast, *in* M. P. A. Jackson, D. G. Roberts, and S. Snelson, eds.,

Salt tectonics: a global perspective: AAPG Memoir 65, p. 177-198.
Simmons, G. R., 1992, The regional distribution of salt in the northwestern Gulf of Mexico: styles of emplacement and implications for early tectonic history: Ph.D. dissertation, Texas A&M University, College Station, Texas, 180 p.
Sumner, H. S., B. A. Robison, W. K. Dirks, and J. C. Holliday, 1991, Structural style of salt/minibasin systems: lower shelf and upper slope, central offshore Louisiana (abs.): Gulf Coast Association of Geological Societies Transactions, v. 41, p. 582.
Vendeville, B. C., and M. P. A. Jackson, 1992a, The rise of diapirs during thin-skinned extension: Marine and Petroleum Geology, v. 9, p. 331-352.
Vendeville, B. C., and M. P. A. Jackson, 1992b, The fall of diapirs during thin-skinned extension: Marine and Petroleum Geology, v. 9, p. 354-371.
Weimer, P., and R. T. Buffler, 1992, Structural geology and evolution of the Mississippi Fan fold belt, deep Gulf of Mexico: AAPG Bulletin, v. 76, p. 225-251.
West, D. B., 1989, Model for salt deformation on deep margin of Gulf of Mexico Basin: AAPG Bulletin, v. 73, p. 1472-1482.
Winker, C. D., 1982, Cenozoic shelf margins, northwestern Gulf of Mexico Basin: Gulf Coast Association of Geological Societies Transactions, v. 32, p. 427-448.
Worrall, D. M., and S. Snelson, 1989, Evolution of the northern Gulf of Mexico, with emphasis on Cenozoic growth faulting and the role of salt, *in* A. W. Bally and A. R. Palmer, eds., The geology of North America; an overview: Geological Society of America Decade of North American Geology, v. A, p. 97-138.
Wu, S., 1993, Salt and slope tectonics, offshore Louisiana: Ph.D. dissertation, Rice University, Houston, Texas, 251 p.
Wu, S., A. W. Bally, and C. Cramez, 1990a, Allochthonous salt, structure and stratigraphy of the northeastern Gulf of Mexico, part II: structure: Marine and Petroleum Geology, v. 7, p. 334-370.
Wu, S., P. R. Vail, and C. Cramez, 1990b, Allochthonous salt, structure and stratigraphy of the northeastern Gulf of Mexico, part I: stratigraphy: Marine and Petroleum Geology, v. 7, p. 318-333.

## ABOUT THE AUTHOR

**Barry C. McBride**

Barry McBride is a geologist at HS Resources Inc. He received his Ph.D. in geology at the University of Colorado, his M.S. degree in geology from Western Michigan University, and his B.S. degree in geology from Furman University. From 1988 to 1993 he worked for Mobil Exploration and Producing U.S. Inc. and Mobil Global Interpretation Support as an exploration geologist. His research interests focus on petroleum exploration, complex structures, and petroleum systems.

# Salt-Sediment Interaction, Northern Green Canyon and Ewing Bank (Offshore Louisiana), Northern Gulf of Mexico[1]

**Mark G. Rowan and Paul Weimer[2]**


## ABSTRACT

Structural and sequence stratigraphic interpretations of two-dimensional seismic and well data from northern Green Canyon and Ewing Bank were integrated to evaluate how salt deformation influenced the distribution of Pliocene–Pleistocene facies in time and space. Two techniques were employed. First, twelve palinspastic maps of near-surface structure were constructed. These were combined with maps of interpreted depositional environments to show how shallow salt diapirism created bathymetric relief that influenced the configuration of sediment transport systems and depocenters through time. Second, tectonostratigraphic packages comprising multiple sequences were defined based on external geometry. Different stacking patterns of these packages characterize four types of minibasins, each with a distinct history of salt evacuation from underlying salt stocks and sheets. Interpreted seismic facies were analyzed within this minibasin framework to evaluate how deep-salt withdrawal influenced the distribution of depositional systems.

The results show that both structural and sedimentological variables influenced lithofacies development. External factors dictated the volume and type of systemwide clastic input. Regional factors, such as nearby salt structures and the position of deltas, controlled the dispersal of clastics. Local factors, such as the thickness of underlying salt, influenced minibasin-specific evolution. These factors interacted at three scales: (1) a broad transition from sand-rich ponded settings to shale-dominated bypass settings during the Pliocene–Pleistocene, (2) fluctuations over periods of several sequences that created highly variable stratigraphic stacking patterns, and (3) a progression from ponded to bypass facies within individual sea level cycles. Analysis of these various factors can improve the prediction of reservoir distribution within slope minibasins, and thereby reduce the risk in subsalt and deep-water exploration.


## INTRODUCTION

Sedimentation and salt deformation are intricately linked in the northern Gulf of Mexico. Interactions between the two processes are reciprocal and take numerous forms. Sedimentary loading drives salt mobilization and affects the shapes of passive diapirs and allochthonous sheets. In turn, salt deformation influences both the stratigraphic architecture and the spatial and temporal distribution of lithofacies. We address in this paper the control exerted by salt tectonics on slope sedimentation.

The manner in which salt influences sedimentation can be divided into two broad categories. First, salt flow into near-surface diapirs and sheets creates sea-floor topography that determines the location and configuration of depocenters and sediment transport systems (e.g., Satterfield and Behrens, 1990; Sumner et al., 1991; Swanson and Bryant, 1992; Apps et al., 1994; Mahaffie, 1994; Pratson and Ryan, 1994; Risch et al., 1994; Rowan et al., 1994, 1995b; Weimer et al., 1994, 1998, this issue; Yeilding et al., 1995; Lee et al., 1996, Winker, 1996). Second, salt withdrawal from beneath minibasins provides much of the accommodation for

---



[1]Manuscript received August 19, 1996; revised manuscript received March 24, 1997; final acceptance November 19, 1997.

[2]Department of Geological Sciences and Energy and Minerals Applied Research Center (EMARC), Campus Box 399, University of Colorado, Boulder, Colorado 80309-0399.

This work represents the culmination of a 4 year, integrated structural and stratigraphic research program that would not have been possible without the contributions of many companies and individuals. Specifically, we thank Halliburton (Bob Graebner, John Anderson, and Gary White) for the bulk of the seismic data; TGS and Exxon for additional seismic data (through Peter Flemings at Pennsylvania State University); Paleo-Data (Art Waterman) and Micro-Strat for biostratigraphic data; and CogniSeis (Roy Kligfield and Bob Ratliff), GeoQuest, and Zeh for software. Funding was provided through the EMARC Gulf of Mexico industrial consortium (Agip, Amoco, Anadarko, BHP, BP Exploration, Burlington Resources, CXY Energy, Chevron, CNG, Conoco, Enterprise, Exxon, Marathon, Mobil, Occidental, PanCanadian, Pennzoil, Petrobras, Phillips, Shell, Texaco, Total, Union-Pacific, and UNOCAL) and by the Department of Energy (through Peter Flemings at Pennsylvania State University). We also acknowledge the donors of the Petroleum Research Fund, administered by the American Chemical Society, for partial support of this research. We thank Jory Pacht and Hongxing Ge for careful and valuable reviews, Kevin Biddle for his editorial comments, and Brad Prather for a copy of his manuscript. We especially acknowledge our graduate students for their vital contributions: Barry McBride, Peter Varnai, Fadjar Budhijanto, Zurilma Acosta, Rafael Martinez, and Alonso Navarro.

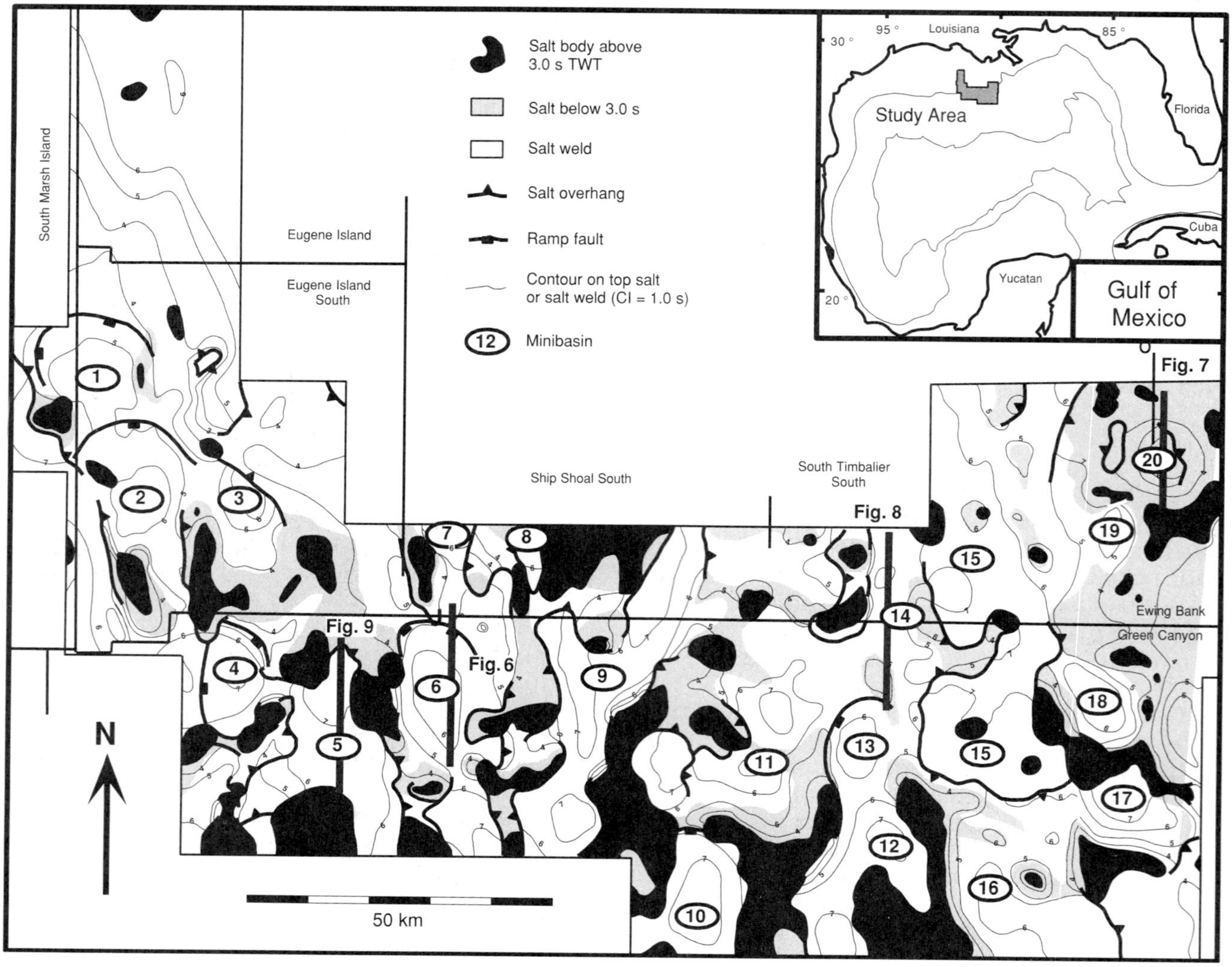


**Figure 1—Map of study area (located in inset) showing structure of multilevel salt system (from Rowan, 1995). Locations of seismic profiles illustrated in other figures and minibasins referred to in text and Table 1 are indicated.**

slope turbidite systems (e.g., Sumner et al., 1991; Pulham, 1993; Weimer et al., 1994, 1998, this issue; Mahaffie et al., 1995; Prather et al., 1998).

In this paper, we examine both ways in which salt deformation controls sedimentation by adopting two different approaches: one for the near-surface effects and one for the deeper subsurface effects of salt evolution. We first evaluate the influence of salt-controlled bathymetry on sediment transport and deposition through a series of twelve palinspastic maps of the study area for the Pliocene–Pleistocene. Each map shows the distribution of depositional environments, interpreted for the maximum lowstand of each sequence, overlain on a reconstructed map of near-surface salt bodies, faults, and folds. This technique addresses only the surface expression of salt diapirs and sheets, and does not reflect the type of salt system or the thickness and evacuation of salt underlying slope minibasins. To evaluate these deeper subsurface controls, four kinds of tectonostratigraphic packages comprising multiple sequences are identified based on their external form. Different stacking patterns of these packages define four types of slope minibasins: two result from varying evolutionary scenarios of salt evacuation from salt stocks, and two are related to salt withdrawal from underlying salt sheets. Interpreted depositional environments are then evaluated within this minibasin classification framework to understand the links between deep-salt deformation and sedimentation.

The primary objectives of our investigation are to (1) determine whether salt evolution controlled sedimentation in any systematic way and (2) develop guidelines to aid interpreters in predicting lithofacies and reservoir distribution in both time and

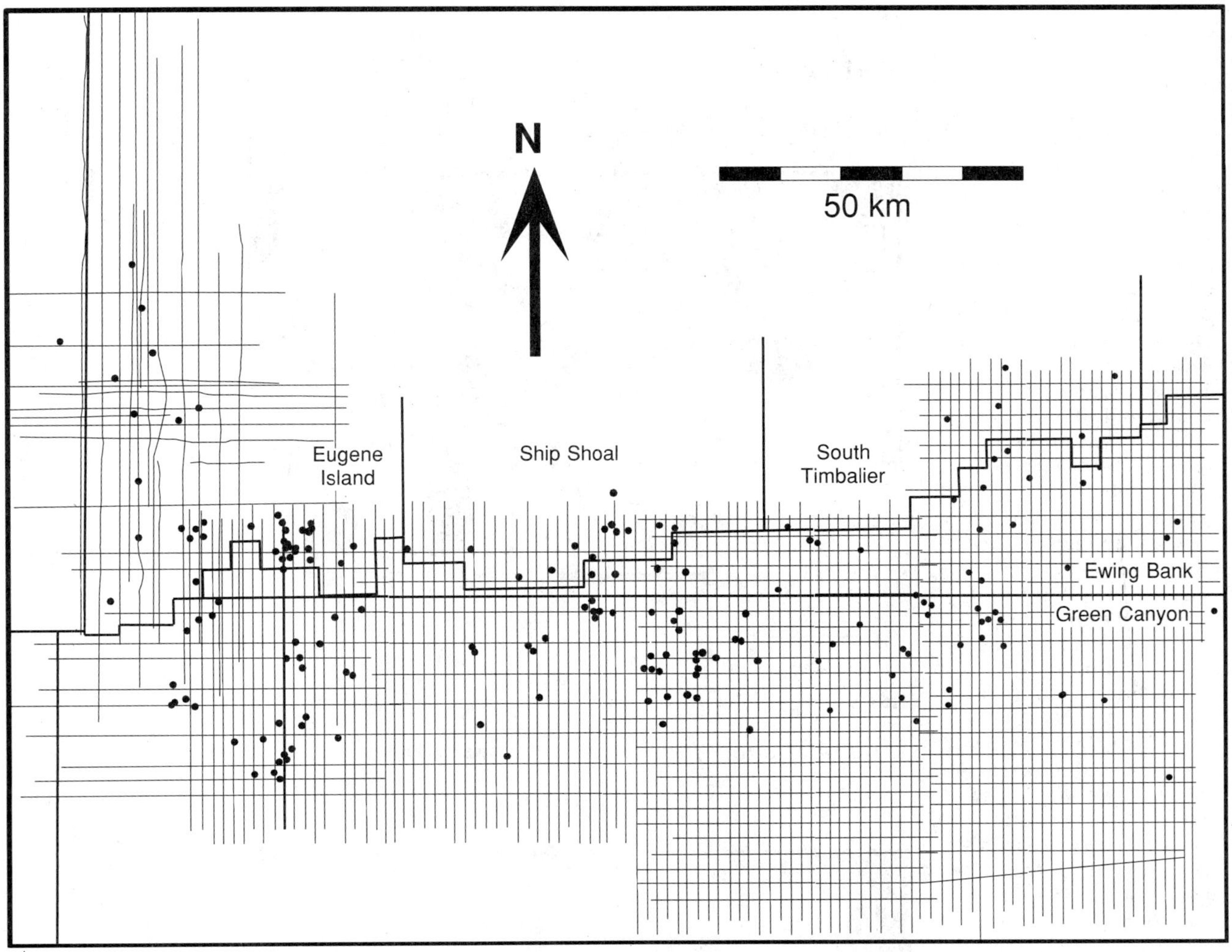


**Figure 2—Map showing two-dimensional seismic data grid and location of wells with biostratigraphic data used in this study.**

space. Our results suggest that a variety of factors, both structural and sedimentological, influence lithofacies development. Complex interaction of these factors at different temporal and spatial scales produces unique stratigraphic histories within individual minibasins. We demonstrate that the influence of salt deformation on sedimentation is neither simple nor systematic; nevertheless, there are meaningful patterns and relationships between the two processes. A better understanding of the various factors will enable geoscientists to identify scenarios in which sand-rich lithofacies are more probable, and thus reduce risk in reservoir prediction in deep-water environments.

Our study area covers northern Green Canyon, most of Ewing Bank, southern Eugene Island, and the southern edges of the southern additions of Ship Shoal, South Timbalier, and Grand Isle (Figure 1). This paper is one of several in this issue that summarize the results of an integrated structural, stratigraphic, biostratigraphic, and petroleum systems investigation of this region. The structural setting and evolution are presented in McBride et al. (1998, this issue), and the sequence stratigraphic evolution is described in Weimer et al. (1998, this issue). The primary seismic data set consists of more than 10,000 km of two-dimensional (2-D) profiles with a 1.6 km spacing between dip (north-south) lines and a 3–5 km spacing between strike (east-west) lines (Figure 2). This is supplemented by 2500 km of more widely spaced 2-D profiles in southern Eugene Island (Figure 2). Age and lithology control are provided by logs and biostratigraphic data from approximately 180 wells (Figure 2) [for details, see Weimer et al. (1998, this issue) and Villamil et al. (1998, this issue)].

## SALT DEFORMATION, SEA-FLOOR TOPOGRAPHY, AND DEPOSITION

In this section, we examine the influence of salt tectonics on sedimentation patterns by reconstructing,

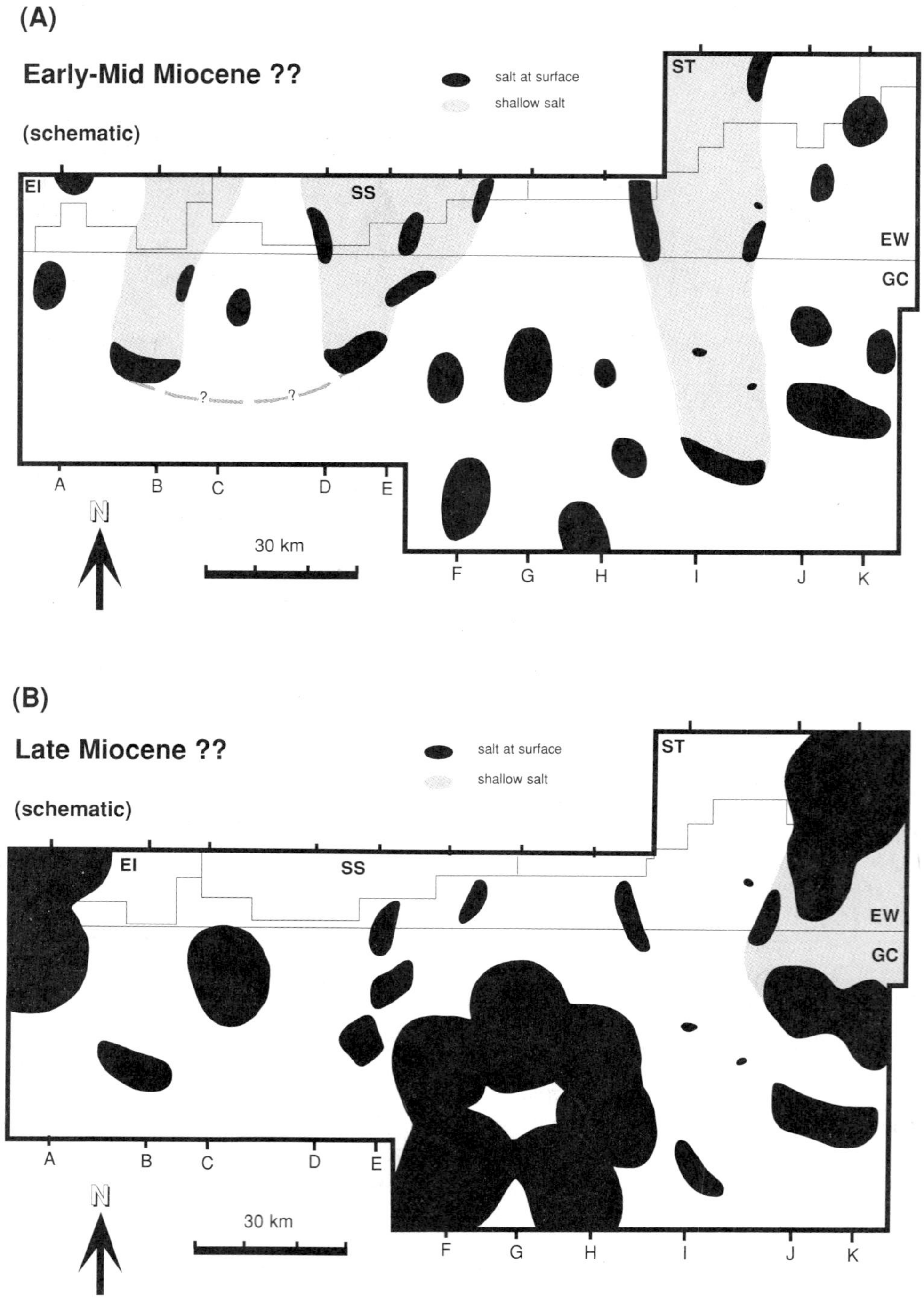


Figure 3—(A and B) Conceptual, schematic reconstructions of Miocene(?) salt evolution; (C–D) palinspastic reconstructions of Pliocene–Pleistocene near–sea-floor structure (salt, faults, and folds) derived from analysis of seismic data (construction method illustrated in Figure 4). The maps do not include the Eugene Island portion of the study area. A–K indicate restored two-dimensional profiles used to constrain the reconstructions (from McBride, 1997). Superimposed on the structural geometries are interval thicks derived from isochron maps and interpreted depositional environments at the base of each sequence (both from Weimer et al., 1998, this issue). Note that restorations, facies interpretations, and isochron data are schematic in the western half of the study area in (C–E) due to minimal deep well control. EI = Eugene Island, SS = Ship Shoal, ST = South Timbalier, EW = Ewing Bank, GC = Green Canyon.

in map view, the sea-floor effects of salt flow and related deformation for discrete time intervals during the Pliocene–Pleistocene. We first describe the technique of structural reconstruction in map view, and then discuss the results in terms of both structural and stratigraphic evolution. The type of salt system and the stage of formation or evacuation are not considered, only the presence or absence of salt diapirs, faults, and folds that probably had bathymetric expression. The underlying assumption is that active, salt-related deformation affects the sea floor in the form of active fault scarps, subsiding basins, and bathymetric highs over salt diapirs and contractional folds. Other factors influence bathymetry, such as shelf progradation and submarine slope failure, but are not addressed here.

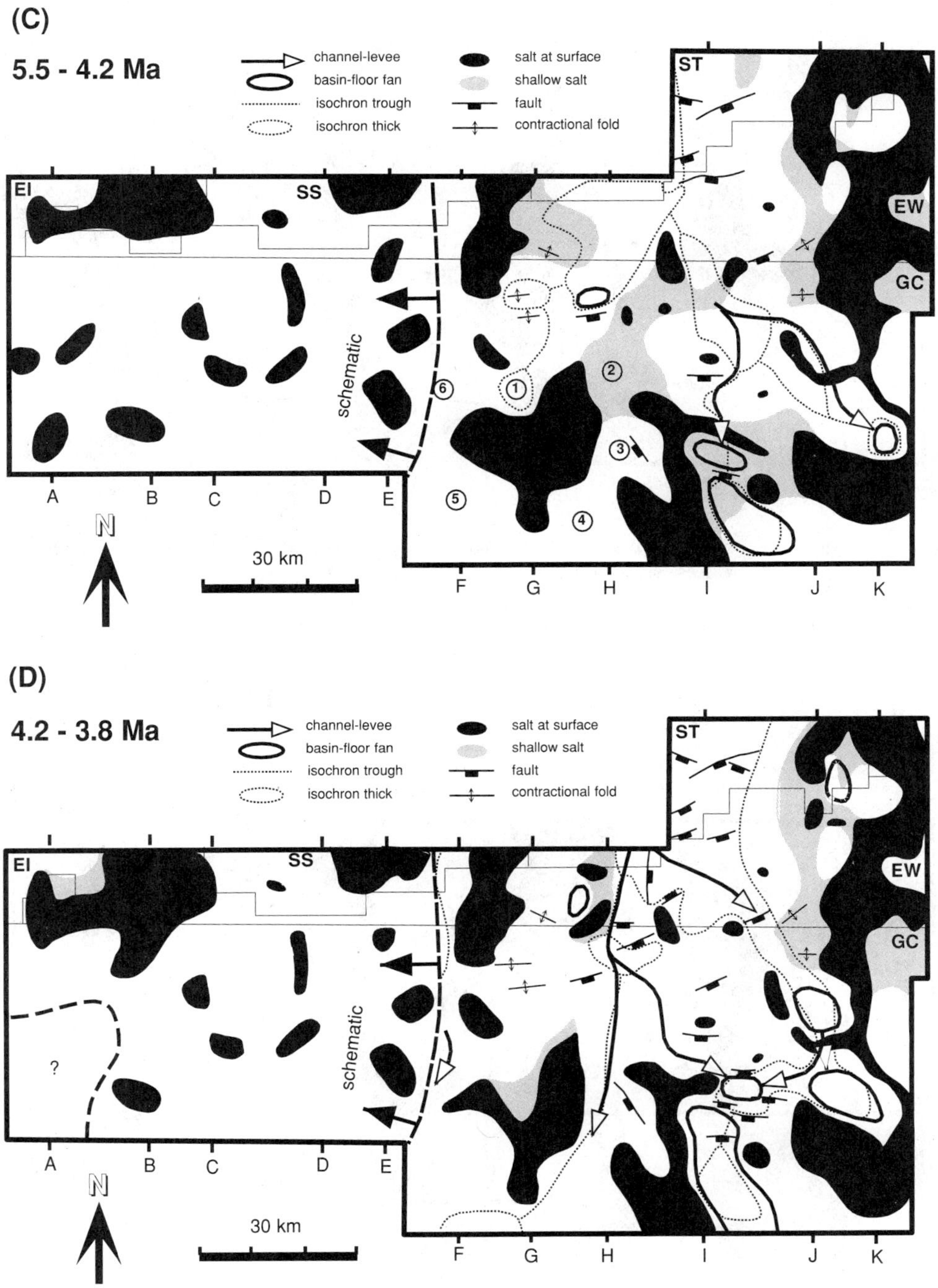


**Figure 3—Continued.**

## Method

Palinspastic maps of near-sea-floor structure were constructed for ten intervals delineated by interpreted sequence boundaries: 5.5–4.2, 4.2–3.8, 3.8–3.0, 3.0–1.4, 1.4–1.1, 1.1–0.8, 0.8–0.7, 0.7–0.5, 0.5–0.4, and 0.4–0 Ma (Figure 3C–L). The horizon interpretations were based on stratigraphic studies of the area (Acosta, 1994; Martinez, 1994; Navarro, 1994; Budhijanto, 1995; Varnai, 1996; Weimer et al., 1998, this issue. In addition, two schematic maps are illustrated for earlier, unconstrained times (Figure 3A, B). The salt distribution shown in these Miocene(?) maps is not the true geometry at any specific time, merely a conceptual model of how salt might have evolved.

For each stratigraphic interval, every profile in the survey was examined and its interpretation

**Figure 3—Continued.**

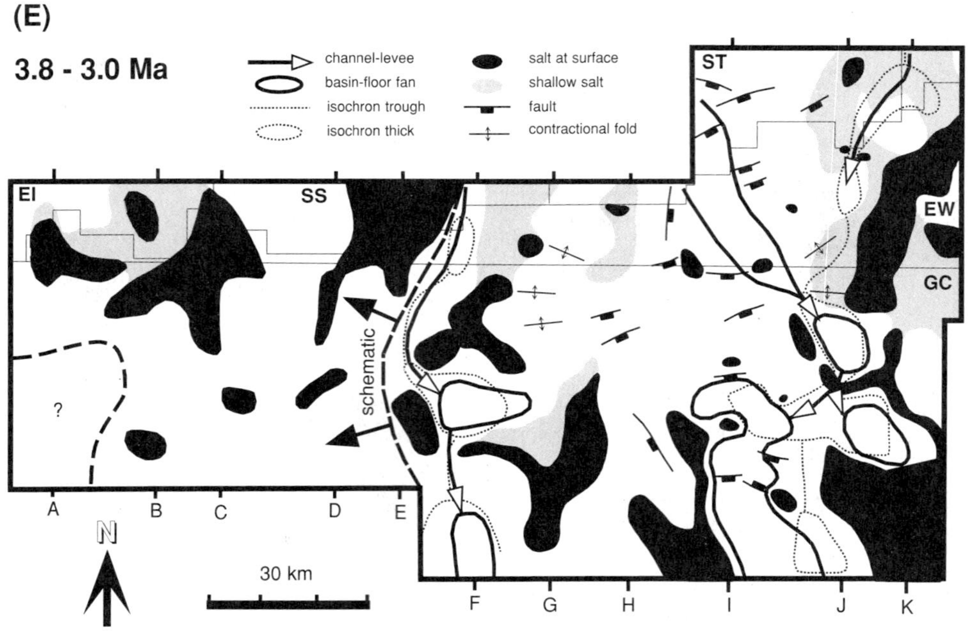


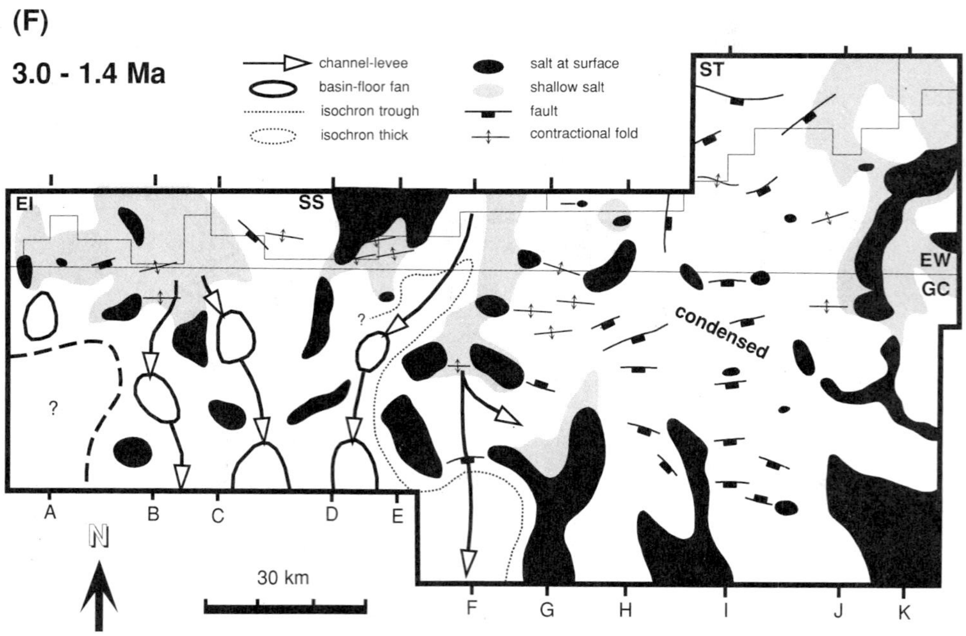


mentally restored to construct the maps using the following criteria. If an interval onlapped salt (Figure 4, locations a) or was truncated by an overhanging salt body (Figure 4, location b), salt was assumed to have sea-floor expression at that time, with a plan-view geometry determined by mapping the onlap or truncation pattern. A salt body and corresponding bathymetric high were also interpreted if the sequence thinned locally immediately above salt (Figure 4, location c). In contrast, if an interval did not thin noticeably, but still occurred within about 800 ms of the top salt or a salt weld over a large area (Figure 4, location d), the area was mapped as shallow salt; these regions probably did not influence the sea floor. If an interval was observed to thin over a contractional fold (Figure 4, location e), the structure was assumed to have bathymetric expression at that time. Likewise, faults were mapped as active during a given sequence if there was significant thickening across

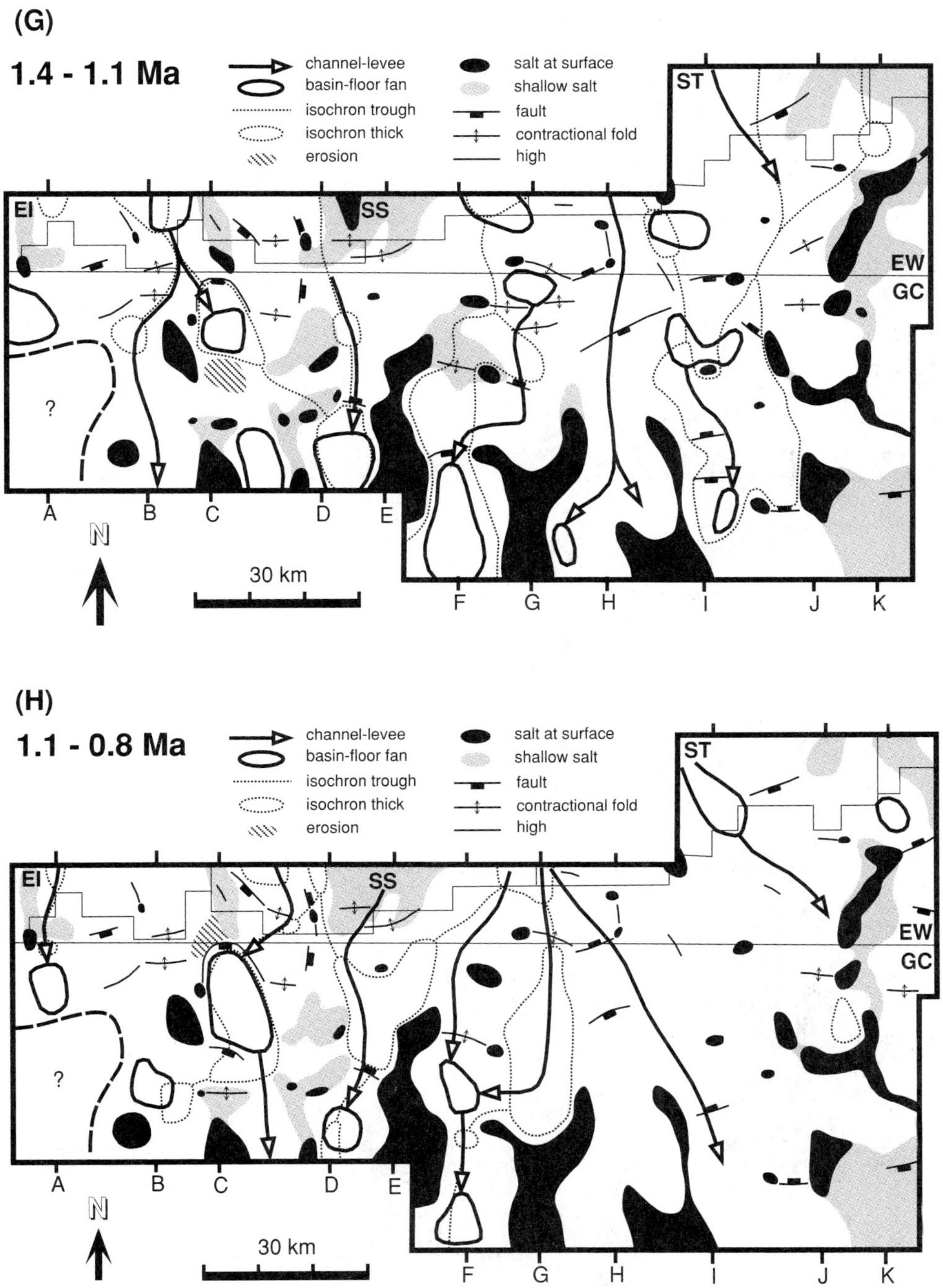


Figure 3—Continued.

the fault (Figure 4, location f). One important aspect that is not shown is the presence or absence of underlying salt that would provide accommodation for sedimentation. Areas between salt diapirs are underlain by either salt or salt welds, and thus may or may not have had bathymetric expression as subsiding basins.

Map construction was in large part a subjective process; however, the results are consistent with a series of eleven north-south profiles that were restored quantitatively (McBride, 1997; McBride et al. 1998, this issue). One component of deformation included in the profile restorations, but not on the maps, is lateral translation. Extension on normal faults resulted in differential downslope translation of suprasalt strata. This would impact the restored position of features such as the landward margins of some salt bodies; however, because extension in the study area was generally minor, with a maximum of about 12% (Rowan et al., 1995a; McBride et al., 1998,

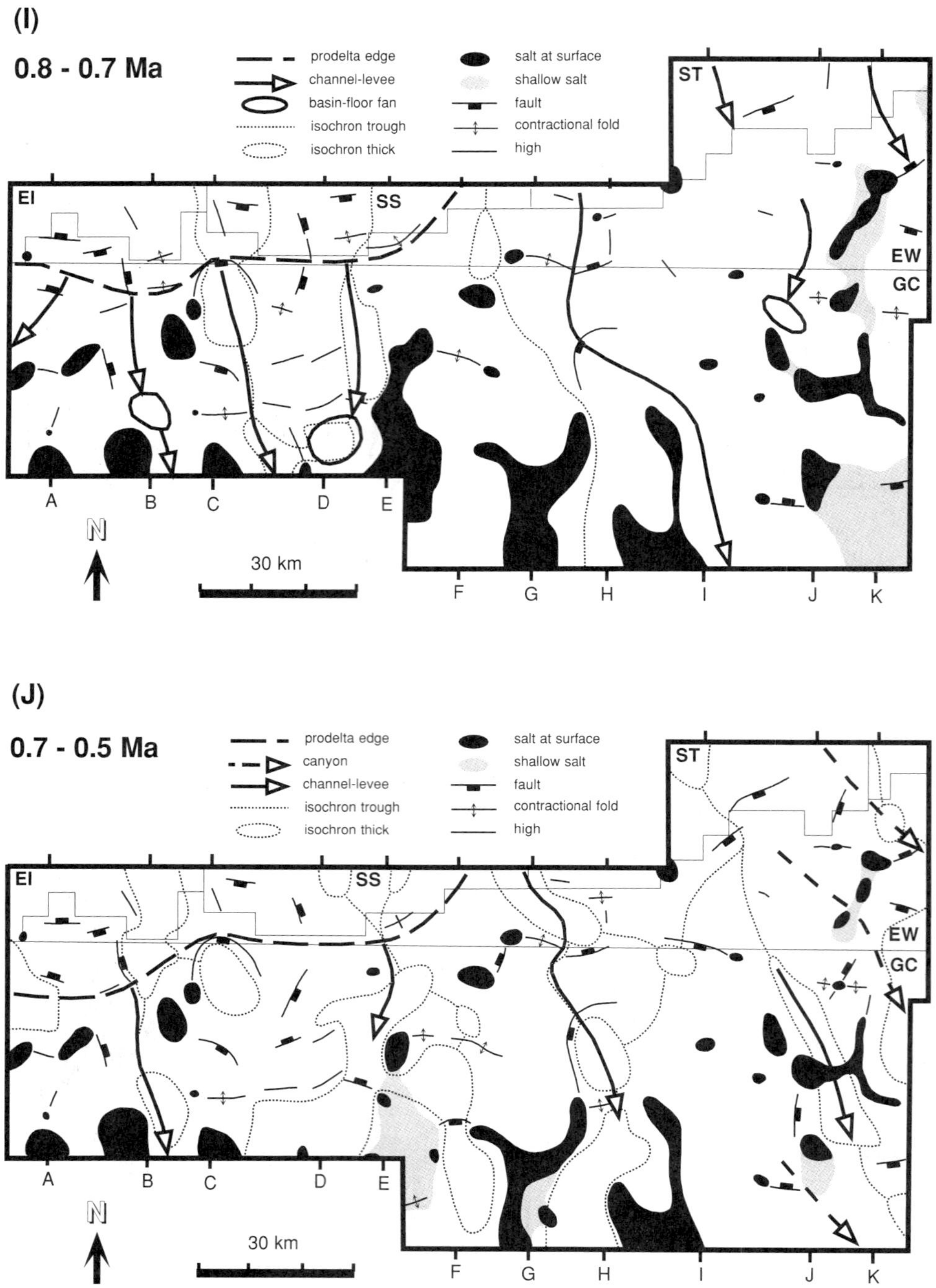


**Figure 3—Continued.**

this issue), this limitation should have insignificant impact on the palinspastic reconstructions. Also, because stratigraphic features were mapped in their present locations, they were more easily compared to the structural reconstructions if translations were not restored.

Two other features are illustrated on the reconstructed maps (Figure 3). First, broad areas and troughs of thicker strata taken from isochron maps for each interval (Weimer et al., 1998, this issue) show areas of increased sedimentation. These may correspond to areas of more rapid subsidence or may represent thicker lobes deposited in uniformly subsiding regions. Second, interpreted sediment transport pathways and depocenters from Weimer et al. (1998, this issue) are illustrated. Depocenters are depicted where basin-floor fans were interpreted and transport pathways follow

Figure 3—Continued.

(K)

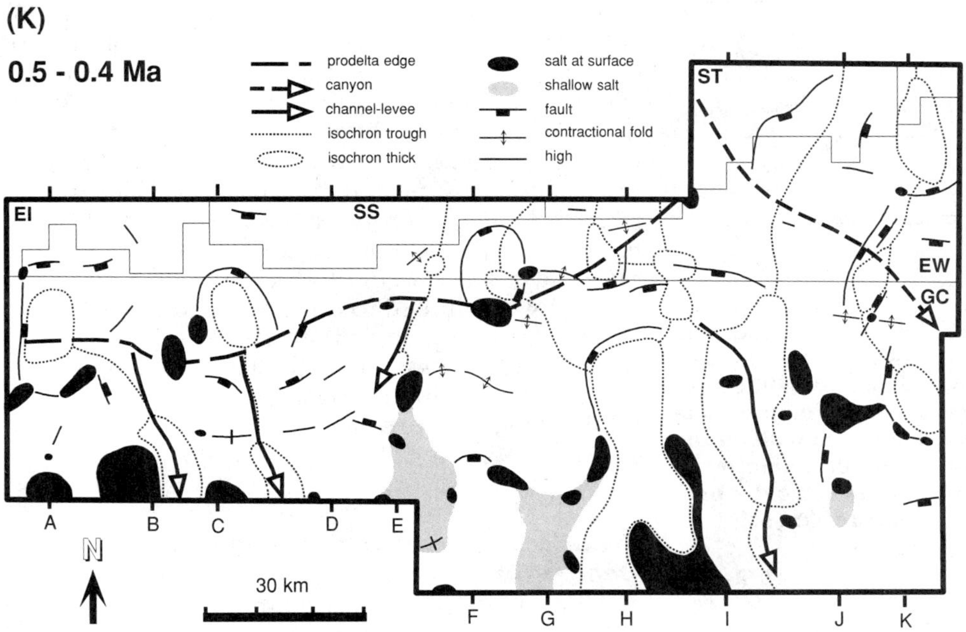


(L)

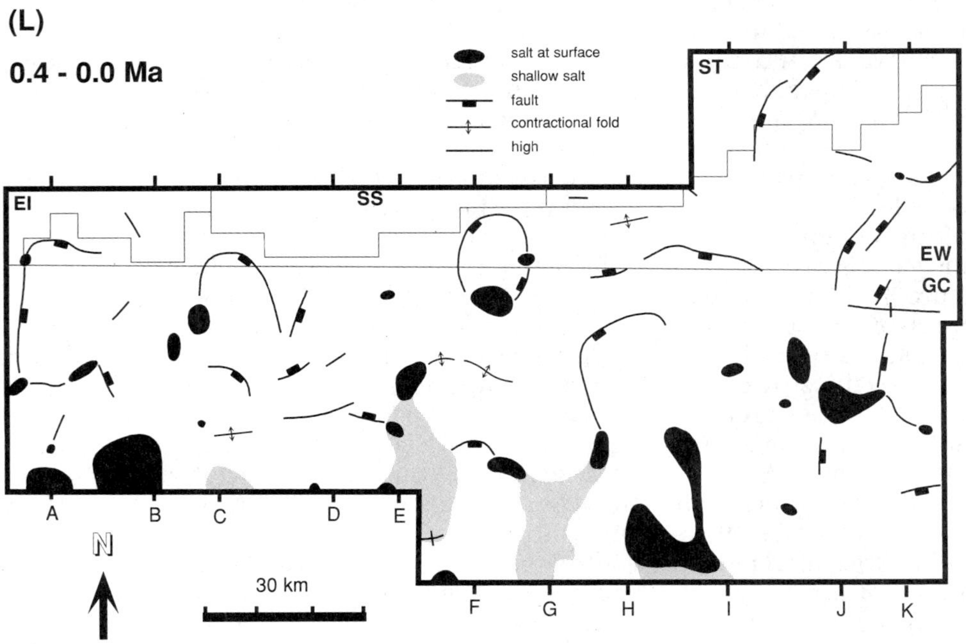


areas where channel-levee systems were observed. The facies shown on the palinspastic maps represent only the lowest facies within each sequence. Furthermore, we emphasize that terms such as "basin-floor fan" are used as descriptions of interpreted seismic facies and do not necessarily signify anything about grain size, depositional processes, or sand-body geometries. Some of these facies are corroborated by well-log data (see Weimer et al., 1998, this issue), but others are interpretive.

## Results

### *Structural Evolution*

The schematic maps (Figure 3A, B) and the oldest constrained reconstruction (Figure 3C) show

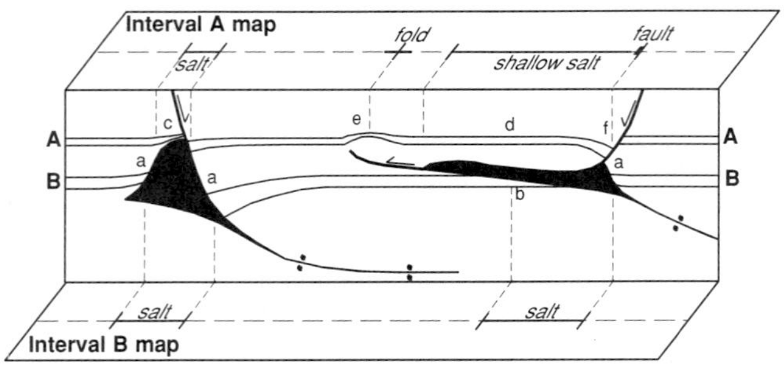


**Figure 4—Palinspastic map construction method. The middle panel shows a schematic cross section with two stratigraphic intervals (A and B) affected by salt tectonics. The upper and lower panels show the map projections of interpreted near-surface deformation along the illustrated profile (locations a–f are described in text). Interval maps are constructed by connecting the projected patterns from all dip and strike profiles. Salt is in black; welds are indicated by pairs of dots.**

how we envision the pre-Pliocene evolution [similar scenarios have been discussed by Rowan et al. (1995b) and Snyder and Nugent (1995)]. Three shallow salt sheets (shaded areas in Figure 3A) gradually segmented into diapirs that advanced upward and laterally through time. The two western areas actually may have been parts of one larger sheet. At the same time, isolated stocks (Figure 3A) grew and expanded in all directions to form several salt-stock canopies (Figure 3B). Eventually, collapse of the bathymetrically elevated stocks led to basin initiation and displacement of salt into diapirs located along the margins of the canopies or over saddles between adjacent stocks (Figure 3C). For example, the large canopy in the center of the study area (Figure 3B) collapsed to form a series of six linked minibasins located exactly where the stocks had been (locations 1–6, Figure 3C). Thus, salt-related bathymetric highs and subsiding minibasins have effectively switched places through time.

A general trend of decreasing near-surface salt and corresponding bathymetric expression took place during the Pliocene–Pleistocene (Figure 3C–L). Diapirs and sheets in the north were gradually buried by shelf progradation, and salt bodies in the south decreased in areal size as basins expanded. This progression through time is directly analogous to the spatial progression observed on the present-day slope, with more salt and smaller, isolated basins basinward and less salt influence landward. The general trend was interrupted locally by the emplacement of salt sheets and tongues, for example at the southeastern corner of the study area during the early Pliocene (Figure 3C–E).

Features other than salt also affected sea-floor topography, although probably to a lesser extent. Contractional anticlines and associated rejuvenated diapirs (Rowan, 1995; Vendeville and Nilsen, 1995) were prevalent, especially in the northern one-half of the study area, but became less common with time (Figure 3C–L). In addition, three types of growth faults presumably affected sea-floor topography: (1) counterregional (landward-dipping) faults were rare, minor, and most active during the Pleistocene (Figure 3G–L), (2) roho faults [basinward-dipping listric faults that sole into salt sheets or subhorizontal welds; see Diegel et al. (1995) and Schuster (1995)] were limited to the eastern one-third of the study area and primarily to the Pliocene (Figure 3C–F), and (3) ramp faults [arcuate faults that wrap around the updip and lateral margins of elliptical minibasins; see Rowan (1995)] were most active during the Pleistocene (Figure 3G–L).

### *Evolution of Sediment Transport and Deposition Patterns*

A good, although not exact, correspondence exists among reconstructed near-surface structural geometry, isopach thicks, and facies distribution. Channel-levee systems generally followed lows between paleobathymetric highs and aligned with areas or troughs of thicker strata (Figure 3C–K). Basin-floor fans were usually associated with isopach thicks and occurred in five general structural settings: (1) minibasins almost completely surrounded basinward and laterally by salt highs (e.g., southeastern corner of study area during the early Pliocene; Figure 3C–E), (2) minibasins located between several salt highs (many examples), (3) lows between contractional anticlines (e.g., northern edge of central Green Canyon between 1.4 and 1.1 Ma; Figure 3G), (4) hanging walls of arcuate ramp faults (e.g., in northern Green Canyon just south of Eugene Island/Ship Shoal boundary between 1.1 and 0.8 Ma; Figure 3H), and (5) hanging walls of roho faults (e.g., southeastern part of study area between 4.2 and 3.0 Ma; Figure 3D, E). However, there also were interpreted basin-floor fans with no obvious structure-related bathymetric control (e.g., in South Timbalier South between 1.1 and 0.8 Ma; Figure 3H), and many apparent bathymetric lows or basinward barriers to flow were not associated with ponding and fan deposition.

The palinspastic maps show qualitatively only the near-surface effects of salt deformation. An equally important role was played by the amount of salt withdrawal from beneath minibasins (this is addressed more thoroughly in the following section). For example, the old sheet in the eastern part of the study area (eastern shallow salt in Figure 3A) was evacuated by the end of the Miocene (McBride, 1997). The resulting lack of differential subsidence during the Pliocene–Pleistocene is reflected in the relative absence of interpreted

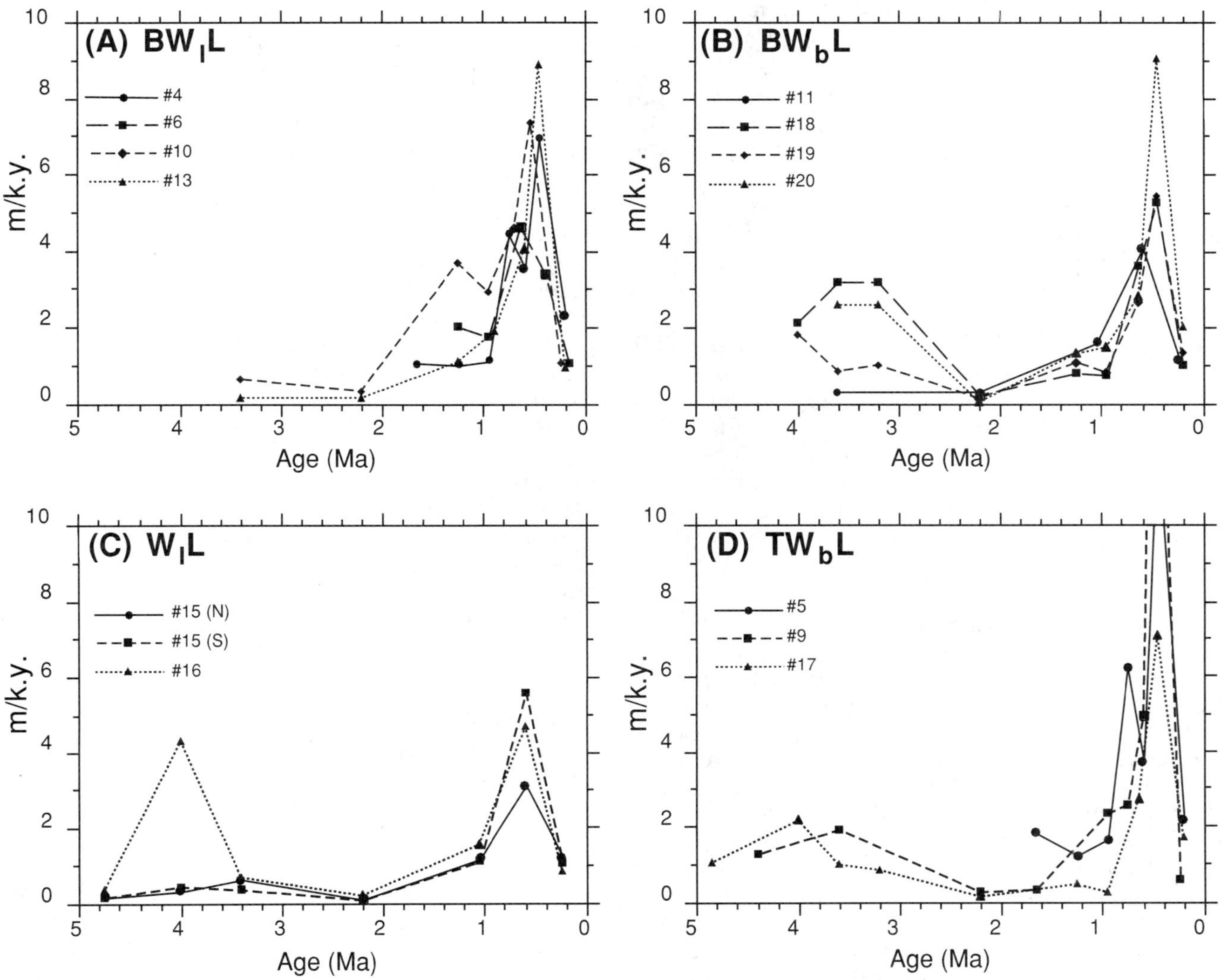


**Figure 5—Plots of sediment accumulation rates (in m/k.y.) vs. age for fourteen minibasins, derived from restored profiles (McBride, 1997). Rates are plotted at the midpoints of each time interval (e.g., 4.2–3.8 Ma interval is plotted at 4.0 Ma). Minibasin numbers correspond to those shown in Figure 1. $BW_lL$, $BW_bL$, $W_lL$, and $TW_bL$ are notations for minibasin types illustrated in Figures 6–9 and are discussed in the text.**

basin-floor fans in this area (Figure 3C–L). In contrast, most other minibasins were underlain by salt that was continuously evacuated throughout much of the Pliocene-Pleistocene (McBride, 1997; McBride et al., 1998, this issue). Consequently, depocenters and associated basin-floor fans were often superimposed for prolonged periods. Some of the minibasins in the western half of the study area have had basin-floor fans deposited intermittently from at least 3.0 to 0.7 Ma (Figure 3F–I).

Basin-floor fans had variable distribution in time. They were common prior to 0.8 Ma (Figure 3C–H), rare in the 0.8–0.7 Ma sequence (Figure 3I), and completely absent after 0.7 Ma (Figure 3J–L). The decrease in occurrences of basin-floor fans may have been related to a surge of sediment input at about 0.8 Ma (Figure 5) that has been ascribed to drainage capture within the ancestral Mississippi River system (Pulham, 1993; Prather et al., 1998, personal communication). Alternatively, a dramatic change in the periodicity of glacial episodes at about 0.8 Ma (J. T. Andrews, 1997, personal communication) somehow may be related to the increased sedimentation rates. This increase may have had two effects, both of which would have diminished the likelihood of ponding and sand deposition: (1) the shelf margin would have advanced a significant distance to the south, creating a steeper slope gradient and higher energy environment, and (2) sedimentation may have been rapid enough to smooth out much of the bathymetric expression associated with salt.

## SALT EVACUATION, SUBSIDENCE, AND DEPOSITIONAL ENVIRONMENTS

We have shown that sediment transport and deposition were controlled, at least in part, by bathymetric highs related to salt diapirism. Bathymetric lows caused by salt withdrawal also were important, and evacuation was probably influenced by the type of underlying salt and its evolutionary stage. To evaluate this relationship, we examine the stratal fill of various minibasins and identify the three-dimensional geometry and vertical stacking of tectonostratigraphic packages. We define a tectonostratigraphic package as consisting of multiple depositional sequences (sensu Mitchum et al., 1977) that have the same external form. The stacking patterns are related to different types of allochthonous salt systems and serve as a framework for analysis of the temporal distribution of seismic facies within each minibasin.

### Tectonostratigraphic Packages

Four types of tectonostratigraphic packages are defined on the basis of their three-dimensional external form: (1) layers, which are sheetlike sequences with constant thickness in all directions (denoted "L"), (2) wedges, which are sequences that thicken in one direction, but remain a constant thickness in the other direction (denoted "$W_l$" or "$W_b$," depending on whether the wedge thickens landward or basinward, respectively), (3) troughs, which are sequences that have an axis of maximum thickness (denoted "T"), and (4) bowls, which are sequences that thin in all directions from a point of maximum thickness (denoted "B"). The character of the internal stratal fill is not considered here, but is used in identifying seismic facies (Weimer et al., 1998, this issue). Also, we do not include external forms characteristic of progradation (e.g., sigmoidal geometry). None of these forms are observed, because of the bathyal setting of the study area throughout most of the Pliocene-Pleistocene and the small scale of individual minibasins. Only in the younger sequences (<0.8 Ma) in the northern part of the study area can clear clinoforms be identified, but these occur within sheetlike tectonostratigraphic packages.

Minibasins are characterized by four different stacking patterns of these tectonostratigraphic packages. The stacking patterns are designated according to the vertical succession of tectonostratigraphic forms: (1) $BW_lL$, which is a deep bowl overlain by a landward-thickening wedge and then a layer (Figure 6), (2) $BW_bL$, which is a deep bowl overlain by a basinward-thickening wedge and then a layer (Figure 7), (3) $W_lL$, which is adjacent landward-thickening wedges overlain by a layer (Figure 8), and (4) $TW_bL$, which is a deep trough overlain by a basinward-thickening wedge and then a layer (Figure 9). The top layer in each pattern is not always observed, but is included as the conceptual final stage because any given slope minibasin eventually will be filled and capped by shelf sequences of near constant thickness as progradation of the margin continues.

We emphasize that this is not an exhaustive list of minibasin types. Our goal here is to describe the patterns observed within the study area, not to present a theoretical classification of minibasin styles for the entire northern Gulf of Mexico. For example, minibasins that include progradational wedges are not included, and all local thickness changes are ascribed to salt tectonics. We suggest that these four stacking patterns can be related directly to different types of allochthonous salt systems. The geometry and evolution of minibasins recorded by the stacking patterns are consistent with existing models for either salt stocks (Rowan, 1995) or salt sheets fed by counterregional feeders (Schuster, 1995).

#### *Salt Stocks*

Bulb-shaped salt stocks (Figure 10A) expand radially during growth and collapse once the source is depleted or cut off (Rowan, 1995). The initial phase of basin growth is dominated by vertical subsidence into the underlying salt stock (Figure 10B). Associated stratal fill is bowl-shaped, with the thickest accumulation in the center of the basin and onlap onto the basin margins. The suprasalt strata eventually touch down, forming a weld at some point beneath the basin. Remaining salt is evacuated through asymmetric collapse, resulting in wedge-shaped stratal fill. Rotation may be landward, with the formation of a ramp fault (Figure 10C), or basinward, with the formation of rollover above a counterregional weld (Figure 10D). These two end members correspond, respectively, to the $BW_lL$ and $BW_bL$ minibasin types. The transition from bowl-shaped strata to either landward- or basinward-thickening wedges is interpreted to represent the approximate time of initial weld formation beneath some portion of the minibasin.

#### *Salt Sheets*

Evacuation of salt sheets fed by counterregional (landward-dipping) feeders (Figure 11A) follows two end-member scenarios (Schuster, 1995). In roho systems, families of basinward-dipping, listric growth faults sole into the salt layer and accommodate significant downslope translation of the overburden (Figure 11B, C). Strata thicken landward in

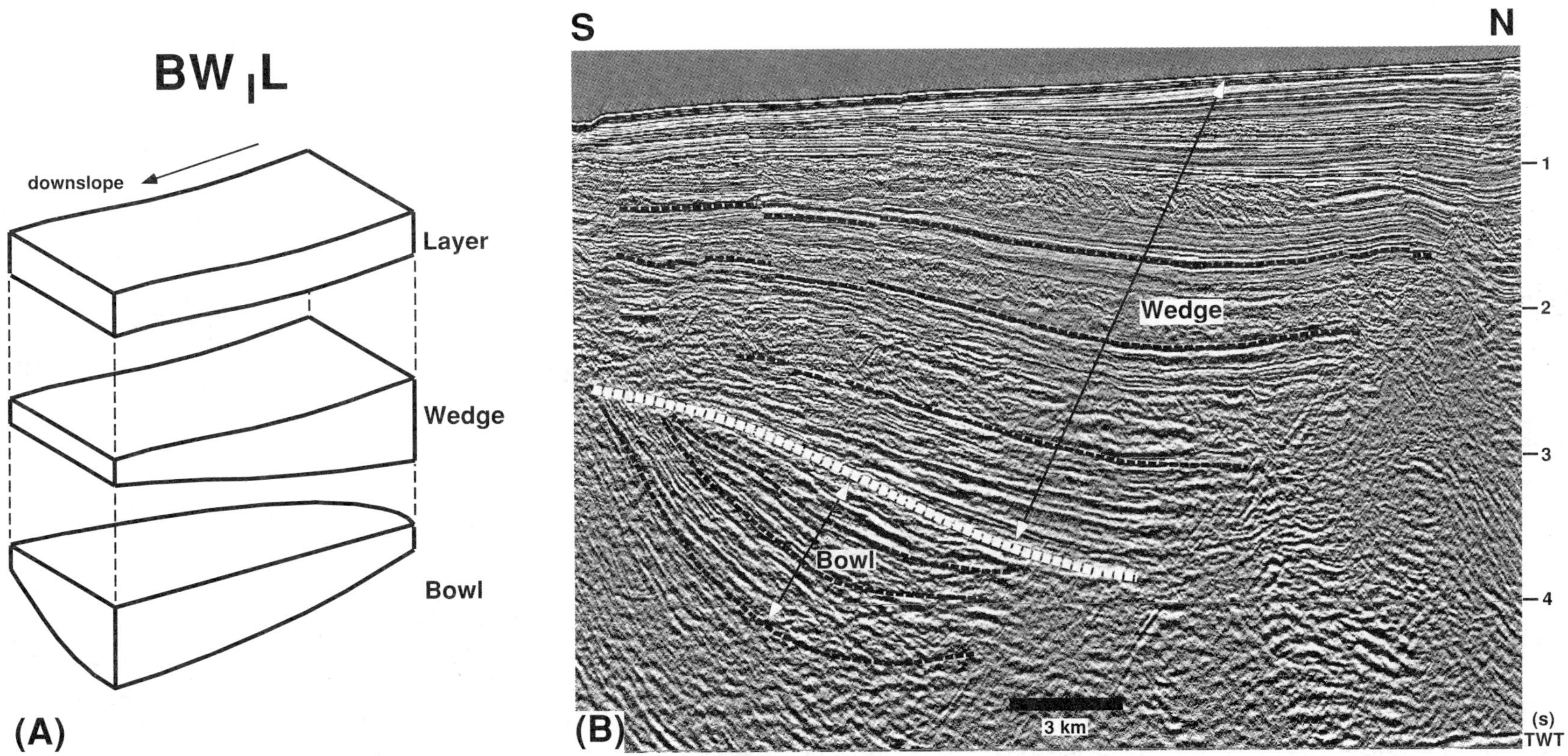


Figure 6—BW$_I$L minibasin type. (A) Diagram showing transition from bowl to landward-thickening wedge to layer; (B) representative seismic profile through minibasin 6 (located on Figure 1). Dashed white line in (B) marks the boundary between bowl and wedge packages; dashed black lines show selected horizons defining the stratal geometry. No layer package is present. Bowl shape is not apparent on two-dimensional profile, but is confirmed in three-dimensional profile. Data courtesy of Halliburton.

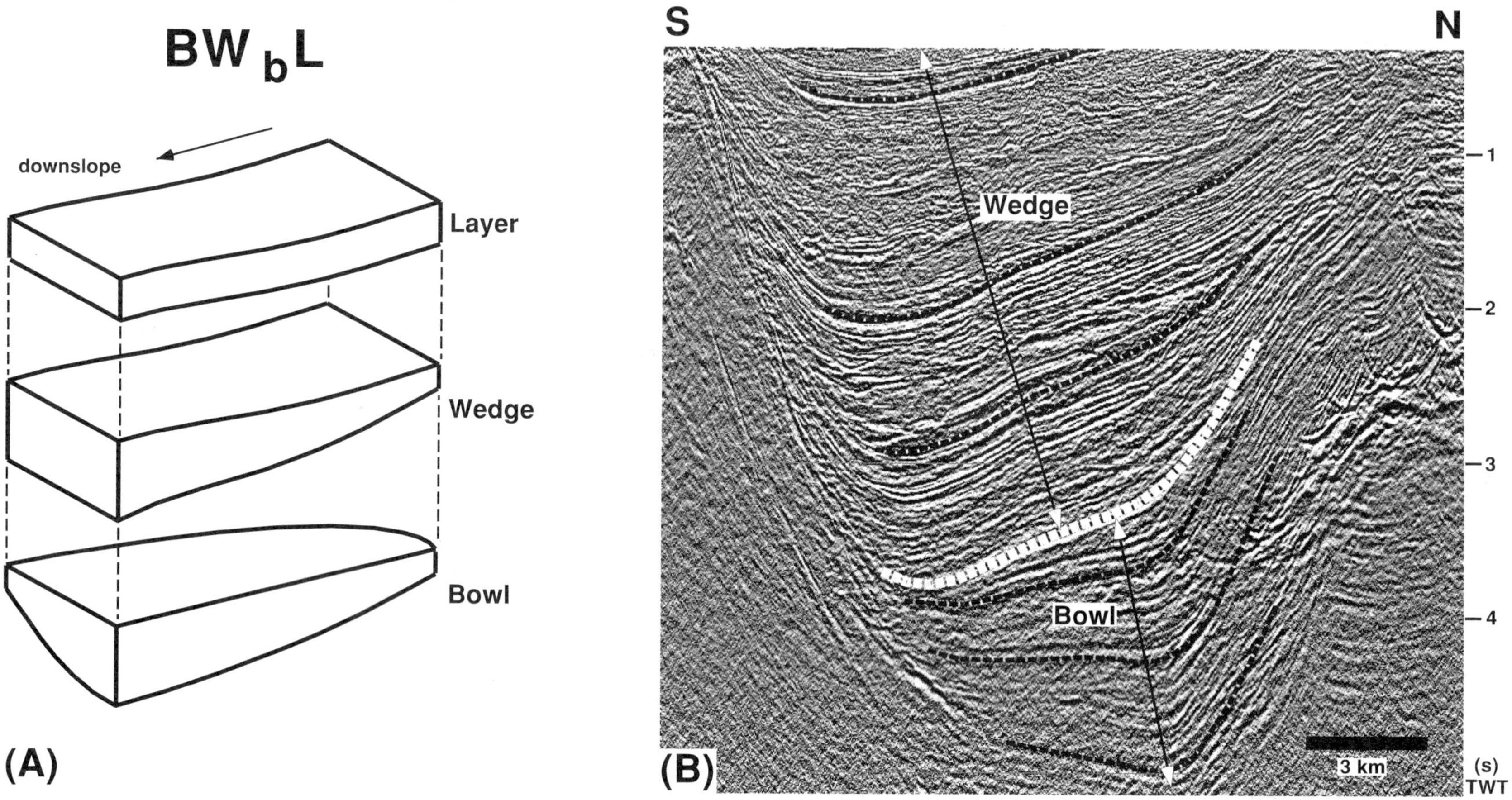


Figure 7—$BW_bL$ minibasin type. (A) diagram showing transition from bowl to basinward-thickening wedge to layer; (B) representative seismic profile through minibasin 20 (located on Figure 1). Dashed white line in (B) marks the boundary between bowl and wedge packages; dashed black lines show selected horizons defining the stratal geometry. No layer package is present. Bowl shape is not apparent on two-dimensional profile but is confirmed in three-dimensional profile. Data courtesy of Halliburton.

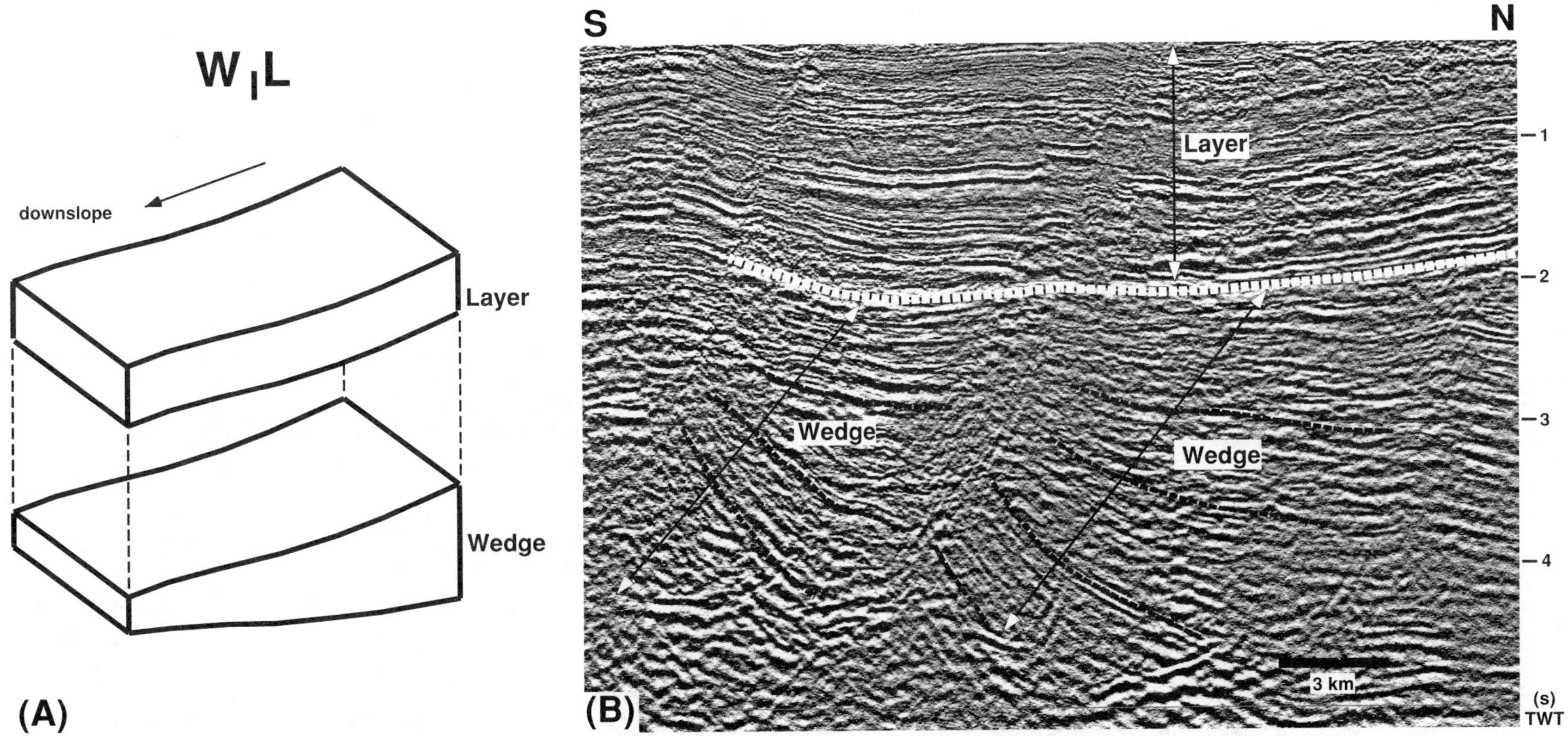


Figure 8—W$_I$L minibasin type. (A) Diagram showing transition from landward-thickening wedge to layer; (B) representative seismic profile through minibasin 14 (located on Figure 1). Dashed white line in (B) marks the boundary between multiple landward-thickening wedges and layer; dashed black lines show selected horizons defining the stratal geometry. Data courtesy of Halliburton.

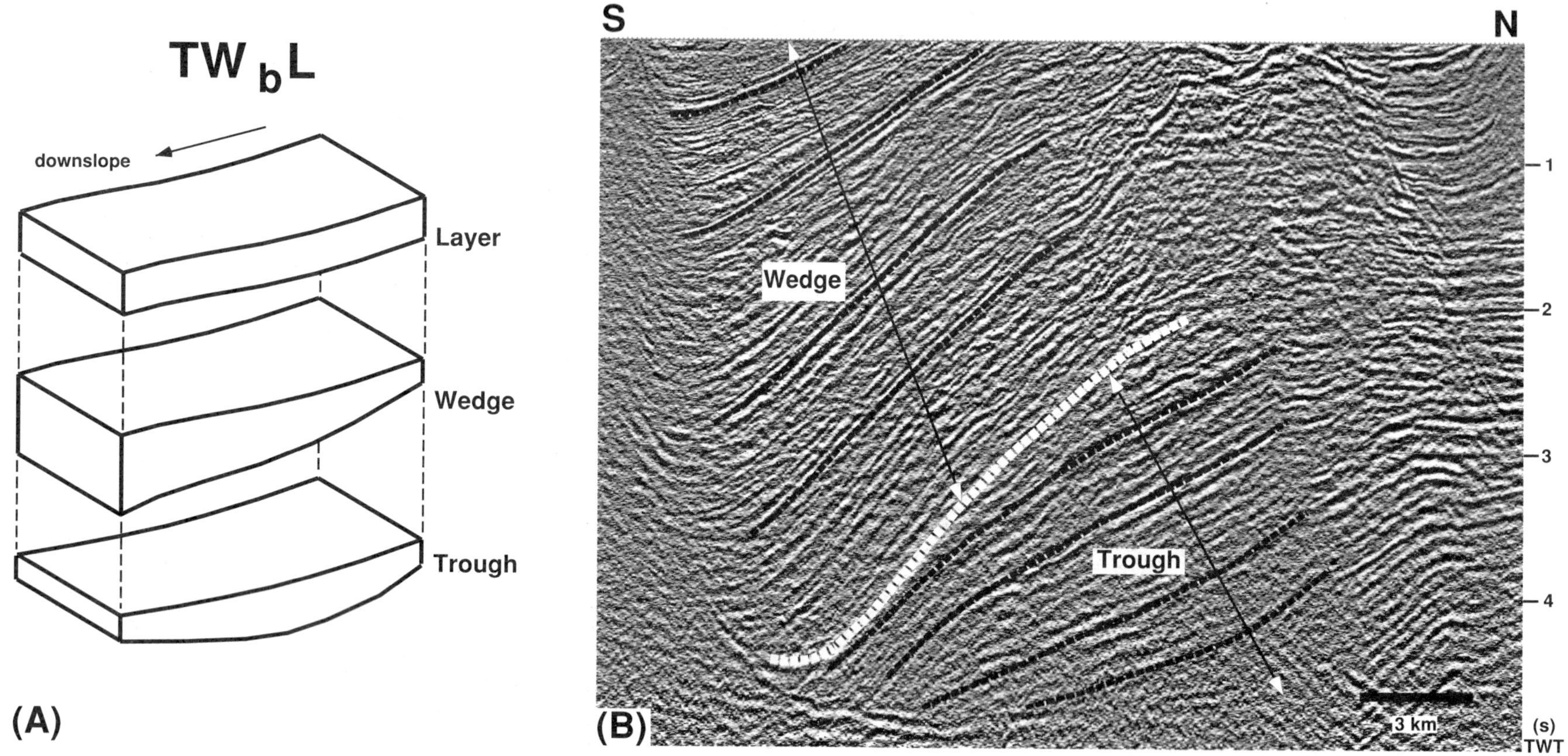


Figure 9—$TW_bL$ minibasin type. (A) Diagram showing transition from trough to basinward-thickening wedge to layer; (B) representative seismic profile through minibasin 5 (located on Figure 1). Dashed white line in (B) marks the boundary between trough and wedge packages; dashed black lines show selected horizons defining the stratal geometry. No layer package is present. Trough appears identical to bowl-shaped packages (Figures 6, 7) in two-dimensional, but is distinguished based on three-dimensional geometry. Data courtesy of Halliburton.

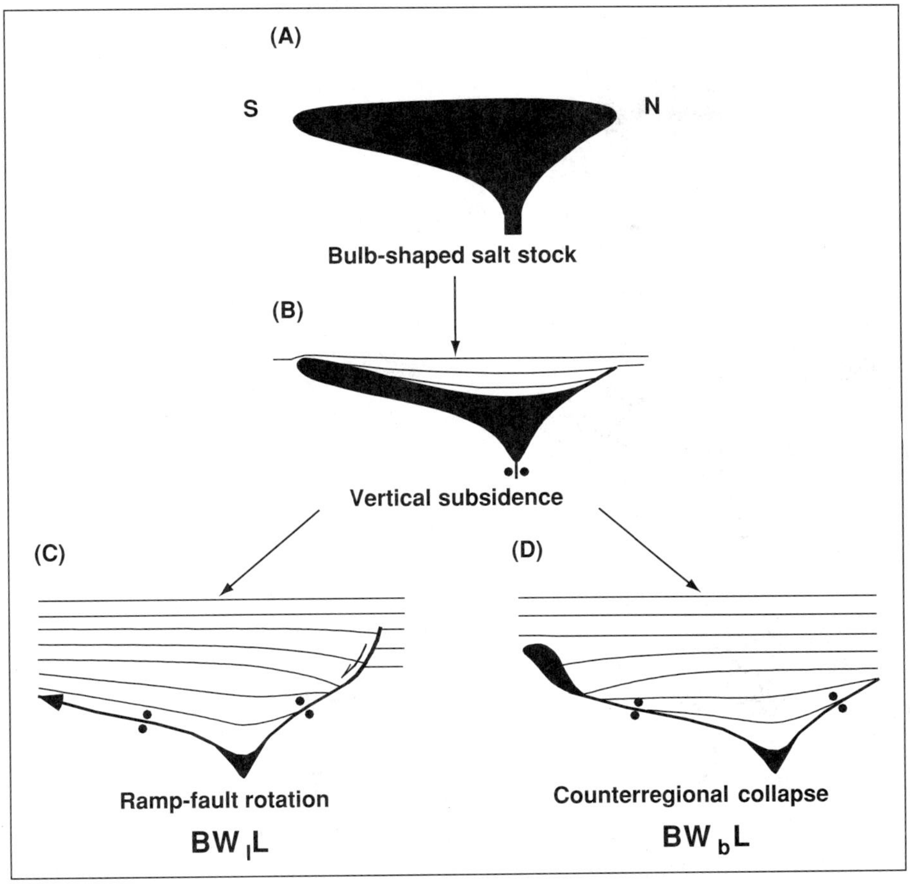


**Figure 10—(A) Evacuation of a bulb-shape salt stock (B) via vertical subsidence to either (C) ramp-fault rotation or (D) counterregional collapse. The two end-members correspond, respectively, to BW$_l$L (C) and BW$_b$L (D) minibasin types. Salt welds indicated by pairs of dots.**

the hanging walls of these faults, thus forming W$_l$L minibasins (Figure 11C). In stepped counterregional systems, salt collapses progressively basinward, driving salt into south-leaning diapirs and creating large, monoclinal rollovers (Figure 11D, E). Early suprasalt strata subside vertically into the salt sheet without significant translation (Figure 11D). Because the sheet is subhorizontal, the basin will tend to be more elongate (trough-shaped) than elliptical (bowl-shaped). Asymmetric collapse proceeds once suprasalt strata touch down, and stratal packages in this stage thicken progressively basinward (Figure 11E). Thus, stepped counterregional systems correspond to the TW$_b$L minibasin type. As in the case of BW$_l$L and BW$_b$L minibasins, the transition to a wedge-shaped geometry is interpreted to represent the approximate time of initial weld formation along the salt sheet beneath the trough.

### *Distribution*

All four minibasin types are found widely distributed within the study area (Figure 12). Not all minibasins have been classified; some of the basins are covered by seismic data with poor resolution, some basins extend out of the study area, and other basins have variable and complex geometries that do not fit any of the simple end-member models. The latter minibasins are located above the southern terminations of shallow salt sheets along the northern margin of the map, where complex salt geometries include rejuvenated diapirs, contractional folds and thrusts, and listric faults with multiple orientations.

Six minibasins have been identified as BW$_l$L types and six have been classified as BW$_b$L minibasins (Figure 12). There is no obvious pattern to their distribution, although both types generally occur in clusters representing old salt-stock canopies. BW$_l$L minibasins are always bounded to the north by highly arcuate ramp fault systems (Figure 12, minibasin 1), with the landward-thickening wedges in the hanging walls of these faults. Basinward, these minibasins may be bounded by salt bodies (Figure 12 minibasins 4 and 10) or not (Figure 12, minibasins 1, 2, 6, and 13). In contrast, BW$_b$L minibasins are always bounded basinward by prominent salt bodies. More importantly, BW$_b$L minibasins do not have ramp faults around their northern margins; instead, their landward margins are usually overridden by either shallow salt sheets (Figure 12, minibasins 3 and 20) or thrust faults where shallow sheets merge with the old stocks (Figure 12, minibasins 11 and 18).

Figure 11—(A) Evacuation of a salt sheet into (B and C) a roho system or (D and E) a stepped counterregional system. The two end members (from Schuster, 1995) correspond, respectively, to (C) $W_lL$ and (E) $TW_bL$ minibasin types. Salt welds indicated by pairs of dots.

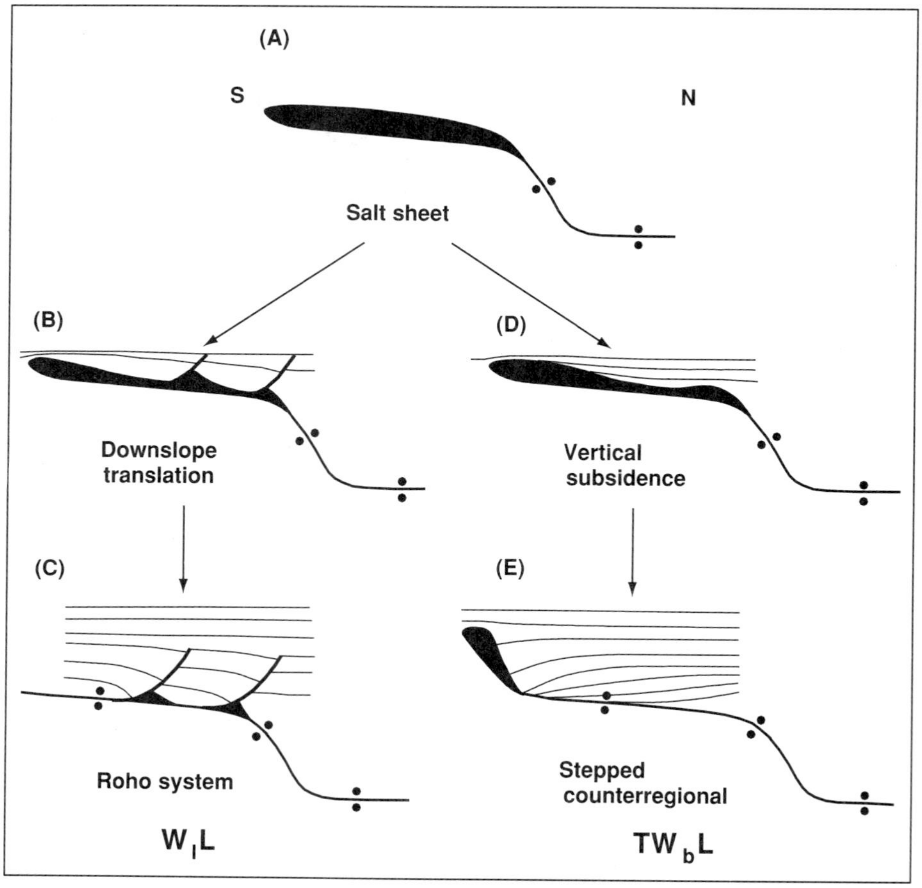


Several factors probably control whether a salt stock evacuates to form a $BW_lL$ or a $BW_bL$ minibasin: (1) the pattern of sediment dispersal into the basin and the distribution of sedimentary loading, which would determine where the initial weld forms, and possibly influence the subsequent direction of minibasin rotation, (2) the geometry of flanking salt diapirs established during early evacuation, which may affect later asymmetric collapse, and (3) the presence or absence of shallow salt sheets immediately north of the minibasin. The latter factor may be the most important because impingement of shallow sheets would inhibit formation of ramp faults and associated landward rotation of minibasins. This effect is observed in minibasin 20, which is in reality a $BW_lW_bL$ minibasin. Initial rotation was landward in the hanging wall of a ramp fault, but emplacement of a shallow sheet from the northeast terminated the faulting and was followed by the major phase of basinward rotation (McBride et al., 1998, this issue).

$W_lL$ minibasins, which are those derived from roho-style evacuation of salt sheets, occur primarily in the eastern part of the study area (Figure 12). The composite system to the east actually comprises three separate sheets: one deep (Figure 12, minibasin 15) and two at intermediate levels (Figure 12, minibasins 14 and 16). $TW_bL$ minibasins (Figure 12, minibasins 5, 9, and 17) are not common, but are widely distributed. The first two (minibasins 5 and 9) actually may represent a single remnant sheet that is continuous beneath shallower salt (see Figure 3A), in which case the isolated salt stock represented by minibasin 6 would be rooted in this sheet. The factors that determine whether a sheet evolves into a roho or stepped counterregional system are uncertain. F. A. Diegel (cited in Schuster, 1995) proposed that the downdip length of the sheet may be paramount; long sheets with multiple feeders would have enough gravitational head to develop downslope translation and roho faults, whereas shorter sheets may collapse basinward. However, because some sheets combine both styles, other factors, such as the distribution of sedimentary loading, also probably are important (F. A. Diegel, 1996, personal communication).

The factors that determine whether stocks or sheets form in the first place are also uncertain. Ultimately, the difference between the two types of systems is that stocks grow symmetrically and counterregional feeders grow asymmetrically. Thus, the symmetry of sedimentary loading above the source salt layer may have been the dominant factor. Sedimentary loading may have varied locally due to irregular sediment dispersal through and around

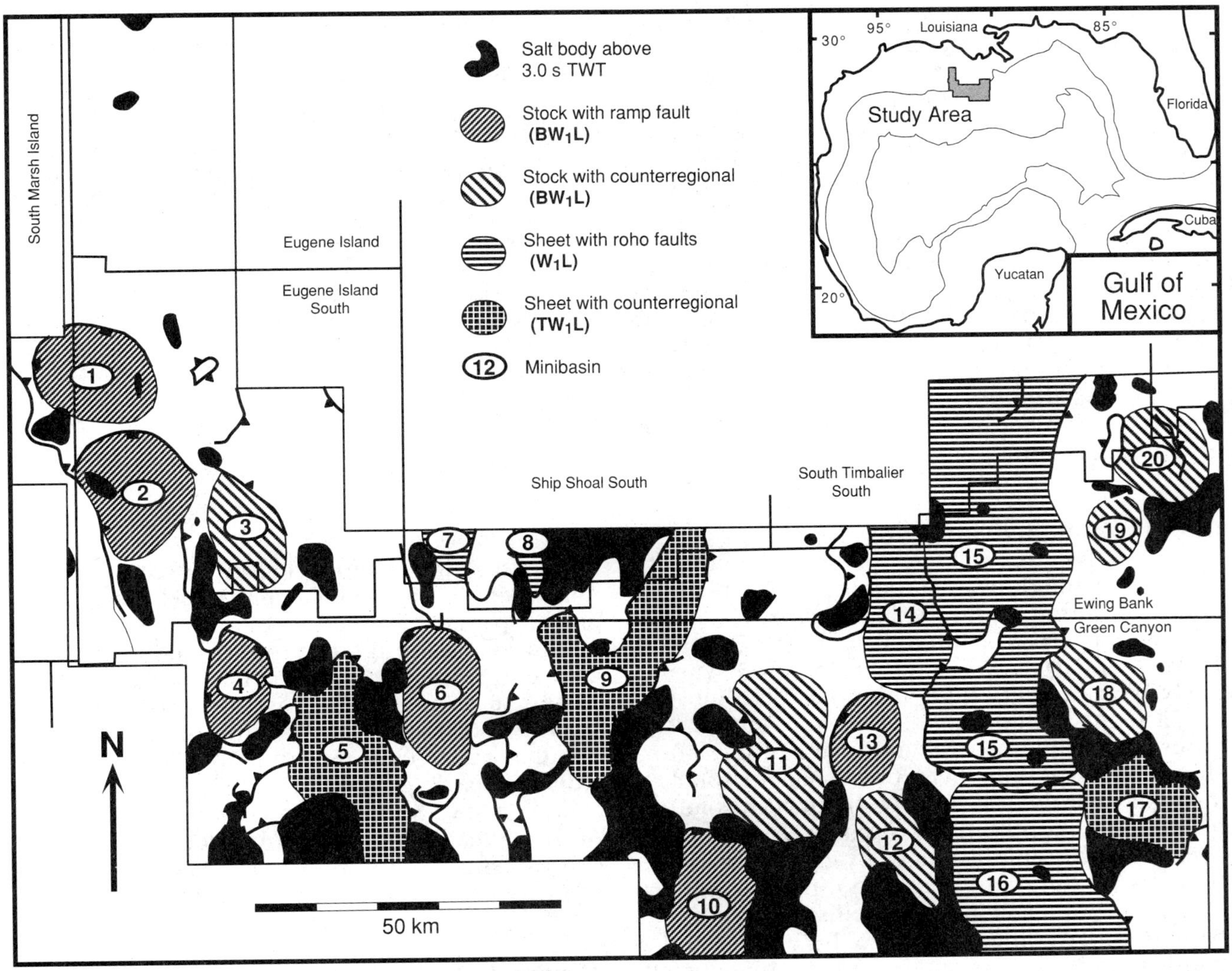


**Figure 12—Map showing distribution of minibasin types superimposed on salt diapir geometry. Minibasin numbers correspond to those in the text and Table 1. Note clusters (canopies) of evacuated salt stocks (s 1–4 in the northwest, s 10–13 in the south-central part of the study area, and s 18–20 in the northeast). White areas comprise either suprasalt strata above shallow sheets or portions of minibasins that could not be classified due to incomplete coverage.**

more landward salt highs. Loading also would have varied regionally: first, as a function of the distance from major deltas; second, dependent on the distribution and thickness of original salt basins. More landward salt diapirs may have grown asymmetrically in response to progradational loading, whereas more basinward diapirs may have been insulated from this loading and grown vertically.

### *Timing*

We have suggested that the transition from bowl- or trough-shaped tectonostratigraphic packages to landward- or basinward-thickening wedges may mark the time of initial salt weld formation beneath growing minibasins. Table 1 lists these transition ages for the thirteen minibasins that display such shifts. The values range from 4.2 to 0.5 Ma and show no obvious patterns. Ages are randomly distributed both parallel and perpendicular to the shelf margin, and for stocks, the minibasin type ($BW_1L$ or $BW_bL$) has no effect on the time of transition. The transition ages for $TW_bL$ minibasins are 3.0 Ma or older, which may be real and related to the deep (older) level of the associated sheets or simply a function of the small sample size.

Table 1 also lists the timing of salt weld formation determined from sequential restorations of the eleven profiles located on the maps of Figure 3 (McBride, 1997; McBride et al., 1998, this issue). Ranges in age represent both uncertainties related to resolution limits of the method and different ages derived for basin margins and centers (diachronous weld formation). There is generally a good correspondence between the transition ages and the timing of weld formation. Where the ages are different (minibasins 6 and 10), the transition age is older, and it may be that weld formation, in fact, was earlier in other parts of the basins not traversed by the restored profiles. Thus, we conclude that the age of transition from a bowl- or trough-shaped geometry to a wedge-shaped geometry marks the approximate timing of initial weld formation beneath minibasins. This is important because it provides an estimate of the timing of earliest possible petroleum migration into the minibasin from below the salt (see also McBride et al., 1998, this issue).

**Table 1. Minibasin Timing**

| Minibasin Type | Minibasin Number | Transition Age (Ma) | Weld Age (Ma) |
|---|---|---|---|
| $BW_lL$ | 2 | 1.9 | 2.6–1.4 |
| ($B \rightarrow W_l$ transition) | 4 | 0.5 | 0.7–0.5 |
| | 6 | 1.4 | 0.5–0.3 |
| | 10 | 3.0 | 0.8–0.5 |
| | 13 | 0.5 | 0.5–0.4 |
| $BW_bL$ | 11 | 4.2 | 5.5–4.2 |
| ($B \rightarrow W_b$ transition) | 12 | 3.0 | – |
| | 18 | 3.0 | 3.0–1.4 |
| | 19 | 3.8 | – |
| | 20 | 1.4 | 3.0–1.4 |
| $TW_bL$ | 5 | 3.0 | >1.9 |
| ($T \rightarrow W_b$ transition) | 9 | 3.0 | 4.2–3.0 |
| | 17 | 3.8 | >3.4 |

## Depositional Environments

The simplified evolution of interpreted depositional environments for each minibasin is shown in Figure 13 (derived from Acosta, 1994; Martinez, 1994; Navarro, 1994; Budhijanto, 1995; Varnai, 1996; Weimer et al., 1998, this issue). The illustrated patterns are complex, and we examine four types of possible control and systematic relationships. Note that only the depositional environment interpreted at the base of each sequence is plotted; for example, although most basin-floor fans are overlain in the same sequence by channel-levee systems (with overbank deposits), unchannelized mud-rich flows, and a condensed section, only the basin-floor fan is indicated. Also, because of uncertainties in well-to-seismic ties and seismic correlation, ages are considered to have a maximum error range of plus-or-minus one sequence (although any such errors are thought to be rare).

### *Temporal Patterns*

Depositional environments in most minibasins tend to gradually evolve from basin-floor fans and channel-levee systems to unchannelized mud-rich flows and prodelta sediments (Figure 13); however, the timing of these transitions varies dramatically. For example, some minibasins (14, 17, and 18) have no basin-floor fans in sequences younger than 3.0 Ma, whereas others (4, 5, 6, 10, 15, and 20) have fans in sequences younger than 1.0 Ma; moreover, there are numerous exceptions to the general trend. Eleven of the eighteen minibasins plotted display at least one transition in the opposite direction. The only systematic relationship is that there are no basin-floor fans after 0.7 Ma, as was shown on the reconstructed maps (Figure 3).

### *Spatial Patterns*

Careful comparison of Figure 13 with the map of minibasin distribution (Figure 12) shows no geographic pattern to the evolution of depositional environments. Although the oldest basin-floor fans appear to be confined to the eastern part of the study area (minibasins 14, 16, 17, and 18), this is an artifact of the lack of deep age control in western minibasins. There also is no clear pattern in the dip direction, as would be expected in a fill-and-spill scenario of basin evolution (e.g., Satterfield and Behrens, 1990). Some of the oldest basin-floor fans occur to the south (minibasins 16 and 17), whereas some of the youngest occur in northern minibasins (15 and 20).

### *Minibasin Type*

There is no clear association between minibasin type and the evolution of interpreted depositional environments (Figure 13). The patterns for each type are broadly similar. The only discernible relationship is that some evacuated stocks (minibasins 13, 19, and 20) may have unchannelized mud-rich flows at the base of the sequence prior to 3.0 Ma, whereas evacuated sheets are characterized only by basin-floor fans and channel-levee systems at this time. Initial deposition above sheets may have been older, such that early overbank deposits occurred prior to 5.5 Ma, the oldest correlated horizon.

### *Evolutionary Stage*

The transition ages between tectonostratigraphic packages are indicated in Figure 13. Again, there is no clear relationship to the evolution of depositional environments. Although geometric transitions coincide in eight cases with facies shifts that follow the general trend (e.g., from basin-floor fans

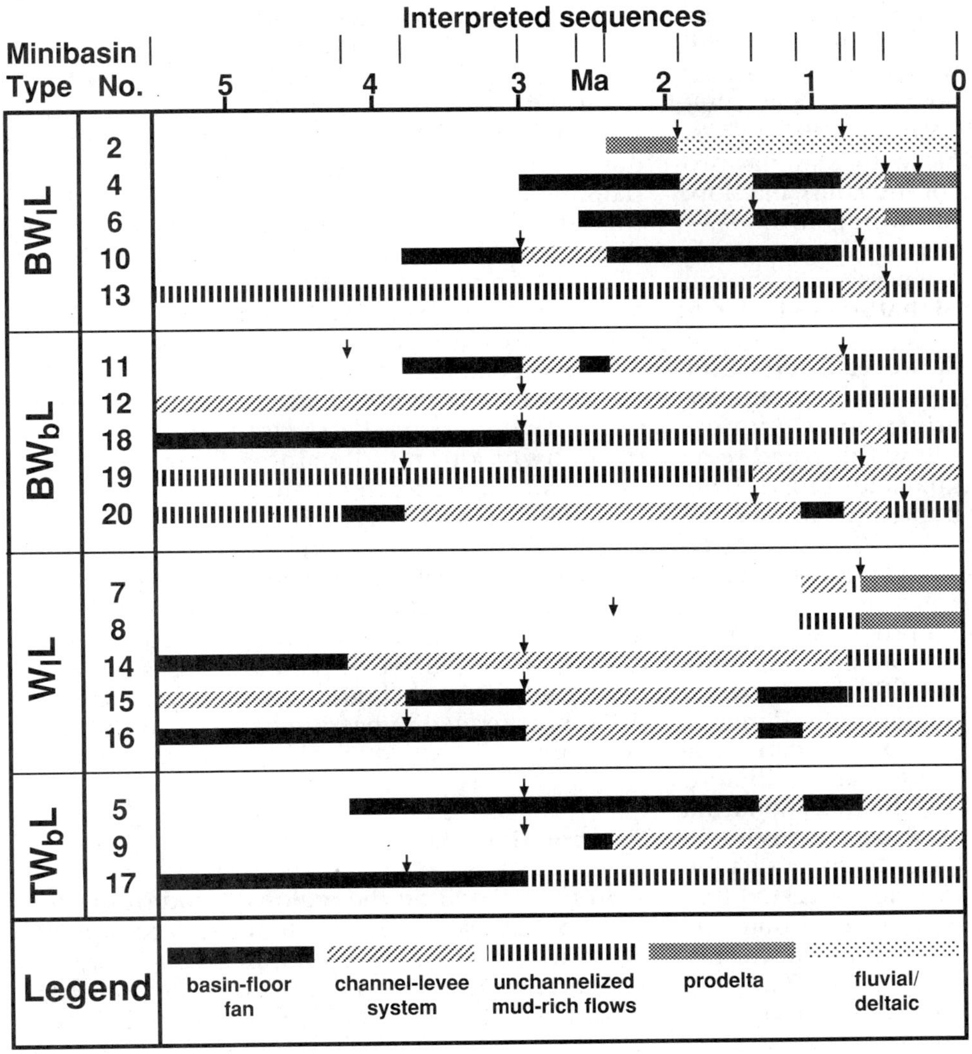


**Figure 13**—Chart summarizing the evolution of interpreted depositional environments within the different minibasins (facies data from Weimer et al., 1998, this issue). Within each minibasin, each sequence (shown at top) is coded with a patterned bar representing only the facies interpreted to occur at the base of the sequence; intrasequence facies variations are not shown. Arrows show the ages of transition from one tectonostratigraphic package to another; where only one arrow exists for the $BW_IL$, $BW_bL$, and $TW_bL$ minibasin types, the transition to a layer package is not observed. Minibasin number refers to numbers shown in Figures 1 and 12 and listed in Table 1.

to channel-levee systems), there are twelve instances of no facies change and one place where the facies shifts in the opposite direction (minibasin 6, from channel-levee to basin-floor fan). The transition from bowl- and trough-shaped packages to wedges apparently has little effect on depositional environment. Basin-floor fans are often found above these transitions; i.e., after initial weld formation (minibasins 5, 6, 9, 10, 11, 17, and 20). Accommodation could have been provided by growth faulting in the absence of salt, or by bathymetric lows surrounded by diapirs fed by salt from adjacent areas. We are not implying, however, that the presence or absence of underlying salt has no impact. We pointed out in the discussion of the reconstructed maps that (1) superimposing basin-floor fans over multiple sequences requires adequate underlying salt for accommodation and (2) Pliocene–Pleistocene basin-floor fans were relatively rare in minibasin 15 because underlying salt within this roho system had been evacuated by the end of the Miocene.

## DISCUSSION

There is a general stacking pattern of interpreted depositional environments in any given minibasin within the study area (Figures 3, 13) (see also Weimer et al., 1998). Deeper sequences are more likely to have basin-floor fans, overlying intervals often have well-developed channel-levee complexes, younger sequences are usually dominated by unchannelized mud-rich flows and slide/slump facies, and the youngest sediments (<0.8 Ma) in northern minibasins are characterized by clinoforms and topsets. There is a corresponding evolution in lithofacies and sand content as determined from well logs; deeper sequences tend to be more sand rich, with the sand occurring as sheets or amalgamated channels, and overlying strata have fewer and thinner sands interpreted as channel-fill, levee, and related deposits (Acosta, 1994; Martinez, 1994; Navarro, 1994; Weimer et al., 1994, 1998, this issue; Budhijanto, 1995; Varnai, 1996). These results are broadly compatible with a similar progression noted in other areas of the Louisiana slope (Pulham, 1993; Holman and Robertson, 1994; Yeilding and Apps, 1994), which has been termed a transition from "ponded" to "bypass" facies assemblages by Shell geologists (Mahaffie et al., 1995; Prather et al., 1998).

Figure 13 clearly illustrates, however, that the overall ponded to bypass stacking pattern is a simplification of a more complex stratigraphic evolution. Although some minibasins follow this trend (e.g., minibasin 14), most switch back and forth from ponded (basin-floor fan) to bypass (channel-levee and overbank) settings. Analysis of the data shows that a variety of simple models and relationships do not explain adequately the distribution of depositional environments: (1) there is no progressive downslope pattern in which more landward basins fill earlier and are then bypassed, (2) depositional environment is clearly not a function of sedimentation rate (Figure 5), (3) there is no systematic relationship between salt deformation and sediment transport and deposition, and (4) the transition from ponded to bypass facies was not synchronous in the study area. This latter point appears to contradict the results of Prather et al. (1998), who suggested that the transitions from ponded to bypass environments were nearly synchronous over large areas, occurring during two main episodes from 2.0 to 1.8 Ma and from 1.2 to 1.0 Ma; however, comparison of their results with ours is not straightforward. Our data, in fact, may fit with their model because we have addressed only the lowest facies in each sequence, whereas they defined facies assemblages. For example, a bypass facies assemblage may have a basin-floor fan at the base of the sequence (B. E. Prather, 1997, personal communication).

Our results suggest a general model in which there are actually three different scales of stratigraphic stacking that interact during the history of each minibasin. These scales are differentiated in terms of both time and areal extent: (1) a large-scale trend that is regional and characterizes the Pliocene–Pleistocene, (2) intermediate-scale effects that produce local variations over multiple sequences, and (3) small-scale stacking that occurs within each minibasin during each depositional sequence.

## Pliocene–Pleistocene (Regional) Scale

Although stratigraphic stacking within minibasins is complex, there was nevertheless a broad evolution from ponded to bypass settings and from sand-rich to shale-rich facies throughout the Pliocene–Pleistocene. Two key factors probably contributed to this pattern. First, the thickness of salt underlying minibasins gradually diminished due to sedimentary loading and evacuation into flanking diapirs and sheets (McBride et al., 1998, this issue). The resulting decrease of both subsidence (accommodation) and nearby diapirism (Figure 3) impacted sea-floor topography by reducing bathymetric relief, and sedimentation correspondingly evolved from ponded to bypass processes (see also Pulham, 1993; Mahaffie et al., 1995). Second, the overall progradation of the shelf meant that any given area gradually progressed from lower and middle slope environments to upper slope water depths before being buried beneath shelf and marginal-marine sediments. As a result, the regional slope gradient increased through time until the shelf margin prograded into the area (Villamil et al., 1998, this issue). Lower gradients characteristic of the lower to middle slope were more conducive to ponding and sand deposition (Figure 3C–H); higher gradients in the upper slope encouraged bypassing of sands to the south into the middle and lower slope (Figure 3I–L).

## Intermediate (Local) Scale

Superimposed on the regional trend from ponded to bypass settings are numerous examples of fluctuations within given minibasins (Figure 13). Similar interruptions to the larger scale ponded-to-bypass evolution have been observed to the east in the greater Mars basin in Mississippi Canyon (M. Mahaffie, 1996, personal communication). These fluctuations are occasionally similar between like minibasins. For example, minibasins 4 and 6 show almost identical stratigraphic stacking patterns (Figure 13). These two basins are both in the western part of the study area, equidistant from the shelf, and underlain by evacuated salt stocks (although the timing of transition from bowl- to wedge-shaped geometries differed). Such similarities are the exception rather than the rule; as discussed, there are no systematic patterns to the variations in stratigraphic stacking.

We suggest that the diverse evolutionary histories are due to numerous factors, both structural and sedimentological, that interact in complex ways to determine lithofacies at any time in each minibasin. These factors can be divided into three categories: external, regional, and local (Figure 14). Although they are first addressed here because their complicated interactions have the greatest influence at the local scale, we stress that they impact stratigraphic stacking patterns at all scales.

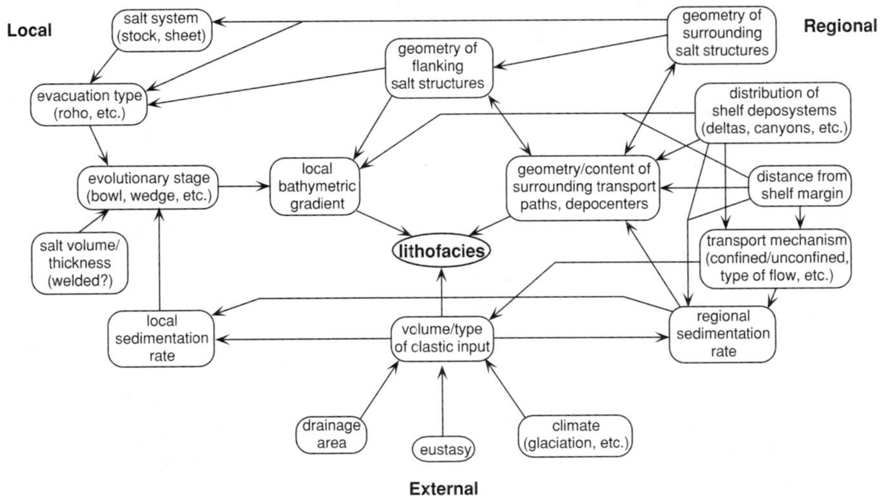


**Figure 14—Diagram showing some of the complex interactions among various structural and sedimentological factors that influence lithofacies development in a minibasin. Listed factors are divided into external, regional, and local groups, and are not intended to be exhaustive. See text for discussion.**

## *External Factors*

External factors are those that largely determine the nature of the clastic input into the shelf/slope system (Figure 14). These factors include the size of the drainage area captured by the ancestral Mississippi River, which increased periodically during the Pliocene–Pleistocene (Pulham, 1993; Prather et al., 1998); continental climatic fluctuations (primarily glacial); and eustasy. Such factors control, in part, the volume and timing of sediment input into the transport/deposition system, as well as the grain size distribution. Their impact is basinwide and thus should affect all minibasins during any interval. These factors may explain some of the changes that appear to have occurred throughout the study area at the same time; for example, the shift from more sand-rich strata prior to 3.0 Ma to more sand-poor strata after 1.4 Ma (Weimer et al., 1994, 1998, this issue), or the transition from ponded to bypass facies assemblages (Prather et al., 1998).

## *Regional Factors*

Regional factors are defined as those that influence the geometry and content of surrounding transport pathways and depocenters (Figure 14). This geometry is crucial in that sedimentation in a given minibasin is affected by nearby bathymetric lows, especially those updip. Conditions otherwise may be perfect for ponding of sheet sands, but if sands are trapped in landward basins or diverted away from the minibasin, only shale-dominated facies will be deposited. The surrounding basin/trough morphology is controlled primarily by the geometry of nearby salt structures, namely growing diapirs and sheets or stocks from which salt is being evacuated. Of course, these in turn are affected by sedimentary loading, so that interactions are mutual.

Equally important is the content of surrounding sedimentological systems. Even if the geometry of basins and troughs is ideal for funneling sands into a ponded minibasin, actual sand deposition is dependent on there being sufficient sand in the system at the time. This is controlled partly by external factors and partly by regional factors such as the distance of the minibasin from the shelf margin, the distribution of deltas and canyons at the shelf edge, and the types of failure/transport mechanisms (e.g., shelf-margin collapse, delta-related gravity flows, and episodic floods).

## *Local Factors*

Superimposed on both external and regional factors are those specific to each minibasin. These local factors influence the local bathymetric gradient (Figure 14); i.e., whether there is enough relief to favor ponded deposition, or whether sea-floor topography has been smoothed enough to favor bypass. These factors are primarily structural, related to the geometry of the salt system and related faults. The local factors include the type of underlying salt body (sheet or stock), the style of evacuation (roho vs. stepped counterregional, or ramp-fault rotation vs. counterregional collapse) (Figures 10, 11), the evolutionary stage (bowl, wedge, etc.), and the amount of salt remaining to be evacuated. Although these factors do not correlate directly to lithofacies (Figure 13), they can be important. For example, once a minibasin forming over a preexisting salt stock subsides enough to expel most of the salt and create a salt weld, ponding becomes more difficult because of the lack of accommodation and further subsidence; however, bathymetric relief may be maintained through faulting or because of flanking diapirs sourced from adjacent minibasins still underlain by sufficient salt, such that ponding is still possible.

## Sequence (Minibasin) Scale

We have already alluded to the stacking of seismic facies within individual sequences. Most sequences within a given minibasin are not characterized by just one facies, but instead show an upward progression that mimics the larger scale evolution from ponded to bypass settings. For example, many sequences are interpreted to have basin-floor fans (amalgamated sheet sands) immediately overlying the lower sequence boundary, followed by channel-levee systems, and then by unchannelized mud-rich flows and slumps/slides, and finally capped by a condensed section of hemipelagic sediments (Weimer et al., 1998, this issue). This pattern may be purely sedimentological in nature, but also may reflect in part structural controls.

### *Sedimentological Controls*

The vertical stacking of lithofacies within a single depositional sequence is fairly common. Stacking was first noted in multifold seismic data by Mitchum (1985), and in field exposures by Mutti (1985). Since then, more detailed studies have recognized similar facies stacking patterns in turbidite systems within sedimentary basins of different tectonic settings around the world (e.g., foreland basins: Mutti and Normark, 1987, 1991; Mutti et al., 1994; transform margins: Hewlett and Jordan, 1993; nonsalt-related divergent margins: Rossen and Sickafoose, 1994; Clemenceau, 1995; Brown et al., 1996; salt-related margins: Vail and Wornardt, 1991; Carminatti and Scarton, 1991; Pacht et al., 1992; Mitchum et al., 1993; Yeilding and Apps, 1994; unconfined deep-basin submarine fans: Weimer, 1990) or in generalized depositional models based on many different stratigraphic basins (Vail, 1987; Posamentier et al., 1991; Richards and Bowman, in press).

In general, these workers attributed the vertical facies stacking to decreases in both grain size and flow volumes during deposition of the sequence. The most sand-rich gravity flows are deposited in the basin at the beginning of a sea level cycle (i.e., during the lowstand), giving rise to sheetlike geometries. Transition to more mud-rich gravity flows leads to development of channel-levee systems, where flows tend to be more channelized and organized. Gradual decrease of sediment volume within individual gravity flows during a sea level cycle results in more hemipelagic sediments (mixed pelagic and gravity flows) and eventually a condensed section characterized by pelagic mud-rich sediments or foraminifera-rich oozes. Slides can develop anywhere within a sequence, originating from local slope failures, oversteepened delta fronts, or nearby salt-related bathymetric highs.

For the Pliocene–Pleistocene of the Gulf of Mexico, we interpret these vertical facies changes to be controlled by relative changes in sea level. During relative lowstands, delta systems are closer to the shelf edge, and more sediment is transported into slope minibasins. During times of relatively higher sea level, the shelf is flooded, and the amount of sediment transported to the slope area is significantly reduced. Although salt plays a dominant role in creating the minibasin where sediment was deposited, the actual facies changes within one depositional sequence, both vertically and laterally, probably are controlled more by sedimentological factors.

### *Structural Controls*

Having stated that facies stacking within a single depositional sequence is primarily a sedimentological feature, one of us (Rowan) suggests that the dynamics of salt withdrawal from beneath the minibasin also may play a role. The rate of basin subsidence is controlled by several variables. First, the driving force for salt flow into flanking diapirs is the differential fluid pressure set up by the large variance in thickness between the overburden above the deep salt and that above the shallow diapirs or sheets (Vendeville and Jackson, 1992). Countering this are viscous forces associated with salt flow along the narrowing source layer and the diapir conduits (Vendeville and Jackson, 1992). Thus, it is primarily a combination of the differential thickness of overburden already present and the thickness of underlying salt that determines the rate of salt evacuation. The implication is that salt flow from beneath a minibasin varies only gradually through time, and does not fluctuate significantly with high-frequency variations in added load.

A near-constant or smoothly varying subsidence rate has important ramifications for sedimentation and suggests a conceptual model for some structural control of facies evolution during one sea level cycle (Figure 15). We assume a constant rate of subsidence above an adequate thickness of salt, and the starting point is a minibasin with significant bathymetric relief at the time of the lower sequence boundary (Figure 15A). This relief enables ponding during the subsequent relative lowstand and deposition of basin-floor fans (Figure 15B). If the sedimentation rate is high relative to the subsidence rate, bathymetric relief will decrease during the lowstand as the minibasin fills, resulting in a transition to channel-levee systems (Figure 15C) and unchannelized mud-rich flows (Figure 15D). As sedimentation slows during relative sea level rise, it is eventually overtaken by the constant subsidence rate such that minibasin relief starts to increase again (Figure 15E). Recent unconsolidated sediments on the basin flanks are uplifted and tilted, resulting in local slumping into the minibasin (slumps and slides are more often found in the mid to upper parts of individual sequences; Weimer et al., 1998, this issue). Further subsidence leads to increased relief that is draped by condensed pelagic shales during the highstand (Figure 15F), reestablishing the starting configuration (Figure 15A) for the next phase of lowstand sedimentation.

A similar model has been formulated by Shell geologists, in which decreasing rates of sedimentation relative to subsidence produce a transition from ponded to bypass settings during one sea level cycle (B. E. Prather, 1997, personal communication). As we have already discussed, however, sedimentological factors are also important in determining facies stacking within a given sequence. Unfortunately, the relative magnitude of sedimentological and structural factors cannot be identified from the available data. In addition, minibasin- and sequence-scale evolution is superimposed upon local- and regional-scale patterns such that the net stratigraphic stacking in any given minibasin is complex and varies from that observed in adjacent minibasins.

## CONCLUSIONS

Salt deformation and sedimentation in slope minibasins of the northern Gulf of Mexico are intricately linked. Allochthonous salt evolution has been driven and controlled to a large degree by the enormous progradational clastic load deposited throughout the Cenozoic in the northern Gulf of Mexico. In turn, evolving salt geometries dictate the morphology of minibasins and sediment transport pathways; i.e., the template in which stratigraphic stacking patterns develop.

We have shown that the distribution of interpreted seismic facies in time and space is compatible with the reconstructed history of salt tectonics. Salt deformation controls the stratigraphic architecture in two ways: (1) Salt diapirism creates bathymetric relief that defines minibasin locations and shapes as well as the geometry of topographic lows that link and feed the minibasins, and (2) evacuation of deep salt provides much of the accommodation for sediment deposition. Minibasins that are partially confined by surrounding salt bodies and associated faults and that are underlain by

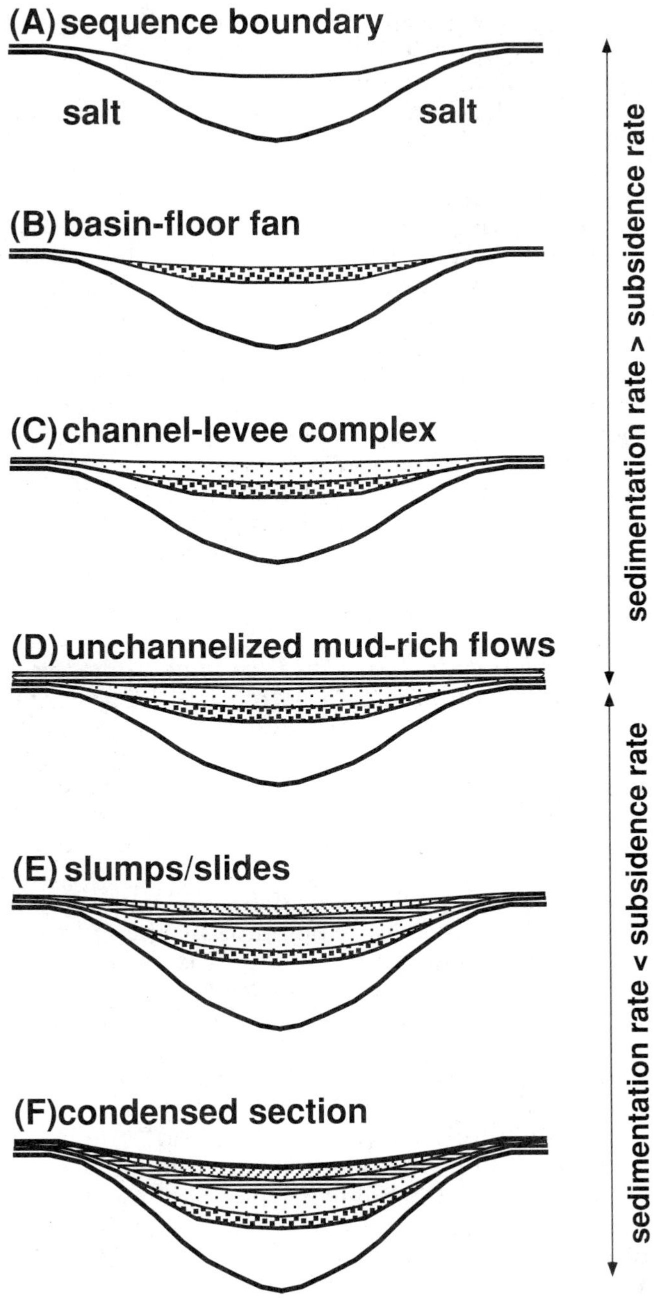


**Figure 15—Simplified schematic diagram showing possible structural influence on facies development during one sea level cycle within a slope minibasin underlain by salt and flanked by growing diapirs (heavy line is top salt, white indicates older minibasin fill). Salt withdrawal and subsidence are assumed to be constant with time. Variations in sedimentation rate result first in decreasing bathymetric relief during the lowstand (A–D), and then in increasing relief (D–F). Sedimentation shifts correspondingly from ponded basin-floor fans to bypass channel-levee systems and unchannelized mud-rich flows, to slumping as relief increases, and finally to condensed deposition during the highstand. Note that this stacking pattern is probably controlled primarily by sedimentological factors (see text). Figure not to scale.**

sufficient salt are the preferred sites for repeated stacking of ponded, sand-rich facies; however, there are many other variables that influence lithofacies development through time, such that otherwise appropriate areas may be sand starved, and less suitable environments may be filled by basin-floor fans. These variables include external factors that define the clastic input into the entire depositional system, regional factors that control the distribution of these clastics, and local factors that determine specific deposition in a given minibasin. The various factors interact at three different, superimposed time scales: (1) a broad transition from ponded to bypass settings during the Pliocene–Pleistocene, (2) fluctuations from ponded to bypass and back that occur over periods representing multiple sequences, and (3) a shift from ponded to bypass environments that occurs during each sea level cycle in most minibasins. The result is a complex interaction that usually produces different stratigraphic stacking patterns in adjacent minibasins.

Although the relationships between salt deformation and sedimentation are complicated, the kind of analysis presented here provides a better understanding of the principal variables that control lithofacies development. Of the many factors shown in Figure 14, some are more important than others. We have identified three questions that are critical and that should be addressed.

(1) Where and when was there adequate sand in the system? The answer depends on the external factors and regional factors, such as the distance from the shelf margin and the distribution of deltas and shelf-margin canyons. In other words, this is simply the regional stratigraphic history that can be derived from published studies and regional analyses of well logs.

(2) Was there adequate salt beneath the minibasin (or fault growth) to provide accommodation for ponding? This answer can be derived from quantitative cross section restoration that incorporates isostasy and paleobathymetry (e.g., Rowan, 1993; McBride et al., 1998, this issue). Alternatively, analysis of tectonostratigraphic packages within a minibasin may be used to infer the timing of initial weld formation as proposed in this paper.

(3) What was the subregional evolution of salt bodies and minibasins? The evolution of the area surrounding a minibasin should be reconstructed in four dimensions (at least schematically) to understand the history of sediment transport and deposition patterns. The area should not only include landward diapirs, sheets, and minibasins, but also should extend laterally and basinward to understand the influence of nearby structures. Such efforts are especially critical in subsalt exploration because the three-dimensional morphology of salt bodies and minibasins prior to sheet emplacement would have been very different from that observed today.

Of course, addressing these issues will not provide a unique and correct interpretation of the stratigraphic history of a given minibasin; however, by answering these questions, geologists can make improved predictions about facies distribution in space and time, and can identify scenarios in which sand-rich lithofacies are more probable. This type of analysis will allow the associated risks to be assessed more accurately and reduced, which is especially important in the frontier subsalt and deep-water Gulf of Mexico plays.

## REFERENCES CITED

Acosta, Z., 1994, Sequence stratigraphy of Plio–Pleistocene sediments in north-central Green Canyon and western Ewing Bank, northern Gulf of Mexico: M.S. thesis, University of Colorado, Boulder, Colorado, 187 p.

Apps, G. M., F. J. Peel, C. J. Travis, and C. A. Yeilding, 1994, Structural controls on Tertiary deep water deposition in the northern Gulf of Mexico, *in* P. Weimer, A. H. Bouma, and B. F. Perkins, eds., Submarine fans and turbidite systems: SEPM Gulf Coast Section 15th Annual Research Foundation Conference, p. 1-7.

Brown, L. F., Jr., J. M. Benson, G. J. Brink, S. Doherty, A. Jollands, E. H. A. Jungslager, J. H. G. Kennan, A. Muntingh, and N. J. S. van Wyk, 1996, Sequence stratigraphy in offshore South African divergent basins—an atlas on exploration for Cretaceous lowstand traps by Soekor (Pty) Ltd.: AAPG Studies in Geology 41, 184 p.

Budhijanto, F. M., 1995, Sequence stratigraphy of Plio–Pleistocene strata of the northeastern Green Canyon and eastern Ewing Bank areas, northern Gulf of Mexico: M.S. thesis, University of Colorado, Boulder, Colorado, 154 p.

Carminatti, M., and J. C. Scarton, 1991, Sequence stratigraphy of the Oligocene turbidite complex of the Campos Basin, offshore Brazil: an overview, *in* P. Weimer and M. H. Link, eds., Seismic facies and sedimentary processes of submarine fans and turbidite systems: New York, Springer-Verlag, p. 241-246.

Clemenceau, G. R., 1995, Ram/Powell field: Viosca Knoll Block 912, deepwater Gulf of Mexico, *in* R. D. Winn, Jr., and J. M. Armentrout, eds., Turbidites and associated deep-water facies: SEPM Core Workshop 20, p. 95-129.

Diegel, F. A., J. F. Karlo, D. C. Schuster, R. C. Shoup, and P. R. Tauvers, 1995, Cenozoic structural evolution and tectono-stratigraphic framework of the northern Gulf Coast continental margin, *in* M. P. A. Jackson, D. G. Roberts, and S. Snelson, eds., Salt tectonics: a global perspective: AAPG Memoir 65, p. 109-151.

Hewlett, J. S., and D. W. Jordan, 1993, Stratigraphic and combination traps within a seismic sequence framework, Miocene Stevens turbidites, Bakersfield arch, California, *in* P. Weimer and H. W. Posamentier, eds., Siliciclastic sequence stratigraphy: AAPG Memoir 58, p. 135-162.

Holman, W. E., and S. S. Robertson, 1994, Field development, depositional model, and production performance of the turbiditic "J" sands at prospect Bullwinkle, Green Canyon 65 field, outer-shelf Gulf of Mexico, *in* P. Weimer, A. H. Bouma, and B. F. Perkins, eds., Submarine fans and turbidite systems: SEPM Gulf Coast Section 15th Annual Research Foundation Conference, p. 139-150.

Lee, G. H., J. S. Watkins, and W. R. Bryant, 1996, Bryant Canyon fan system: an unconfined, large river-sourced system in the northwestern Gulf of Mexico: AAPG Bulletin, v. 80, p. 340-358.

Mahaffie, M. J., 1994, Reservoir classification for turbidite intervals at the Mars discovery, Mississippi Canyon 807, Gulf of Mexico, *in* P. Weimer, A. H. Bouma, and B. F. Perkins, eds., Submarine fans and turbidite systems: SEPM Gulf Coast Section 15th Annual Research Foundation Conference, p. 233-244.

Mahaffie, M. J., R. S. Barnard, J. W. Kendrick, and L. B. Franceware, 1995, The interaction between accommodation space and sediment supply on depositional fill patterns within the greater Mars basin, deep-water Gulf of Mexico, U.S.A. (abs.): AAPG Annual Convention Program, p. 59A.

Martinez, R. E., 1994, Sequence stratigraphy of late Pliocene and Pleistocene sediments of northwestern Green Canyon area, northern Gulf of Mexico: M.S. thesis, University of Colorado, Boulder, Colorado, 138 p.

McBride, B. C., 1997, The geometry and kinematics of alloch thonous salt systems and their impact on petroleum systems, northern Gulf of Mexico Basin: studies in three- and four-dimensional analysis: Ph.D. dissertation, University of Colorado, Boulder, Colorado, 185 p.

McBride, B. C., M. G. Rowan, and P. Weimer, 1998, The evolution of allochthonous salt systems, northern Green Canyon and Ewing Bank (offshore Louisiana), northern Gulf of Mexico: AAPG Bulletin, v. 82, no. 5B, p. 1013-1036.

Mitchum, R. M., Jr., 1985, Seismic stratigraphic recognition of submarine fans, *in* O. R. Berg and D. G. Woolverton, eds., Seismic stratigraphy II: AAPG Memoir 39, p. 117-136.

Mitchum, R. M., Jr., P. R. Vail, and J. B. Sangree, 1977, Seismic stratigraphy and global changes of sea level, part 6: stratigraphic interpretation of seismic reflection patterns in depositional sequences: AAPG Memoir 26, p. 117-133.

Mitchum, R. M., Jr., J. B. Sangree, P. R. Vail, and W. W. Wornardt, 1993, Recognizing sequences and systems tracts from well logs, seismic data, and biostratigraphy: examples from the late Cenozoic of the Gulf of Mexico, *in* P. Weimer and H. W. Posamentier, eds., Siliciclastic sequence stratigraphy: recent developments and applications: AAPG Memoir 58, p. 163-197.

Mutti, E., 1985, Turbidite systems and their relations to depositional sequences, *in* G. G. Zuffa, ed., Provenance of arenites: Dordrecht, Reidel, p. 65-93.

Mutti, E., and W. R. Normark, 1987, Comparing modern and ancient turbidite systems: problems and concepts, *in* J. Leggett and G. G. Zuffa, eds., Marine and clastic sedimentology: concepts and case studies: London, Graham and Trotman, p. 1-38.

Mutti, E., and W. R. Normark, 1991, An integrated approach to the study of turbidite systems, *in* P. Weimer and M. H. Link, eds., Seismic facies and sedimentary processes of submarine fans and turbidite systems: New York, Springer-Verlag, p. 75-106.

Mutti, E., G. Davoli, S. Mora, and L. Papani, 1994, Internal stacking patterns of ancient turbidite systems from collisional basins, *in* P. Weimer, A. H. Bouma, and B. F. Perkins, eds., Submarine fans and turbidite systems: SEPM Gulf Coast Section 15th Annual Research Foundation Conference, p. 257-268.

Navarro, A. F., 1994, Sequence stratigraphy of Pleistocene sediments of northwestern Green Canyon area, northern Gulf of Mexico: M.S. thesis, University of Colorado, Boulder, Colorado, 108 p.

Pacht, J. A., B. Bowen, B. L. Shaffer, and W. R. Pottorf, 1992, Systems tracts, seismic facies, and attribute analysis within a sequence stratigraphic framework—example from the offshore Louisiana Gulf Coast, *in* E. G. Rhodes and T. Moslow, eds., Marine clastic reservoirs: New York, Springer-Verlag, p. 21-39.

Posamentier, H. W., R. D. Erskine, and R. M. Mitchum, Jr., 1991, Models for submarine fan deposition within a sequence stratigraphic framework, *in* P. Weimer and M. H. Link, eds., Seismic facies and sedimentary processes of submarine fans and turbidite systems: New York, Springer-Verlag, p. 127-136.

Prather, B. E., J. R. Booth, G. S. Steffens, and P. A. Craig, 1998, Classification, lithologic calibration, and stratigraphic succession of seismic facies in intraslope basins, deep-water Gulf of Mexico: AAPG Bulletin, v. 82, no. 5A, p. 701-728.

Pratson, L. F., and W. B. F. Ryan, 1994, Pliocene to Recent infilling and subsidence of intraslope basins, offshore Louisiana: AAPG Bulletin, v. 78, p. 1483-1506.

Pulham, A. J., 1993, Variations in slope deposition, Plio-Pleistocene, offshore Louisiana, north-east Gulf of Mexico, *in* P. Weimer and H. W. Posamentier, eds., Siliciclastic sequence stratigraphy: recent developments and applications: AAPG Memoir 58, p. 199-234.

Richards, M., and M. Bowman, in press, Sequence stratigraphy of deep-marine clastic systems: Marine and Petroleum Geology.

Risch, D. L., A. N. Chowdhury, and A. E. Hannan, 1994, Regional depositional history of the Miocene-Pleistocene Louisiana slope, Green Canyon-Mississippi Canyon: Transactions of the Gulf Coast Association of Geological Societies, v. 44, p. 611-617.

Rossen, C., and D. K. Sickafoose, 1994, 3-D seismic expression and architecture of deep-water reservoirs at Ram/Powell field, Viosca Knoll Block 912, Gulf of Mexico, *in* P. Weimer, A. H. Bouma, and B. F. Perkins, eds., Submarine fans and turbidite systems: SEPM Gulf Coast Section 15th Annual Research Foundation Conference, p. 309-310.

Rowan, M. G., 1993, A systematic technique for the sequential restoration of salt structures: Tectonophysics, v. 228, p. 331-348.

Rowan, M. G., 1995, Structural styles and evolution of allochthonous salt, central Louisiana outer shelf and upper slope, *in* M. P. A. Jackson, D. G. Roberts, and S. Snelson, eds., Salt tectonics: a global perspective: AAPG Memoir 65, p. 199-228.

Rowan, M. G., P. Weimer, and P. B. Flemings, 1994, Three-dimensional geometry and evolution of a composite, multi-level salt system, western Eugene Island, offshore Louisiana: Transactions of the Gulf Coast Association of Geological Societies, v. 44, p. 641-648.

Rowan, M. G., B. C. McBride, and P. Weimer, 1995a, The role of extension in the evolution of allochthonous salt, central Louisiana outer shelf and upper slope, *in* C. J. Travis, H. Harrison, M. R. Hudec, B. C. Vendeville, F. J. Peel, and B. F. Perkins, eds., Salt, sediment and hydrocarbons: SEPM Gulf Coast Section 16th Annual Research Foundation Conference, p. 223-233.

Rowan, M. G., P. Weimer, B. C. McBride, P. Varnai, Z. Acosta, F. Budhijanto, R. Martinez, A. Navarro, T. Villamil, and J. C. Fiduk, 1995b, Interactions between salt tectonics and sedimentation in the Louisiana offshore: preliminary results, *in* C. J. Travis, H. Harrison, M. R. Hudec, B. C. Vendeville, F. J. Peel, and B. F. Perkins, eds., Salt, sediment and hydrocarbons: SEPM Gulf Coast Section 16th Annual Research Foundation Conference, p. 235-244.

Satterfield, W. M., and E. W. Behrens, 1990, A late Quaternary canyon/channel system, northwest Gulf of Mexico continental slope: Marine Geology, v. 92, p. 51-67.

Schuster, D. C., 1995, Deformation of allochthonous salt and evolution of related salt/structural systems, eastern Louisiana Gulf Coast, *in* M. P. A. Jackson, D. G. Roberts, and S. Snelson, eds., Salt tectonics: a global perspective: AAPG Memoir 65, p. 177-198.

Snyder, F. C., and J. A. Nugent, 1995, Teak—testing a subsalt hydrocarbon trap geometry, South Timbalier Block 260, Gulf of Mexico, *in* C. J. Travis, H. Harrison, M. R. Hudec, B. C. Vendeville, F. J. Peel, and B. F. Perkins, eds., Salt, sediment and hydrocarbons: SEPM Gulf Coast Section 16th Annual Research Foundation Conference, p. 257-267.

Sumner, H. S., B. A. Robison, W. K. Dirks, and J. C. Holliday, 1991, Structural style of salt/mini-basin systems: lower shelf and upper slope, central offshore Louisiana (abs.): Transactions of the Gulf Coast Association of Geological Societies, v. 41, p. 582.

Swanson, J. P., and W. R. Bryant, 1992, Sediment pathways of the northwest Green Canyon area, offshore Louisiana: Transactions of the Gulf Coast Association of Geological Societies, v. 42, p. 707-715.

Vail, P. R., 1987, Seismic stratigraphy interpretation procedure, *in* A. W. Bally, ed., Atlas of seismic stratigraphy: AAPG Studies in Geology 27, v. 1, p. 1-10.

Vail, P. R., and W. W. Wornardt, 1991, An integrated approach to exploration and development in the 90s: well log-seismic sequence stratigraphy analysis: Transactions of the Gulf Coast Association of Geological Societies, v. 41, p. 630-650.

Varnai, P., 1996, Integrated 2-D and 3-D sequence stratigraphic analysis of Plio-Pleistocene turbidite systems, northern Green Canyon and central Ewing Bank protraction areas, northern Gulf of Mexico: Ph.D. dissertation, University of Colorado, Boulder, Colorado, 503 p.

Vendeville, B. C., and M. P. A. Jackson, 1992, The rise of diapirs during thin-skinned extension: Marine and Petroleum Geology, v. 9, p. 331-353.

Vendeville, B. C., and K. T. Nilsen, 1995, Episodic growth of salt diapirs driven by horizontal shortening, *in* C. J. Travis, H. Harrison, M. R. Hudec, B. C. Vendeville, F. J. Peel, and B. F. Perkins, eds., Salt, sediment and hydrocarbons: SEPM Gulf Coast Section 16th Annual Research Foundation Conference, p. 285-295.

Villamil, T., C. Arango, P. Weimer, A. Waterman, M. G. Rowan, P. Varnai, A. J. Pulham, and J. R. Crews, 1998, Biostratigraphic techniques for analyzing benthic biofacies, stratigraphic condensation, and key surface identification, Pliocene and Pleistocene sediments, northern Green Canyon and Ewing Bank (offshore Louisiana), northern Gulf of Mexico: AAPG Bulletin, v. 82, no. 5B, p. 961-985.

Weimer, P., 1990, Sequence stratigraphy, seismic geometries, and depositional history of the Mississippi Fan, deep Gulf of Mexico: AAPG Bulletin, v. 74, p. 425-453.

Weimer, P., P. Varnai, A. Navarro, Z. Acosta, F. Budhijanto, R. Martinez, M. G. Rowan, B. McBride, and T. Villamil, 1994, Sequence stratigraphy of Neogene turbidite systems, Green Canyon and Ewing Bank, northern Gulf of Mexico: preliminary results, *in* P. Weimer, A. H. Bouma, and B. F. Perkins, eds., Submarine fans and turbidite systems: SEPM Gulf Coast Section 15th Annual Research Foundation Conference, p. 383-399.

Weimer, P., P. Varnai, F. M. Budhijanto, Z. M. Acosta, R. E. Martinez, A. F. Navarro, M. G. Rowan, B. C. McBride, T. Villamil, C. Arango, and A. J. Pulham, 1998, Sequence stratigraphy of Pliocene and Pleistocene turbidite systems, northern Green Canyon and Ewing Bank (offshore Louisiana), northern Gulf of Mexico: AAPG Bulletin, v. 82, no. 5B, p. 918-960.

Winker, C. D., 1996, High-resolution seismic stratigraphy of a late Pleistocene submarine fan ponded by salt-withdrawal mini-basins on the Gulf of Mexico continental slope: Offshore Technology Conference Program with Abstracts, p. 619-628.

Yeilding, C. A., and G. M. Apps, 1994, Spatial and temporal variations in the facies associations of depositional sequences on the slope: examples from the Miocene-Pleistocene of the Gulf of Mexico, *in* P. Weimer, A. H. Bouma, and B. F. Perkins, eds., Submarine fans and turbidite systems: SEPM Gulf Coast Section 15th Annual Research Foundation Conference, p. 425-437.

Yeilding, C. A., C. J. Travis, D. M. Urban, and E. J. Ekstrand, 1995, Geologic evolution of the Late Jurassic-Recent Bogart Basin, northeastern Gulf of Mexico: an integrated study of depositional architecture and salt tectonics in a continental slope setting (abs.): AAPG Annual Convention Program, p. 107A.

## ABOUT THE AUTHORS

### Mark G. Rowan

Mark Rowan is a graduate of Caltech (1976), University of California, Berkeley (1982), and the University of Colorado (1991), where he is currently an assistant research professor in structural geology. He has also worked in the petroleum industry, both as an exploration geologist for Sohio and as a consultant with Geo-Logic Systems and Alastair Beach Associates. His research interests focus on the three-dimensional geometry and evolution of both extensional and contractional structures, with a recent emphasis on salt tectonics.

### Paul Weimer

Paul Weimer is an associate professor and director of the Energy and Minerals Applied Research Center at the University of Colorado at Boulder. His research interests are in sequence stratigraphy and basin analysis, three-dimensional seismic interpretation, and reservoir geology.

# The Effect of Allochthonous Salt on the Petroleum Systems of Northern Green Canyon and Ewing Bank (Offshore Louisiana), Northern Gulf of Mexico[1]

**Barry C. McBride,[2] Paul Weimer,[3] and Mark G. Rowan[3]**

## ABSTRACT

The northern Green Canyon/Ewing Bank region, northern Gulf of Mexico basin, contains the Oxfordian-Neogene (.), Tithonian-Neogene (.), Albian-Neogene (.), Turonian-Neogene (.), and Eocene-Neogene (.) petroleum systems. The systems encompass 42 fields or discoveries in the study area and include four subsalt discoveries. Essential elements of the systems include source shales of Oxfordian, Tithonian, Albian, Turonian, and Eocene age; Neogene siliciclastic turbidite reservoirs; allochthonous salt; and overburden strata ranging in age from Jurassic to Quaternary.

The petroleum systems of the area are significantly affected by the evolution of allochthonous salt. The high thermal conductivity of salt retards the thermal maturation of subsalt petroleum source rocks and causes late generation and migration from them. Most traps were formed during the Pliocene-Pleistocene, and the generation-migration-accumulation of petroleum ranges from early Miocene to the Holocene. The critical moment of peak oil generation for each source varies spatially and temporally as a function of the overlying sediments and allochthonous salt evolution.

The impermeability of salt prevents vertical petroleum migration and causes migration pathways to be deflected laterally up the dip of base salt. Where salt welds form, petroleum migration is unimpeded and continues vertically. By integrating predictions of potential source rocks, structural restorations, thermal maturation modeling, regional salt maps, and petroleum systems logic, we can determine petroleum migration pathways and zones of concentration. All 42 fields or discoveries within the study area are associated with predicted zones of paleosubsalt petroleum concentration. Present-day salt geometries do not delineate many of these zones because of salt weld formation during the Pleistocene. This generation, migration, and accumulation technique enables geoscientists to focus their exploration efforts toward areas with a greater probability of success.

## INTRODUCTION

Although the northern Gulf of Mexico Basin is one of the most extensively studied and explored sedimentary basins in the world, no consensus exists regarding the mechanics of its petroleum systems. This lack of consensus is primarily the result of the basin's large size, thickness of strata, and complex structural evolution. In the offshore regions, much of the data needed for a petroleum systems analysis are difficult to compile because of the relatively shallow well penetrations (source rocks occur at much greater depths), the ambiguity of deep structure, difficult stratigraphic correlations, and the



[1]Manuscript received September 30, 1996; revised manuscript received October 1, 1997; final acceptance November 19, 1997.

[2]Energy and Minerals Applied Research Center, Department of Geological Sciences, Campus Box 399, University of Colorado, Boulder, Colorado 80309-0250. Present address: HS Resources Inc., 1999 Broadway, Suite 3600, Denver, Colorado 80202.

[3]Energy and Minerals Applied Research Center, Department of Geological Sciences, Campus Box 399, University of Colorado, Boulder, Colorado 80309-0250.

We thank the following companies and their representatives for their tremendous support and input as part of the Gulf of Mexico research consortium at the University of Colorado: AGIP, Amoco, Anadarko, BHP, BP Exploration, Burlington Resources, Canadian Occidental, Chevron USA, CNG, Conoco, Enterprise, Exxon, Marathon, Mobil, Occidental, PanCanadian, Pennzoil, Petrobras, Phillips Petroleum, Shell Oil Company, Texaco, Total, Union Pacific, and Unocal. We extend special thanks to Halliburton Geophysical (now Western Geophysical), specifically to Bob Graebner, John Anderson, and Gary White, for the use of seismic data used in the project. We are indebted to PaleoData Inc. and Art Waterman for use of biostratigraphic data. Software donations from GeoQuest, CogniSeis Development Inc., Landmark, Platte River Associates, and Zeh Graphics are essential to the success of our program and very much appreciated. Special thanks are in order to Bob Ratliff, Jay Leonard, China Leonard, Doug Waples, Fred Meissner, Al Koch, Thane McCulloh, Andy Pulham, Tom McKenna, and George Gail for their support, advice, discussions, training, and input during the development of this research over the years. We appreciate reviews by Mike Lewan, Bill Galloway, an anonymous reviewer, and especially Chris Travis for his particularly thorough review. We thank Jonathan Irick, Jennifer Crews, and Ryan Crow for drafting. Acknowledgment is made to the Donors of the Petroleum Research Fund, administered by the American Chemical Society, for partial support for this research.

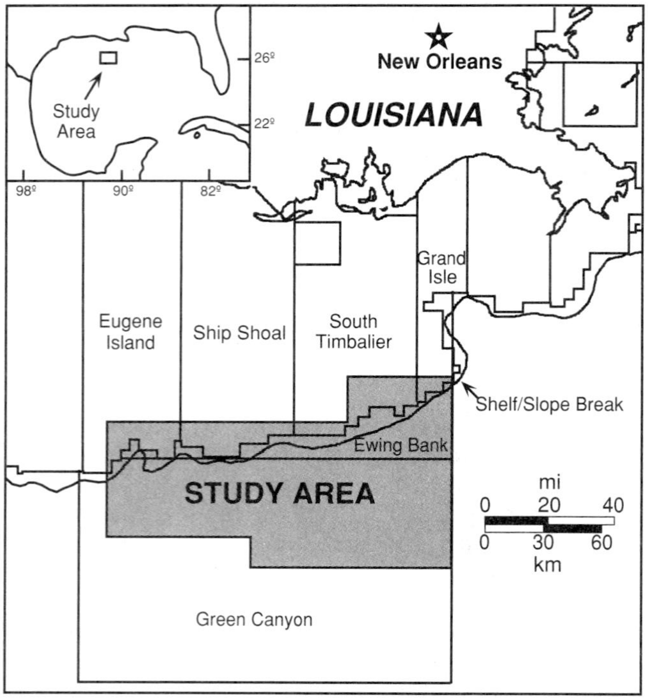


**Figure 1—Location map of study area, northern Green Canyon and Ewing Bank, northern Gulf of Mexico.**

generally poor understanding of the evolution of specific allochthonous salt systems.

The northern Green Canyon/Ewing Bank area, offshore Louisiana (Figure 1), is a location where improved understanding of the structural evolution of allochthonous salt features now allows us to conduct petroleum systems analysis with greater confidence (McBride et al., 1998, this issue). This region is a developing frontier exploration petroleum province with 38 fields or discoveries in suprasalt Pliocene–Pleistocene turbidite-related reservoirs (Figure 2; Table 1) (Weimer et al., 1998a, this issue). The region also includes active subsalt exploration with four subsalt discoveries: Mahogany (Ship Shoal 349), Agate (Ship Shoal 361), Teak (South Timbalier 260), and North Lobster (South Timbalier 308) in Miocene(?), turbidite-related reservoirs (Figure 2; Table 1). The area has experienced a complex structural history with multiple phases of extrusion and evacuation of allochthonous salt sheets and canopies (e.g., Rowan, 1995; McBride et al., 1998, this issue). Although the area is considered to have excellent potential for additional discoveries in the suprasalt, subsalt, and deeper water regions, relatively little has been published about the petroleum source rocks and their timing of maturation and expulsion, or secondary migration pathways. In particular, the effect of the evolution of allochthonous salt on the expulsion-migration subsystems has not been addressed.

Understanding the evolution of allochthonous salt is critical to petroleum systems analysis in the Gulf of Mexico because it affects (1) retardation of thermal maturation in subsalt petroleum source rocks and (2) subsalt petroleum migration pathways. Thermal maturation of subsalt source rocks is retarded due to the significantly higher thermal conductivity of salt compared to laterally adjacent siliciclastic strata; therefore, allochthonous salt transfers heat more effectively from deeper strata and keeps the underlying strata anomalously cool (O'Brien and Lerche, 1988; Mello et al., 1994; Waples, 1994a). Subsalt petroleum migration pathways are affected because the impermeable salt deflects migration pathways laterally up the dip of the base salt near the salt/sediment interface. By integrating predictions of potential source rocks, structural restorations, thermal maturation modeling, and petroleum systems logic (Smith, 1994), it is possible to delineate spatially and temporally thermal maturity, petroleum migration pathways, and, most importantly, zones of focused petroleum migration and accumulation through time.

The purpose of this paper is to illustrate how an integrated structural and stratigraphic interpretation and quantification of the incremental evolution of allochthonous salt constrains maturation modeling and petroleum systems analyses within the Gulf of Mexico. The analysis of the expulsion-migration subsystems of each petroleum system in this manner enables an exploration program to focus its efforts on regions with a much greater probability of petroleum accumulation and discoveries. In this paper, we intend to stimulate thinking regarding the implementation of petroleum system analysis toward exploration in the Gulf of Mexico, as well as toward other salt-related basins in the world.

## SEDIMENTARY BASIN SETTING

The northern Green Canyon/Ewing Bank study area is contained within the north-central portion of the Gulf of Mexico Basin. The basin is approximately 1500 km (930 mi) in diameter, with Upper Triassic–Holocene sedimentary rocks reaching thicknesses of 10–15 km (6–9 mi) (Salvador, 1991a).

### Structural Setting and Evolution

The Gulf of Mexico Basin formed as a consequence of rifting and attendant crustal stretching and thinning during the Late Triassic–Early Jurassic as the North American plate separated from the African and South American plates (e.g., Pilger, 1981; Pindell, 1985; Sawyer et al., 1991). Beta

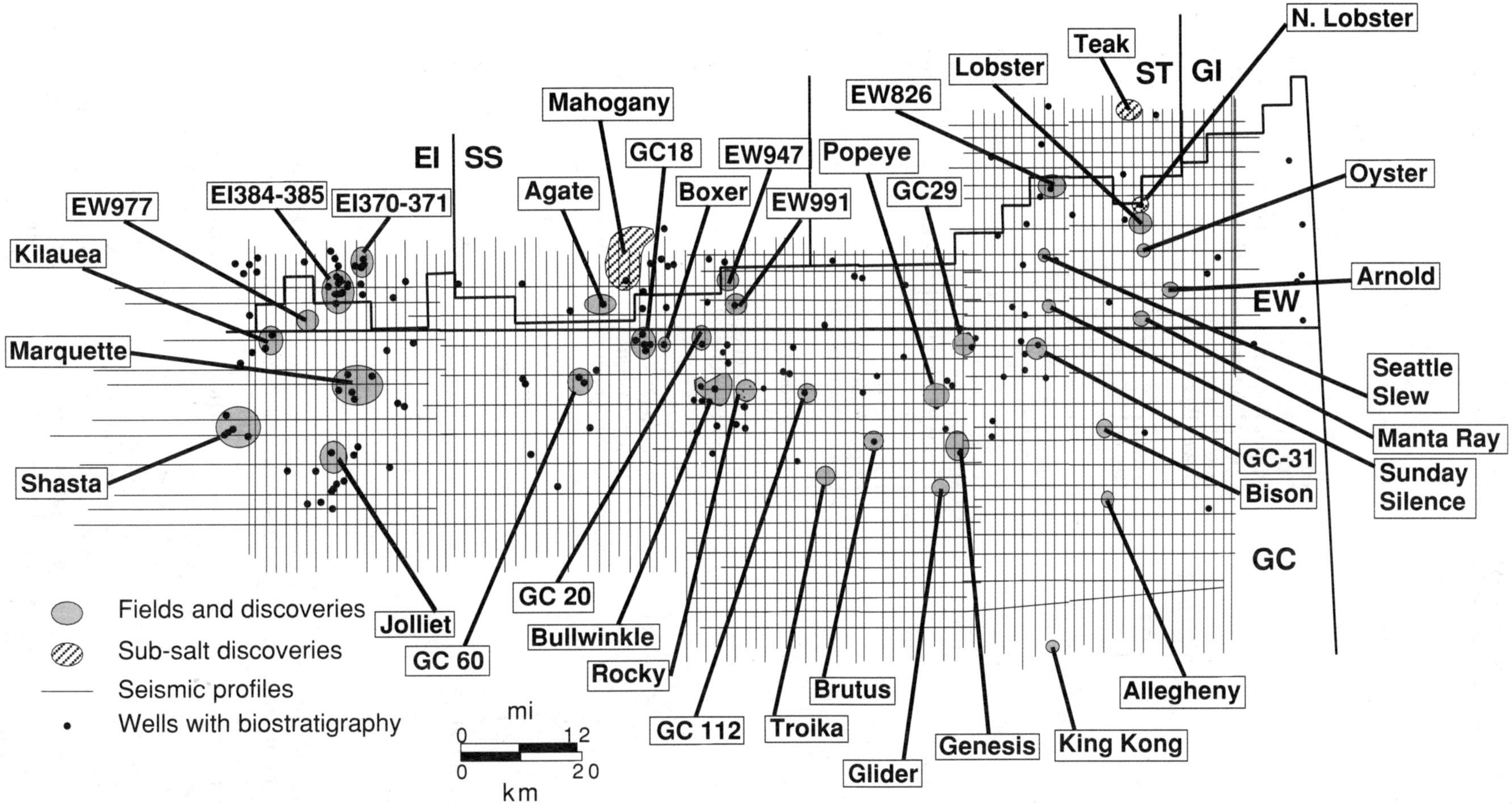


**Figure 2—Map showing distribution of two-dimensional seismic data, wells with biostratigraphy, and discoveries or fields in study area. See Table 2 for summary of fields.**

**Table 1. Fields and Discoveries of the Northern Green Canyon/Ewing Bank Study Area**

| Name | Location | Date | Reservoir | Water Depth (m) |
|---|---|---|---|---|
| Agate | SS361 | 1996 | Miocene(?) | 112 |
| Allegheny | GC254 | 1985 | Lower Pliocene | 983 |
| Arnold | EW963 | 1996 | Lower Pliocene | 530 |
| Bison | GC166 | 1986 | Lower Pliocene | 726 |
| Boxer | GC19 | 1980 | Upper Pliocene | 252 |
| Brutus | GC158 | 1986 | Upper Pliocene/lower Pleistocene | 866 |
| Bullwinkle | GC65/109 | 1983 | Upper Pliocene/lower Pleistocene | 354 |
| EI370/371 | EI370/371 | 1977 | Lower Pleistocene | 125 |
| EI384/385 | EI384/385 | 1977 | Lower Pleistocene | 140 |
| | | 1974 | | |
| EI380 | EI380 | 1977 | Lower Pleistocene | 100 |
| EW826 | EW826 | 1985 | Lower Pleistocene | 153 |
| EW947 | EW947 | 1984 | Lower Pleistocene | 150 |
| EW977 | EW977 | 1996 | Lower Pleistocene | 140 |
| EW991 | EW991 | 1996 | Upper Pliocene | 228 |
| Fuji | GC506 | 1995 | Upper Pliocene | 1220 |
| GC6 | GC6 | 1985 | Pleistocene | 190 |
| GC18 | GC18 | 1987 | Upper Pliocene | 229 |
| GC20 | GC20 | 1989 | Miocene(?) | 243 |
| GC29 | GC29 | 1986 | Upper Pliocene/Pleistocene | 463 |
| GC31 | GC31 | 1985 | Pliocene | 625 |
| GC60 | GC60 | 1983 | Pliocene(?) | 274 |
| Genesis | GC205/161 | 1988 | Upper Pliocene | 823 |
| Glider | GC248 | 1997 | Lower Pliocene | 1000 |
| Jolliet | GC184 | 1981 | Pleistocene | 466 |
| King Kong | GC472/473 | 1989 | Lower Pliocene | 1160 |
| Lobster | EW873 | 1991 | Lower Pliocene | 216 |
| Mahogany | SS349/359 | 1993 | Middle Miocene(?) | 114 |
| Manta Ray | EW1006 | 1988 | Lower Pliocene | 548 |
| Marquette | GC52/53 | 1984 | Pleistocene | 186 |
| Morpeth | EW921 | 1993 | Pliocene | 512 |
| North Lobster | EW829/ST308 | 1985 | Miocene(?) | 173 |
| Oyster | EW917 | 1996 | Upper Pliocene | 366 |
| Popeye | GC116/72-73 | 1985 | Upper Pliocene/lower Pleistocene | 640 |
| Rocky | GC110 | 1983 | Pliocene | 401 |
| Seattle Slew | EW914/915 | 1994 | Upper Pliocene/lower Pleistocene | 310 |
| Shasta | GC136 | 1981 | Pleistocene | 291 |
| Starfish | EW963 | 1996 | Lower Pliocene | 457 |
| Stellaria | GC112/113 | 1997 | Lower Pliocene (?) | 623 |
| Sunday Silence | EW958-959/1002-1003 | 1985 | Lower Pliocene | 588 |
| Teak | ST260 | 1994 | Miocene(?) | 82 |
| Troika | GC200/244 | 1994 | Lower/upper Pliocene | 815 |

(crustal extension) values have been proposed to range from 1.0 to 4.5 throughout the basin (Sawyer et al., 1991), and from 3.2 to 4.2 within the study area. This means the thickness of the crust has thinned from 69 to 76%. Mesozoic subsidence was greatest within the central portion of the basin due to thermal effects; Cenozoic subsidence was greatest along the northern margin of the basin due to sedimentary loading beneath regional depocenters.

Within the northern Green Canyon/Ewing Bank study area, the present-day geometry is dominated by allochthonous salt and associated evacuation structures, such as extensional growth faults, salt welds, and contractional folds (Figures 3, 4). The time structure map (Figure 3) is a composite map; where there is a discrete thickness of salt, the top salt has been mapped, and where salt is believed evacuated, the resultant salt weld has been mapped. The study area comprises a structurally complex, multilevel salt system consisting of a series of allochthonous, rootless salt bodies and salt welds that are the result of an interplay between salt-stock canopies and salt sheets (Rowan, 1995; McBride et al., 1998, this issue).

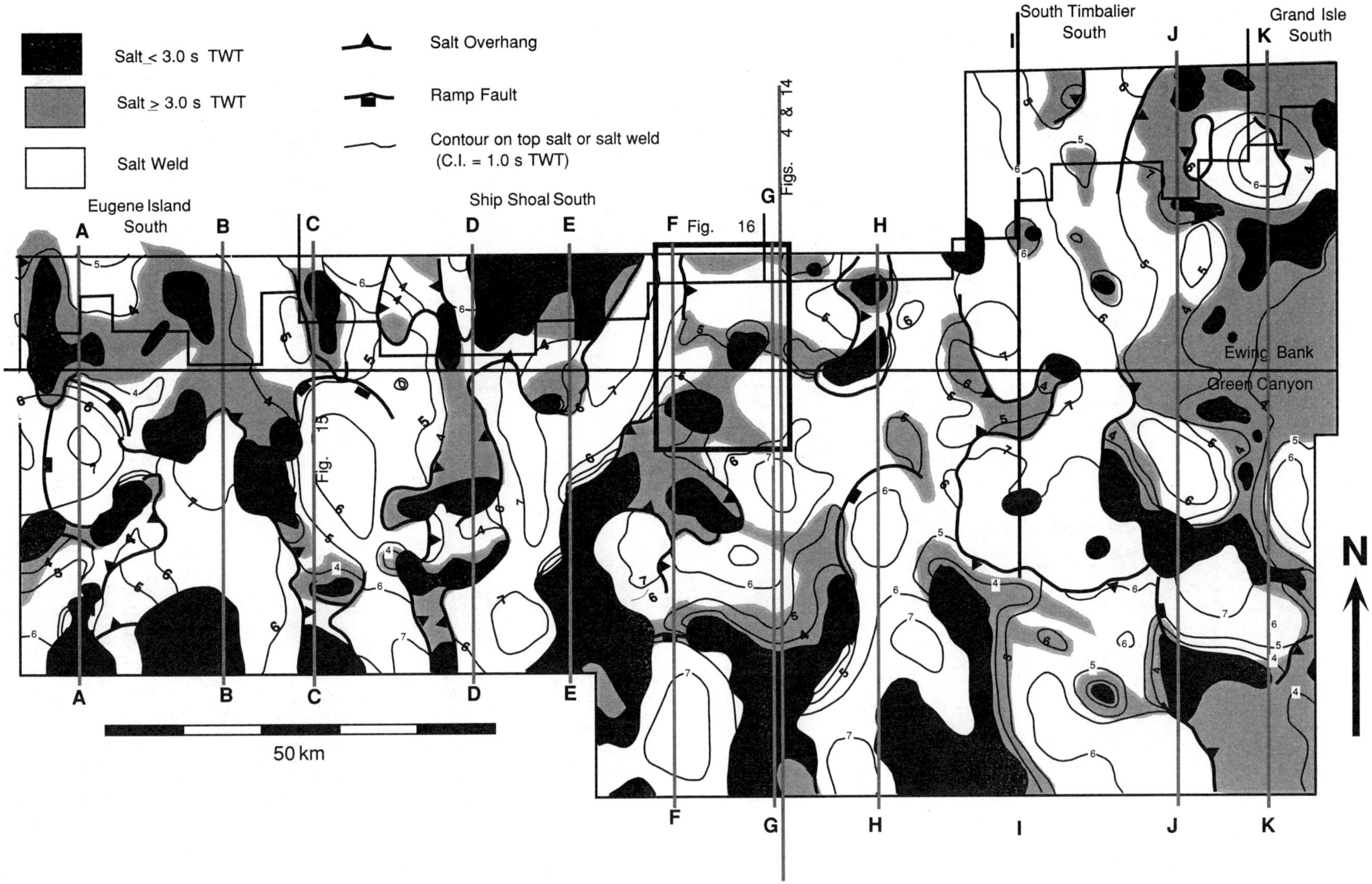


Figure 3—Structure contour map on top salt or equivalent salt weld (contour interval = 1.0 s two-way traveltime). The area is divided into salt bodies (shallower than 3.0 s two-way traveltime), thin salt (deeper than 3.0 s two-way traveltime), and salt welds (salt generally less than 100 ms in thickness). The map also shows areas where shallow salt overhangs deeper levels, and faults that arc around the lows. A–K indicate the locations of restored seismic profiles; locations of other figures are also shown. Map modified from Rowan (1995).

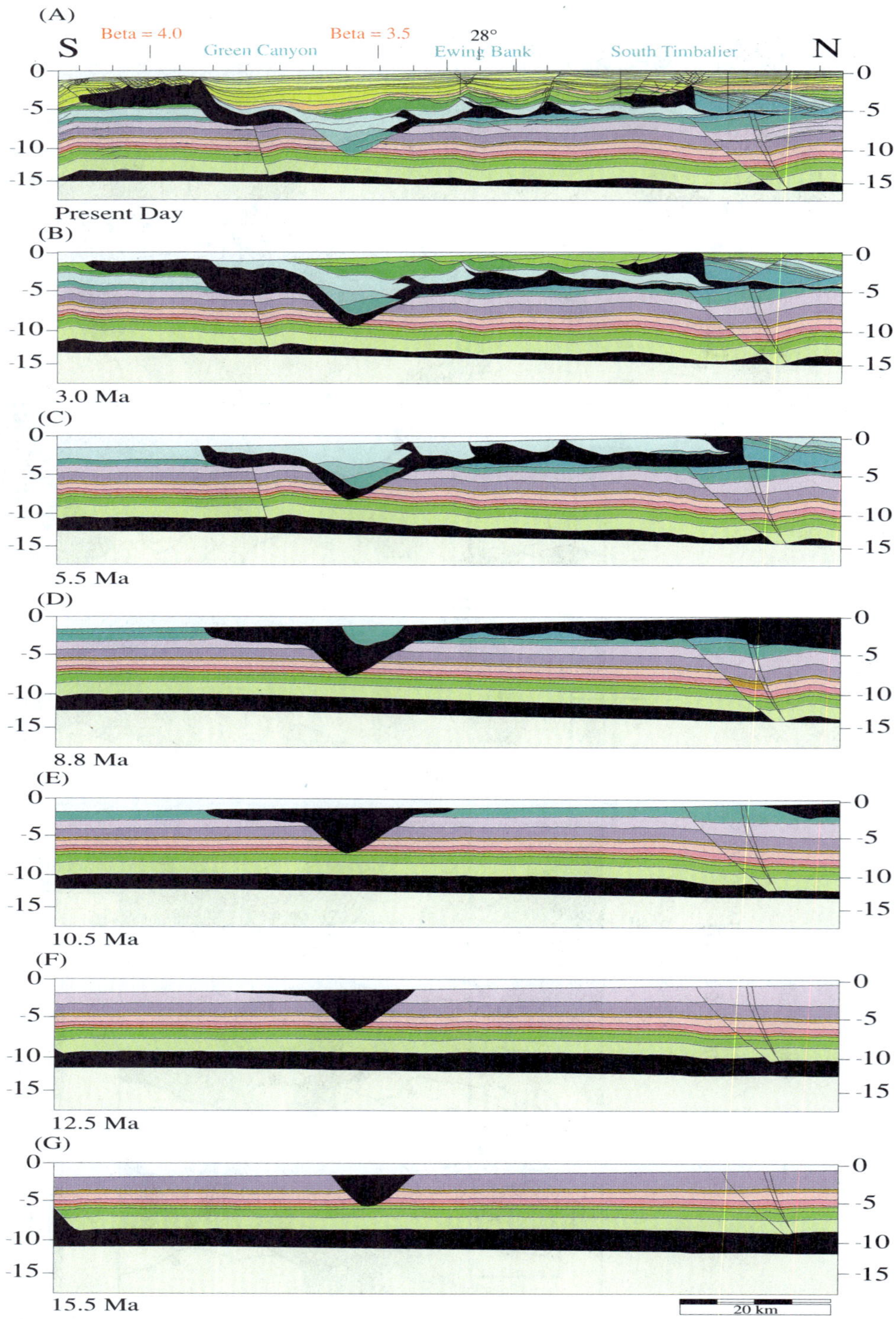


**Figure 4—Green Canyon/Ewing Bank portion of megaregional cross section and sequential restorations (1:1 scale): (A) present-day geometry; (B) 3.0 Ma restoration; (C) 5.5 Ma restoration; (D) 8.8 Ma restoration; (E) 10.5 Ma restoration; (F) 12.5 Ma restoration; (G) 15.5 Ma restoration. Salt is black; water is shaded. Restorations were done using GeoSec. Location shown in Figure 3. Modified from McBride (1998, this issue). Refer to Figure 5 for color legend.**

Restorations of a north-south cross section across the study area illustrate the timing of allochthonous salt extrusion and amalgamation to form a composite salt-stock canopy/salt-sheet canopy that was subsequently loaded and evacuated (Figure 4) (McBride, 1998, this issue). This cross section was extracted from a megaregional cross section extending from onshore Mississippi to the central Gulf of Mexico (~590 km; 365 mi) (McBride, 1998). Working at a megaregional scale provided better control on interpreting depth to basement, deep structures, and stratigraphic thicknesses.

The major structural feature of the study area during the early–middle Miocene was a bulb-shaped

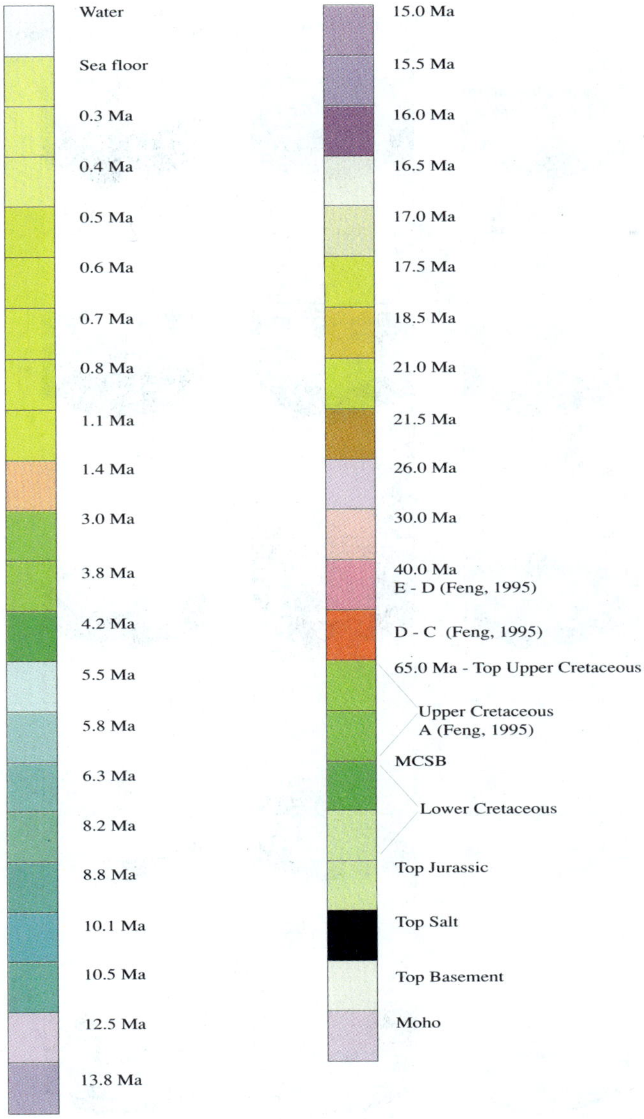


**Figure 5—Stratigraphic column for megaregional cross section (Figure 4), northern Gulf of Mexico Basin. Letters refer to the stratigraphic sequences defined by Feng (1995) for the deep Gulf of Mexico.**

salt stock in the central portion of the section that had been growing at least since 40 Ma (Figure 4G). This salt stock is believed to have been sourced from the autochthonous Louann Salt, with its feeder stem just out of the plane of section. A relatively rapid pulse of lateral spreading during the middle Miocene produced a prominent overhang along the southern margin of the salt stock and a relatively minor one along its northern margin by approximately 10.5 Ma (Figure 4E, F). By the late Miocene (8.8 Ma, Figure 4D), the salt stock had amalgamated with an allochthonous salt sheet flowing from the north to form a composite salt-stock/salt-sheet canopy. The similarity in timing between the spreading of the salt stock and the basinward advance of the salt sheet suggests a similar origin, as a result of either a regional reduction in sedimentation rates or a regional acceleration of salt flow due to updip loading, or both.

Evacuation of the allochthonous salt began during the late Miocene and culminated during the Pliocene–Pleistocene as the major clastic depocenter sourced from the Mississippi Delta prograded into the study area (Figure 4A–C). During this period, allochthonous salt continued to flow basinward in advance of the prograding depocenter until the salt became encased by sediments sometime during the Pleistocene (Figure 4A). Contractional features developed where the advancing salt sheet amalgamated with the salt stock. These structures are partially compensating extension above the salt sheet farther to the north.

The structural restorations illustrate that salt within the study area has undergone a dynamic evolution, with dramatic thickness variations occurring both laterally and temporally. Salt thicknesses were constrained by the restoration method of Rowan (1993). The two-dimensional (2-D) technique makes no assumptions about the kinematics of salt deformation, and salt area is not necessarily maintained. Salt is assumed to flow into or out of the section because of the three-dimensional (3-D) kinematics of salt deformation. In addition, salt can leave the system entirely by dissolution. The method incorporates decompaction, isostasy, thermal subsidence, and paleobathymetry, and its accuracy has been demonstrated by Rowan et al. (1996). Ignoring these factors has a major impact on the restored thickness of salt through time, and thus on the timing of salt weld formation (Figure 6). The absence of properly restored salt thicknesses through time in turn leads to incorrect modeling of the timing of subsalt thermal maturation and subsalt petroleum migration pathways.

Restoration results are largely determined by the interpreted stratigraphic correlations. Restorations to times older than the deepest strata penetrated by wells are less reliable. In the interpretation shown in Figure 4A, for example, the oldest strata in the central minibasin were assumed to be younger than the youngest adjacent subsalt strata. The resulting restorations show that the entire central minibasin was originally filled with salt (Figure 4E). If the oldest strata in the minibasin were coeval with the shallowest subsalt strata, however, the restoration would show that upward and lateral growth of the salt stock occurred as a result of loading and evacuation of the stock rather than as a result of flow from the deep source. Thus, the depicted thicknesses of salt in the stock (Figures 4C–G) are the maximum possible, and may be overestimates of the thickness variation through time.

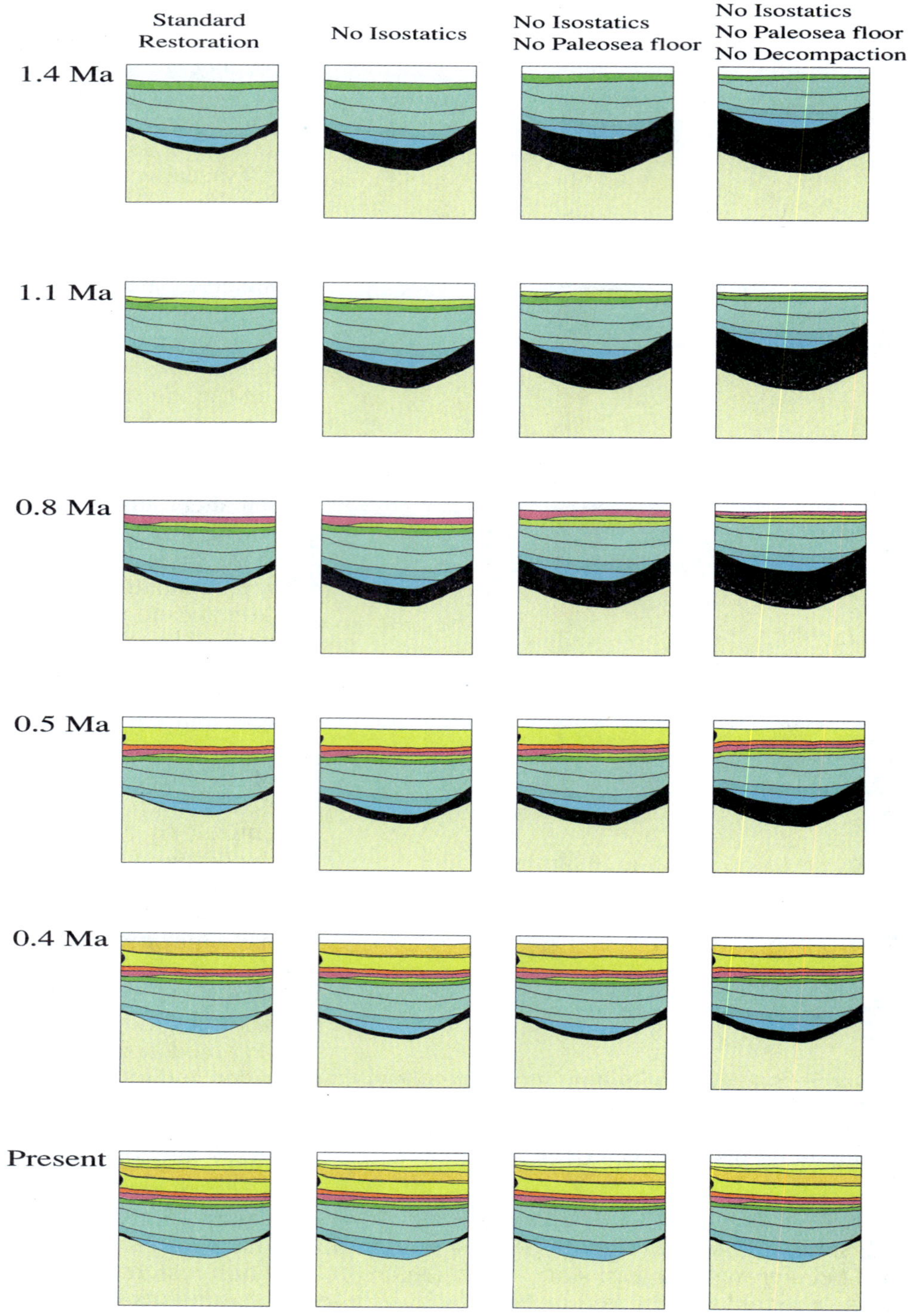


**Figure 6—Multiple structural restorations (using GeoSec) illustrating technique sensitivity to essential factors that need to be considered during restoration. Present-day geometry is the same in each column; restorations vary depending on which factors (isostatics, paleobathymetry, decompaction) are incorporated. Ignoring these factors causes reconstructed salt thicknesses to be too thick and changes the timing of salt weld formation, two critical parameters in conducting thermal modeling and petroleum migration analyses. Salt is black; water is shaded.**

## Stratigraphic Section

Mesozoic synrift and postrift strata comprise nonmarine siliciclastics, overlain by evaporites, marine carbonates, and shales (e.g., McFarlan and Menes, 1991; Salvador, 1991b; Sohl et al., 1991). Cenozoic strata record a history of thick, siliciclastic wedges, prograding primarily along the basin's northwestern and northern margins, and major episodes of allochthonous salt extrusion and evacuation (e.g., Worrall and Snelson, 1989; Galloway et al., 1991; Diegel et al., 1995; Peel et al., 1995). Numerous regional unconformities and depositional hiatuses subdivide the Mesozoic and Cenozoic. The following description of the regional stratigraphy of the northern Gulf of Mexico is summarized from Salvador (1987, 1991a), Piggott and Pulham (1993), and Weimer et al. (1994, 1998b, this issue) and illustrated in Figure 7.

Following the deposition of synrift Upper Triassic–Lower Jurassic sediments, extensive, thick deposits of Louann Salt (>4 km [2.5 mi] locally) accumulated throughout much of the basin during the Callovian. Thickness variations in salt were probably related to the basement rift geometry of fault-bounded grabens. The basement geometry shown in Figure 4 was taken directly from the regional depth-to-basement map of Sawyer et al. (1991). This surface was smoothed and does not show any fault blocks associated with initial rifting. Basement faults undoubtedly exist along this profile, but they cannot be resolved with present seismic or gravity data sets because of the great depth to basement and the interference of shallow salt. Trudgill et al. (1995) suggested that basement faults beneath the Perdido fold belt controlled original salt thickness variations.

From the Late Jurassic–Early Cretaceous, the basin became rimmed by broad, stable, carbonate shelves, banks, and ramps; meanwhile, pelagic carbonate ooze and shale were deposited in the deeper portions of the basin. Following the widespread middle Cretaceous unconformity (middle Cretaceous sequence boundary) (Buffler, 1991), siliciclastic sediments began dominating the stratigraphy with the development of the Woodbine and Tuscaloosa deltas. Chalk deposition then developed during the Late Cretaceous to form the Austin and Selma chalks. By the end of the Cretaceous, the sediments were predominantly shales and silts.

During the Cenozoic, thick siliciclastic depocenters began prograding into the northwestern and northern portions of the basin (Winker, 1982) in response to the Sevier and Laramide orogenies in the western United States and Mexico. These sediments are in excess of 10 km (6 mi) thick locally. The study area during this time consisted primarily of slope turbidite systems deposited in bathyal water depths, with Pliocene–Pleistocene strata reaching thicknesses of up to 6000 m (20,000 ft). These sediments were shale dominated, with sands developing locally within channel-levee systems and basin-floor fans.

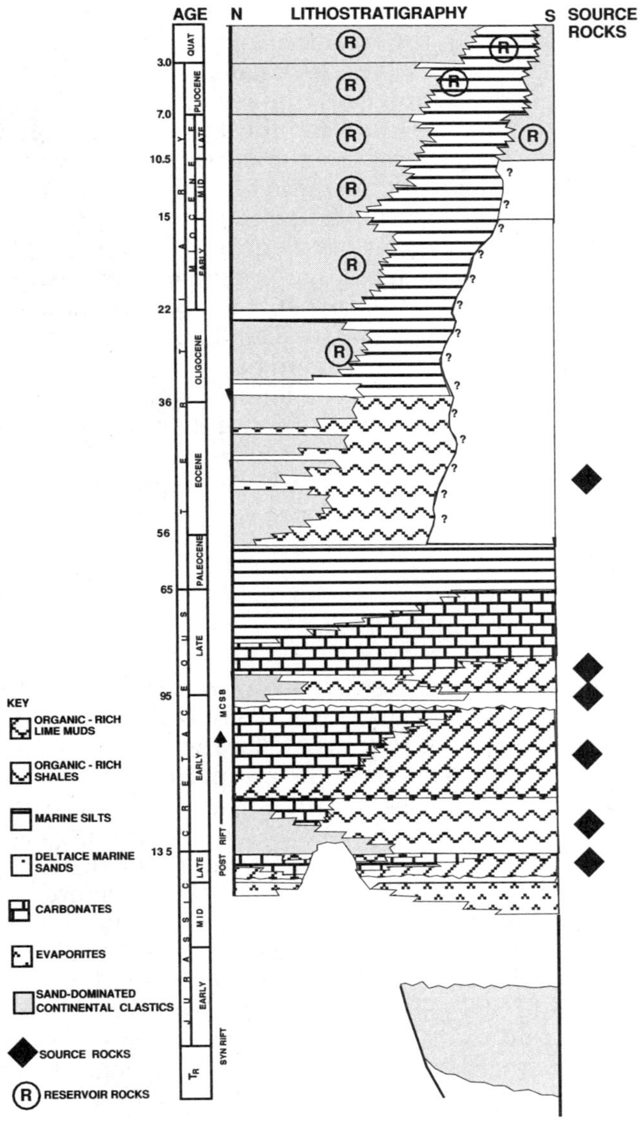


**Figure 7—Generalized stratigraphic framework of the northern Gulf of Mexico Basin with reservoirs and probable source intervals highlighted. Modified from Piggott and Pulham (1993). MCSB = middle Cretaceous sequence boundary.**

## PETROLEUM SYSTEMS

A petroleum system consists of all the geologic elements and processes responsible for source rock maturation and petroleum generation, migration, and accumulation (Magoon, 1989; Demaison and Huizinga, 1991; Magoon and Dow, 1994). Discussion of all the petroleum systems of the northern Gulf of Mexico Basin is beyond the scope of this paper; however, the northern Green Canyon/Ewing Bank study area is large enough to constrain key petroleum system components believed to be representative of a much larger portion of the basin. Figure 8 summarizes the timing of the key events for the various petroleum systems thought to be present within the Green Canyon/Ewing Bank area. The main focus of this paper is the effects of salt evolution on the expulsion-migration subsystem. A more detailed discussion of these elements is presented in following sections.

Specific names for individual petroleum systems are not given in the traditional manner (i.e., source rock and major reservoir rock names) because of the diversity of reservoir ages, the regional variations in reservoir nomenclature, and the apparent mixing of petroleum originating from multiple sources within individual reservoirs. Furthermore, all the numerous reservoirs are Neogene in age; therefore, the individual petroleum systems are consolidated and simplified based on the age of the potential source rocks and reservoir rocks. Hence, the study area contains the Oxfordian–Neogene (.), Tithonian–Neogene (.), Albian–Neogene (.), Turonian–Neogene (.), and Eocene–Neogene (.) petroleum systems. Because none of these source rock intervals have been penetrated within the study area, the level of certainty for these petroleum systems has been assigned as hypothetical (see Magoon and Dow, 1994, for explanation of these symbols).

## Petroleum Source Rocks

Mesozoic and Cenozoic source rocks of the Oxfordian, Tithonian, Albian, Turonian, and Eocene are believed to be responsible for the majority of oil and gas accumulations in the northern offshore Gulf of Mexico Basin (Nunn and Sassen, 1986; Curtis, 1989; Bissada et al., 1990; Sassen and Chinn, 1990; Kennicutt et al., 1993; Piggott and Pulham, 1993; Sassen et al., 1993; Wenger et al., 1994) (Figure 7). Although these strata have not been penetrated by exploration drilling in the offshore, they have been identified onshore and within Deep-Sea Drilling Project (DSDP) sites 535 and 540 and have been correlated to known petroleum accumulations using biomarkers, whole-rock pyrolysis, and gas chromatograms (Herbin et al., 1983; Patton et al., 1983; Oehler, 1984; Sassen et al., 1988; Sassen and Moore, 1988; Walters and Dusang, 1988; Sassen, 1990; Sassen and Chinn, 1990; Wenger et al., 1990, 1994; Piggott and Pulham, 1993); therefore, deposition of these source rocks is assumed to be widespread throughout the northern Gulf of Mexico Basin. Biomarker compounds and geochemical analyses of crude oils from deep-water seeps and produced oils are consistent with deeply buried Mesozoic carbonate and siliciclastic source rocks (Nunn and Sassen, 1986; Bissada et al., 1990; Piggott and Pulham, 1993; Sassen et al., 1993) and Eocene siliciclastic source rocks (Wenger et al., 1994). These sources were deposited in marine environments, and most of the kerogen is of type II or type II-S (sulfur rich). Where penetrated, the source rock ages correspond to global oceanic anoxic events that are associated worldwide with organic-rich source rock deposition and petroleum accumulations (Ulmishek and Klemme, 1990).

Offshore drilling has not encountered any effective Cenozoic oil source rocks; shales within producing intervals are invariably organically lean and thermally immature, and contain only gas-generating type III kerogen (Dow, 1984; Nunn and Sassen, 1986; Curtis, 1989; Bissada et al., 1990). Such information has been used to argue that older, more deeply buried source rocks were necessary to generate petroleum found within relatively young, shallowly buried reservoirs (e.g., Curtis, 1989; Bissada et al., 1990). This model created a paradox of how to explain late expulsion and long-distance vertical migration of petroleum out of the deeply buried source rocks and into young reservoirs and traps (Curtis, 1989). Modeling suggested deeply buried source rocks should have matured much earlier (i.e., Mello and Karner, 1996). Although much of the shallow gas within Pleistocene reservoirs has been attributed to biogenic sources based on its isotopically light composition, thermogenic methane has also been found within Pleistocene, Pliocene, and Miocene reservoirs (Rice, 1980; Rice and Claypool, 1981; Curtis, 1989).

## Maturity Modeling

Maturity modeling was conducted across the study area along the regional cross section (Figure 4A) to determine the lateral and temporal effects of allochthonous salt flow and evacuation on the timing and distribution of subsalt source rock maturities. Thirteen pseudo-wells were created along the profile for one-dimensional (1-D) modeling using the software package BasinMod. Input parameters used for modeling are listed in Tables 2 and 3.

### *Burial History*

The burial history of the stratigraphic section was modeled across the study area. Present-day thicknesses were measured directly from the cross section (Figure 4A). Variations in salt thickness and water depth through time were measured using the appropriate structural restorations (Figure 4B–G). No significant periods of erosion were incorporated within the models. Petrophysical parameters necessary for calculating compaction, porosity, and thermal conductivity were calculated automatically by BasinMod as a function of rock type specified for each horizon (Table 2). Compaction was calculated using the equation of Baldwin and Butler (1985). One-dimensional Airy isostatics and thermal subsidence calculations were incorporated in the construction of the restorations.

Overpressures were not incorporated within the models in this study. To incorporate overpressures properly into the burial history requires fluid-flow

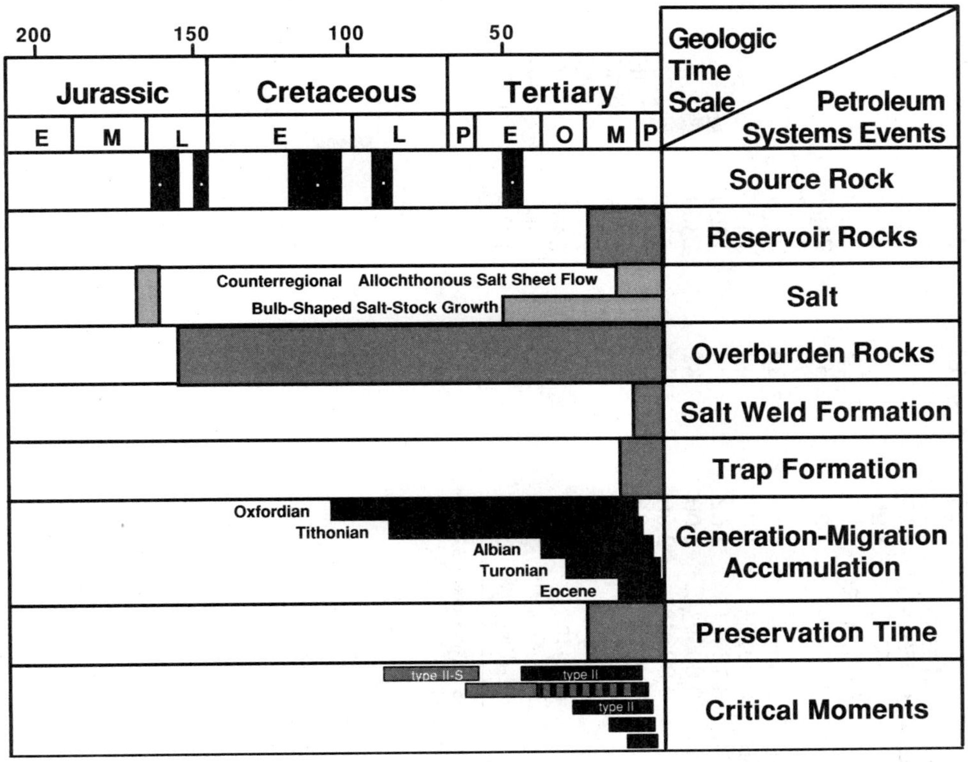


**Figure 8—Events chart for the petroleum systems within the northern Green Canyon/Ewing Bank study area showing the temporal relationships of the essential elements and processes. Preservation time represents time since initial critical moment of peak generation. Critical moments represent the time each source rock experienced peak oil generation (0.90–0.95 %$R_o$) as shown in Figures 12, 13, and 16.**

modeling, which is beyond the scope of this study. Overpressure zones can vary spatially and temporally. Presence of overpressure may increase the thermal gradient because overpressured strata in the Gulf of Mexico are commonly more porous than normally pressured rocks, and therefore have lower thermal conductivity. The overpressure zone acts as an insulator, causing deeper strata to achieve hotter temperatures than normally pressured strata (Mello and Karner, 1996).

### *Present-Day Temperatures*

A present-day mean annual sea-floor temperature at 96 m (314 ft) below sea level of 17.8°C (64.0°F) (Cathles and Nunns, 1991) was used to calculate the mean annual sea-floor temperature at each of the 1-D modeling locations. Temperatures were decreased by 4°C/100 m (2.2°F/100 ft) for the first 200 m (656 ft) below sea level, and 2°C/100 m (1.1°F/100 ft) at depths greater than 200 m (656 ft), with a minimum value of 4°C (39°F) (Waples et al., 1992a; Waples, 1994a). Present-day temperature profiles were obtained by using corrected bottom-hole temperatures (BHT) of wells in and near the study area using bottom-hole temperature correction curves for the Gulf Coast (F. Meissner, 1993, personal communication). Thirty-one BHT measurements were used with depths ranging from 837 m (2747 ft) to 4205 m (13,794 ft), and temperature correction factors ranging from 1.13 to 1.42. These temperature values were not used as true input data for the modeling, but to constrain sedimentary radiogenic heat values and present-day heat flow following the method of Waples et al. (1992a).

Although the BHT correction method used is common, BHT data overall constitute a low-quality data set of temperature observations. A multitude of BHT correction techniques have been published (e.g., Horner, 1951; Middleton, 1979; Deming, 1989; Hermanrud and Shen, 1989), and the degree of bias and error inherent in these measurements has been analyzed (Hermanrud et al., 1990). Even though the method used in this study may give temperatures that are too low by 5–10°C (9–18°F), these errors can be partly eliminated by using data from as many wells as possible, and using the deepest measurements available (Hermanrud et al., 1990). Consequently, we used 33 BHT measurements from eight wells in Ewing Bank and northern Green Canyon. Although more sophisticated BHT correction techniques or drill-stem test temperatures can considerably reduce errors (Hermanrud et al., 1990; Waples et al., 1992a), we believe the quality of the data set and correction technique used in this study are acceptable for the objectives of this paper.

Thermal conductivities were calculated automatically by the thermal modeling software (BasinMod)

**Table 2. Input Parameters for Thermal Maturation Modeling Using BasinMod**

| Formation | Lithology* | Initial Porosity (%) | Compaction Factor (FM)** | Exponential Factor (SC)** | Density ($g/cm^3$) | Grain Size (mm) | Matrix Conductivity (W/mC) | Matrix Conductivity Correction | Heat Capacity ($kJ/m^3C$) | Heat Generation ($mW/m^3$) |
|---|---|---|---|---|---|---|---|---|---|---|
| Recent | 50SH-45SLT-5SS | 57 | 2.27 | 0.45 | 2.62 | 0.03222 | 1.87 | 0 | 2382 | 4.4 |
| 0.5 Ma | 50SH-45SLT-5SS | 57 | 2.27 | 0.45 | 2.62 | 0.03222 | 1.87 | 0 | 2382 | 3.3 |
| 1.1 Ma | 50SH-40SLT-10SS | 56 | 2.25 | 0.44 | 2.62 | 0.05644 | 1.99 | 5 | 2390 | 2.2 |
| 3.0 Ma | 50SH-25SLT-25SS | 55 | 2.18 | 0.42 | 2.62 | 0.1291 | 2.35 | 20 | 2412 | 1.8 |
| 3.8 Ma | 50SH-25SLT-25SS | 55 | 2.18 | 0.42 | 2.62 | 0.1291 | 2.35 | 20 | 2412 | 1.8 |
| 4.2 Ma | 50SH-25SLT-25SS | 55 | 2.18 | 0.42 | 2.62 | 0.2391 | 2.35 | 20 | 2412 | 1.36 |
| 5.5 Ma | 50SH-30SLT-20SS | 55 | 2.21 | 0.43 | 2.62 | 0.10488 | 2.23 | 15 | 2405 | 1.36 |
| 8.8 Ma | 50SH-30SLT-20SS | 55 | 2.21 | 0.43 | 0.43 | 0.10488 | 2.23 | 15 | 2405 | 1.8 |
| 10.1 Ma | 50SH-30SLT-20SS | 55 | 2.21 | 0.43 | 2.62 | 0.10488 | 2.23 | 15 | 2405 | 1.8 |
| 10.5 Ma | 50SH-40SLT-10SS | 56 | 2.25 | 0.44 | 2.62 | 0.05644 | 1.99 | 5 | 2390 | 1.8 |
| 12.5 Ma | 50SH-40SLT-10SS | 56 | 2.25 | 0.44 | 2.62 | 0.05644 | 1.99 | 5 | 2390 | 1.8 |
| 15.5 Ma | 50SH-40SLT-10SS | 56 | 2.25 | 0.44 | 2.62 | 0.05644 | 1.99 | 5 | 2390 | 2.2 |
| 21.5 Ma | 55SLT-40SH-5SS | 56 | 2.25 | 0.44 | 2.624 | 0.004285 | 1.92 | 35 | 2437 | 2.2 |
| 30.0 Ma | 70SH-25SLT-5SS | 58 | 2.31 | 0.47 | 2.612 | 0.001427 | 1.77 | -70 | 2272 | 1.36 |
| Eocene | 60SH-40SLT | 58 | 2.32 | 0.47 | 2.616 | 0.001731 | 1.7 | -40 | 2320 | 1.36 |
| Paleocene | 60SH-40SLT | 58 | 2.32 | 0.47 | 2.616 | 0.001731 | 1.7 | -40 | 2320 | 1.36 |
| Late Cretaceous | 80LS-20SH | 60 | 1.68 | 0.27 | 2.696 | 0.120113 | 2.62 | 244 | 2500 | 0.66 |
| Turonian | 60SH-40LS | 60 | 2.04 | 0.39 | 2.64 | 0.006931 | 2.06 | 32 | 2300 | 1.36 |
| Cenomanian | 60SH-40LS | 60 | 2.04 | 0.39 | 2.64 | 0.006931 | 2.06 | 32 | 2300 | 1.36 |
| Albian | 60SH-40LS | 60 | 2.04 | 0.39 | 2.64 | 0.006931 | 2.06 | 32 | 2300 | 1.36 |
| Early Cretaceous | 60SH-30LS-10SLT | 59 | 2.11 | 0.41 | 2.64 | 0.0049 | 1.97 | 14 | 2305 | 0.66 |
| Tithonian | 60SH-30LS-10SLT | 59 | 2.11 | 0.41 | 2.64 | 0.0049 | 1.97 | 14 | 2305 | 1.36 |
| Kimmeridgian | 60SH-30LS-10SLT | 59 | 0.41 | 0.41 | 2.64 | 0.0049 | 1.97 | 14 | 2305 | 1.36 |
| Oxfordian | 60SH-40LS | 60 | 2.04 | 0.39 | 2.64 | 0.006931 | 2.06 | 32 | 2300 | 1.36 |
| Louann Salt | Salt | 0 | 0 | 0 | 0 | 1e-006 | 6.5 | 0 | 1750 | 0.012 |

*This column lists the percentage and then rock type. For example, 50SH = 50% shale. SH = shale, SLT = siltstone, SS = sandstone, LS = limestone.
**Common equations for porosity as a function of depth: FM = Falvey and Middleton (1981), SC = Sclater and Christie (1980).

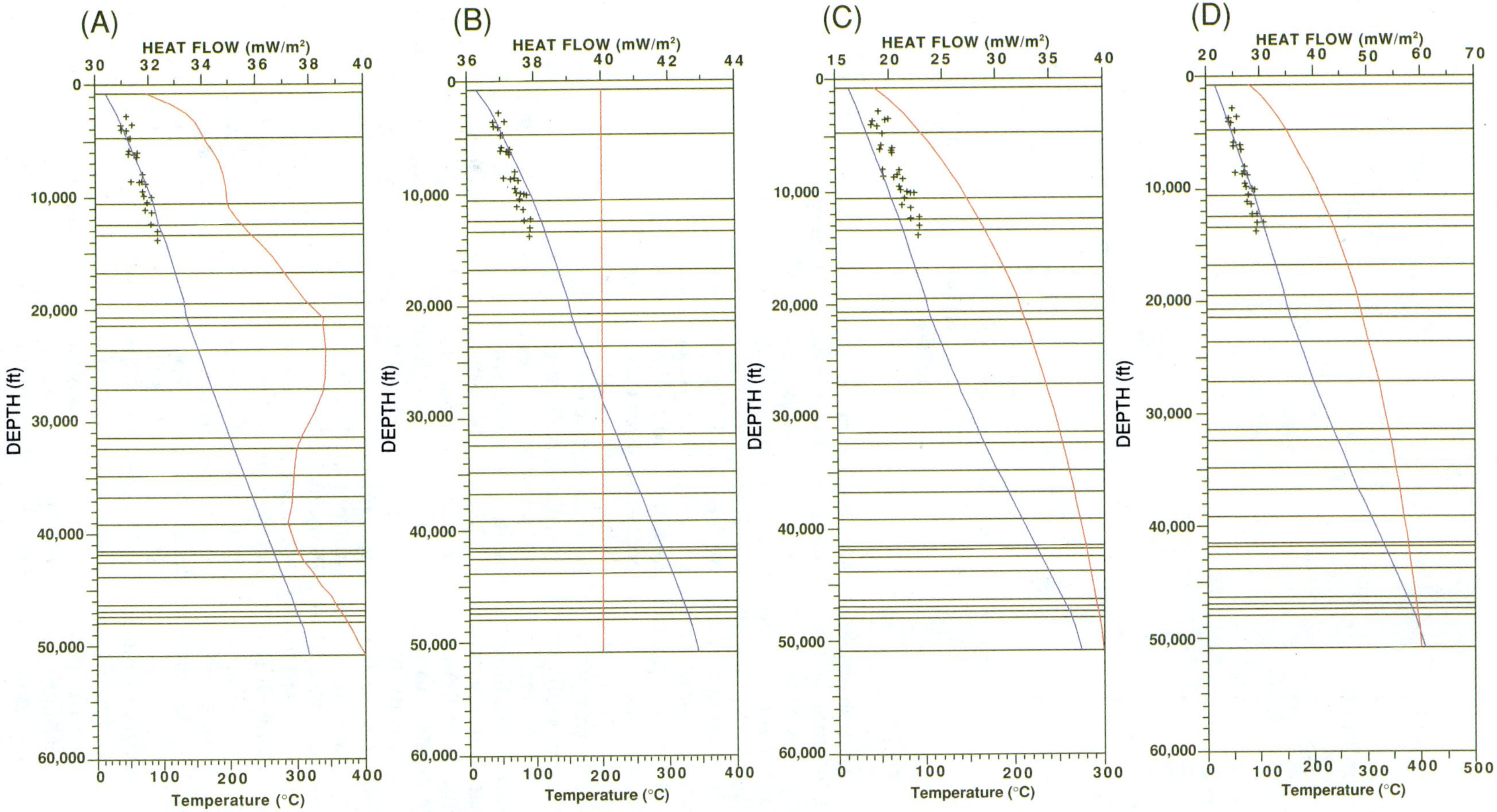


**Figure 9**—Plots of measured and predicted present-day subsurface temperatures and heat flow in the northern Green Canyon/Ewing Bank study area compared to different heat flow models. + = corrected bottom-hole temperature (BHT) data are from wells located in ST316, EW952, EW994, GC21, GC23, GC67, GC70, and GC154. (A) Transient heat flow model with additional radiogenic heat sourced from stratigraphic section produces optimized model; (B) steady-state heat flow model calculates subsurface temperatures that are too hot; (C) transient heat flow model without radiogenic heat sourced from stratigraphic section calculates subsurface temperatures that are too cool; (D) calculated transient heat flow from BHT without radiogenic heat sourced from stratigraphic section calculates heat flow values that are too high. Red lines represent modeled present-day subsurface temperature profiles. Blue lines represent modeled present-day subsurface heat flow profiles. Models generated using software package BasinMod.

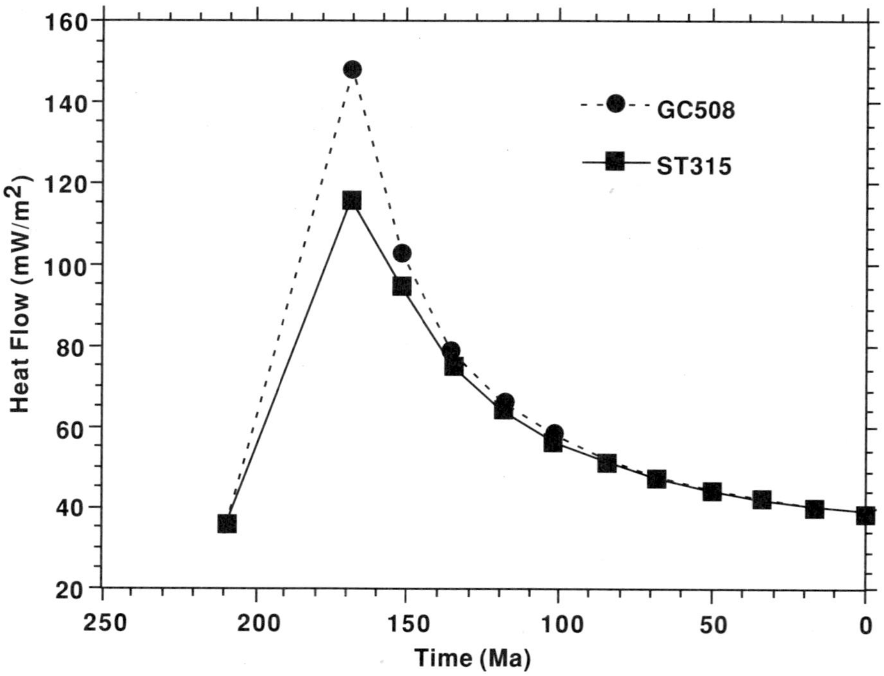


**Figure 10—Spectrum of paleoheat flow curves used in the thermal modeling. Heat flow history determined as a function of beta using BasinMod software. Beta for pseudowell GC508 is 4.15. Beta for pseudowell ST315 is 3.30.**

for each rock unit based on its lithology, porosity, and temperature (Waples et al., 1992a) (Table 2). Rock types were estimated based on nearby well penetrations, regional net sand trends, and regional stratigraphic trends (Salvador, 1991a; Piggott and Pulham, 1993; Pulham, 1993; 1996, personal communication). A value of 6.5 W/m$^3$ was used for the thermal conductivity of salt. No measured thermal conductivity values were available for this study area.

Several heat transfer assumptions were used within the modeling runs (modified from Waples et al., 1992a). (1) Heat was assumed to come from the basement (pre-Middle Jurassic Louann Salt) and radiogenic heat sources within the sedimentary section. (2) The basement heat source was not differentiated between radiogenic heat from the basement and heat from mantle convection. (3) Heat transfer was by conduction and by water expelled vertically during compaction. (4) Transient thermal conditions were used to model heat flow at the base of the sedimentary section to the sea floor. (5) All present-day heat flows were modeled at 40 mW/m$^2$, and paleoheat flows varied through time (40–150 mW/m$^2$) as a function of the estimated local beta value for the opening of the Gulf of Mexico at about 190 Ma (Sawyer et al., 1991).

Present-day heat flow measurements made along the sea floor west of the study area range from 19 to 104 mW/m$^2$, dependent on the thickness and proximity of underlying salt (Nagihara et al., 1993). Likewise, surface heat flow measurements made in northern Louisiana and adjacent Mississippi range from 39.7 to 83.7 mW/m$^2$ as a function of underlying salt geometries (Smith et al., 1981). Nunn et al. (1984) predicted a value of 41.8 mW/m$^2$ onshore in north Louisiana. Sclater and Nagihara (1991) noted the distinctive contrast between high heat flow values collected along the sea floor over areas of shallow salt vs. relatively uniform and low heat flow values over salt-free areas and between salt. These workers suggested these lower values (~40 mW/m$^2$) were more representative of the geothermal heat flux through the lithosphere; therefore, 40 mW/m$^2$ was used as the present-day heat flow for all of the maturity modeling in this study.

Historically, radiogenic heat production of sedimentary rocks and its contribution to heat flow density has been considered insignificant (Hermanrud, 1993); however, work by Keen and Lewis (1982) and McKenna and Sharp (1998) indicated that radiogenic-heat production from sediment can contribute from 15 to 26% of the overall surface heat flow density. Although measurements of radiogenic heat production in Gulf of Mexico sediments are sparse, radiogenic-heat values were assigned to the stratigraphic column as a function of rock type based on published values by Epp et al. (1970), Keen and Lewis (1982), Pollack et al. (1993), Nagihara et al. (1996), and McKenna and Sharp (1998) (Table 2).

Based on this information, we tried to optimize the data set following the method of Waples et al. (1992a, b) by constructing a temperature profile that fit the surface temperature and BHT and honored the 40 mW/m$^2$ heat flow value, while applying a

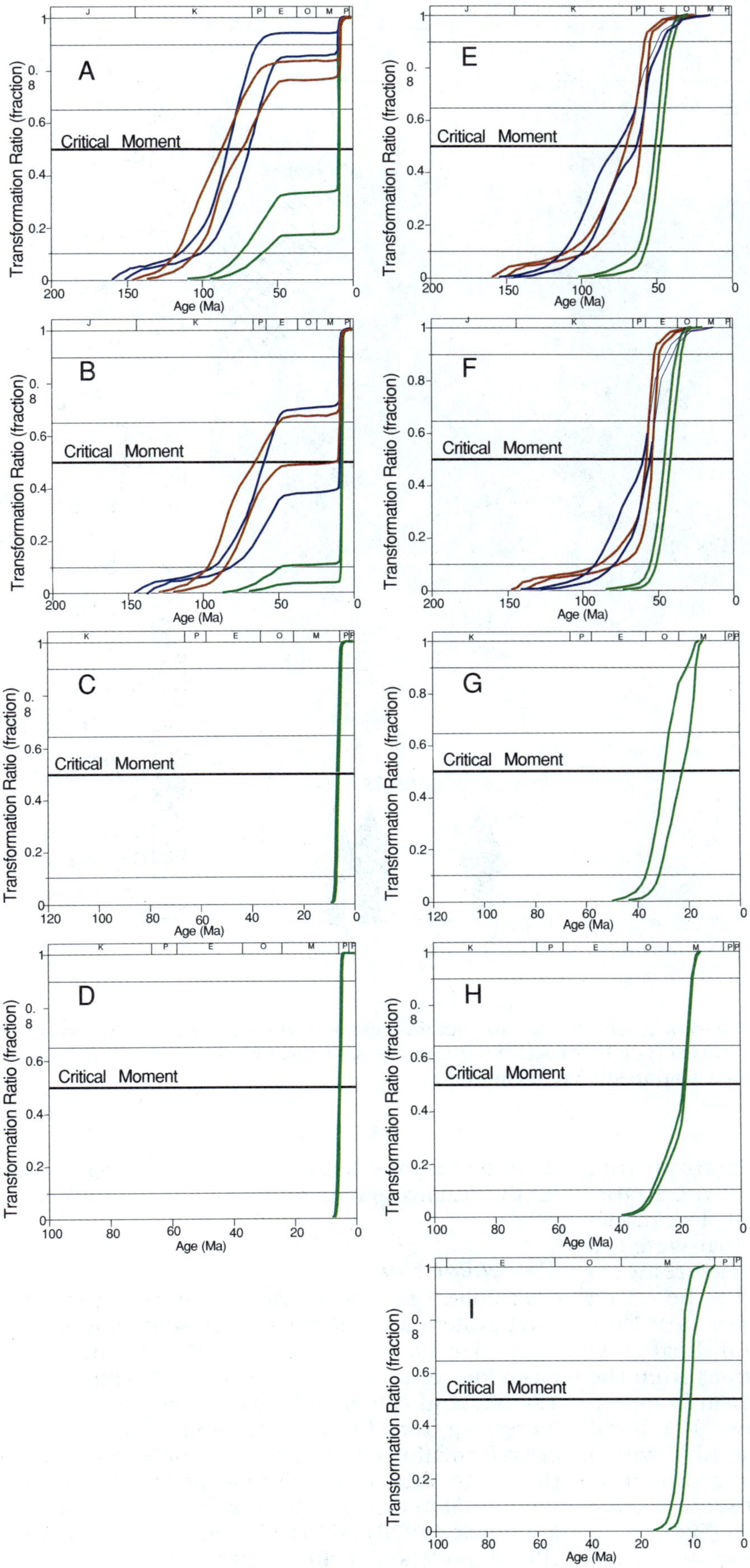


**Figure 11—Transformation ratio calculations (from BasinMod) of potential source rocks in Green Canyon 112 (A–D) and Ewing Bank 950 (E–I) pseudo-wells: (A) GC112 Oxfordian, (B) GC112 Tithonian, (C) GC112 Albian, (D) GC112 Turonian, (E) EW950 Oxfordian, (F) EW950 Tithonian, (G) EW950 Albian, (H) EW950 Turonian, and (I) EW950 Eocene. Green lines represent tops and bottoms of standard type II kerogen source rocks, red lines represent tops and bottoms of type II-S (Monterey) kerogen source rocks, blue lines represent tops and bottoms of type II-S (Kirkuk) kerogen source rocks (Table 3) (Tissot et al., 1987). Locations are shown in Figure 14.**

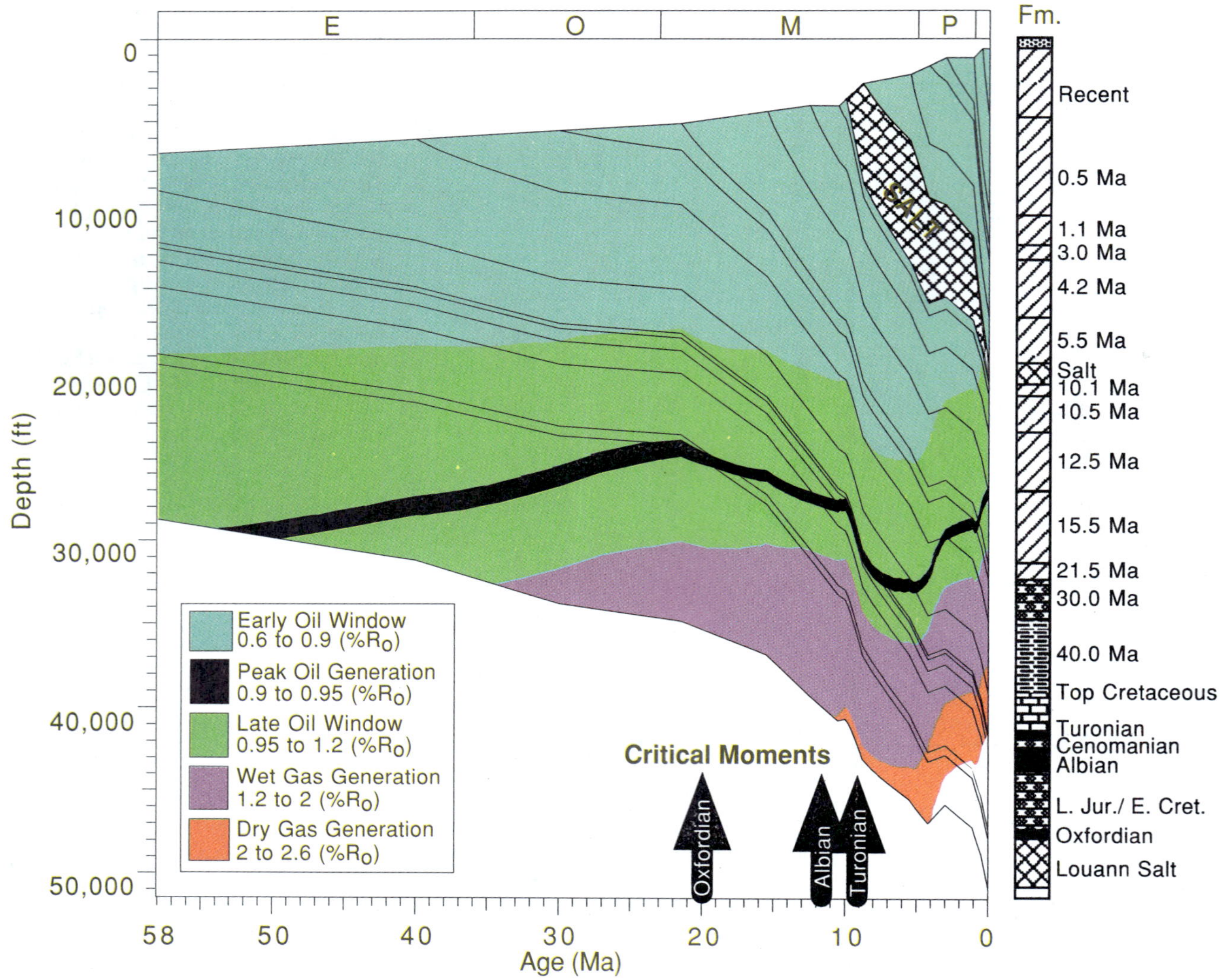


**Figure 12—Geohistory plot (from BasinMod) of Ewing Bank 950 pseudo-well with thermal maturation overlay. Potential Mesozoic and early Cenozoic standard type II kerogen source rocks pass through the peak oil generation window (~0.90–0.95 % $R_o$) from 25 to 5 Ma. Location shown in Figure 14.**

transient heat flow model and maintaining acceptable radiogenic-heat values for the stratigraphy (Figure 9A). A steady-state heat flow model calculated subsurface temperatures that were too hot and did not match the BHT measurements (Figure 9B). Conductivities needed to be raised approximately 25% for the steady-state heat flow model to match the BHT. A transient heat flow model without radiogenic-heat sources from the sedimentary section calculated subsurface temperatures that were too cool (Figure 9C). Basal heat flow values calculated from the BHT with a transient heat flow model without radiogenic-heat sources from the sedimentary section were too high compared to published values (Figure 9D). We consider the parameters used to constrain Figure 9A to be the most reasonable, and used these values for thermal maturation modeling.

### *Paleoheat Flow*

Paleoheat flow was constrained using the structural evolution of the basin as presented by Sawyer et al. (1991). Sawyer et al. (1991) proposed the basin formed as a result of rifting between 210 and 170 Ma, and calculated beta values progressively increasing from 1.0 (no extension) to 4.5 (oceanic crust formation) from onshore to offshore, respectively. Paleoheat flow values were calculated using the BasinMod rifting heat flow option that uses the McKenzie (1978) and Jarvis and McKenzie (1980) rifting models. For this study, heat flow from the

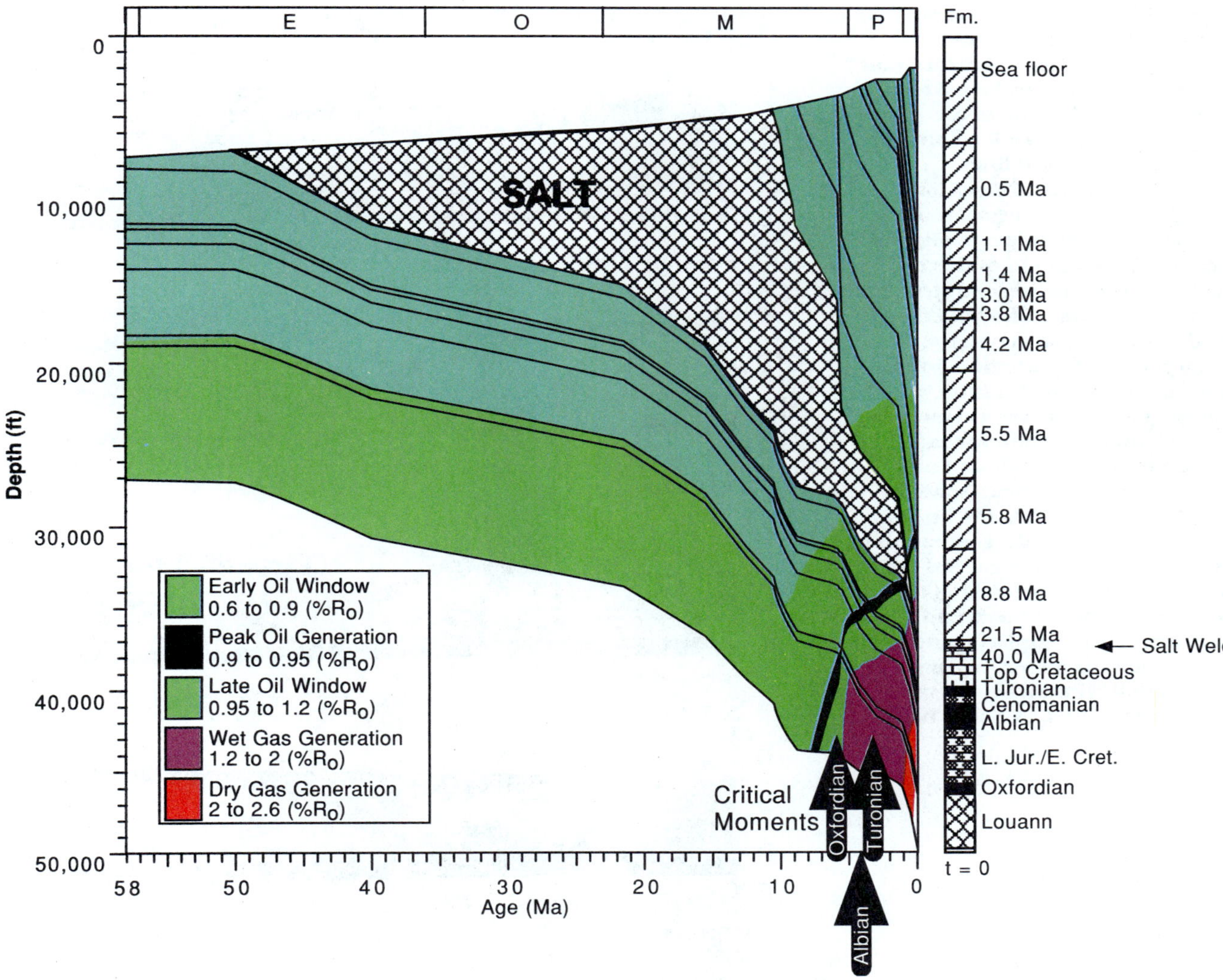


**Figure 13—Geohistory plot (from BasinMod) of Green Canyon 112 pseudo-well with thermal maturation overlay. Potential Mesozoic standard type II kerogen source rocks are in the peak oil generation window (~0.90 –0.95 %$R_o$) from 5 to 3 Ma. Thermal history and timing of generation are significantly different from those in Ewing Bank 950 pseudo-well (Figure 12). Location shown in Figure 14.**

basement is modeled at the base of the Louann Salt, with a sudden increase to a maximum during the rifting, followed by an exponential decay to the present-day value of 40 mW/m$^2$ (Figure 10). Paleoheat flows for the individual 1-D models were varied as a function of the beta values mapped by Sawyer et al. (1991).

Neither vitrinite-reflectance ($R_o$) values nor any other thermal-indicator data were available for this study; therefore, calculated paleotemperatures, heat flows, or thermal maturity cannot be constrained or optimized in a manner similar to the temperature profile and BHT values discussed. Miocene and younger sediments, however, have been reported to be thermally immature (Dow, 1984; Nunn and Sassen, 1986; Curtis, 1989; Piggott and Pulham, 1993), in agreement with the modeling results of this study.

### *Thermal Maturity Timing*

The results of thermal maturation modeling depend on many variables and factors, some of which are poorly constrained. Thus, the calculated timing of events (e.g., peak oil generation) may be inaccurate; however, we are more concerned here with the patterns, both spatial and temporal, produced by allochthonous salt evolution on maturation. Because the variables and modeling methods are internally consistent, we consider the shifting patterns shown here to be realistic.

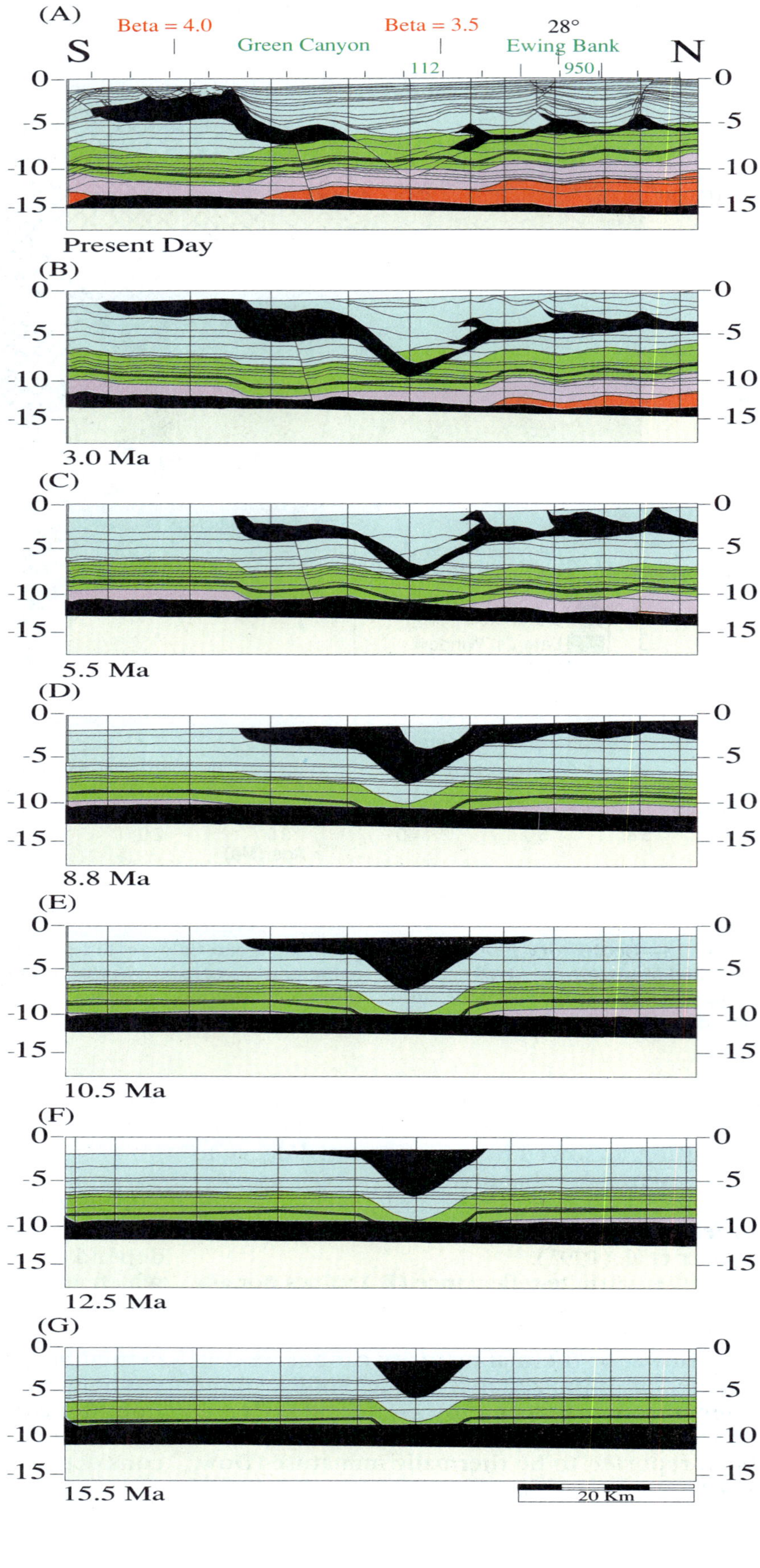


**Figure 14**—Structural restorations (identical to Figure 4) with superimposed thermal maturation windows across northern Green Canyon from 15.5 Ma to the present. Section location shown on Figure 3; vertical lines indicate locations of BasinMod one-dimensional thermal models, including those shown in Figures 12 and 13. Combination of thermal maturation modeling and structural restorations illustrates the effect of the original salt geometry, evolution, and evacuation on the subsalt thermal maturation. Consequently, the critical moment of peak oil generation for each of the potential Mesozoic and early Cenozoic sources varies spatially and temporally along the line of section. For example, generation is retarded beneath the central salt stock until weld formation (15.5–3.0 Ma), and at the southern end of section after sheet emplacement (5.5–0 Ma). Color legend for thermal maturation windows is the same as in Figures 12 and 13. Salt is black.

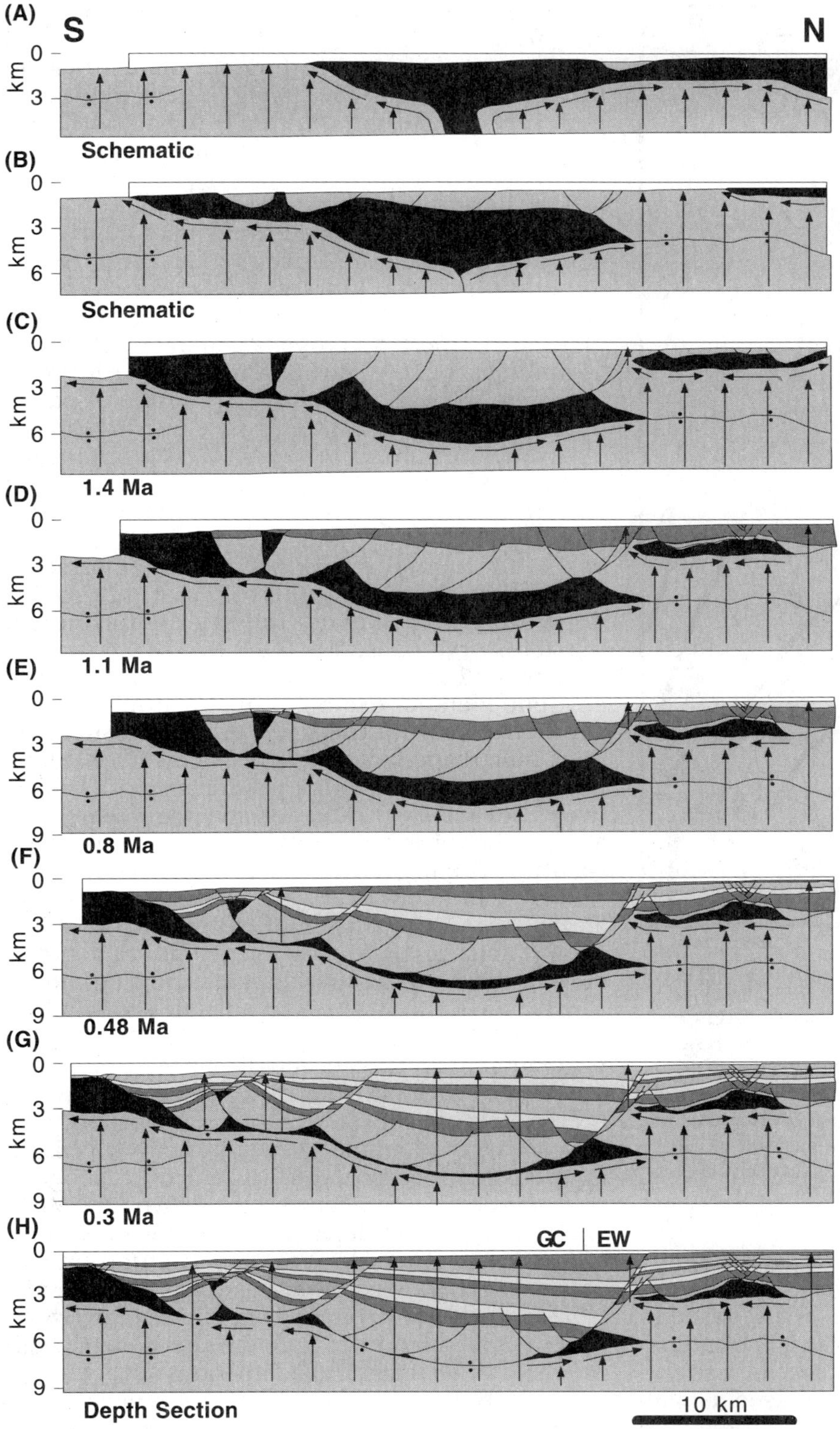


**Figure 15—1:1 depth section and sequential restorations of seismic profile C from regional study area (see Figure 3 for location). The restorations show the evolution of a central salt stock into a bowl-shaped minibasin, and how this evolution influenced petroleum migration pathways through time. Arrows represent petroleum migration pathways at the time of each restoration. Migration is believed to be vertical until it reaches the base of salt (in black), at which point it is deflected up the dip along the base of salt. Vertical migration is believed to resume when and where salt welds form. The evolution of salt concentrates petroleum migration in some regions through time, whereas salt shields suprasalt sediments from petroleum migration for significant amounts of time. Changes in length of suprasalt section show amounts of extension and contraction.**

Thermal maturity of the stratigraphic section was calculated using a kinetic model from the Lawrence Livermore National Laboratory (Sweeney and Burnham, 1990). This model calculates multiple parallel reactions as organic matter transforms to petroleum. The model calculates vitrinite-reflectance (%$R_o$) values that are used to map the thermal maturity windows of oil

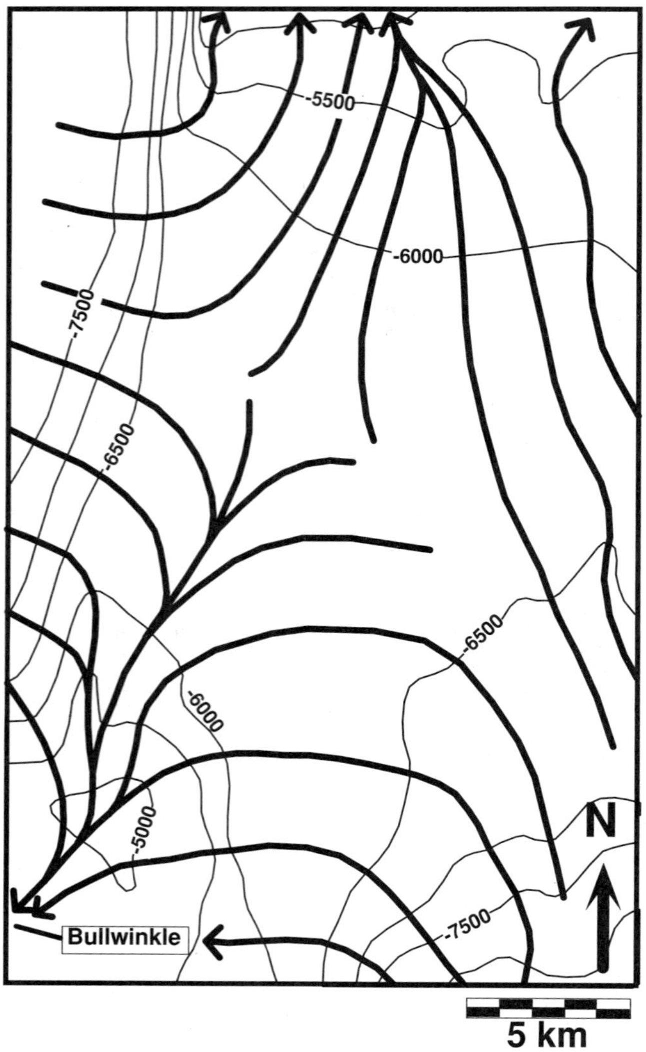


**Figure 16—Structure contour map of present-day base of salt or equivalent salt weld in 3-D structural restoration study area (contour interval = 600 m, 2000 ft) illustrating possible control on subsalt petroleum migration pathways prior to salt weld formation. Flow paths are assumed to be up the dip of the base salt surface. See Figure 3 for location. Modified from McBride et al. (1995).**

(0.6–1.2 %$R_o$), peak oil (0.9–0.95 %$R_o$), wet gas (1.2–2.0 %$R_o$), and dry gas (2.0–2.6 %$R_o$) used in this study (F. Meissner, 1993, personal communication). Standard type II kerogen and its associated default kinetic parameters were used as well (Table 3). In addition, transformation ratios were calculated for the individual source rocks using standard type II kerogens for the Eocene, Albian, and Turonian source rocks and type II-S kerogens for the Tithonian and Oxfordian source rocks (Table 3). The type II-S kerogens were modeled using *n*-component kinetics determined for the Monterey and Kirkuk (Iraq) formations (Tissot et al., 1987) because no data have been published for the potential type II-S Jurassic source rocks in the Gulf of Mexico. The timing of peak petroleum generation (critical moment) for a particular source rock is defined as when the transformation ratio equals 0.5.

The thermal maturation modeling indicates that potential Jurassic source rocks may have been generating petroleum since the Late Cretaceous, depending on the kinetics of the source rocks (Figure 11A, B, E, F). The modeling indicates Oxfordian type II-S kerogens generate petroleum as much as 90 m.y. earlier than standard type II kerogens, depending on the magnitude of paleoheat flow and allochthonous salt history (Figure 11A, E). The timing of the critical moment for Oxfordian type II-S kerogens is most sensitive to variations in the paleoheat flow history rather than allochthonous history (Figure 11A, E); however, the timing of peak petroleum generation varies spatially among the other source rocks, depending on the overlying salt evolution and burial history (Figure 11B, C, F, I).

Figures 12 and 13 are geohistory plots that represent end-member examples of allochthonous salt geometries within the study area (thin sheet vs. thick bulb-shaped salt stock, respectively) that later underwent complete or partial evacuation. The relatively late timing of maturation and petroleum generation of the subsalt type II kerogen source rocks is a result of the high thermal conductivity of salt (O'Brien and Lerche, 1984, 1988; Blackwell and Steele, 1989; Yu and Lerche, 1992; Mello et al., 1994; Pepper and Yu, 1995). Allochthonous salt acts as a radiator that cools underlying strata, creating anomalously low temperatures in subsalt strata while anomalously heating suprasalt strata. The thermal maturation retardation of subsalt strata enables deep Mesozoic and Cenozoic potential source rocks (Oxfordian, Tithonian, Albian, Turonian, and Eocene) within the study area to be actively generating petroleum before, during, and after Neogene reservoir deposition (Figures 11–13). Work by Mello and Karner (1996) determined significantly older (Cretaceous–Paleogene) ages for maturation and generation of Mesozoic source rocks. Those results were obtained by integrating the effects of overpressures, but did not incorporate the effects of the high thermal conductivity of allochthonous salt.

The critical moment of petroleum generation for each of the potential sources varies throughout the study area (Figure 14). This variation is a direct result of the difference in age and maximum thickness of the original overlying allochthonous salt bodies, as well as variations in the timing and degree of salt evacuation. Because the timing of the critical moment of peak oil generation for a particular

**Table 3. Kinetic Parameters for Standard Type II and Type II-S Kerogens***

| | Standard Type II** | Type II-S† Monterey | Initial Potential (mg/g OC) | Type II-S† Kirkuk | Initial Potential (mg/g OC) |
|---|---|---|---|---|---|
| Oil potential from kerogen (mg/g TOC) | 350 | 350 | | 350 | – |
| Gas potential from kerogen (mg/g TOC) | 65 | – | | – | – |
| Oil to gas cracking efficiency (%) | 50 | – | | – | – |
| A factor, hydrocarbon generation (1/s) | $3.012 \times 10^{13}$ | $1.5 \times 10^{14}$ | | $1.5 \times 10^{14}$ | – |
| A factor, oil cracking (1/s) | $1.015 \times 10^{12}$ | – | | – | – |
| Activation energy for oil cracking (kcal/mol) | 54 | – | | – | – |
| Activation energy for hydrocarbon generation (kcal/mol); fraction†† | | | | | |
| 40 | – | 0.0030 | 2.0 | 0.0260 | 12.8 |
| 42 | – | – | | 0.0183 | 9.0 |
| 44 | – | 0.0137 | 9.0 | 0.0098 | 4.8 |
| 46 | – | 0.0222 | 14.5 | 0.0154 | 7.6 |
| 48 | – | 0.3106 | 203.3 | 0.0601 | 29.6 |
| 49 | 0.05 | – | – | – | – |
| 50 | 0.20 | 0.4349 | 284.7 | 0.7903 | 389.0 |
| 51 | 0.50 | – | – | – | – |
| 52 | 0.20 | 0.1448 | 94.8 | 0.061 | 30.0 |
| 53 | 0.05 | – | – | – | – |
| 54 | – | 0.052 | 34.0 | 0.0130 | 6.4 |
| 56 | – | 0.0104 | 6.8 | 0.0061 | 3.0 |
| 58 | – | 0.0038 | 2.5 | – | – |
| 60 | – | 0.0046 | 3.0 | – | – |

*OC = organic carbon, TOC = total organic carbon.
**Lawrence Livermore National Laboratory kinetics (Burnham and Sweeney, 1989).
†Type II-S kinetics (Tissot et al., 1987).
††Fraction of the kerogen having bonds with a particular activation energy.

source rock varies so dramatically across the study area, the different source intervals may all be within the peak oil generation window at the same time, but in different locations (Figure 14). Figure 14 was constructed by posting the output from thirteen BasinMod 1-D thermal maturation models onto a portion of the regional cross section restorations (from Figure 4) and connecting zones of equivalent maturity. The overlying salt geometry was taken under consideration during interpolation of maturity boundaries. The combination of structural restorations with 1-D thermal modeling nicely illustrates salt-influenced retardation of subsalt maturation, its lateral variations, and its rapid reequilibration once a salt weld forms (Figure 14). Because multiple source rocks may be within the peak oil generation window at the same time, it is likely that multiple sources contribute petroleum to the same reservoir rocks within any particular accumulation.

## Reservoir Rocks

Neogene siliciclastic turbidite systems comprise the reservoir rocks within the petroleum systems of northern Green Canyon/Ewing Bank (e.g., Cook and D'Onfro, 1991; Brannon et al., 1993; Holman and Robertson, 1994; Rafalowski et al., 1994; Harrison et al., 1995; Snyder and Nugent, 1995; Weimer et al., 1998b, this issue). These sediments were deposited within bathyal water depths in intraslope basins controlled primarily by active salt tectonics (Weimer et al., 1994, 1998b, this issue; Rowan and Weimer, 1998, this issue). Seismic and geologic facies show abrupt variations both laterally and temporally as a function of the structural controls on sediment pathways across the slope, and the positioning of shallow/marginal marine depocenters.

Most of the reservoirs occur within Pliocene and lower Pleistocene sands deposited in a variety of environments (Weimer et al., 1994, 1998b, this issue). Although the sediments are generally all fine grained, differences in geometries and rock types exist among these turbidite systems. Detailed sequence stratigraphic interpretations of this area were published by Acosta (1994), Martinez (1994), Navarro (1994), Budhijanto (1995), and Varnai (1996). Older reservoirs may exist within the study area, but wells have penetrated only to the upper middle Miocene.

### Level of Certainty

The level of certainty for a petroleum system represents the degree to which the available geological and geochemical data support the hypothesis that petroleum within the reservoir rocks originated from the designated source rocks (Magoon and Dow, 1994). The three levels of certainty are (1) known (!), (2) hypothetical (.), and (3) speculative (?) (see Magoon and Dow, 1994). General agreement exists that the source of oil for the Neogene reservoirs in the northern Green Canyon/Ewing Bank area is not from interbedded shales, but rather must be derived from depths well below current production (e.g., Bissada et al., 1990; Sassen et al., 1993). Because the proposed deep source rocks have not been penetrated in the offshore nor confirmed to be the actual sources for petroleum, the level of certainty for the Oxfordian-Neogene, Tithonian-Neogene, Albian-Neogene, Turonian-Neogene, and Eocene-Neogene petroleum systems has been assigned to hypothetical (.). In addition, it is not possible to determine the actual thickness, richness, thermal maturity, or areal extent of any of the individual sources responsible for generating the petroleum found within the study area; however, because of the reasonably well-established correlations between offshore oils and onshore source rocks of Mesozoic and Eocene age, the level of certainty for these petroleum systems could perhaps be raised to known (!).

## MIGRATION AND ACCUMULATION

### Petroleum Migration Pathways

Petroleum migration pathways in the Gulf of Mexico are believed to be predominantly vertical, driven by buoyancy and accentuated by compaction, dewatering, overpressures, microfractures, and faulting (Dow, 1984; Nunn and Sassen, 1986; Curtis, 1989; Hanor and Sassen, 1990; Demaison and Huizinga, 1991; Pratsch, 1991; Piggott and Pulham, 1993; Sassen et al., 1993; Waples, 1994b; Pepper and Yu, 1995; Mello and Karner, 1996). In the Green Canyon/Ewing Bank area, Neogene strata are dominated by fine-grained turbidite systems, whereas the deeper section we consider to be dominated by fine-grained, bathyal to distal mudstones and micrites. Significant fluid flow through such fine-grained rocks occurs only if large fluid-potential gradients are present (England et al., 1991). Significant overpressures were generated during the Cenozoic in this region due to rapid deposition of extremely thick sediments (Piggott and Pulham, 1993; Mello and Karner, 1996). Because the rate of deposition within a sedimentary basin usually changes slowly in a lateral direction, overpressure gradients are primarily vertical (England et al., 1991).

The combination of buoyancy and overpressure causes petroleum migration to be predominantly vertical in mudstones and other low-permeability strata (England et al., 1991). Overpressuring also results in low effective stress that promotes low sealing capacities and the creation of vertically oriented tension fractures, both of which further enhance vertical migration (Phillips, 1972; Meissner, 1978; du Rouchet, 1981; Piggott and Pulham, 1993; F. Meissner, 1993, personal communication; Mello and Karner, 1996). The vertical fractures occur when pore fluid pressure exceeds the least compressive stress ($s_3$, horizontal) and causes the rock to undergo tensile failure (Phillips, 1972; Meissner, 1978; Mello and Karner, 1996).

Concentration of petroleum along some Gulf of Mexico fault zones clearly shows that faults can play a major role in collecting and focusing petroleum migration (i.e., Vicksburg flexure in south Texas Frio-Vicksburg play); however, such occurrences commonly are associated with laterally extensive reservoir (carrier) beds offset by faulting and disrupting lateral petroleum flow. Regionally widespread, coarse-grained carrier beds that could have deflected regional petroleum migration pathways from vertical are most likely not present in the study area. Sandstone reservoirs, such as sheet sands and channel-fill sands, tend to be isolated and areally restricted. Some faults appear to serve as local migration conduits based on the occurrence of petroleum seeps at the sea floor along fault traces (Sassen et al., 1993) and apparent four-dimensional seismic anomalies (Anderson et al., 1991a, b, 1994). Faults may provide local zones of enhanced permeability, but they are unlikely to significantly deflect vertical petroleum migration flow paths because faults in the study area tend to be relatively steep and to access small drainage areas.

In contrast, allochthonous salt, which is impermeable, is believed to form major barriers to vertical petroleum migration (House and Pritchett, 1995; McBride et al., 1995; Pepper and Yu, 1995). The impermeability of salt prevents petroleum migration through it, causing petroleum migration pathways to be deflected laterally beneath salt (House and Pritchett, 1995; McBride et al., 1995; Pepper and Yu, 1995). Hydrofracturing associated with overpressures has no effect on salt because of the ability of salt to flow in a ductile manner.

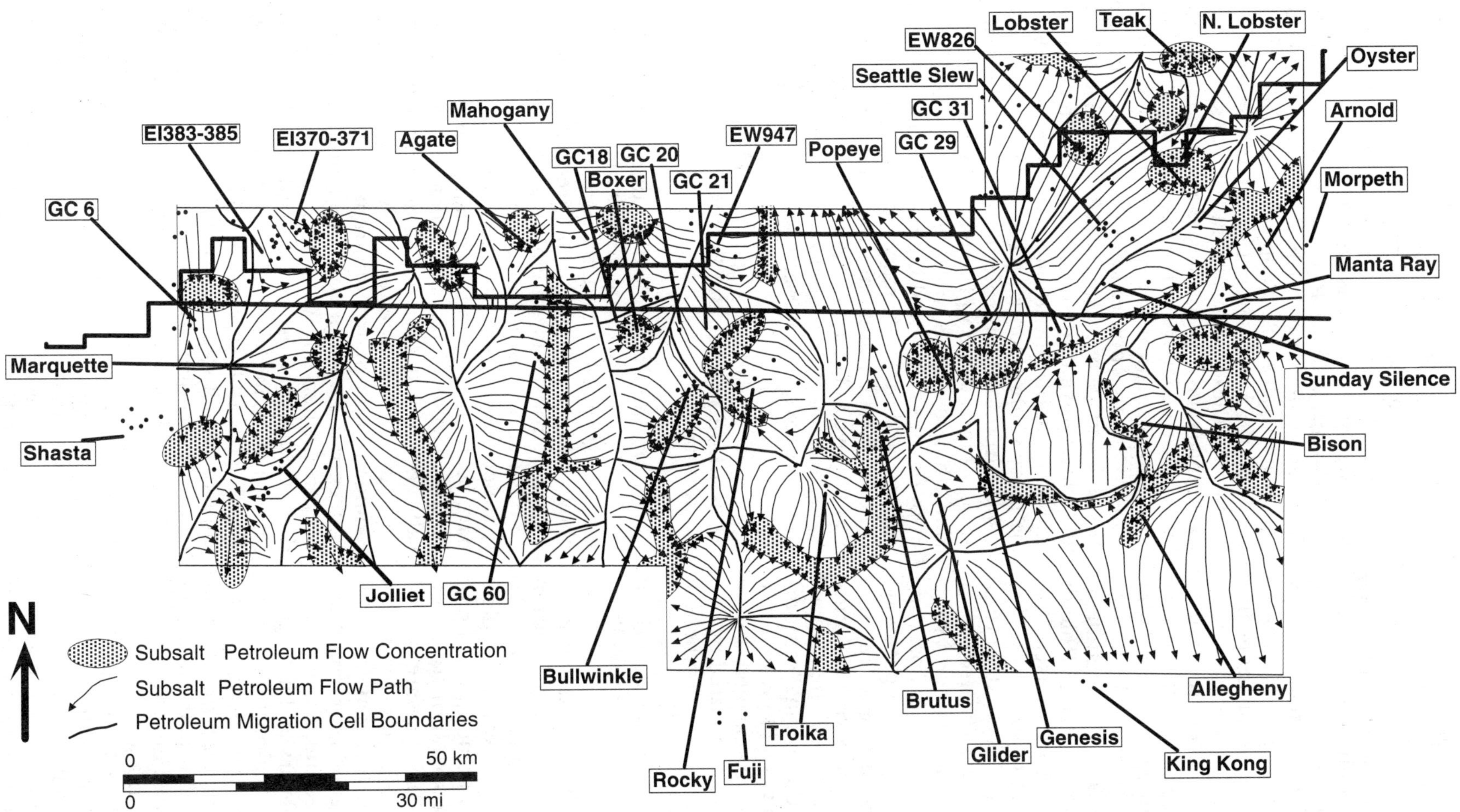


Figure 17—Reconstructed subsalt petroleum migration pathways map. Arrows show flow paths up the dip of interpreted base salt. Zones of flow concentration highlighted with flow cells outlined by heavy lines. The map represents diachronous allochthonous salt distribution and does not represent a discrete moment in time. Locations of fields and discoveries also are shown to illustrate their correspondence to original subsalt petroleum migration concentrations. See text for discussion.

## Allochthonous Salt and Petroleum Migration Pathways

Subsalt petroleum migration has been proposed to be vertical until it reaches the base of salt, at which point it is deflected laterally up the dip of the base salt (House and Pritchett, 1995; McBride, 1995; McBride et al., 1995; Pepper and Yu, 1995) (Figures 15, 16). Because of the variability of salt geometries through time, it is critical to understand the structural evolution of allochthonous salt to reconstruct ancient subsalt petroleum migration pathways and current accumulations. By integrating structural restorations of salt (2-D and 3-D), regional salt maps, thermal maturation modeling (1-D), and petroleum systems logic (Smith, 1994), potential petroleum migration pathways and zones of focused flow through time can be reconstructed.

By careful integration of eleven structural restorations across the study area (Figure 3, profiles A–K) and the regional top salt/salt weld structure map (see Figure 3), we interpreted past and present subsalt petroleum migration pathways and zones of petroleum concentration. The first step was to construct a base-salt structural form map. Where salt has been evacuated, the mapped weld represents the base of the original salt bodies. Where allochthonous salt is still present, the base of salt could sometimes be imaged, but in other places the base of salt was estimated by projection and interpolation. These eleven structural restorations (Figure 3, profiles A–K) were constructed using the same methodology described for Figure 4 and are discussed in detail by McBride et al. (1998, this issue).

The base-salt form map was then used to predict potential regional subsalt petroleum migration pathways prior to salt weld formation (Figure 17). Migration pathway arrows were drawn perpendicular to structural contours and point up the dip of the interpreted base salt (orthocontour, see MacKenzie and Quigley, 1988). There are several limitations and potential sources of error in this map. First, the map is based on the geometry of the base salt and thus is only as good as the base-salt structural interpretation. Second, the base-salt geometry is assumed to be primary and not significantly affected by subsequent deformation. Finally, and most importantly, the map does not represent a discrete moment in time, but rather is diachronous across the region due to variations in the timing of both salt extrusion and salt weld formation. Thus, the deflection of petroleum migration pathways shown in any given place is what would have occurred when salt was present at that location, which varies across the area from roughly the early Miocene (21.5 Ma) to the present. To map the petroleum migration pathways at a discrete moment in time requires 3-D structural restorations (see McBride, 1997).

Figure 17 depicts 32 subregional, subsalt cells of petroleum migration (areas of fetch) and concentration. Within these cells, major potential subsalt migration pathways (arteries) and zones of focused flow can be delineated. The structural style of the salt system (salt-stock canopies vs. salt sheets) (see Rowan, 1995; Schuster, 1995) has the greatest impact on these pathways. Individual salt stocks cause migration to disperse as a result of the elliptical, conical geometry of the base-salt surface; however, amalgamation of two or more stocks into a canopy creates domelike geometries along the base-salt surface between the salt stocks. Fluid flow is focused toward these culminations. In contrast, the base salt of allochthonous salt sheets tends to be more subhorizontal, with regional dips landward; therefore, petroleum migration beneath sheets tends to be unfocused. The merging of allochthonous sheets and salt-stock canopies, however, creates ridges and highs along the base-salt surface, and these are more likely to concentrate fluid flow. The potential subsalt migration pathways shown in Figure 17 assume hydrostatic conditions, and full-scale fluid-flow modeling is needed to evaluate whether these, in fact, are viable.

## Allochthonous Salt and Petroleum Accumulation

The locations of 42 fields or discoveries in the northern Green Canyon and Ewing Bank region may be explained, in part, by being located within, above, or in proximity to the predicted zones of subsalt petroleum migration concentration (Figure 17). Numerous additional zones where subsalt petroleum migration has been concentrated also have been modeled, but have not yet been tested. Because Figure 17 assumes a regional presence of salt, however, it does not accurately depict the true migration scenario throughout time. Remigration of petroleum from subsalt zones of concentration to suprasalt sediments would have occurred when salt welds formed, creating windows through the original allochthonous salt bodies and allowing vertical migration to resume. Thus, Figure 17 is supplemented by a map of present-day subsalt migration pathways (Figure 18). Figure 18 is noticeably different from Figure 17 because of extensive salt weld formation throughout the region. Where salt still exists, regional subsalt petroleum migration pathways are the same as depicted in Figure 17; where salt welds are present, petroleum migration is considered vertical and flow paths are not shown. Salt weld surfaces are not uniformly transmissible to vertical

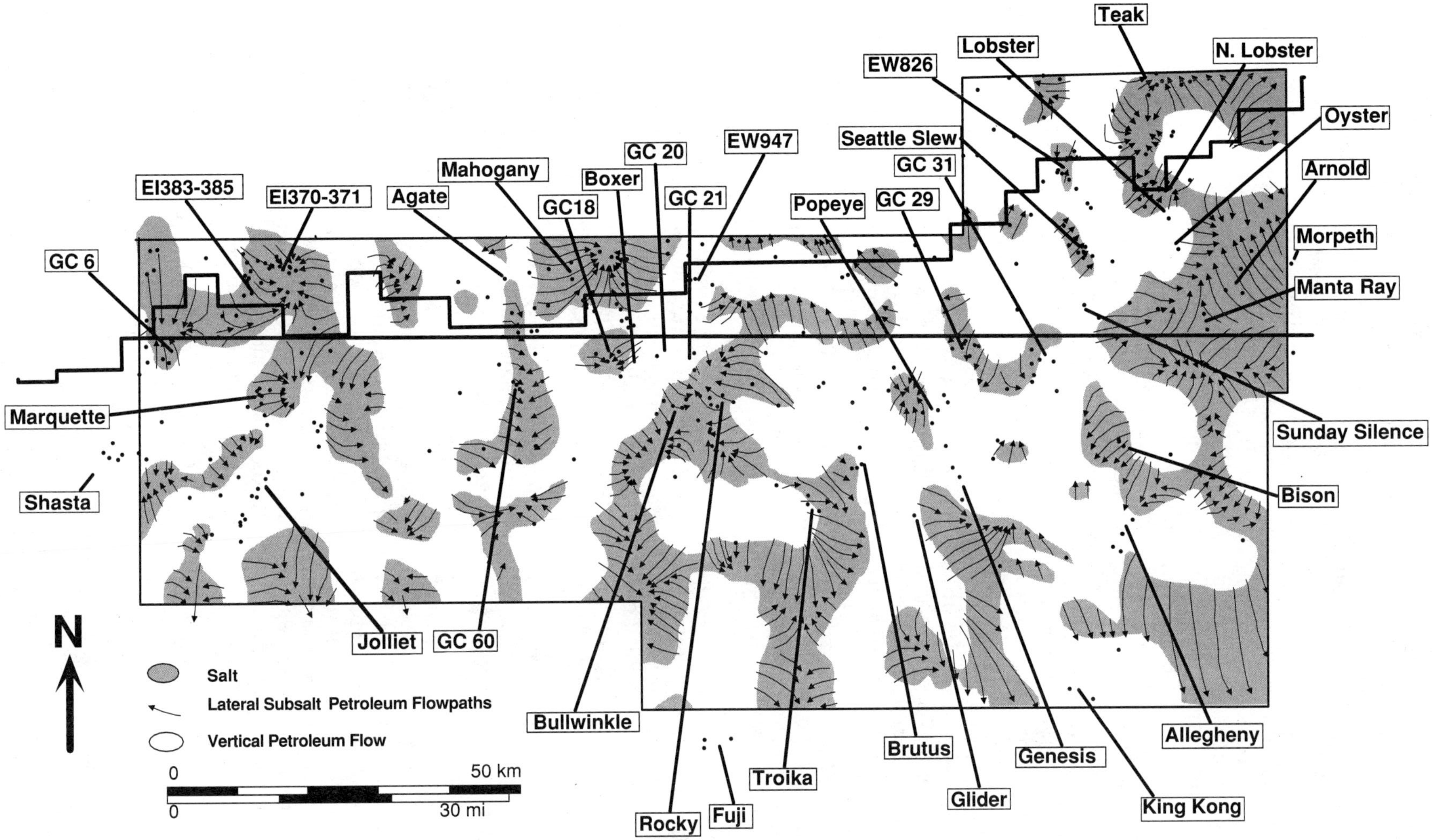


Figure 18—Present-day subsalt petroleum migration pathways map. Present-day salt distribution is shaded, with arrows showing subsalt flow paths. White represents salt welds, which are regions of vertical petroleum migration. Locations of fields and discoveries are shown to illustrate spatial positioning near edges of allochthonous salt (suprasalt) or where flow is concentrated (subsalt). See text for discussion.

migration, and therefore may alter the interpreted migration pathways.

Many of the zones of subsalt petroleum migration concentration illustrated in Figure 17 no longer exist because of the dynamic evolution of salt (see Figure 18). Nevertheless, numerous areas still remain as zones of subsalt petroleum migration concentration from deep source rocks currently generating petroleum. The locations of most suprasalt fields tend to be along the flanks of present-day allochthonous salt (Figure 18) and near zones of concentrated subsalt flow (Figure 17). We presume that fields have not been found above areally extensive allochthonous salt because these sediments are still shielded by the salt from receiving petroleum-generated subsalt. Likewise, zones of subsalt concentration still covered by allochthonous salt are potential areas of subsalt entrapment.

Although beyond the scope of this study, volumetrics of the amount of petroleum generated within each of the individual flow cells could be estimated by conducting more extensive thermal maturation modeling in 2-D or 3-D, and by using mass balance/genetic yield potential calculations for the various source beds (Waples, 1979, 1981; MacKenzie and Quigley, 1988); however, because there are no penetrations of these source rocks, the necessary input parameters for source richness, thicknesses, and genetic yield would have to be estimated. In addition, more extensive structural restorations in 3-D would be required to constrain the allochthonous salt thicknesses through time. Nevertheless, the exercise would be beneficial in ranking the individual cells based on the predicted volumes of generated petroleum, migration distances, and timing relationships relative to salt tectonics.

Figures 17 and 18 depict only petroleum migration pathways and concentration; other factors such as the presence of high-quality reservoir rocks and adequate structural trapping configurations are needed for commercial discoveries. The combination of petroleum migration pathway maps, geologic facies maps, depth-structure maps, and thermal modeling is essential to high-grade areas of interest and focus exploration efforts. The petroleum flowpath maps provide a means of estimating the area of fetch and quantities of petroleum generation and concentration. In addition, these maps provide information that could be used to influence bidding and leasing strategies.

## CONCLUSIONS

The northern Green Canyon/Ewing Bank region, northern Gulf of Mexico Basin, is believed to contain the Oxfordian–Neogene (.), Tithonian–Neogene (.), Albian–Neogene (.), Turonian–Neogene (.), and Eocene–Neogene (.) petroleum systems. The systems encompass 42 fields or discoveries in the study area and include four subsalt discoveries. Essential elements of the systems include source rocks of Oxfordian, Tithonian, Albian, Turonian, and Eocene age; Neogene siliciclastic turbidite reservoir rocks; allochthonous salt as migration barriers; and overburden strata ranging in age from Late Jurassic to Quaternary. Although the petroleum systems cover a much larger area regionally, analyses were restricted to the local study area and focused on the expulsion-migration subsystems.

The petroleum systems of the study area were significantly affected by the evolution of allochthonous salt. Assessing these evolving petroleum systems requires a means of quantifying allochthonous salt flow (changing salt thicknesses through time). The integration of sequence stratigraphy and structural restorations of allochthonous salt provides the necessary foundation for conducting thermal maturation modeling and petroleum migration analyses. Sequence stratigraphy constrains the geometries and ages of the incremental structural restorations. The incremental structural restorations, in turn, constrain salt thickness variations through time and the timing of salt weld formation. The high thermal conductivity of salt retards the thermal maturation of subsalt source rocks and causes late petroleum generation below thick salt bodies. The fine-grained strata of the region, in association with hydrocarbon buoyancy, faults, fractures, and overpressures, causes fluid flow to be primarily vertical; however, the impermeability of salt prevents vertical petroleum migration and causes migration pathways to be deflected laterally up the dip of the base of allochthonous salt. Where salt withdrawal eventually formed a salt weld, vertical petroleum migration is unimpeded; therefore, the petroleum systems of the region are a hybrid of vertical and lateral drainage associated with low and high impedance to flow. Lateral and vertical migration pathways varied spatially and temporally throughout the basin's evolution.

Most known traps were formed during the Pliocene–Pleistocene, and the generation-migration-accumulation of petroleum started during the Late Cretaceous and continues to the present; however, the critical moment of peak oil generation for each of the potential source rocks varies throughout the study area as a function of the overlying allochthonous salt evolution. Because the critical moment of peak oil generation varies spatially, overlaps in the timing of generation exist among the different source rocks. This relationship enhances the prospectivity of this area because it allows late generation in multiple source rocks, and early migration and accumulation into numerous porous Neogene reservoirs.

By integrating predictions of potential source rocks, structural restorations, thermal maturation modeling, regional salt maps, and petroleum systems logic, it is possible to delineate potential petroleum migration pathways and their variations through time, and to identify potential zones of focused petroleum flow through time. Major allochthonous salt sheets and canopies existed during periods of peak oil generation (late Miocene–late Pliocene). The base-salt geometry during these times influenced regional subsalt petroleum migration pathways and created zones of petroleum concentration beneath salt. All 42 fields or discoveries within the study area are located directly in, above, or very close to the zones of subsalt petroleum migration concentration. Most of these fields are within suprasalt reservoirs that get charged with remigrated petroleum once salt welds form above zones of subsalt petroleum concentration. Present-day salt distributions and geometries do not delineate many of these zones of ancient subsalt petroleum migration concentration because of the extensive formation of salt welds during the Pleistocene; however, many prospective zones of subsalt petroleum concentration have remained active to the present because salt welds have not yet formed above them.

The proposed methodology of evaluating the effect of allochthonous salt evolution on petroleum systems sheds light on the expulsion-migration subsystems. Sequential structural restorations, when used for petroleum systems analysis, can provide substantial benefits to explorationists in addition to validating interpretations. The overall technique enables geoscientists to focus their exploration efforts toward areas with the greatest probability of exploration success. Although the method delineates zones of petroleum concentration, reservoirs and traps still need to be present and mapped within these areas for commercial discoveries.

## REFERENCES CITED


Acosta, Z. M., 1994, Sequence stratigraphy of Plio–Pleistocene sediments in north-central Green Canyon and western Ewing Bank, northern Gulf of Mexico: M.S. thesis, University of Colorado, Boulder, Colorado, 187 p.

Anderson, R. N., W. He, M. A. Hobart, C. R. Wilkinson, and H. R. Nelson, 1991a, Active fluid flow in the Eugene Island area, offshore Louisiana: The Leading Edge, v. 10, no. 4, p. 12–17.

Anderson, R. N., L. M. Cathles, and H. R. Nelson, 1991b, "Data Cube" depicting fluid flow history in Gulf Coast sediments: Oil & Gas Journal, v. 89, no. 44, p. 60–65.

Anderson, R. N., P. Flemings, S. Losh, J. Austin, and R. Woodhams, 1994, Gulf of Mexico growth fault drilled, seen as oil, gas migration pathway: Oil & Gas Journal, v. 92, no. 23, p. 97–104.

Baldwin, B., and C. O. Butler, 1985, Compaction curves: AAPG Bulletin, v. 69, p. 622–625.

Bissada, K. K., B. J. Katz, S. C. Barnicle, and D. J. Schunk, 1990, On the origin of hydrocarbons in the Gulf of Mexico Basin—a reappraisal, *in* D. Schumacher and B. F. Perkins, eds., Gulf coast oils and gases, their characteristics, origin, distribution, and exploration and production significance: Gulf Coast Section SEPM Foundation Ninth Annual Research Conference, p. 163–171.

Blackwell, D. D., and J. L. Steele, 1989, Thermal conductivity of rocks: measurement and significance, *in* N. D. Naeser and T. H. McCulloh, eds., Thermal history of sedimentary basins: methods and case histories: New York, Springer-Verlag, p. 13–36.

Brannon, D., A. Chedburn, W. J. Schneider, and M. Sheedlo, 1993, Jolliet field revisited: producibility of upper slope turbidites in Green Canyon 184 (abs): AAPG Annual Convention Program, v. 2, p. 79–80.

Budhijanto, F. M., 1995, Sequence stratigraphy of Plio–Pleistocene sediments, northeastern Green Canyon and eastern Ewing Bank, Gulf of Mexico: M.S. thesis, University of Colorado, Boulder, Colorado, 140 p.

Buffler, R. T., 1991, Seismic stratigraphy of the deep Gulf of Mexico Basin and adjacent margins, *in* A. Salvador, ed., The Gulf of Mexico Basin: Decade of North American Geology, Geological Society of America, p. 353–387.

Burnham, A. K., and J. J. Sweeney, 1989, A chemical kinetic model of vitrinite maturation and reflectance: Geochimica et Cosmochimica Acta, v. 53, p. 2649–2657.

Cathles, L. M., and A. G. Nunns, 1991, A temperature probe survey on the Louisiana shelf: effects of bottom-water temperature variations: AAPG Bulletin, v. 75, p. 180–186.

Cook, D., and P. D'Onfro, 1991, Jolliet field thrust fault structure and stratigraphy, Green Canyon Block 184, offshore Louisiana: Gulf Coast Association of Geological Societies Transactions, v. 41, p. 100–121.

Curtis, D. M., 1989, A conceptual model for sources of oils in Gulf Coast Cenozoic reservoirs: Gulf Coast Association of Geological Societies Transactions, v. 39, p. 37–56.

Demaison, G. J., and B. J. Huizinga, 1991, Genetic classification of petroleum systems: AAPG Bulletin, v. 75, p. 1626–1643.

Deming, D., 1989, Application of bottom-hole temperature corrections in geothermal studies: Geothermics, v. 18, p. 775–786.

Diegel, F. A., J. F. Karlo, D. C. Schuster, R. C. Shoup, and P. R. Tauvers, 1995, Cenozoic structural evolution and tectonostratigraphic framework of the northern Gulf Coast continental margin, *in* M. P. A. Jackson, D. G. Roberts, and S. Snelson, eds., Salt tectonics: a global perspective: AAPG Memoir 65, p. 109–151.

Dow, W. G., 1984, Oil source beds and oil prospect definition in the upper Tertiary of the Gulf Coast: Gulf Coast Association of Geological Societies Transactions, v. 34, p. 329–339.

du Rouchet, J. H., 1981, Stress field, a key to oil migration: AAPG Bulletin, v. 65, p. 74–85.

England, W. A., A. L. Mann, and D. M. Mann, 1991, Migration from source to trap, *in* R. K. Merril, ed., Source and migration processes and evaluation techniques: AAPG Treatise of Petroleum Geology, Handbook of Petroleum Geology, p. 23–46.

Epp, D., P. J. Grim, and M. G. Langseth, Jr., 1970, Heat flow in the Caribbean and Gulf of Mexico: Journal of Geophysical Research, v. 75, p. 5655–5669.

Falvey, D. A., and M. F. Middleton, 1981, Passive continental margins: evidence for a prebreakup deep crustal metamorphic subsidence mechanism: Oceanologica Acta SP, p. 103–114.

Feng, J., 1995, Post mid-Cretaceous seismic stratigraphy and depositional history, deep Gulf of Mexico: Ph.D. dissertation, University of Texas at Austin, Austin, Texas, 253 p.

Galloway, W. E., D. G. Bebout, W. L. Fisher, J. B. Dunlap, R. Cabera-Castro, J. E. Lugo-Rivera, and T. M. Scott, 1991, Cenozoic, *in* A. Salvador, ed., The Gulf of Mexico Basin: Geological Society of America, The Geology of North America, v. J, p. 245–324.

Hanor, J. S., and R. Sassen, 1990, Evidence for large-scale vertical and lateral migration of formation waters, dissolved salt, and crude oil in the Louisiana gulf coast, *in* D. Schumacher and B. F. Perkins, eds., Gulf Coast oils and gases, their characteristics, origin, distribution, and exploration and production significance: Gulf Coast Section SEPM Foundation Ninth Annual Research Conference, p. 283–296.

Harrison, H., D. Moore, and P. Hodgkins, 1995, The Mahogany subsalt discovery: a unique hydrocarbon play, offshore Louisiana: Gulf Coast Section SEPM Foundation Sixteenth Annual Research Conference, p. 95-97.

Herbin, J. P., G. Deroo, and J. Roucache, 1983, Organic geochemistry of lower Cretaceous sediments from Site 535, Leg 77: Florida straits: Initial Reports of the Deep-Sea Drilling Project Leg 77, p. 459-474.

Hermanrud, C., 1993, Basin modelling techniques—an overview, *in* A. G. Dore, J. H. Auguston, C. Hermanrud, D. S. Stewart, and O. Sylta, eds., Basin modelling: advances and applications: Norwegian Petroleum Society (NPF) Special Publication 3, p. 1-34.

Hermanrud, C., and P. Y. Shen, 1989, Virgin rock temperatures from well logs—accuracy analysis for some advanced inversion models: Marine and Petroleum Geology, v. 6, p. 360-363.

Hermanrud, C., S. Cao, and I. Lerche, 1990, Estimates of virgin rock temperature derived from BHT measurements: bias and error: Geophysics, v. 55, p. 924-931.

Holman, W. E., and S. S. Robertson, 1994, Field development, depositional model, and production performance of the turbiditic "J" sands at prospect Bullwinkle, Green Canyon 65 field, outer-shelf Gulf of Mexico, *in* P. Weimer, A. H. Bouma, and B. F. Perkins, eds., Submarine fans and turbidite systems: Gulf Coast Section SEPM Foundation Fifteenth Annual Research Conference, p. 139-150.

Horner, D. R., 1951, Pressure buildups in wells: Third World Petroleum Congress Proceedings, The Hague, v. 2, p. 503-521.

House, W. M., and J. A. Pritchett, 1995, Fluid migration and formation pressures associated with allochthonous salt sheets in the northern Gulf of Mexico: Gulf Coast Section SEPM Foundation Sixteenth Annual Research Conference, p. 121-124.

Jarvis, G. T., and D. P. McKenzie, 1980, Sedimentary basin formation with finite extension rates: Earth and Planetary Science Letters, v. 48, p. 42-52.

Keen, C. E., and T. Lewis, 1982, Measured radiogenic heat production in sediments from continental margin of eastern North America: implications for petroleum generation: AAPG Bulletin, v. 66, p. 1402-1407.

Kennicutt, M. C., A. G. Requejo, R. Sassen, and J. M. Brooks, 1993, Oil formation in the northern Gulf of Mexico: geochemical evidence (abs): AAPG Annual Convention Program, v. 2, p. 128.

MacKenzie, A. S., and T. M. Quigley, 1988, Principles of geochemical prospect appraisal: AAPG Bulletin, v. 72, p. 399-415.

Magoon, L. B., 1989, The petroleum system—status of research and methods, 1990: U.S. Geological Survey Bulletin 1912, 88 p.

Magoon, L. B., and W. G. Dow, eds., 1994, The petroleum system—from source to trap: AAPG Memoir 60, 655 p.

Martinez, R. E., 1994, Sequence stratigraphy of Plio-Pleistocene sediments, northwestern Green Canyon, Gulf of Mexico: M.S. thesis, University of Colorado, Boulder, Colorado, 138 p.

McBride, B. C., 1995, Evaluation of subsalt petroleum potential using structural restorations (abs.): AAPG Annual Convention Program, v. 4, p. 62A.

McBride, B. C., 1997, The geometry and evolution of allochthonous salt and its impact on petroleum systems, northern Gulf of Mexico Basin: studies in three- and four-dimensional analysis: Ph.D. dissertation, University of Colorado, Boulder, Colorado, 276 p.

McBride, B. C., M. G. Rowan, P. Weimer, and R. Ratliff, 1995, Three-dimensional restoration of allochthonous salt structures and its impact on understanding reservoir trends and petroleum migration: preliminary results from Green Canyon and Ewing Bank, northern Gulf of Mexico: Gulf Coast Section SEPM Foundation Sixteenth Annual Research Conference, p. 171-175.

McBride, B. C., M. G. Rowan, and P. Weimer, 1998, The evolution of allochthonous salt systems, northern Green Canyon and Ewing Bank (offshore Louisiana), northern Gulf of Mexico: AAPG Bulletin, v. 82, no. 5B, p. 1013-1036.

McFarlan, E., Jr., and L. S. Menes, 1991, Lower Cretaceous, *in* A. Salvador, ed., The Gulf of Mexico Basin: Geological Society of America, The Geology of North America, Boulder, Colorado, v. J, p. 181-204.

McKenna, T. E., and J. M. Sharp, Jr., 1998, Radiogenic heat production in sedimentary rocks of the Gulf of Mexico Basin (south Texas): AAPG Bulletin, v. 82, no. 3, p. 484-496.

McKenzie, D. P., 1978, Some remarks on the development of sedimentary basins: Earth and Science Planetary Letters, v. 40, p. 25-32.

Meissner, F. F., 1978, Petroleum geology of the Bakken Formation, Williston basin, North Dakota and Montana: The economic geology of the Williston basin: Montana Geological Society 1978 Williston Basin Symposium, Billings, Montana, p. 207-227.

Mello, U. T., and G. D. Karner, 1996, Development of sediment overpressure and its effect on thermal maturation: application to the Gulf of Mexico Basin: AAPG Bulletin, v. 80, p. 1367-1396.

Mello, U. T., R. N. Anderson, and G. D. Karner, 1994, Salt restrains maturation in subsalt plays: Oil & Gas Journal, January 31, v. 92, no. 5, p. 101-108.

Middleton, M. F., 1979, A model for bottom-hole temperature stabilization: Geophysics, v. 44, p. 1458-1462.

Nagihara, S., L. E. Beckley, E. W. Behrens, and J. G. Sclater, 1993, Characteristics of heat flow through diapiric salt structures on the Texas continental slope: Gulf Coast Association of Geological Societies Transactions, v. 43, p. 269-279.

Nagihara, S., J. G. Sclater, J. D. Phillips, E. W. Behrens, T. Lewis, L. A. Lawver, Y. Nakamura, J. Garcia-Abdeslem, and A. E. Maxwell, 1996, Heat flow in the western abyssal plain of the Gulf of Mexico: implications for thermal evolution of the old oceanic lithosphere: Journal of Geophysical Research, v. 101, p. 2895-2914.

Navarro, A., 1994, Sequence stratigraphy of Plio-Pleistocene sediments, northwestern Green Canyon and western Ewing Bank, Gulf of Mexico: M.S. thesis, University of Colorado, Boulder, Colorado, 150 p.

Nunn, J. A., and R. Sassen, 1986, The framework of hydrocarbon generation and migration, Gulf of Mexico continental slope: Gulf Coast Association of Geological Societies Transactions, v. 36, p. 257-262.

Nunn, J. A., A. D. Scardina, and R. H. Pilger, Jr., 1984, Thermal evolution of the north-central Gulf Coast: Tectonics, v. 7, p. 723-740.

O'Brien, J. J., and I. Lerche, 1984, The influence of salt domes on paleotemperature distributions: Geophysics, v. 49, p. 2032-2043.

O'Brien, J. J., and I. Lerche, 1988, Impact of heat flux anomalies around salt diapirs and salt sheets in the Gulf Coast on hydrocarbon maturity: Gulf Coast Association of Geologic Societies, v. 58, p. 231-243.

Oehler, J. H., 1984, Carbonate source rocks in the Jurassic Smackover trend of Mississippi, Alabama, and Florida, *in* J. G. Palacas, ed., Petroleum geochemistry and source rock potential of carbonate rocks: AAPG Studies in Geology 18, p. 63-70.

Patton, J. W., P. W. Choquette, G. K. Guennel, A. J. Kaltenback, and A. Moore, 1983, Organic geochemistry and sedimentology of lower to mid-Cretaceous deep-sea carbonates, Sites 535 and 540, Leg 77: Initial Reports of the Deep-Sea Drilling Project Leg 77, p. 417-443.

Peel, F. J., C. J. Travis, and J. R. Hossack, 1995, Genetic structural provinces and salt tectonics of the Cenozoic offshore U.S. Gulf of Mexico: a preliminary analysis, *in* M. P. A. Jackson, D. G. Roberts, and S. Snelson, eds., Salt tectonics: a global perspective: AAPG Memoir 65, p. 153-175.

Pepper, A. S., and Z. Yu, 1995, Influence of an inclined salt sheet on petroleum emplacement in the Pompano field area, offshore Gulf of Mexico Basin: Gulf Coast Section SEPM Foundation Sixteenth Annual Research Conference, p. 197-205.

Phillips, W. J., 1972, Hydraulic fracturing and mineralization: Journal of the Geological Society of London, v. 128, p. 337-359.

Piggott, N., and A. J. Pulham, 1993, Sedimentation rate as the control on hydrocarbon sourcing, generation and migration in the deepwater Gulf of Mexico: Gulf Coast Section SEPM Foundation Fourteenth Annual Research Conference, p. 179-191.

Pilger, R. H., Jr., 1981, The opening of the Gulf of Mexico; implications for the northern Gulf Coast: Gulf Coast Association of Geological Societies Transactions, v. 3, p. 377–381.

Pindell, J. L., 1985, Alleghenian reconstruction and subsequent evolution of the Gulf of Mexico, Bahamas, and proto-Caribbean: Tectonics, v. 4, p. 1–39.

Pollack, H. N., S. J. Hurter, and J. R. Johnson, 1993, Heat flow from the Earth's interior: analysis of the global data set: Reviews of Geophysics, v. 31, p. 231–245.

Pratsch, J. C., 1991, Vertical hydrocarbon migration: a major exploration parameter: Journal of Petroleum Geology, v. 14, no. 4, p. 429–444.

Pulham, A. J., 1993, Variations in slope deposition, Pliocene–Pleistocene, offshore Louisiana, northeast Gulf of Mexico, *in* P. Weimer and H. W. Posamentier, eds., Siliciclastic sequence stratigraphy: recent developments and applications: AAPG Memoir 58, p. 199–234.

Rafalowski, J. W., B. W. Regel, D. J. Jordan, and D. O. Lucidi, 1994, Green Canyon block 205 lithofacies, seismic facies, and reservoir architecture, *in* P. Weimer, A. H. Bouma, and B. F. Perkins, eds., Submarine fans and turbidite systems: Gulf Coast Section SEPM Foundation Fifteenth Annual Research Conference, p. 293–306.

Rice, D. D., 1980, Chemical and isotope evidence of the origins of natural gases in offshore Gulf of Mexico: Gulf Coast Association of Geological Societies Transactions, v. 30, p. 203–213.

Rice, D. D., and G. E. Claypool, 1981, Generation, migration, and resource potential of biogenic gas: AAPG Bulletin, v. 65, p. 5–25.

Rowan, M. G., 1993, A systematic technique for the sequential restoration of salt structures: Tectonophysics, v. 228, p. 331–348.

Rowan, M. G., 1995, Structural styles and evolution of allochthonous salt, central Louisiana outer shelf and upper slope, *in* M. P. A. Jackson, D. G. Roberts, and S. Snelson, eds., Salt tectonics: a global perspective: AAPG Memoir 65, p. 199–228.

Rowan, M. G., and P. Weimer, 1998, Salt-sediment interaction, Northern Green Canyon and Ewing Bank (offshore Louisiana), northern Gulf of Mexico: AAPG Bulletin, v. 82, no. 5B, p. 1055–1082.

Rowan, M. G., T. Villamil, P. B. Flemings, and P. Weimer, 1996, Use of cross-section restoration to determine paleobathymetry and sea-floor paleotopography in the Gulf of Mexico Basin: Geology, v. 24, p. 299–302.

Salvador, A., 1987, Late Triassic–Jurassic paleogeography and origin of the Gulf of Mexico: AAPG Bulletin, v. 71, p. 419–451.

Salvador, A., ed., 1991a, The Gulf of Mexico Basin: Geological Society of America, The Geology of North America, v. J, 568 p.

Salvador, A., 1991b, Triassic–Jurassic, *in* A. Salvador, ed., The Gulf of Mexico Basin: Geological Society of America, The Geology of North America, v. J, p. 131–180.

Sassen, R., 1990, Geochemistry of carbonate source rocks and crude oils in Jurassic salt basins of the Gulf Coast, *in* J. Brooks, ed., Classic petroleum provinces: Geological Society Special Publication 50, p. 265–277.

Sassen, R., and E. W. Chinn, 1990, Implications of lower Tertiary source rocks in south Louisiana to the origin of crude oil, offshore Louisiana, *in* D. Schumacher and B. F. Perkins, eds., Gulf coast oils and gases, their characteristics, origin, distribution, and exploration and production significance: Gulf Coast Section SEPM Foundation Ninth Annual Research Conference, p. 175–179.

Sassen, R., and C. H. Moore, 1988, Framework of hydrocarbon generation and destruction in eastern Smackover trend: AAPG Bulletin, v. 72, p. 649–663.

Sassen, R., R. S. Tye, E. W. Chinn, and R. C. Lemoine, 1988, Origin of crude oil in the Wilcox trend of Louisiana and Mississippi: evidence of long-range migration: Gulf Coast Association of Geological Societies Transactions, v. 38, p. 27–34.

Sassen, R., J. M. Brooks, M. C. Kennicutt, I. A. MacDonald, and N. L. Guinasso, 1993, How oil seeps, discoveries relate in deepwater Gulf of Mexico: Oil & Gas Journal, April 19, v. 91, no. 16, p. 64–69.

Sawyer, D. S., R. T. Buffler, and R. H. Pilger, Jr., 1991, The crust under the Gulf of Mexico Basin, *in* A. Salvador, ed., The Gulf of Mexico Basin: Geological Society of America, The Geology of North America, v. J, p. 53–72.

Schuster, D. C., 1995, Deformation of allochthonous salt and evolution of related salt/structural systems, eastern Louisiana Gulf Coast, *in* M. P. A. Jackson, D. G. Roberts, and S. Snelson, eds., Salt tectonics: a global perspective: AAPG Memoir 65, p. 177–198.

Sclater, J. G., and P. A. F. Christe, 1980, Continental stretching: an explanation of the post-mid-Cretaceous subsidence of the Central North Sea basin: Journal of Geophysical Research, v. 85, p. 3711–3739.

Sclater, J. G., and S. Nagihara, 1991, Comment on "Heat flow in the Caribbean and Gulf of Mexico" by D. Epp, P. J. Grim, and M. G. Langseth, Jr.: Journal of Geophysical Research, v. 96, p. 21,807–21,810.

Smith, D. L., W. T. Dees, and D. W. Harrelson, 1981, Geothermal conditions and their implications for basement tectonics in the Gulf Coast margin: Gulf Coast Association of Geological Societies Transactions, v. 31, p. 181–190.

Smith, J. T., 1994, Petroleum system logic as an exploration tool in a frontier setting, *in* L. B. Magoon and W. G. Dow, eds., The petroleum system—from source to trap: AAPG Memoir 60, p. 25–49.

Snyder, F. C., and J. A. Nugent, 1995, Teak-testing a subsalt hydrocarbon trap geometry, South Timbalier block 260, Gulf of Mexico: Gulf Coast Section SEPM Foundation Sixteenth Annual Research Conference, p. 257–267.

Sohl, N. F., R. E. Martínez, P. Salerón-Ureña, and F. Soto-Jaramillo, 1991, Upper Cretaceous, *in* A. Salvador, ed., The Gulf of Mexico Basin: Geological Society of America, The Geology of North America, v. J, p. 205–244.

Sweeney, J. J., and A. K. Burnham, 1990, Evaluation of a simple model of vitrinite reflectance based on chemical kinetics: AAPG Bulletin, v. 74, p. 1559–1570.

Tissot, B. P., R. Pelet, and P. Ungerer, 1987, Thermal history of sedimentary basins, maturation indices, and kinetics of oil and gas generation: AAPG Bulletin, v. 71, p. 1445–1466.

Trudgill, B. D., et al., 1995, The structural geometry and evolution of the salt-related Perdido fold belt, Alaminos Canyon, northwestern deep Gulf of Mexico: Gulf Coast Section SEPM Foundation Sixteenth Annual Research Conference, p. 275–284.

Ulmishek, G., and H. D. Klemme, 1990, Depositional controls, distribution, and effectiveness of the world's petroleum source rocks: U.S. Geological Survey Bulletin 1931, 59 p.

Varnai, P., 1996, Integrated 2-D and 3-D sequence stratigraphic analysis of Plio-Pleistocene turbidite systems, northern Green Canyon and central Ewing Bank protraction areas, northern Gulf of Mexico: Ph.D. dissertation, University of Colorado, Boulder, Colorado, 502 p.

Walters, C. C., and D. D. Dusang, 1988, Source and thermal history of oils from Lockhart Crossing, Livingston Parish, Louisiana: Gulf Coast Association of Geological Societies Transactions, v. 38, p. 37–44.

Waples, D. W., 1979, Simple method for oil source bed evaluation: AAPG Bulletin, v. 63, p. 239–245.

Waples, D. W., 1981, Organic geochemistry for exploration geologists: Minneapolis, Burgess Publishing Company, 151 p.

Waples, D. W., 1994a, Maturity modeling: thermal indicators, hydrocarbon generation, and oil cracking, *in* L. B. Magoon and W. G. Dow, eds., The petroleum system—from source to trap: AAPG Memoir 60, p. 285–306.

Waples, D. W., 1994b, Modeling of sedimentary basins and petroleum systems, *in* L. B. Magoon and W. G. Dow, eds., The petroleum system—from source to trap: AAPG Memoir 60, p. 307–322.

Waples, D. W., H. Kamata, and M. Suizu, 1992a, The art of maturity modeling, part 1: finding a satisfactory geologic model: AAPG Bulletin, v. 76, p. 31–46.

Waples, D. W., M. Suizu, and H. Kamata, 1992b, The art of maturity modeling, part 2: alternative models and sensitivity analysis: AAPG Bulletin, v. 76, p. 47–66.

Weimer, P., P. Varnai, Z. M. Acosta, F. M. Budhijanto, R. E. Martinez, A. F. Navarro, M. G. Rowan, B. C. McBride, and T. Villamil, 1994,

Sequence stratigraphy of Neogene turbidite systems, northern Green Canyon and Ewing Bank, northern Gulf of Mexico: Gulf Coast Section SEPM Fifteenth Annual Research Conference, p. 383–396.

Weimer, P., J. R. Crews, R. S. Crow, and P. Varnai, 1998a, Atlas of petroleum fields and discoveries, northern Green Canyon, Ewing Bank, and Southern Ship Shoal and South Timblier areas (offshore Louisiana), northern Gulf of Mexico: AAPG Bulletin, v. 82, no. 5B, p. 878–917.

Weimer, P., P. Varnai, F. M. Budhijanto, Z. M. Acosta, R. E. Martinez, A. F. Navarro, M. G. Rowan, B. C. McBride, T. Villamil, C. Arango, and A. J. Pulham, 1998b, Sequence stratigraphy of Pliocene and Pleistocene turbidite systems, northern Green Canyon and Ewing Bank (offshore Louisiana), northern Gulf of Mexico: AAPG Bulletin, v. 82, no. 5B, p. 918–960.

Wenger, L. M., R. Sassen, and D. Schumacher, 1990, Molecular characteristics of Smackover, Tuscaloosa, and Wilcox-reservoired oils in the eastern Gulf Coast, *in* D. Schumacher and B. F. Perkins, eds., Gulf Coast oils and gases, their characteristics, origin, distribution, and exploration and production significance: Gulf Coast Section SEPM Foundation Ninth Annual Research Conference, p. 37–57.

Wenger, L. M., L. R. Goodoff, O. P. Gross, S. C. Harrison, and K. C. Hood, 1994, Northern Gulf of Mexico: an integrated approach to source, maturation, and migration: First Joint AAPG/AMGP Hedberg Research Conference Program, 6 p.

Winker, C. D., 1982, Cenozoic shelf-margins, northwestern Gulf of Mexico: Gulf Coast Association of Geological Societies Transactions, v. 32, p 427–448.

Worrall, D. M., and S. Snelson, 1989, Evolution of the northern Gulf of Mexico, with emphasis on Cenozoic growth faulting and the role of salt, *in* A. W. Bally and A. R. Palmer, eds., The geology of North America; an overview: Geological Society of America Decade of North American Geology, v. A, p. 97–138.

Yu, A., and I. Lerche, 1992, Thermal impact of salt: simulation of thermal anomaly in the Gulf of Mexico: Pure and Applied Geophysics, v. 138, no. 2, p. 180–192.

## ABOUT THE AUTHORS

### Barry C. McBride

Barry McBride is a geologist at HS Resources Inc. He received his Ph.D. in geology at the University of Colorado, his M.S. degree in geology from Western Michigan University, and his B.S. degree in geology from Furman University. From 1988 to 1993 he worked for Mobil Exploration and Producing U.S. Inc. and Mobil Global Interpretation Support as an exploration geologist. His research interests focus on petroleum exploration, complex structures, and their petroleum systems.

### Paul Weimer

Paul Weimer is an associate professor and director of the Energy and Minerals Applied Research Center at the University of Colorado at Boulder. His research interests are in sequence stratigraphy and basin analysis, three-dimensional seismic interpretation, and reservoir geology.

### Mark G. Rowan

Mark Rowan is a graduate of Caltech (1976), University of California, Berkeley (1982), and the University of Colorado (1991), where he is currently an assistant research professor in structural geology. He has also worked in the petroleum industry, both as an exploration geologist for Sohio and as a consultant with Geo-Logic Systems and Alastair Beach Associates. His research interests focus on the three-dimensional geometry and evolution of both extensional and contractional structures, with a recent emphasis on salt tectonics.

**Qualifications:** Any person engaged in the practice or teaching of geology may apply for Active membership, provided the applicant holds a Bachelor's, Master's or Doctoral Degree in geological science from a college of acceptable standards and, in addition, has had three (3) years of experience in the practice or teaching of geology. Credit for experience can be counted as follows: Master's Degree—one year; Doctor's Degree—two years. The Executive Committee may waive degree or current professional activity requirements if in its judgement an applicant has adequate professional experience and has attained standing in the profession. (If less than 3 years' experience, apply for Junior.)

# Application for Active Membership


**The American Association of Petroleum Geologists, Inc.**
**An International Geological Organization**

P.O. Box 979 • Tulsa, OK 74101-0979 USA
Membership Phone: (918) 560-2615 • Membership Fax: (918) 560-2694


❑ Ms. ❑ Mr. ❑ Dr.

PLEASE TYPE OR PRINT (LAST NAME) (FIRST) (MIDDLE)

To the Executive Committee, The American Association of Petroleum Geologists:

I hereby apply for Active Membership in the American Association of Petroleum Geologists (AAPG).

BIRTH DATE PLACE CITIZENSHIP

MAILING ADDRESS

EMPLOYER POSITION OR TITLE

ADDRESS

( ) ( )

PHONE FAX

E-MAIL ADDRESS

I have read the Constitution, including the Code of Ethics, and the Bylaws of the Association (see December *Bulletin*). I recognize the need to enforce and protect a meaningful Code of Ethics, and I will hold blameless any member of the Association who may, without malice, bring charges against me of a violation of the Code of Ethics. If admitted to membership, I agree to abide by the provisions of the Constitution and Bylaws now in effect and as they may be hereafter amended; and I further agree that said Constitution and Bylaws, including the Code of Ethics, shall be the sole measure of my rights, privileges, and obligations in the Association. I further understand and agree that, without limitation, violation by me of any provision of the Code of Ethics, as now existing, or as may be amended hereafter, shall be sufficient grounds for my expulsion or suspension as may be decided by the Executive Committee. I authorize the Association to include information about me in its publications.

SIGNATURE OF APPLICANT DATE

**For full membership complete the above and attach the following information:**

1. A listing of all college work by date, name of school, field of major study, degree received and date degree received.
2. A photocopy of official documents verifying this training.
3. A listing of all experience since high school by date, nature of work, and for whom. Resumé or c.v. acceptable if it contains this information.
4. A listing of three Active AAPG Members who have known you for at least one year and who are willing to sponsor your membership. Give name, position and company, address and phone number.
5. Payment of US $72 which covers a $10 non-refundable application fee and $62 for dues. If mailing address is outside North America, include additional $10 mailing charge. VISA, MC, AmX

**Other membership classes:**

**Junior:** *(see qualifications, above)* Comply with 1 and 3; list one Active AAPG Member as sponsor; remit $30 ($40 if outside North America).

**Student:** Comply with 1 (include date degree expected); attach an endorsement by a faculty sponsor; remit $10 ($20 if outside North America).

**Associate:** *(not qualified for other classes but employment is associated with geology)* Comply with 1, 2, and 3; list one Active AAPG Member as sponsor; remit $62 ($72 if outside North America).

***This page may be copied.***

AAPG An International Geological Organization

# Bulletin

Volume 82, Number 5B • May 1998, Part B

REGULAR FEATURES



**ON COVER—Seismic chair display of northern Green Canyon (northern Gulf of Mexico) showing distribution of the Bullwinkle reservoir (yellow amplitude) and boreholes from two producing wells. View is to the northeast. East of the reservoir are two prominent allochthonous salt stocks that formed from an evacuated salt sheet. Strata consist of Pliocene and Pleistocene bathyal turbidite systems deposited in intraslope mini-basins. Image is a voxel-based, 3-D seismic cube displayed in VoxelGeo (courtesy of Paradigm Geophysical, formerly CogniSeis Development). 3-D seismic data are courtesy of Halliburton (now Western Geophysical).** ***Figure by Robert Ratliff, Paul Weimer, and Barry McBride.***

ISBN: 0-89181-819-7

ARTICLES



**Acknowledgments—**

*AAPG thanks the following contributors for financial support of this issue of the Bulletin.*

**Geco-Prakla ◆ GeoQuest**

**Conoco ◆ Exxon Exploration**

**Paradigm Geophysical**

**Petroleum Research Fund of the American Chemical Society**

**Platte River Associates ◆ AAPG Foundation**

# Previews